Organization of Insect Societies

Organization of Insect Societies

FROM GENOME TO SOCIOCOMPLEXITY

Edited by

Jürgen Gadau and Jennifer Fewell

WITH A FOREWORD BY EDWARD O. WILSON

HARVARD UNIVERSITY PRESS

Cambridge, Massachusetts

London, England

2009

Library of Congress Cataloging-in-Publication Data
Organization of insect societies : from genome to sociocomplexity /
edited by Juergen Gadau and Jennifer Fewell ; with a foreword by Edward O. Wilson.
p. cm.
Includes bibliographical references and index.
ISBN 978-0-674-03125-8
1. Insect societies.   2. Insects—Evolution.   I. Gadau, Juergen.
II. Fewell, Jennifer.
QL496.O74 2009
595.717'82—dc22   2008031365

Contents

Foreword

EDWARD O. WILSON

I AM PLEASED to take this opportunity to pay a double tribute. The first is to Bert Hölldobler, a great scientist whose contributions to the study of social insects, and especially the wonderfully paradigmatic ants, has been profound in both content and influence on other investigators. His observations and experiments have proceeded magisterially from field to laboratory to synthesis and theoretical reasoning. During the four decades of our friendship, he has greatly influenced my own life and work.

The second tribute is to the distinguished international group of biologists who in this volume chronicle the growth of insect sociobiology during the past half century. The authors of each chapter of *Organization of Insect Societies: From Genome to Sociocomplexity* are among the most productive researchers on the subjects they address. The overall picture they assemble is of a discipline that has grown exponentially in recent years, and at publication time shows no sign of slacking off. The book is thus at once a history, a dispatch from the research front, and a vision as accurate as can be made of future advances in the study of social insects.

Having been an active researcher myself for over six decades (since I purchased William Morton Wheeler's 1910 *Ants: Their Structure, Development and Behavior* as a teenager and began my first collection of ants), I can fairly say that I have been attentive to original scientific reports spanning more than half and perhaps three quarters of all the knowledge of social insects existing today. As such, I hope it will not seem overly presumptuous to offer three meta-generalizations about the status of the subject.

To begin, we have been blessed by the extraordinary observability of social insects. Entomologists, unlike students of the social behavior of primates and birds for example, do not as a rule have to sit for months or years in remote localities to acquire a large database. Colonies of ants, termites, and bees of most species, when uprooted and taken to a laboratory and given simple artificial nests, quickly resume a normal repertory of social behavior. The culture of honey bees, perfected since ancient times, requires only small adjustments to make these insects open to minute scientific observation.

The sanguinity of social insects in captivity also makes it possible to manipulate colonies easily for experiments on social organization. The worker population can be trimmed to reduce colony size, chosen to obtain particular age-size distributions, deprived of a queen or brood or given new ones, and so forth. They can be induced repeatedly to emigrate from one nest site to another. There are few possible such combinations that have not been tried in one context or other.

The second overarching advantage is the importance easily demonstrated of the study of social insects to general biology. A major theme of biology in the present century, perhaps *the* major theme, is the nature and evolutionary origin of the transitions across levels of organization. The most transparent of the transitions, greater than that, for example, from molecule to cell or species to ecosystem, is organism to superorganism, the level reached when societies are tightly bound by altruism and division of labor. One of the major advances of the past half century has been the demonstration, well illustrated in the present volume, of how the transition is made through the emergence of colony-level traits. The process of emergence, driven by colony-level natural selection, can be described with some exactitude through studies of easily observed traits of the participant organisms.

Finally, to achieve the many advances on colony organization reported in the present volume has required researchers to draw on almost all other branches of biology. It is now routine to bring in techniques and ideas from genomics and behavioral genomics to ecology and paleontology. The study of social insects has thereby become one of the most mature and inclusive of all biology's many disciplines.

Organization of Insect Societies

PART ONE

Transitions in Social Evolution

JÜRGEN GADAU

TRANSITIONS FROM SOLITARY to social living and between alternative social phenotypes did not happen just once or in the distant past, as did the transition from prokaryotes to eukaryotes. Such transitions continue to be observed today in facultative eusocial halictine bees that switch between a solitary and social lifestyle and in ant species with colonies headed by a single or multiple queens depending on the population studied. The widespread occurrence of such transitions raises the question of whether these transitions are major or minor. Social insects have two important advantages for someone interested in studying the mechanistic basis of evolutionary transitions from solitary to social living, or from one social phenotype to another. First, because these transitions can be studied in extant species, detailed mechanistic analyses of the underlying genetic, physiological, or developmental changes responsible for these transitions are possible. Second, because there are many major and minor transitions that happen in phylogenetically independent groups, comparative studies enable us to understand the selective forces shaping these transitions.

Throughout his scientific career, Bert Hölldobler was interested in both proximate and ultimate questions concerning the evolution of sociality. Beginning with his PhD thesis and continuing throughout his career, Hölldobler has explored the evolution of queen number and the evolutionary implications of multiple queen societies in ants. Ants are especially rich in the variation of the number and mode of reproduction, and are arguably the most successful social insect group in terms of species number, distribution, and ecological dominance. In ants we discover a dazzling amount of variation from the standard model of one, singly mated queen. Polygynous societies with multiple reproductive individuals can develop by starting

1

together as pleometrotic foundress associations or by adding additional queens later. Some ants, like *Platythyrea punctata*, have lost queens completely and reproduction has been taken over by clonally reproducing workers through thelytoky. In the ponerine ant *Diacamma*, reproduction depends on the possession of intact gemmulae, structures that are thought to be homologous to wings. In many invasive ant species, queen number has become inflationary, with sometimes thousands of reproducing queens. These are just some of the extremes, and ants have realized almost all imaginable variations. Hölldobler regarded the huge leaf cutter or army ant colonies as the final transition in social evolution to true superorganisms. In these species, with the degeneration of ovaries in workers, selection now predominantly acts on the level of a colony.

The following chapters explore the proximate and ultimate mechanisms of minor and major transitions within all major social insect clades. One major theme of these chapters is the key requisite of knowledge of the phylogenetic relationships of the clades of social insects. Boomsma, Kronauer, and Pedersen review our current knowledge about the evolution of mating frequency, and Heinze and Foitzik do the same for queen number. Both focus on transitions between different classes of mating frequencies or queen numbers per colony in order to understand the selective forces behind this transition. In Chapters 3 and 4, Rueppell and Cole deal with colony demography in bees and ants, respectively, and lay out the basic methods to study this important but so far neglected research area. Cole's analysis makes especially clear that the selective unit shaping important life history parameters in ants is the colony rather than an individual queen or worker. Rueppell draws attention to the perplexing differences in the mean age between queens and workers that otherwise do not differ in their genotype, and summarizes our current knowledge about the mechanistic and evolutionary explanations. Brent and Korb both focus on termites. Brent summarizes the current research on caste determination in termites and asks for more in-depth studies of basal termite lineages. Korb introduces an interesting interpretation of the factors that were important for the evolution of eusociality in termites, emphasizing ecological factors over kin selection. Finally, Wcislo and Tierney explain why communal behavior should be considered an ultimate social phenotype equivalent to eusociality rather than an intermediate stage in the evolution of sociality.

CHAPTER ONE

The Evolution of Social Insect
Mating Systems

JACOBUS J. BOOMSMA

DANIEL J. C. KRONAUER

JES S. PEDERSEN

THIS BRIEF REVIEW is about the causes and consequences of mating systems in the eusocial Hymenoptera: ants, bees, and wasps. Forces of natural and sexual selection have shaped mating behavior and genital morphology over evolutionary time within the constraints that specialization of the queen caste has allowed. Consequences are expressed in the genetic variation of queen offspring, which may significantly affect the way in which reproductive conflicts are expressed and regulated. Although early studies emphasized that both aspects are crucial, in the last quarter century there have been many consequential reconstructions of mating behavior based on (mother-) offspring analysis with genetic markers, but relatively few studies on the actual causes of variation in social insect mating systems. One reason for this emphasis is that Hamiltonian predictions developed in the 1960s and 1970s focused on the importance of relatedness and so the tools to estimate relatedness developed in a spectacular way. Other reasons are that aspects of sexual selection in eusocial Hymenoptera are very hard to study in the field (and close to impossible in the laboratory), and that it seemed unlikely that sexual selection was very important overall, because male ornaments and sperm displacement devices are absent and male fighting rare. We begin this chapter with a review of the now extensive literature on genetic marker studies of queen mating frequency, emphasizing mostly recent developments, as this area has been thoroughly and repeatedly reviewed elsewhere. We then address recent developments in the study of genitalia, sperm, accessory gland material, and sperm storage organs. These approaches have opened various windows for future sexual selection studies that may help to redress some of the imbalances in the evolutionary study of causes and consequences of social insect mating systems. In conclusion, we formulate three themes where future studies can be expected to be particularly fruitful.

Background

Multiple queen mating in eusocial Hymenoptera was considered a potentially serious problem for the evolution and maintenance of worker helping behavior by kin selection, but only when it appeared that it evolved before the irreversible establishment of sterile worker castes (Hamilton 1964; Starr 1979; Trivers and Hare 1976; Wilson 1971). This notion inspired a series of early reviews of the sociobiology of mating systems in ants, bees, and wasps (Page and Metcalf 1982; Cole 1983; Starr 1984; Hölldobler and Bartz 1985), setting the standard for all subsequent work. While the first three of these studies addressed the possibility that multiple mating could be relatively common, the Hölldobler and Bartz review focused primarily on the unique characteristics of social insect mating systems and established the following: (1) ant males hatch with a fixed amount of sperm, which imposes significant constraints on mating system evolution; (2) there are two major and fundamentally different mating syndromes ("female calling" and "male aggregation"), which tend to evolve under different conditions; and (3) in at least some species, both queens and males can mate multiply.

How far the study of social insect mating systems has progressed since the 1980s becomes apparent by comparing the data and tools that were available then and now. Although they cite the pioneering studies by Crozier (1979, 1980), Pamilo and Varvio-Aho (1979), Pearson (1983), and Ward (1983) on the derivation and practical use of relatedness parameters, the Hölldobler and Bartz synthesis is essentially based on a clever combination of meticulous field observations and dissection studies. The first data tables presented in the other three reviews (Page and Metcalf 1982; Cole 1983; Starr 1984) gave unprecedented insight into the variation of mating systems across taxa, but also exposed how much these records relied on occasional dissections and anecdotal field observations—the data available at that time had hardly ever been collected with the explicit purpose of studying the evolution of social insect mating systems. As we write, two decades later, an entire generation of students has been educated with the necessity of studying the consequences of mating systems (i.e., the genetic structure of colonies, with DNA microsatellite markers) (Queller, Strassman, and Hughes, 1993) and advanced statistical programs to estimate queen mating frequencies and relatedness among queen offspring (Goodnight and Queller 1998, 1999; Moilanen, Sundström, and Pedersen 2004; Wang 2004). The accumulation of genetic marker studies on social insect mating systems reflects the substantial growth of this field (Figure 1.1).

More detailed reviews on social insect mating systems that capitalized on further conceptual and technical developments saw the light in the mid-1990s (Boomsma and Ratnieks 1996; Bourke and Franks 1995; Crozier and Pamilo 1996; Schmid-Hempel 1998). A now larger but still very limited comparative data set of genetic marker studies allowed Boomsma and Ratnieks (1996) to infer that obligate multiple mating of queens was likely

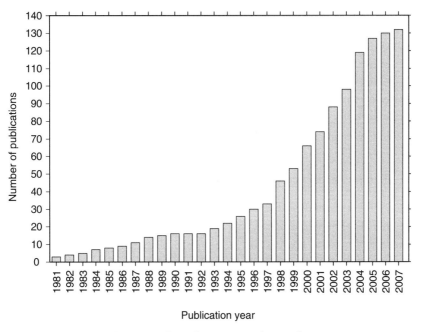

Publication year

Figure 1.1. The cumulative number of genetic marker studies on Hymenopteran queen mating frequencies ($N = 132$) included in our online appendix as of 6 January 2007. Studies before 1982 are combined.

to be an evolutionary derived trait, so that multiple mating was unlikely to have been a constraint for the early evolution of eusociality. A more recent review by Strassmann (2001) even emphasized that multiple mating is surprisingly rare and evaluated why this might be the case. In a contemporaneous review, Crozier and Fjerdingstad (2001) concluded that although many more reliable data on social insect mating systems had become available by the end of the 1990s, scientists were not much closer to an integrated understanding of the various selection forces that affect social insect mating systems. In fact, these authors argued that there may not be a single, prominent explanation for facultative or obligate multiple mating across the different taxa of ants, bees, and wasps.

In our present review we do not aim to repeat all the conceptual advances that were established earlier, but to take a tangential view at the present state of the field. We focus on four particular aspects: (1) the recent explosion of high quality data based on DNA microsatellite analysis and the development of software to analyse these data; (2) a novel appraisal of some of the leading hypotheses based on these recent developments; (3) the recent discovery of mating plugs and sperm-length variation and the new light that this sheds on the likelihood for facultative and obligate multiple mating to evolve; and (4) the identification of some new questions that need to be addressed. We also introduce a new web-based tool that will, we hope, facilitate the future study of social insect mating systems; that is, a complete phylogenetically organized database of all genetic marker studies currently available, subject to regular updating when new material appears in print (see below).

The Increase in Genetic Marker Studies and Analytic Tools

Our data set contains estimates of queen mating frequency in 173 species of ants ($N=89$), social bees ($N=57$), and social wasps ($N=27$) and covers a total of 64 genera. The data include (mostly older) allozyme studies, a limited number of dominant marker studies, and (more recent) studies using highly variable co-dominant microsatellite loci. As the size of this table surpasses the space available in this chapter, we have chosen to present the comparative data set as an online appendix, and to present a summary here. Our table includes data on typical colony queen number ($m=$ exclusive monogyny; $m-p=$ facultative polygyny; $p=$ obligate polygyny), typical mature colony size (median of mature colony worker population expressed as power of 10), mode of colony founding (i=independent founding; p=dependent founding by single or multiple queens, including social parasites; b=colony budding where workers accompany queens in polygynous species; f=colony fission where workers accompany a single queen in monogynous species), the number of single-mother broods or spermathecae analyzed in studies of queen mating frequency (N), the mating system category according to Boomsma and Ratnieks (1996, after merging their two intermediate categories): s=singly mated, $s-m$=facultatively multiply mated [usually ≥50% singly mated with a variable minority of queens mated to 2–5 males], m=obligately multiply mated [almost always ≥ 2 and often ≥ 5 matings per queen]), and the methods that were used to estimate the mean observed number of matings (k_{obs}) and the mean effective number of matings (m_e). The latter estimates are based on one of three types of pedigree analyses, using respective estimators of Starr (1984), ps; Pamilo (1993), pp; and Nielsen, Tarpy, and Reeve (2003), pn, or on relatedness estimates, using codominant (rc) or dominant (rd) markers. In some cases paternity estimates were maximum estimates assuming that multiple mating would be detected in the next group or colony sampled (e.g., Boomsma and Ratnieks 1996), or on an estimated upper 95% confidence limit of a binomial distribution (e.g., Villesen et al. 2002).

In Figure 1.2, we present a summary of our online data set as a single figure, organized according to the presently known phylogenetic relationships. A number of previously identified trends (Boomsma and Ratnieks 1996) seem to hold up in this much larger data set, but a number of novel insights have also emerged. (1) When multiple species of the same genus were analyzed, they almost always fell in the same mating category (i.e., exclusively singly mated, facultatively multiply mated, or exclusively multiply mated). This confirms that the three mating system categories are a reasonable overall classification and that genera across the ants, bees, and wasps tend to belong to only one of these mating system categories. (2) The inference that obligate multiple mating of queens is characteristic for many taxa with large colony size has been further corroborated. Examples now include the *Apis* honey bees, the *Atta* and *Acromyrmex* leaf-cutter ants, at least some of the *Vespula* wasps (for references, see earlier reviews cited above), as well as all investigated North American *Pogonomyrmex* harvester ants (Cole and Wiernasz 2000; Gadau et al. 2003; Helms Cahan et al. 2002; Rheindt et al. 2004; Volny and Gordon 2002; Wiernasz, Perroni, and Cole 2004) and *Eciton, Neivamyrmex, Aenictus,* and *Dorylus*

M Q W

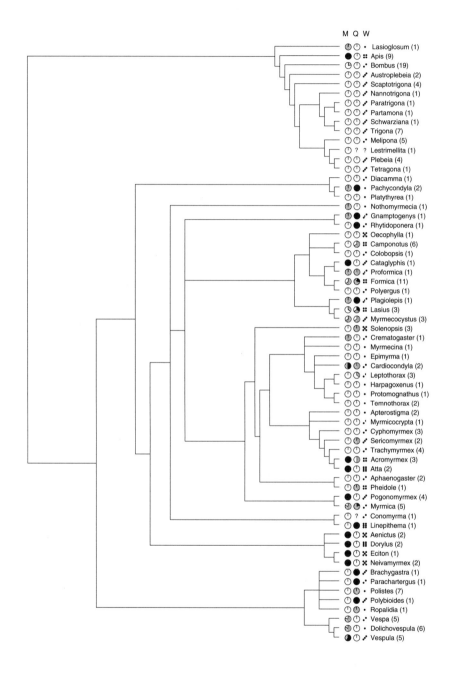

Lasioglossum (1)
Apis (9)
Bombus (19)
Austroplebeia (2)
Scaptotrigona (4)
Nannotrigona (1)
Paratrigona (1)
Partamona (1)
Schwarziana (1)
Trigona (7)
Melipona (5)
Lestrimellita (1)
Plebeia (4)
Tetragona (1)
Diacamma (1)
Pachycondyla (2)
Platythyrea (1)
Nothomyrmecia (1)
Gnamptogenys (1)
Rhytidoponera (1)
Oecophylla (1)
Camponotus (6)
Colobopsis (1)
Cataglyphis (1)
Proformica (1)
Formica (11)
Polyergus (1)
Plagiolepis (1)
Lasius (3)
Myrmecocystus (3)
Solenopsis (3)
Crematogaster (1)
Myrmecina (1)
Epimyrma (1)
Cardiocondyla (2)
Leptothorax (3)
Harpagoxenus (1)
Protomognathus (1)
Temnothorax (2)
Apterostigma (2)
Myrmicocrypta (1)
Cyphomyrmex (3)
Sericomyrmex (2)
Trachymyrmex (4)
Acromyrmex (3)
Atta (2)
Aphaenogaster (2)
Pheidole (1)
Pogonomyrmex (4)
Myrmica (5)
Conomyrma (1)
Linepithema (1)
Aenictus (2)
Dorylus (2)
Eciton (1)
Neivamyrmex (2)
Brachygastra (1)
Parachartergus (1)
Polistes (7)
Polybioides (1)
Ropalidia (1)
Vespa (5)
Dolichovespula (6)
Vespula (5)

army ants (Denny et al. 2004, Kronauer et al. 2004; Kronauer, Johnson, and Boomsma 2007). However, obligate multiple mating also occurs in at least single species of the ant genera *Cataglyphis* (Pearcy et al. 2004) and *Cardiocondyla* (Lenoir et al. 2007), which have rather small colonies, suggesting that additional factors may also play a role. The unique intranidal mating system of *Cardiocondyla* ants may be one such factor (Heinze and Hölldobler 1993; reviewed in Boomsma, Baer, and Heinze 2005). The same comparison also reinforces the notion that, without exception so far, obligate multiple mating is associated with monogyny (Table 1.1). Direct evidence for this was recently obtained from a derived polygynous army ant that had reverted to low queen-mating frequencies (Kronauer and Boomsma 2007b) and from a polygynous social parasite of *Acromyrmex* leaf-cutting ants that made a similar shift in mating system (sumner et al. 2004). (3) Rapid reversal from obligate multiple mating to almost exclusively single mating is apparently possible in evolutionary young inquiline social parasites. This has now been demonstrated for a host-parasite sister species pair of *Acromyrmex* leafcutter ants and has been predicted to also apply to inquilines in other polyandrous clades (Sumner et al. 2004). (4) A number of new genera with facultative multiple mating have been identified: *Cardiocondyla* (Schrempf et al. 2005), *Crematogaster* (Heinze et al. 2000), *Gnamptogenys* (Giraud et al. 2000), *Lasioglossum* (Paxton et al. 2002), *Myrmecocystus* (Kronauer, Gadau, and Hölldobler 2003), *Nothomyrmecia* (Sanetra and Crozier 2001), *Pachycondyla* (Kellner et al. 2007), *Plagiolepis* (Trontti et al. 2006), *Proformica* (Fernández-Escudero, Pamilo, and Seppä 2002), and *Vespa* (Foster

Figure 1.2. A simplified composite phylogeny for the genera of social Hymenoptera listed in the online appendix. The phylogeny is based on the same sources as in Hammond and Keller (2004), supplemented with recent or more detailed studies on Meliponini (Costa et al. 2003), Formicidae (Brady et al. 2006; Moreau et al. 2006), army ants and relatives (Brady 2003), Formicoxenini (Beibl et al. 2005), Attini (Villesen et al. 2002), and Vespinae (Wenseleers et al. 2005). There are two columns with small pie diagrams on the mating system class (M) and the number of queens per colony (Q). For the M-column, the pies show the proportion of species being obligately singly mated (white), facultatively multiply mated (grey), and obligately multiply mated (black). The pies in the Q-column have similar meaning: obligately monogynous (white), facultatively polygynous (grey), and obligately polygynous (black). The third column (W) indicates the median mature colony size (worker number) with a dot for each power of 10. The number in brackets behind each genus name is the number of species studied.

Table 1.1. A summary of studies demonstrating obligate effective multiple
mating in recently studied ants

Species	k_{obs}	m_e	Colony size	Reference
Aenictus laeviceps	17.8	18.8	5	Kronauer et al. (2007)
Dorylus molestus	17.8	15.9	6	Kronauer et al. (2004); Kronauer et al. (2006)
Eciton burchellii	12.9	12.9	5	Kronauer et al. (2006)
Neivamyrmex nigrescens	14.9	12.8	5	Kronauer et al. (2007)
Pogonomyrmex badius	11.0	6.7	3	Rheindt et al. (2004)
Pogonomyrmex barbatus	—	3.1; 3.3	4	Helms Cahan et al. (2002); Volny and Gordon (2002)
Pogonomyrmex rugosus	6.0	4.7	3–4	Gadau et al. (2003)
Cardiocondyla elegans	4.5	4.4	2	Lenoir et al. (2007)
Cataglyphis cursor	5.6	—	3	Pearcy et al. (2004)

Notes: This table includes studies that could not be included in Strassmann 2001.

All these species are exclusively monogynous. Only studies with several colonies per species are included. The observed (k_{obs}) and effective (m_e) queen mating frequencies are given together with median mature colony sizes (worker power as power of 10) and references (see online appendix for more detailed information).

et al. 1999; Takahashi, Akimoto, et al. 2004, Takahashi, Nakamura et al. 2004), although data are too sparse to exclude as yet that some of these may belong to the obligate multiple mating class. Overall, facultative multiple mating is about equally frequent in exclusively monogynous, facultatively polygynous, and obligatory polygynous species. This underlines that facultative multiple mating is not just a precursor of obligate multiple mating, but most likely a fundamentally different mating system that is maintained by another combination of selection forces than obligate multiple mating (see also Boomsma and Ratnieks 1996).

The comparative data of our online appendix obviously await formal analysis with independent contrasts methods (Purvis and Rambault 1995), but this is premature as accurate molecular phylogenies are still lacking for most groups. At some point in the not too distant future, such analysis will allow the different relevant predictor variables for multiple queen mating to be entered in a single or very few multiple regression analyses. The compilation of our large database and the heterogeneities in the source papers that we encountered clearly suggest that the field would benefit from further standardization of analyses and presentation of results. While the now ample availability of DNA microsatellite markers has enormously increased the power of paternity analyses, results have also become more easily affected by occasional null or dropout alleles, mutations, reading errors in laboratories, copy paste errors during data handling, the occasional offspring from a second nest queen, or a few unrelated individuals that joined a colony. Some of these errors can lead to serious overestimates of both the mean absolute number of matings (k_{obs}) and the mean effective number of matings (m_e). In addition, methods for the estimation of standard errors and 95% confidence limits have recently become available, facilitating the statistical comparison of mating systems across species or populations. The software program MATESOFT (Moilanen, Sundström, and Pedersen 2004) combines the most advanced algorithms for estimating mating system parameters and also offers a series of error checks that significantly reduce the risk of spurious overestimations of paternity. A recent example of the potential of this program for unravelling complex data sets consisting of a mother and daughter queen each mated to >10 unrelated males is given in Kronauer et al. (2004). The program COLONY (Wang 2004) uses a likelihood method to infer full-sib and half-sib families from genetic marker data and identifies genotyping errors (Kronauer and Boomsma 2007a).

Progress in Understanding Mating System Evolution

Males and queens are committed to each other for life in all advanced social insects, as queens do not remate later in life and males share paternity when inseminating the same female (Boomsma, Baer, and Heinze 2005; Kronauer and Boomsma 2007). Nonetheless, mating with multiple males has generally been assumed to be costly for queens, so that clear compensating fitness benefits have been hypothesized to exist. Recent studies (Crozier and Fjerdingstad 2001; Sumner et al. 2004) have narrowed a large variety of possible working hypotheses for these benefits down to the following: more stored sperm; lower costs due to reproductive conflict over sex allocation; reduced load due to diploid male production; and enhanced colony performance owing to higher genetic diversity, either by genetic polyethism or by improved resistance to diseases.

Studies of the past few years have indicated that the first hypothesis may well be conceptually problematic because it does not take into account that sperm length (and thus sperm mass) is highly variable, both within and between species of the same clade (Baer 2003; Baer, Nash and Boomsma 2008), and that storing more sperm from a single male could also be achieved by reducing sperm length (Baer, Nash, and Boomsma 2008; Boomsma, Baer, and Heinze 2005). The second hypothesis cannot generally apply because it requires that worker-controlled relatedness-induced split sex ratios give queens that mate with $(n+1)$ males a direct fitness benefit compared to queens that mate with (n) males (Ratnieks and Boomsma 1997). This seems to work very well in some *Formica* ants (Sundström and Ratnieks 1998), but apparently does not apply in *Lasius* ants (Fjerdingstad, Gertsch, and Keller, 2002). In addition, this hypothesis is unlikely to apply in taxa with obligate multiple mating as variation in relatedness asymmetry across colonies becomes very low when all queens mate with many males (Boomsma and Ratnieks 1996; Ratnieks and Boomsma 1997).

The third (diploid male load) hypothesis assumes that there is limited genetic variation for the complementary sex determination (CSD) locus (Crozier and Page 1985; Crozier and Pamilo 1996; Pamilo et al. 1994), so that a certain proportion of matched matings is unavoidable, resulting in diploid offspring that were intended to become female workers but develop a sterile male phenotype instead. Under single mating, all queens mated to an unmatched male will have no fitness load at all, while a minority of queens will have 50% of their worker brood failing so that their colonies inevitably die. In contrast, obligate multiple mating implies that many colonies may have a single matched mating, but that the fitness load remains limited because the majority of unmatched fathers will sire normal brood. This hypothesis, which is theoretically well documented (reviewed in Crozier and Pamilo 1996), predicts that multiple mating will be favored when the fitness price of diploid males is primarily paid when colonies have reached their final size, whereas single mating is favored when fitness load reduces early colony growth. The diploid male load hypothesis is, therefore, a rather peculiar "genetic diversity" hypothesis (Crozier and Fjerdingstad 2001), because diversity among workers owing to multiple paternity does not have a consistently positive effect (as in the final hypotheses discussed below), but instead a fitness effect that depends on idiosyncratic details of the biology (Ratnieks 1990).

The diploid male hypothesis has been difficult to test because diploid males may be recognized by workers and removed before they become adult, while still imposing a fitness cost of unknown magnitude. However,

Tarpy and Page (2001, 2002) recently showed that the genetic load from diploid males can indeed have a significant effect in the honey bee, consistent with theory. Furthermore, based on a study of multiple mating in army ants, Kronauer, Johnson, and Boomsma (2007) argued that diploid male load may be particularly important in social insects that multiply by colony fission, because the large investment (accompanying workers in a swarm) in a single daughter colony needs to be made before the new queen mates. This hypothesis would predict that in-depth studies of fitness load due to diploid male brood might be particularly likely to give significant effects in polyandrous species that multiply by colony fission in addition to suffering from genetic erosion because of small effective population sizes (Crozier and Pamilo 1996).

Over recent years, new evidence has accumulated in support of the genetic diversity hypotheses. The standard version of this idea (coined "genetic polyethism" by Crozier and Fjerdingstad [2001]; Oldroyd and Fewell 2007) has met with recent support in honey bee (Tarpy 2003; Jones et al. 2004; Mattila and Seeley 2007) and harvester ant (Cole and Wiernasz 1999) research. In addition, the special version—that genetic diversity enhances resistance against disease (Hamilton 1987; Sherman, Seeley and Reeve 1988)—has been supported by recent evidence in leaf-cutting ants (Hughes and Boomsma 2004, 2006) and honey bees (Palmer and Oldroyd 2003; Tarpy and Seeley 2006; Seeley and Tarpy 2007) and by a comparative study (Brown and Schmid-Hempel 2003). However, it has also become clear that model study systems of ants, bees, and wasps might not be as comparable as they were previously assumed to be. For example, a recent review (Boomsma, Schmid-Hempel, and Hughes 2005) provided comparative evidence that disease pressure in annual and perennial social insects may be fundamentally different.

The overall picture seems to reinforce the conclusion by Crozier and Fjerdingstad (2001) that we are unlikely to find a single major selection force favoring multiple queen mating that is almost universally valid. Several factors may play a role in most of the clades where multiple mating evolved and some of them may have been important for the origin of multiple mating or for its later maintenance, but possibly not for both. The single overall trend that remains robust is that obligate multiple mating is typically found in large-colony genera with monogynous queens across ants, bees, and wasps (Boomsma and Ratnieks 1996; Cole 1983). Also, the newly discovered cases of obligate multiple mating in *Cataglyphis* and

Cardiocondyla ants concern monogynous species, although these species do not have particularly large colonies (Table 1.1). New insights that have the potential to apply more generally are, first, that colony fission may be a major general factor promoting multiple queen mating (e.g., in combination with diploid male load), because multiple mating ensures that the risk of failure is minimized for each of the single or very few extremely valuable daughter queens that honey bees, army ants, and any other such system produce (Kronauer, Johnson, and Boomsma 2007). Second, the reversal from obligate multiple mating in *Acromyrmex* inquiline social parasites (Sumner Hughes, Pedersen and Boomsma 2004) is predicted to apply to other inquilines that have evolved in clades with obligate multiple mating, because genetic diversity advantages of costly multiple mating do not apply in social parasites. This result further suggests that mating systems may change rapidly after dramatic life history changes between sister species and faster than many morphological traits. A recent confirmation of this notion was obtained by Kronauer and Boomsma (2007b), who showed that the only known polygynous army ant has reverted to facultative multiple mating from obligate multiply mating monogynous ancestors.

Mating Plugs and
Multiple Mating by Males

A surprising finding of the last few years has been the discovery of mating plugs in bumblebees (Baer, Morgan, and Schmid-Hempel 2001), fire ants (Mikheyev 2003), queenless *Dinoponera* ants (Monnin and Peeters 1998), and (most likely) attine ants (Baer and Boomsma 2004), adding to a similar previous observation for stingless bees (Kerr et al. 1962). All these records concern social insects with obligatory single mating of queens or gamergates and in the bumblebee case it has been experimentally demonstrated that the plugs are effective in preventing additional matings (Baer, Morgan, and Schmid-Hempel 2001). Prior to these recent findings, mating plugs were hardly considered in social insects, perhaps because they had been shown to serve another function in the honey bee (Woyciechowski, Kabat and Krol 1994). However, it is now clear that the occurrence of mating plugs must be considered as a primary hypothesis to explain that the majority of eusocial insects have maintained exclusively singly mated queens (Figure 1.2; see also Strassmann 2001). Should future studies show that mating plugs are indeed common, this would imply that the mating systems of the eusocial Hymenoptera are much more male-controlled than we have so far acknowledged (Boomsma, Baer, and Heinze 2005). If that is the case, then the next set of questions need to address how mating plug functions were modified in the various clades where facultative and obligate multiple mating evolved.

There is now fairly solid evidence that males of *Atta* and *Acromyrmex* leaf-cutting ants can mate multiply (Baer and Boomsma 2004). The direct evidence in these cases comes from dissections of mature males where the accessory testes have a constriction that separates the first ejaculate from the remaining sperm to be used in later ejaculates. Adding this evidence to earlier data on multiply mating males in *Pogonomyrmex* harvester ants (Hölldobler 1976; Wiernasz et al. 2001), *Myrmica* ants (Woyciechowski 1990), and *Formica* ants (Fortelius 1994), it seems clear that obligate single mating by males may well be restricted to the eusocial Hymenoptera with either high degrees of sexual dimorphism (with males being much smaller than queens) or reproduction by colony fission and extremely male-biased sex ratios (e.g., honey bees and army ants; Boomsma, Baer, and Heinze 2005).

Sperm Variation and Sperm Storage Costs

A recent study in the bumblebee *Bombus terrestris* has shown that sperm length has a significantly heritable component (Baer, De Jong, et al. 2006). Whether this heritability has any adaptive significance remains to be seen as *B. terrestris* normally has obligate single mating of queens. However, sperm in *B. hypnorum*, where facultative multiple mating of queens occurs, is significantly longer than sperm of *B. terrestris*, which could indicate that sperm competition in the bursa copulatrix has selected for longer sperm that are more efficient in reaching the queen spermatheca (Baer et al. 2003). This would be at least partly consistent with a heuristic model for sperm length evolution presented by Boomsma, Baer, and Heinze (2005), a though recent evidence on sperm length across the in attine fungus-growing ants indicates that sexual dimorphism is a more direct overall predictor of sperm length across genera (Baer et al. 2008).

Artificial insemination of bumblebees that normally have single mating queens results in a significant reduction of colony fitness when sperm and semen from multiple males are mixed (Baer and Schmid-Hempel 2001); however, there is no evidence for any such effects in honey bees, where queens always mate with many males. If anything the effect is opposite, as colonies with a high genetic diversity among workers tend to do better (Tarpy 2003; Tarpy and seeley 2006). A recent field study on incipient colonies of *Atta* leaf-cutting ants showed that there are significant immunity costs for storing more than average numbers of sperm, and that the immune defense of newly inseminated queens was further negatively affected by the number of fathers that contributed to the insemination (Baer, Armitage, and Boomsma 2006). These results suggest that some forms of antagonistic co-evolutionary arms races between mating partners may occur in eusocial Hymenoptera, as they do in other insects (Chapman et al. 2003). However, the evolution of sexually antagonistic traits is constrained and the effects of such sexual conflicts are predicted to be relatively weak in perennial species such as *Atta* leaf-cutting ants and at best moderately strong in annual societies such as bumblebees (Baer and Boomsma 2004; Boomsma, Baer, and Heinze 2005). Even in annual bumblebee societies the sexual brood is preceded by several sterile worker broods (Duchateau, Velthuis, and Boomsma 2004), which implies that selection for male traits in the sperm or seminal fluid that harm the queen must necessarily be weaker than, for example, in *Drosophila* or other promiscuous nonsocial insects (Baer and Boomsma 2004; Boomsma, Baer, and Heinze 2005).

Conclusion: Where to Go from Here

The evolutionary study of social insect mating systems has realized significant advances since the pioneering insights by the first researchers (Page and Metcalf 1982; Cole 1983; Starr 1984; Hölldobler and Bartz 1985) who organized the overall concepts and data. The data on mating frequencies currently available (Figure 1.2; online appendix) are more substantial than comparable data sets for any other insect family (Simmons 2001). However, our understanding of the evolutionary dynamics of social insect mating has been constrained by the difficulty of doing large-scale laboratory experiments. In addition, the forces of sexual selection that operate in social insects are rather different from those in nonsocial insects, so that most conceptual advances have been achieved independent of the mainstream studies in sexual selection (Boomsma, Baer, and Heinze 2005).

It is important to realize that the somewhat idiosyncratic nature of social insect mating systems not only imposes constraints, but also provides opportunities for innovative studies that cannot easily be pursued in other groups of insects. Based on the topics reviewed here, we identified the following three areas as being particularly promising for research efforts in the coming years: (1) As genus and subfamily phylogenies become increasingly available, reconstructing the evolutionary transitions among the three classes of mating systems (Figure 1.1) will become more feasible. Ultimately, it will be these transitions that will be most informative for inferring the costs and benefits of mating systems in an overall phylogenetic analysis using independent contrasts (as, for example, implemented in CAIC; Purvis and Rambaut 1995). Such solid, albeit indirect, evidence will continue to be important because direct measurements of life-time costs and benefits will remain very difficult. (2) Studies should concentrate more on the spectacular aspects of social insect mating systems and try to devise experimental techniques to study them. For example, queens of many ants store sperm for many years and manage to keep it viable (Tschinkel 1987), while other ant species (e.g., *Cardiocondyla*) have evolved intranidal mating systems that are functionally reminiscent to those of nonsocial insects and vertebrates where males monopolize groups of females (Boomsma, Baer, and Heinze 2005; Heinze and Hölldobler 1993). Furthermore, research into the multitude of unknown chemical compounds in mating plugs (Baer, Morgan, and Schmid-Hempel 2001) and seminal fluids that may manipulate sexual behavior would be beneficial. (3) More studies are needed on the variation in mating system parameters both within and across populations of the same species and on their environmental and morphological correlates. The average number of references per species studied in our online appendix table is at present only 1.2 and almost no record has more than five independent studies supporting it. A more thorough appreciation of the natural population-wide variation in mating system traits will be essential to prioritize hypothesis-testing in the few species of social insects that are amenable for experimental work on mating system evolution.

Acknowledgments

We thank Jürgen Gadau and two anonymous reviewers for comments on previous versions of this manuscript. This work was supported by a grant from the Danish National Research Foundation (JJB and JSP) and the Danish Research Training Council (DJCK).

Links

Appendix: http://sols.asu.edu/publications/pdf/chapter1.pdf. Table of 132 genetic marker studies on hymenopteran queen-mating frequencies including references.

MateBase: http://www.bio.ku.dk/matebase.htm. An online database holding the same information as the Appendix but being kept updated by the authors. We encourage colleagues to let us know about new studies to be included.

Literature Cited

Baer, B., S. A. O. Armitage, and J. J. Boomsma. 2006. "Sperm storage induces an immunity cost in ants." *Nature* 441: 872–875.

Baer, B., and J. J. Boomsma. 2004. "Male reproductive investment and queen mating-frequency in fungus-growing ants." *Behavioral Ecology* 15: 426–432.

Baer, B., G. De Jong, R. Schmid-Hempel, P. Schmid-Hempel, J. T. Høeg, and J. J. Boomsma. 2006. "Heritability of sperm length in the bumblebee *Bombus terrestris.*" *Genetica* 127: 11–23.

Baer, B., E. D. Morgan, and P. Schmid-Hempel. 2001. "A nonspecific fatty acid within the bumblebee mating plug prevents females from remating." *Proceedings of the National Academy of Sciences USA* 98: 3926–3928.

Boris Baer, B., M. B. Dijkstra, U. G. Mueller, D. R. Nash, and J. J. Boomsma. 2008. "Sperm length evolution in the fungus-growing ants." *Behavioral Ecology*, Advance Access published on September 11, 2008. doi:10.1093/beheco/arn112.

Baer, B., and P. Schmid-Hempel. 2001. "Unexpected consequences of polyandry for parasitism and fitness in the bumblebee, *Bombus terrestris.*" *Evolution* 55: 1639–1643.

Baer, B., P. Schmid-Hempel, J. T. Høeg, and J. J. Boomsma. 2003. "Sperm length, sperm storage and mating system characteristics in bumblebees." *Insectes Sociaux* 50: 101–108.

Beibl, J., R. Stuart, J. Heinze, and S. Foitzik. 2005. "Six origins of slavery in formicoxenine ants." *Insectes Sociaux* 52: 291–297.

Boomsma, J. J., B. Baer, and J. Heinze. 2005. "The evolution of male traits in social insects." *Annual Review of Entomology* 50: 395–420.

Boomsma, J. J., and F. L. W. Ratnieks. 1996. "Paternity in eusocial Hymenoptera." *Philosophical Transactions of the Royal Society of London B* 351: 947–975.

Boomsma, J. J., P. Schmid-Hempel, and W. O. H. Hughes. 2005. "Life histories and parasite pressure across the major groups of social insects." In *Insect evolutionary ecology*, ed. F. Fellowes, G. Holloway, and J. Rolff. Wallingford, UK: CABI.

Bourke, A. F. G., and N. R. Franks. 1995. *Social evolution in ants*. Princeton, NJ: Princeton University Press.

Brady, S. G. 2003. "Evolution of the army ant syndrome: The origin and long-term evolutionary stasis of a complex of behavioral and reproductive adaptations." *Proceedings of the National Academy of Sciences USA* 100: 6575–6579.

Brady, S., T. Schultz, B. Fisher, and P. Ward. 2006. "Evaluating alternative hypotheses for the early evolution and diversification of ants." *Proceedings of the National Academy of Sciences USA* 103: 18172–18177.

Brown, M. J. F., and P. Schmid-Hempel. 2003. "The evolution of female multiple mating in social Hymenoptera." *Evolution* 57: 2067–2081.

Chapman, T., G. Arnqvist, J. Bangham, and L. Rowe. 2003. "Sexual conflict." *Trends in Ecology and Evolution* 18: 41–47.

Cole, B. J. 1983. "Multiple mating and the evolution of social behavior in the Hymenoptera." *Behavioral Ecology and Sociobiology* 12: 191–201.

Cole, B. J., and D. C. Wiernasz. 1999. "The selective advantage of low relatedness." *Science* 285: 891–893.

Cole, B. J., and D. C. Wiernasz. 2000. "The nature of ant colony success—response." *Science* 287: 1363b.

Costa, M., M. Del Lama, G. Melo, and W. Sheppard. 2003. "Molecular phylogeny of the stingless bees (Apidae, Apinae, Meliponini) inferred from mitochondrial 16S rDNA sequences." *Apidologie* 34: 73–84.

Crozier, R. H. 1979. "Genetics of sociality." In *Social insects*, ed. H. R. Hermann, 223–286. New York: Academic Press.

———.1980. "Genetical structures of social insects populations." In H. Markl, ed., *Evolution of social behaviour: Hypotheses and empirical tests*, 129–146. Weinheim: Verlag Chemie, Germany.

Crozier, R. H., and E. J. Fjerdingstad. 2001. "Polyandry in social Hymenoptera—disunity in diversity?" *Annales Zoologici Fennici* 38: 267–285.

Crozier, R. H., and R. E. Page Jr. 1985. "On being the right size: Male contributions and multiple mating in social Hymenoptera." *Behavioral Ecology and Sociobiology* 18: 105–115.

Crozier, R. H., and P. Pamilo. 1996. *Evolution of social insect colonies: Sex allocation and kin selection*. Oxford: Oxford University Press.

Denny, A. J., N. R. Franks, S. Powell, and K. J. Edwards. 2004. "Exceptionally high levels of multiple mating in an army ant." *Naturwissenschaften* 91: 396–399.

Duchateau, M. J., H. H. W.Velthuis, and J. J. Boomsma. 2004. "Sex ratio variation in the bumblebee *Bombus terrestris*." *Behavioral Ecology* 15: 71–82.

Fernández-Escudero, I., P. Pamilo, and P. Seppä. 2002. "Biased sperm use by polyandrous queens of the ant *Proformica longiseta.*" *Behavioral Ecology and Sociobiology* 51: 207–213.

Fjerdingstad, E. J., P. J. Gertsch, and L. Keller. 2002. "Why do some social insect queens mate with several males? Testing the sex ratio manipulation hypothesis in *Lasius niger.*" *Evolution* 56: 553–562.

Fortelius, W. 1994. *Reproductive biology in the wood ant genus Formica (Hymenoptera, Formicidae).* Helsinki: University of Helsinki.

Foster, K. R., P. Seppä, F. L. W Ratnieks,. and P. A. Thorén. 1999. "Low paternity in the hornet *Vespa crabro* indicates that multiple mating by queens is derived in vespine wasps." *Behavioral Ecology and Sociobiology* 46: 252–257.

Gadau, J., C. P. Strehl, J. Oettler, and B. Hölldobler. 2003. "Determinants of intracolonial relatedness in *Pogonomyrmex rugosus* (Hymenoptera; Formicidae): Mating frequency and brood raids." *Molecular Ecology* 12: 1931–1938.

Giraud, T., R. Blatrix, C. Poteaux, M. Solignac, and P. Jaisson. 2000. "Population structure and mating biology of the polygynous ponerine ant *Gnamptogenys striatula* in Brazil." *Molecular Ecology* 9: 1835–1841.

Goodnight, K. F., and D. C. Queller. 1998. *Relatedness.* Houston: Goodnight Software.

———.1999. "Computer software for performing likelihood tests of pedigree relationships using genetic markers." *Molecular Ecology* 8: 1231–1234.

Hamilton, W. D. 1964. "The genetical evolution of social behaviour I–II." *Journal of Theoretical Biology* 7: 1–52.

———.1987. "Kinship, recognition, disease, and intelligence: Constraints of social evolution." In Y. Itô, J. L. Brown, and J. Kikkawa, eds., *Animal societies: Theories and facts,* 81–102. Tokyo: Japan Scientific Society Press.

Hammond, R. L., and L. Keller. 2004. "Conflict over male parentage in social insects." *PLoS Biology* 2: e248.

Heinze, J., and B. Hölldobler. 1993. "Fighting for a harem of queens: Physiology of reproduction in *Cardiocondyla* male ants." *Proceedings of the National Academy of Sciences USA* 90: 8412–8414.

Heinze, J., M. Strätz, J. S. Pedersen, and M. Haberl. 2000. "Microsatellite analysis suggests occasional worker reproduction in the monogynous ant *Crematogaster smithi.*" *Insectes Sociaux* 47: 299–301.

Helms Cahan, S., J. D. Parker, S. W. Rissing, R. A. Johnson, T. S. Polony, M. D. Weiser, and D. R. Smith. 2002. "Extreme genetic differences between queens and workers in hybridizing *Pogonomyrmex* harvester ants." *Proceedings of the Royal Society of London B* 269: 1871–1877.

Hölldobler, B. 1976. "The behavioral ecology of mating in harvester ants (Hymenoptera: Formicidae: *Pogonomyrmex*)." *Behavioral Ecology and Sociobiology* 1: 405–423.

Hölldobler, B., and S. H. Bartz. 1985. "Sociobiology of reproduction in ants." *Fortschitte der Zoologie* 31: 237–257.

Hughes, W. O. H., and J. J. Boomsma. 2004. "Genetic diversity and disease resistance in leaf-cutting ant societies." *Evolution* 58: 1251–1260.

———.2006. "Does genetic diversity hinder parasite evolution in social insect colonies?" *Journal of Evolutionary Biology* 19: 132–143.

Jones, J. C., M. R. Myerscough, S. Graham, and B. P. Oldroyd. 2004. "Honeybee nest thermoregulation, diversity promotes stability." *Science* 305: 402–404.

Kellner, J., A. Trindl, J. Heinze, and P. D'Ettore. 2007. "Polygyny and polyandry in small ant societies." *Molecular Ecology.* 16: 2363–2369

Kerr, W. E., R. Zucchi, J. T. Nakadaira, and J. E. Butolo. 1962. "Reproduction in the social bees (Hymenoptera: Apidae)." *Journal of New York Entomological Society* 70: 265–276.

Kronauer, D. J. C., S. M. Berghoff, S. Powell, A. J. Denny, K. J. Edwards, N. R. Franks, and J. J. Boomsma. 2006. "A reassessment of the mating system characteristics of the army ant *Eciton burchellii*." *Naturwissenschaften* 93: 402–406.

Kronauer, D. J. C., and J. J. Boomsma. 2007a. "Do army ant queens re-mate later in life?" *Insectes Sociaux* 54: 20–28.

Kronauer, D. J. C., and J. J. Boomsma. 2007b. "Multiple queens means fewer mates." *Current Biology* 17: R753–R755.

Kronauer, D. J. C., J. Gadau, and B. Hölldobler. 2003. "Genetic evidence for intra- and interspecific slavery in honey ants (genus *Myrmecocystus*)." *Proceedings of the Royal Society of London B* 270: 805–810.

Kronauer, D. J. C., R. A. Johnson, and J. J. Boomsma. 2007. "The evolution of multiple mating in army ants." *Evolution* 61: 413–422.

Kronauer, D. J. C., C. Schöning, and J. J. Boomsma. 2006. "Male parentage in army ants." *Molecular Ecology* 15: 1147–1151.

Kronauer, D. J. C., C. Schöning, J. S. Pedersen, J. J. Boomsma, and J. Gadau. 2004. "Extreme queen-mating frequency and colony fission in African army ants." *Molecular Ecology* 13: 2381–2388.

Lenoir, J. C., A. Schrempf, A. Lenoir, J. Heinze, and J. L. Mercier. 2007. "Genetic structure and reproductive strategy of the ant *Cardiocondyla elegans*: Strictly monogynous nests invaded by unrelated sexuals." *Molecular Ecology* 16: 345–354.

Mattila, H. R., and T. D. Seeley. 2007. "Genetic diversity in honey bee colonies enhances productivity and fitness." *Science* 317: 362–364.

Mikheyev, A. S. 2003. "Evidence for mating plugs in the fire ant Solenopsis invicta." *Insectes Sociaux* 50: 401–402.

Moilanen, A., L. Sundström, and J. S. Pedersen. 2004. "MATESOFT: A program for deducing parental genotypes and estimating mating system statistics in haplodiploid species." *Molecular Ecology Notes* 4: 795–797.

Monnin, T., and C. Peeters. 1998. "Monogyny and regulation of worker mating in the queenless ant *Dinoponera quadriceps*." *Animal Behaviour* 55: 299–306.

Moreau, C., C. Bell, R. Vila, S. Archibald, and N. Pierce. 2006. "Phylogeny of the ants: Diversification in the age of angiosperms." *Science* 312: 101–104.

Nielsen, R., D. R. Tarpy, and H. K. Reeve. 2003. "Estimating effective paternity number in social insects and the effective number of alleles in a population." *Molecular Ecology* 12: 3157–3164.

Oldroyd, B. P., and J. H. Fewell. 2007. "Genetic diversity promotes homeostasis in insect colonies." *Trends in Ecology & Evolution* 22: 408–413.

Page, R. E., Jr., and R. A. Metcalf. 1982. "Multiple mating, sperm utilization, and social evolution." *American Naturalist* 119: 263–281.

Palmer, K. A., and B. P. Oldroyd. 2003. "Evidence for intra-colonial genetic variance in resistance to American foulbrood of honey bees *(Apis mellifera)*: Further support for the parasite/pathogen hypothesis for the evolution of polyandry." *Naturwissenschaften* 90: 265–268.

Pamilo, P. 1993. "Polyandry and allele frequency differences between the sexes in the ant *Formica aquilonia*." *Heredity* 70: 472–480.

Pamilo, P., L. Sundström, W. Fortelius, and R. Rosengren. 1994. "Diploid males and colony-level selection in *Formica* ants." *Ethology, Ecology, and Evolution* 6: 211–235.

Pamilo, P., and S. L. Varvio-Aho. 1979. "Genetic structure of nests in the ant *Formica sanguinea*." *Behavioral Ecology and Sociobiology* 6: 91–98.

Paxton, R. J., M. Ayasse, J. Field, and A. Soro. 2002. "Complex sociogenetic organization and reproductive skew in a primitively eusocial sweat bee, *Lasioglossum malachurum*, as revealed by microsatellites." *Molecular Ecology* 11: 2405–2416.

Pearcy, M., S. Aron, C. Doums, and L. Keller. 2004. "Conditional use of sex and parthenogenesis for worker and queen production in ants." *Science* 306: 1780–1783.

Pearson, B. 1983. "Intra-colonial relatedness amongst workers in a population of nests of the polygynous ant, *Myrmica rubra* Latreille." *Behavioral Ecology and Sociobiology* 12: 1–4.

Purvis, A., and A. Rambaut. 1995. "Comparative analysis by independent contrasts (CAIC): An Apple Macintosh application for analyzing comparative data." *Computer Applications in Biosciences* 11: 247–251.

Queller, D. C., J. E. Strassmann, and C. R. Hughes. 1993. "Microsatellites and kinship." *Trends in Ecology and Evolution* 8: 285–288.

Ratnieks, F. L. W. 1990. "The evolution of polyandry by queens in social Hymenoptera: The significance of the timing of removal of diploid males." *Behavioral Ecology and Sociobiology* 26: 343–348.

Ratnieks, F. L. W., and J. J. Boomsma. 1997. "On the robustness of split sex ratio predictions in social Hymenoptera." *Journal of Theoretical Biology* 185: 423–439.

Rheindt, F. E., J. Gadau, C. P. Strehl, and B. Hölldobler. 2004. "Extremely high mating frequency in the Florida harvester ant *(Pogonomyrmex badius)*." *Behavioral Ecology and Sociobiology* 56: 472–481.

Sanetra, M., and R. H. Crozier. 2001. "Polyandry and colony genetic structure in the primitive ant *Nothomyrmecia macrops*." *Journal of Evolutionary Biology* 14: 368–378.

Schmid-Hempel, P. 1998. *Parasites in social insects*. Princeton, NJ: Princeton University Press.

Schrempf, A., C. Reber, A. Tinaut, and J. Heinze. 2005. "Inbreeding and local mate competition in the ant *Cardiocondyla batesii*." *Behavioral Ecology and Sociobiology* 57: 502–510.

Seeley, T. D., and D. R. Tarpy. 2007. "Queen promiscuity lowers disease within honeybee colonies." *Proceedings of the Royal Society of London Series B* 274: 67–72.

Sherman, P. W., T. D. Seeley, and H. K. Reeve. 1988. "Parasites, pathogens, and polyandry in social Hymenoptera." *American Naturalist* 131: 602–610.

Simmons, L. W. 2001. *Sperm competition and its evolutionary consequences in the insects*. Princeton, NJ: Princeton University Press.

Starr, C. K. 1979. "Origin and evolution of insect sociality: A review of modern theory." In H. R. Hermann, ed., *Social insects*, 35–79. New York: Academic Press.

———.1984. "Sperm competition, kinship, and sociality in the aculaeate Hymenoptera." In R. L. Smith, ed., *Sperm competition and the evolution of animal mating systems*, 427–464. Orlando: Academic Press.

Strassmann, J. 2001. "The rarity of multiple mating by females in the social Hymenoptera." *Insectes Sociaux* 48: 1–13.

Sumner, S., W. O. H. Hughes, J. S. Pedersen, and J. J. Boomsma. 2004. "Ant parasite queens revert to mating singly." *Nature* 428: 35–36.

Sundström, L., and F. L. W. Ratnieks. 1998. "Sex ratio conflicts, mating frequency, and queen fitness in the ant *Formica truncorum*." *Behavioral Ecology* 9: 116–121.

Takahashi, J., S. Akimoto, S. J. Martin, M. Tamukae, and E. Hasegawa. 2004. "Mating structure and male production in the giant hornet *Vespa mandarinia* (Hymenoptera: Vespidae)." *Applied Entomology and Zoology* 39: 343–349.

Takahashi, J., J. S. Nakamura, S. Akimoto, and E. Hasegawa. 2004. "Kin structure and colony male reproduction in the hornet Vespa crabro (Hymenoptera: Vespidae)." *Journal of Ethology* 22: 43–47.

Tarpy, D. R. 2003. "Genetic diversity within honeybee colonies prevents severe infections and promotes colony growth." *Proceedings of the Royal Society of London B* 270: 99–103.

Tarpy, D. R., and R. E. Page Jr. 2001. "The curious promiscuity of queen honeybees *(Apis mellifera)*: Evolutionary and behavioral mechanisms." *Annales Zoologici Fennici* 38: 255–265.

———. 2002. "Sex determination and the evolution of polyandry in honeybees *(Apis mellifera)*." *Behavioral Ecology and Sociobiology* 52: 143–150.

Tarpy, D. R., and T. D. Seeley. 2006. "Lower disease infections in honeybee *(Apis mellifera)* colonies headed by polyandrous vs monandrous queens." *Naturwissenschaften* 93: 195–199.

Trivers, R. L,. and H. Hare. 1976. "Haplodiploidy and the evolution of the social insects." *Science* 191: 249–263.

Trontti, K., N. Thurin, L. Sundström, and S. Aron. 2006. "Mating for convenience or genetic diversity? Mating patterns in the polygynous ant *Plagiolepis pygmaea*." *Behavioral Ecology,* 18: 298–303.

Tschinkel, W. R. 1987. "Fire ant queen longevity and age: Estimation by sperm depletion." *Annals of the Entomology Society of America* 80: 263–266.

Villesen, P., T. Murakami, T. R. Schultz, and J. J. Boomsma. 2002. "Identifying the transition between single and multiple mating in fungus-growing ants." *Proceedings of the Royal Society of London B* 269: 1541–1548.

Volny, V. P., and D. M. Gordon. 2002. "Genetic basis for queen-worker dimorphism in a social insect." *Proceedings of the National Academy of Sciences U.S.A.* 99: 6108–6111.

Wang, J. 2004. "Sibship reconstruction from genetic data with typing errors." *Genetics* 166: 1963–1979.

Ward, P. S. 1983. "Genetic relatedness and colony organization in a species complex of Ponerine ants. II. Patterns of sex ratio investment." *Behavioral Ecology and Sociobiology* 12: 301–307.

Wenseleers, T., N. Badcock, K. Erven, A. Tofilski, F. Nascimento, A. Hart, T. Burke, M. Archer, and F. Ratnieks. 2005. "A test of worker policing theory in an advanced eusocial wasp, *Vespula rufa*." *Evolution* 59: 1306–1314.

Wiernasz, D. C., C. L. Perroni, and B. J. Cole. 2004. "Polyandry and fitness in the western harvester ant, *Pogonomyrmex occidentalis*." *Molecular Ecology* 13: 1601–1606.

Wiernasz, D. C., A. K. Sater, A. J. Abell, and B. J. Cole. 2001. "Male size, sperm transfer, and colony fitness in the western harvester ant, *Pogonomyrmex occidentalis*." *Evolution* 55: 324–329.

Wilson, E. O. 1971. *The insect societies.* Cambridge, MA: Harvard University Press.

Woyciechowski, M. 1990. "Mating behaviour in the ant *Myrmica rubra* (Hymenoptera, Formicidae)." *Acta Zoologica Cracoviensia, Kraków* 33: 565–574.

Woyciechowski, M., L. Kabat, and E. Krol. 1994. "The function of the mating sign in honeybees, *Apis mellifera* L—new evidence." *Animal Behaviour* 47: 733–735.

CHAPTER TWO

The Evolution of Queen Numbers in Ants:
From One to Many and Back

JÜRGEN HEINZE

SUSANNE FOITZIK

ONE OF THE MOST CENTRAL features of Hymenopteran societies is the number of reproductive queens in mature colonies (Buschinger 1968a; Hölldobler and Wilson 1977; Keller 1993). Whether a colony contains one or multiple queens is of fundamental importance for its genetic structure and thus strongly affects the fitness interests of workers concerning, for example, the origin of males, investment in colony maintenance versus reproduction, and sex allocation (Bourke and Franks 1995; Crozier and Pamilo 1996). Though much of the earlier reasoning on social evolution in Hymenoptera was based on the assumption that their societies typically have a single, once-mated reproductive female, the coexistence of multiple fertile queens is widespread in particular among the ants, where about half of all species may have colonies with several reproducing queens or gamergates (mated workers). Well-known examples are invasive ant species such as the pharaoh's ant (*Monomorium pharaonis*) and the Argentine ant *Linepithema humile*, many wood ants (*Formica rufa* group), and the red garden ants (*Myrmica*). Multiple-queening has convergently evolved numerous times in almost every ant subfamily, with the dorylomorph subfamilies ("army ants") probably being the only clade in which colonies usually contain only a single reproductive queen (for exceptions see Rettenmeyer and Watkins 1978; Buschinger, Peeters, and Crozier 1990).

   The transition from the assumedly ancestral single-queening (monogyny) to multiple-queening (polygyny) has been compared with the evolution of insect sociality from solitary species (Keller and Vargo 1993). As in the early stages of the pathway toward obligatory group living, and as in the other major transitions in evolution (Maynard Smith and Szathmáry 1995), individual queens give up their autonomy to cooperate. In such an interaction conflict arises over the egoistic fitness interests of the cooperating queens which forces them to achieve a compromise (e.g., Reeve and Ratnieks 1993). Numerous studies have focused on the ecological causes as well as the social and genetic consequences of variation in queen number within and between ant species.

The Ecology of Queen Number

In principle, stable polygyny can arise through joint colony-founding by multiple queens ("primary polygyny"); the adoption of young, mated queens (normally those of their mothers) in established colonies ("secondary polygyny"); and nest fusion. Of these three pathways, two appear to be extremely uncommon: cooperation among co-foundresses, in most cases, comes to an end once the first workers have eclosed and aggression between them leads to secondary monogyny (Bartz and Hölldobler 1982; Rissing and Pollock 1988; Strassmann 1989; Heinze 1993a; Bernasconi and Strassmann 1999; but see Mintzer 1987; Rissing et al. 1989; Trunzer, Heinze, and Hölldobler 1998; Helms Cahan and Fewell 2004). Nest fusion is restricted primerily to unicolonial species (Passera 1994) or is also associated with subsequent queen elimination (Foitzik and Heinze 1998; Strätz, Strehl, and Heinze 2002). Thus, most cases of polygyny arise through queen adoption.

In this chapter, we focus on secondary polygyny. Queen number in this situation is intricately linked to the dispersal and colony founding strategies of female sexuals (e.g., Rosengren and Pamilo 1983; Rosengren, Sundström, and Fortelius 1993; Keller 1991, 1995), which are reflected in their morphological and physiological adaptations. In obligatorily monogynous species, new colonies are—with few exceptions—founded independently by solitary queens following dispersal and mating flights. In contrast, queens in facultatively polygynous species usually mate in or near their maternal nests, and new colonies may be founded in a dependent way through the fragmentation or budding of established colonies with multiple queens. Because polygyny is often associated with the complete or partial loss of long-range dispersal, the easiest explanation for its occurrence is that its evolution is favored whenever dispersal and solitary founding are exceptionally costly (Cole, this volume). Such conditions might include, among others, strong inter- and intraspecific competition, (e.g., for nest sites) and particularly cold and hot climates (Table 2.1).

Ant colonies may become polygynous when habitats are saturated and empty nest sites are not available for solitary founding. Instead of attempting independent nest foundation, young queens would then benefit from reproducing in the maternal nest. The few vacant nest sites, which may become available over time, are probably more easily located and defended by workers from the maternal nest than by individual queens, which would lead to polydomy and subsequent colony fragmentation. Evidence for the association between polygyny and habitat saturation comes from two sources. First, age of habitat patches in taiga forests was found to be correlated with queen number, perhaps because of variation in nest-site availability and predation rates between different succession stages (Seppä,

Table 2.1. Ecological conditions and morphological, behavioral, and life history traits of queens associated with monogyny or polygyny in ants

Condition/trait	Monogyny	Polygyny
Ecological conditions		
Climate	Tropical—temperate	Extremely dry, hot or cold
Competition	Low to intermediate	Intermediate to high
Population growth	Zero to negative	Positive
Life history		
Reproductive potential	High, late in life	High, early in life
Juvenile mortality	Extremely high	Low to intermediate
Selection on longevity	Extremely high	Low
Behavioral/morphological traits of queens		
Body size relative to workers	Large	Not as large
Fat content	High	Low
Dispersal capability	High	Low
Relatedness to mate	Low	Low to high
Parental care	Shown during colony founding phase	Never expressed

Sources: Hölldobler and Wilson 1990; Keller 1993; Heinze 1993b; Stille 1996; and others.

Sundström, and Punttila 1995). Second, experimental seeding of habitats with artificial nest sites reduced the mean queen number per nest in the facultatively polygynous *Temnothorax longispinosus,* presumably due to the fragmentation of polygynous colonies (Herbers 1986a; Foitzik et al. 2004).

There is as yet only indirect evidence for a negative impact of cold and hot climate on founding success of solitary queens. An association between polygyny and latitude was indicated by a higher proportion of polygynous species in ant communities farther north compared to those in temperate or subtropical climates (Heinze 1993b; Heinze and Hölldobler 1994). However, the ant fauna near the tree-line in both alpine and boreal habitats consists of the same few genera worldwide and data are therefore not phylogenetically independent. Facultative polygyny in the boreal *Formica, Myrmica,* and *Leptothorax* may have arisen independent of cold climate, possibly due to intense intraspecific competition. Hence, more information on the social structure of ant colonies from the tree-line in subantarctic areas are needed to test the hypothesis that cold temperatures are indeed associated with dependent colony founding and polygyny. Interestingly, young queens of the only obligatorily monogynous ant genus in the taiga-tundra ecotone, *Camponotus,* hibernate in the maternal nest before mating and found solitarily in early summer (Eidmann 1943). Like dependent founding, pre-mating hibernation in the maternal nest probably increases the likelihood of surviving the first winter and successfully establishing a new colony.

As endothermic animals it is easy for us to see why solitary hibernation might be more costly when winters are very long and harsh. Ants, however, are ectothermic and spend deep winter in a state of supercooled inactivity (Berman and Zhigulskaya 1995). Workers of boreal *Leptothorax* species can survive temperatures of down to −40°C thanks to the accumulation of high concentrations of glycerol in their hemolymph (Leirikh 1989). Nevertheless, workers survive hibernation less well in isolation than in groups, probably because they have a higher starvation risk. *Leptothorax* are active in the nest at temperatures around 0°C, which are too low for foraging, and during this time some well-nourished workers with full crops provide food to other workers that have fewer reserves. Such food exchange is clearly not possible in isolation (Heinze, Stahl, and Hölldobler 1996). Starvation resistance might also explain why individuals of several insect species, including *Leptothorax* ants, grow to larger average body size at higher latitudes, an ecogeographic trend known as Bergmann's rule

(Heinze et al. 2003). Furthermore, in *T. longispinosus*, queen number and survival rate of workers during winter were positively correlated (Herbers 1986b). Although the causality of this association has not yet been investigated, it is possible that queens themselves have a positive influence on worker survival; one recent study documented that non-laying queens in colonies of *L. gredleri* can significantly boost colony productivity (Heinze and Oberstadt 2003).

A similar reasoning as for cold climates may apply to deserts, semi-deserts, or steppes, where very hot and dry conditions make solitary colony founding exceedingly difficult. Young queens of desert ants, such as *Myrmecocystus* honeypot ants and *Pogonomyrmex* harvester ants, often found their nests after heavy rain falls, which facilitates the digging of deep nest holes. Additionally, queens of many species cooperate during founding, although cooperation in these species is mostly transient and mature colonies are typically monogynous. In several other species from xeric habitats the evidence for an association between hot climate and polygyny is based on the observation that queens lack functional wings and are ergatoid or intermorphic (Briese 1983; Bolton 1986; Tinaut and Heinze 1992; Tinaut and Ruano 1992). Long-range dispersal therefore is impossible, but this does not necessarily mean that queens will return into their maternal nest to become fertile and that colonies will become polygynous. For example, *Cataglyphis cursor* queens attract workers from their maternal nests and found new colonies by budding immediately after mating (Lenoir et al. 1988). The association between wing reduction in queens and polygyny has been demonstrated in only a number of species, such as *Pogonomyrmex imberbiculus* (Heinze, Hölldobler, and Cover 1992) or *Monomorium* spp. (Bolton 1986; Fersch, Buschinger, and Heinze 2000). Briese (1983) suggested that colonies of a certain *Monomorium* species produce wingless female reproductives in response to adverse climatic conditions and winged female sexuals when enough resources are available, but at present the data are not sufficient to show convincingly that these phenomena are indeed adaptations to xeric environments.

Finally, polygyny appears to be favored in disturbed habitats with increasing population sizes. As shown by Tsuji and Tsuji (1996), in such areas early reproduction increases individual fitness more than high productivity later in life. In this situation, queens are under strong selection to produce female and male sexual offspring during their first year of adult life. They can do so only by staying in their maternal nest and taking advantage of the already existing worker force. Polygynous ants can therefore easily spread in man-made and disturbed habitats. Selection for early reproduction probably better explains why invasive ant species are polygynous (Hölldobler and Wilson 1977; Passera 1994; Yamauchi and Ogata 1995) rather than the previously suggested higher risk of queen loss during frequent moving (Hölldobler and Wilson 1977).

Queen Adaptations to Polygyny

The loss of dispersal and solitary founding and the transition to polygyny are associated with numerous social and genetic consequences, which have been discussed elsewhere in detail (e.g., Keller 1993), and are also reflected in morphological specializations of queens (Table 2.1, Figure 2.1). Queens that found dependently are typically smaller than queens from related, monogynous species (Stille 1996, Figure 2.1), and they also have a relatively lower fat content (Keller and Passera 1989). Many facultatively polygynous species are characterized by a pronounced queen polymorphism, with large and well-endowed queens that presumably disperse and found solitarily, and smaller, often wing-reduced queens that found dependently (Buschinger and Heinze 1992; Heinze and Tsuji 1995; Rüppell and Heinze 1999). In such queen-polymorphic species, number and size of queens in a colony are often correlated, giving indirect evidence of presumed alternative reproductive tactics. For example, dealate queens of *Leptothorax* sp. A from Quebec and New England (informally named "*L. ergatogyneus*," Francoeur 2000) are typically found in monogynous nests, whereas wingless, intermorphic queens occur in polygynous nests (Heinze and Buschinger 1989). A similar association has been documented for various other species (Buschinger and Heinze 1992; Rüppell and Heinze 1999; Heinze and Keller 2000; Fersch et al. 2000; Buschinger and Schreiber 2002; Foitzik et al. 2004). Cross-breeding studies and detailed heritability estimates suggest that variation in morphological traits associated with dispersal, be it the presence of wings or wing length, have a genetic basis. Controlled laboratory crosses in *Leptothorax* sp. A and *Myrmecina graminicola* showed that female larvae, which are homozygous for a certain allele, can develop into a winged queen, whereas queen-destined larvae carrying the alternative allele invariably grow to intermorphic, wingless queens (Heinze and Buschinger 1989; Buschinger 2005). The proximate mechanisms underlying queen size polymorphism in *Temnothorax rugatulus*

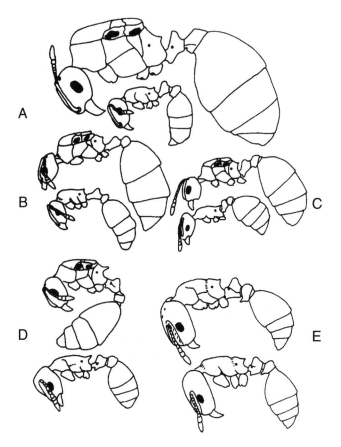

Figure 2.1. Queens and workers of (A) *Tetramorium caespitum* (monogynous), (B) *Temnothorax unifasciatus* and (C) *T. nylanderi* (monogynous), (D) *Leptothorax acervorum* (facultatively polygynous), and (E) the slave-maker *Harpagoxenus sublaevis* (monogynous), demonstrating the larger queen-worker dimorphism in a few monogynous, nonparasitic species.

appear to be more complex, but also involve genetic influences on queen morphology (Rüppell, Heinze, and Hölldobler 2001).

In some species, such as the invasive ants *Monomorium pharaonis, Linepithema humile, Lasius neglectus,* and *Wasmannia auropunctata,* queens invariably appear to be weak dispersers. They carry little fat reserves and are unable to establish a colony without worker assistance (Passera 1994; Hee et al. 2000). For example, queens of *Cardiocondyla minutior* never succeeded in solitary founding and also failed when accompanied by three workers, but quickly managed to start new colonies when assisted by ten or more workers (Alexandra Schrempf, personal observation). Furthermore, female sexuals of many obligatorily polygynous species mate in the maternal nest or its immediate vicinity and pre-mating dispersal has been lost—at least in the female sex. Consequently, matings often involve close relatives. Although genetic variability may be maintained through the adoption of alien males and high queen numbers with consequently low average relatedness among nestmates, in *Linepithema humile* the percentage of inbreeding is not significantly lower than expected under the hypothesis of random mating, suggesting that sib-mating is not avoided (Keller and Fournier 2002).

Finally, queens from obligatorily polygynous species are typically very short-lived compared to queens from monogynous species (Keller and Genoud 1997; Keller 1998). While queens of some monogynous ants are among the invertebrates with the longest life span (up to 25 years and more; Rueppell, this volume), those of *L. humile* or *M. pharaonis* live on average less than one year. As early reproduction counts more than longevity when population size increases, the above-mentioned hypothesis by Tsuji and Tsuji (1996) nicely explains the reduced longevity of queens in polygynous species.

Polygyny therefore appears to evolve toward a life history characterized by weak and nondispersing female sexuals that mate locally and indiscriminately, reproduce early in life, and die soon after laying eggs for a few months. Though exceptions certainly occur, for example in the genus *Formica,* and data are not available on all of these traits, it appears that obligatorily polygynous species with this suite of traits have evolved repeatedly even in predominantly monogynous or at most weakly polygynous genera, as exemplified by *Tetramorium rhenanum* (=*T. moravicum,* Schlick-Steiner et al. 2005), *Lasius sakagamii,* and *Lasius neglectus* (Steiner et al. 2004). Detailed phylogenetic studies on other genera with both monogynous and polygynous taxa will presumably yield many more examples of obligatorily polygynous species being derived from monogynous and facultatively polygynous ancestors.

Reversed Evolution: From Polygyny to Monogyny

The characteristic adaptations of a polygynous queen—weak dispersal capability, intranidal mating strategy, low body reserves, fast reproduction, and short life span—may preclude a reversal to solitary founding. Thus polygyny can turn out to be an evolutionary dead end. However, just as reversals are possible in other major evolutionary transitions, as indicated by the phylogeny of the unicellular Myxozoa, which probably are not protists but degenerate Cnidaria (Siddall et al. 1995), and the fact that sociality has been lost repeatedly in bees (Wcislo and Danforth 1997), we suggest that under certain life histories reversed evolution toward monogyny is possible from facultative polygyny and possibly also from obligatory polygyny. There is evidence for such reversals in slave-making ants (Buschinger, Ehrhardt, and Winter 1980) and the nonparasitic genus *Cardiocondyla*; in both cases, phylogenetic analyses suggest that strict monogyny has evolved from ancestral facultative polygyny (Beibl et al. 2005; Heinze et al. 2005). Because of the very different life histories of social parasites and monogynous *Cardiocondyla*, we discuss these two groups separately.

Queen Number in Social Parasites

The myrmicine tribe Formicoxenini is a hot spot in the evolution of social parasites, with six or more independent origins of slavery alone (Beibl et al. 2005). While all monophyla of slave-makers (*Chalepoxenus, Harpagoxenus, Myrmoxenus, Protomognathus, Temnothorax duloticus,* and a second, as yet undescribed *Temnothorax* slave-maker from North America) are strictly monogynous (Alloway 1979; Buschinger 1966, 1968b; Herbers and Foitzik 2002; Stuart and Alloway 1983; Talbot 1957; Wesson 1939; Wilson 1975), with the exception of *Myrmoxenus algeriana* (Buschinger, Jessen, and Cagniant 1990), at least four of them (*Harpagoxenus, Protomognathus, Temnothorax duloticus, Temnothorax* sp.) parasitize facultatively polygynous *Temnothorax* or *Leptothorax* host species (Alloway et al. 1982; Buschinger 1968a). Assuming that slave-makers indeed evolved from the clade of species serving them as slaves as suggested by Emery's rule (Emery 1909; Bourke and Franks 1991; Buschinger 1990)—and this is highly likely at least for *Temnothorax duloticus* and *T.* sp. (Beibl et al. 2005)—the evolution of slave-making must have been associated with a switch from ancestral facultative polygyny to derived, obligatory monogyny. A similar situation appears to exist in the formicine genus *Rossomyrmex,* monogynous slave-makers of facultatively polygynous *Proformica* spp., though at present it is not absolutely clear whether Emery's rule does apply in this case (Hasegawa, Tinaut, and Ruano 2002). The obligatory myrmicine slave-makers *Strongylognathus, Myrmoxenus,* and *Chalepoxenus* are also monogynous, but predominantly exploit related monogynous slave species.

The reversal to monogyny in the four formicoxenine clades might be caused by the relaxation or absence of those ecological constraints that supposedly forced the slave species to adopt daughter queens, such as nest site limitation (Herbers 1986a) and a high mortality of solitarily hibernating queens (Heinze 1993b; Heinze et al. 1996). Queens of slave-making ants circumvent these constraints in that they never found solitarily but instead invade host nests. In addition, polygynous species often show low queen dispersal as colonies reproduce by budding, which could lead to the local overexploitation of a host population. Thus, polygyny in slave-making ants might be selected against due to the reduction of ecological constraints in the founding phase and the risk of local extinction of the host.

Parasitic founding does not require large body reserves but is nevertheless a risky task leading to a high juvenile mortality of queens that might select for large body size and a prolonged life span. Slave-maker queens have to single-handedly usurp a host colony, kill or expel the resident queen(s), and either drive away all adult host workers or become adopted by them. The queens of those slave-makers which engage in severe fighting with host workers, such as *Harpagoxenus* and *Protomognathus,* are large compared to host queens, whereas *Myrmoxenus* queens, which sneak into the host colony without much fighting, are often smaller than their host queens. On the one hand, there is evidence for strong selection on body size, depending on parasitic founding strategy; on the other hand, data are insufficient to determine whether slave-maker queens indeed are more long-lived than queens of their host species. Both slave-maker and host queens have been kept for ten or more artificial, shortened breeding cycles in the laboratory (Buschinger 1974), but a systematic comparison of polygynous host and monogynous slave-maker species has never been conducted.

In stark contrast to the situation in formicoxenine slave-makers, the facultatively slave-making formicine species *Formica sanguinea* appears to be weakly polygynous (Pamilo 1981; Pamilo and Seppä 1994) and the same has been suggested for a few species of the obligatorily slave-making formicine genus *Polyergus* (Goodloe and Sanwald 1985). In contrast to the slave raids of Formicoxenini, which usually result in the extinction of the raided host colony (Foitzik and Herbers 2001; Fischer-Blass, Heinze, and Foitzik 2006), those of *Formica* and *Polyergus* are less destructive, and polygynous slave-maker colonies probably are less prone to exhaust a local host population (but see Yasuno 1964). Furthermore, queens of these species have been observed to return into their maternal nests after mating and to found new colonies either by usurping a slave colony directly during a slave raid (Mori and Le Moli 1998; Mori et al. 2001) or by budding (Marlin 1968). Such a life history might eventually result in the co-occurrence of two reproductive queens in a single colony.

An interesting pattern emerges in the inquilines, social parasites, which have lost their worker caste (Hölldobler and Wilson 1990). Inquiline queens either kill the slave queen and rapidly start reproducing or allow the host queen to stay alive and reproduce at a moderate pace. In the case of the queen-intolerant inquilines, monogyny is the most common social organization. As in the formicoxenine slave-makers, obligate monogyny appears to have evolved from facultative polygyny at least in *Leptothorax paraxenus, L. wilsoni,* and *L. goesswaldi* (Buschinger 1990; Heinze 1995). Colonies of queen-intolerant parasites are doomed, as dying slave workers are not replaced. Parasite queens are therefore expected to transform all resources available in the slave colony into their own sexual offspring as quickly and completely as possible (e.g., Heinze and Tsuji 1995; Bekkevold and Boomsma 2000). Toleration of other parasite queens or even the adoption of daughter queens into the declining slave colony should be selected against. The highly virulent queen-intolerant parasites are thus expected to be comparatively short-lived, albeit this has not been systematically studied.

In queen-tolerant inquilines, natural selection may act in a different way: because the host queen stays alive and can replenish the stock of slave workers, parasite colonies survive much longer, in the case of large polygynous slave colonies, theoretically indefinitely. Instead of risking the takeover of a new host colony alone, queen-tolerant parasites can remain and reproduce in the maternal nest or spread by budding with the slave colony. The possibility of vertical transmission thus selects for polygyny in the inquiline, and indeed, queen-tolerant inquilines are frequently polygynous, regardless of the queen number of their hosts (e.g., *Kyidris yaleogyna, Leptothorax kutteri, Myrmecia inquilina, Myrmica hirsuta, Plagiolepis xene, Pogonomyrmex colei, Solenopsis daguerrei, Temnothorax minutissimus, Vollenhovia nipponica*) (Buschinger and Linksvayer 2004; Calcaterra et al. 2001; Douglas and Brown 1959; Elmes 1978; Johnson 1996; Kinomura and Yamauchi 1992; Wilson and Brown 1956).

Monogyny in *Cardiocondyla*

The genus *Cardiocondyla* comprises more than 100 species of tiny ants predominantly from the Old World tropics. All studied tropical species are polygynous, and several are notorious tramp species and have become distributed around the world through human commerce (Passera 1994; Heinze et al. 2006). In contrast, a couple of species which live in desert and steppe habitats and other xeric environments in Eurasia appear to be monogynous. Indeed, for *C. batesii* from Southern Spain, *C. elegans* from France, and *C. nigra* from Cyprus, both population genetic and behavioral studies demonstrated strict monogyny (Schrempf, Reber, et al. 2005; Lenoir et al. 2007; Schrempf Alexandra pers. observation), and all excavated colonies of *C. ulianini* from Central Asia contained only a single fertile queen each (Marikovskii and Yakushkin 1974). A phylogeny based on mitochondrial DNA sequences indicates that these monogynous species, together with a number of other Eurasian species with unknown social structure, form a monophyletic group nested in a tree otherwise composed of polygynous taxa (Figure 2.2). The probability that monogyny was the ancestral state in *Cardiocondyla* is negligible based on the available genetic data, suggesting that monogyny evolved at least once from polygyny. In the following paragraphs we investigate which evolutionary pathway *Cardiocondyla* species used in their interesting reversal toward monogyny and which traits are important for adaptation to independent colony foundation.

Cardiocondyla are famous for the occurrence of a peculiar male polymorphism with dispersing, winged males and territorial, wingless males ("ergatoid males," Kugler 1983; Stuart, Francoeur, and Loiselte 1987; Kinomura and Yamauchi 1987; Yamauchi and Kinomura 1993; Boomsma, Kronauer, and Pederson, this volume). Queens mate with winged and/or ergatoid males in the nest and either become fertile in the maternal colony, emigrate with workers, or perhaps join unrelated colonies. Solitary founding by queens appears unlikely at least in the tropical *C. minutior* and *C. obscurior,* and indeed only very few solitary queens have been found in the field (Heinze and Delabie 2005).

Queens of monogynous *C. batesii, C. nigra,* and *C. elegans* mate and hibernate in the maternal nest and shed their wings before dispersing on foot (Heinze et al. 2002; Lenoir et al., 2007; Schrempf, Alexandra pers. observation). Fertile queens are intolerant of each other and do not jointly

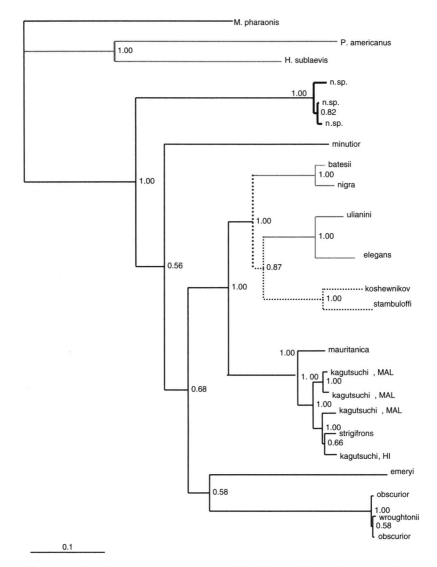

Figure 2.2. Majority rule consensus tree of *Cardiocondyla* ants from complete COI/COII and 16SrRNA sequences, based on 3,940 trees from a Bayesian analysis. The numbers give Bayesian probabilities. Thin lines indicate taxa that are monogynous. Lines in black are taxa that are (facultatively) polygynous. Dotted lines represent species of unknown queen number. A reconstruction of queen numbers at basal nodes of the phylogeny suggests that monogyny is derived and monogynous species evolved once or twice from polygynous ancestors (for details see Heinze et al. 2005).

start new colonies. As queen-worker dimorphism is only slightly more pronounced in monogynous than polygynous *Cardiocondyla* (Seifert 2003), foundresses of monogynous species cannot rely on histolysis of their body tissue to rear their first offspring. Instead, queens forage during the founding period. Such "semi-claustral founding," though often considered a primitive trait, has recently been documented also in the more derived ant subfamilies (Brown 1999; Brown and Bonhoeffer 2002; Johnson 2002) and is probably more widespread, particularly in genera with small colony size and low caste dimorphism. Monogynous *Cardiocondyla* are characterized by a wing dimorphism in queens (Heinze et al. 2002; Seifert 2003; Schrempf and Heinze 2007). In contrast to the established or assumed associations between queen polymorphism and alternative dispersal tactics in facultatively polygynous ants, both long-winged and short-winged *Cardiocondyla* queens found solitarily. Strangely enough, short-winged queens are more successful foundresses, because wing size reduction is associated with the replacement of unnecessary flight muscles by fat (Figure 2.3). Long-winged *C. batesii* queens laid their first eggs significantly later than short-winged queens of *C. batesii* and *C. nigra*, had a much lower egg laying rate and, as a result, had a higher failure rate in colony founding (Schrempf and Heinze 2007).

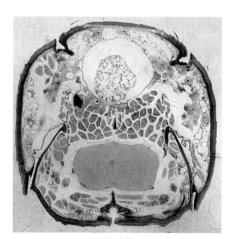 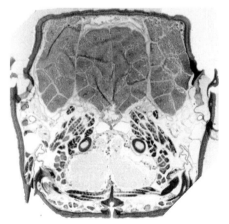

Figure 2.3. Histological section through the alitrunks of a short-winged female sexual (left) and a long-winged female sexual (right) of *Cardiocondyla batesii*, showing that the flight muscles are strongly reduced and presumably replaced by fat in short-winged sexuals (courtesy of B. Lautenschläger and A. Schrempf).

Queens of polygynous *Cardiocondyla* resemble other predominantly or obligatorily polygynous ants in having a very short life span, usually much less than one year (e.g., Heinze, Hölldobler, and Yamauchi 1998; Schrempf, Heinze, and Cremer 2005). In contrast, queens of monogynous *Cardiocondyla* have been observed to survive for two years and longer. A simple comparison of species from nonseasonal and seasonal environments is likely to be imprecise, but queens of monogynous *Cardiocondyla* outlived their polygynous relatives even when their life span was corrected for seasonality by subtracting the supposedly inactive hibernation time (Schrempf and Heinze 2007). Nevertheless, queens of *C. batesii* and *C. nigra* are ephemeral when compared to those of other monogynous ants with similar colony size, such as *Temnothorax* (Plateaux 1986; Keller 1998). This might eventually result in a rather high percentage of orphaned colonies in the field, which could potentially be usurped by young, founding queens (Lenoir et al. 2007).

Monogynous *Cardiocondyla* therefore appear to have reverted to independent founding, albeit in a rather unusual way, with no or almost no long-range dispersal. This leaves the second problem associated with the switch from polygyny to monogyny: inbreeding, which inevitably arises from intranidal mating and monogyny. Inbreeding is generally detrimental in that it increases the homozygosity of lethal or semilethal recessive alleles, but in many ants and other Hymenoptera with single-locus complementary sex determination (sl-CSD) it is also associated with the production of sterile, diploid males (Cook 1993; Cook and Crozier 1995). Genetic studies reveal that more than 80% of all matings in *C. batesii* are between brothers and sisters (Schrempf, Reber, et al. 2005), however, inbreeding apparently does not result in the production of large numbers of diploid males (Schrempf, Aron, and Heinze 2006). Sex in *Cardiocondyla* therefore appears to be determined by other mechanisms than sl-CSD, either by multiple sex loci or by genomic imprinting as in the parasitoid wasp *Nasonia* (Dobson and Tanouye 1998).

Given the life history of monogynous *Cardiocondyla* with wingless males and flightless female sexuals, it is surprising that 20% of all matings in *C. batesii* are not between close relatives. There is some evidence that both sexes can disperse on foot and in this way attempt to reach other colonies. Bolton (1982; pers. comm.) found a male of *C. emeryi* outside of a nest, and J. L. Mercier and colleagues (personal communication,) report the same for *C. elegans*.

At present, the ultimate causes of the transition to monogyny remain obscure. Data on the colony structure of additional species are needed to corroborate the hypothesis that monogyny is indeed associated with life in more seasonal, xeric environments in contrast to what has been observed for other ant taxa. The reversal is made possible by foraging by founding queens and the replacement of wing muscles by fat, which facilitates colony founding by short-winged queens.

Conclusions

These two examples document that a reversal from ancestral polygyny to derived monogyny is possible, although it appears to be associated with novel and unusual colony founding tactics. More cases of such an evolutionary transition might be detected by phylogenetic analyses of ant genera containing both monogynous and polygynous taxa. An indication for the potential occurrence of such an evolutionary reversal could be monogynous species with atypical queen nest foundation habits. A promising candidate is the formicine ant genus *Cataglyphis,* in which both polygynous and monogynous species occur (e.g., Dahbi et al. 1996; De Haro and Cerdá 1984; Wehner, Wehner, and Agosti 1994). As in *Cardiocondyla, Cataglyphis* queens can be short-winged or completely wingless (Tinaut and Ruano 1992). Young queens of monogynous *C. cursor* do not disperse far for mating and are assisted during colony founding by workers from their maternal colonies (Lenoir et al. 1988; Clémencet, Vignier, and Doums 2005).

Evolutionary reversals toward obligatory monogyny might be rare and occur predominantly in special biological settings. In addition, they appear to follow new evolutionary pathways involving novel behavioral and morphological traits. Investigations of these systems might help us to better understand the ecological and genetic causes underlying the evolution of social organizations in ants. The evolution of new traits associated with secondary monogyny can further elucidate how similar selection pressures result in different evolutionary solutions.

Acknowledgments

Supported by Deutsche Forschungsgemeinschaft (He 1623/15–1,2).

Literature Cited

Alloway, T. M. 1979. "Raiding behaviour of two species of slave-making ants, *Harpagoxenus americanus* (Emery) and *Leptothorax duloticus* Wesson (Hymenoptera: Formicidae)." *Animal Behaviour* 27: 202–210.

Alloway, T. M., A. Buschinger, M. Talbot, R. Stuart, and D. Thomas. 1982. "Polygyny and polydomy in three North American species of the ant genus *Leptothorax* Mayr (Hymenoptera: Formicidae)." *Psyche* 89: 249–274.

Bartz, S. H., and B. Hölldobler. 1982. "Colony founding in *Myrmecocystus mimicus* Wheeler (Hymenoptera: Formicidae) and the evolution of foundress associations." *Behavioral Ecology and Sociobiology* 10: 137–147.

Beibl, J., R. J. Stuart, J. Heinze, and S. Foitzik. 2005. "Six origins of slavery in formicoxenine ants." *Insectes Sociaux* 52: 291–297.

Bekkevold, D., and J. J. Boomsma. 2000. "Evolutionary transition to a semel-parous life history in the socially parasitic ant *Acromyrmex insinuator.*" *Journal of Evolutionary Biology* 13: 615–623.

Berman, D. I., and Z. A. Zhigulskaya. 1995. "Cold resistance of the ants of the north-west and north-east of the palaearctic region." *Acta Zoologica Fennica* 199: 73–80.

Bernasconi, G., and J. E. Strassmann. 1999. "Cooperation among unrelated individuals: The ant foundress case." *Trends in Ecology and Evolution* 14: 477–482.

Bolton, B. 1982. "Afrotropical species of the myrmicine genera *Cardiocondyla, Leptothorax, Melissotarsus, Messor* and *Cataulacus* (Formicidae)." *Bulletin of the British Museum (Natural History) Entomology* 45: 307–370.

———. 1986. "Apterous females and shift of dispersal strategy in the *Monomorium salomonis*-group (Hymenoptera: Formicidae)." *Journal of Natural History* 20: 267–272.

Bourke, A. F. G., and N. R. Franks. 1991. "Alternative adaptations, sympatric speciation and the evolution of parasitic, inquiline ants." *Biological Journal of the Linnean Society* 43: 157–178.

———. 1995. *Social evolution in ants.* Princeton: Princeton University Press.

Briese, D. T. 1983. "Different modes of reproductive behaviour (including a description of colony fission) in a species of *Chelaner* (Hymenoptera: Formicidae)." *Insectes Sociaux* 30: 308–316.

Brown, M. J. F. 1999. "Semi-claustral founding and worker behaviour in gynes of *Messor andrei.*" *Insectes Sociaux* 46: 194–195.

Brown, M. J. F., and S. Bonhoeffer. 2002. "On the evolution of claustral colony founding in ants." *Evolutionary Ecology Research* 5: 305–313.

Buschinger, A. 1966. "Untersuchungen an *Harpagoxenus sublaevis* Nyl. (Hym. Formicidae). I. Freilandbeobachtungen zu Verbreitung und Lebensweise." *Insectes Sociaux* 13: 5–16.

————. 1968a. "Mono- und Polygynie bei Arten der Gattung *Leptothorax* Mayr (Hymenoptera Formicidae)." *Insectes Sociaux* 15: 217–226.

————. 1968b. "Untersuchungen an *Harpagoxenus sublaevis* Nyl. (Hymenoptera, Formicidae), III: Kopula, Koloniegründung, Raubzüge." *Insectes Sociaux* 15: 89–104.

————. 1974. "Monogynie und Polygynie in Insektensozietäten." In G.H. Schmidt, ed., *Sozialpolymorphismus bei Insekten*, 862–896. Stuttgart: Wissenschaftliche Verlagsgesellschaft.

————. 1990. "Sympatric speciation and radiative evolution of socially parasitic ants—heretic hypotheses and their factual background." *Zeitschrift für Zoologische Systematik und Evolutionsforschung* 28: 241–260.

————. 2005. "Proof of genetically mediated queen polymorphism in the ant, *Myrmecina graminicola* (Hymenoptera, Formicidae)." *Entomologia Generalis* 27: 185–200.

Buschinger, A., and J. Heinze. 1992. "Polymorphism of female reproductives in ants." In *Biology and Evolution of Social Insects,* ed. J. Billen, 11–23. Leuven: Leuven University Press.

Buschinger, A., and T. A. Linksvayer. 2004. "Novel blend of life history traits in an inquiline ant, *Temnothorax minutissimus*, with a description of the male (Hymenoptera: Formicidae)." *Myrmecologische Nachrichten* 6: 67–76.

Buschinger, A., W. Ehrhardt, and U. Winter. 1980. "The organization of slave-raids in dulotic ants—A comparative study (Hymenoptera; Formicidae)." *Zeitschrift für Tierpsychologie* 53: 245–264.

Buschinger, A., K. Jessen, and H. Cagniant. 1990. "The life history of *Epimyrma algeriana,* a slave-making ant with facultative polygyny (Hymenoptera, Formicidae)." *Zoologische Beiträge Neue Folge* 33: 23–49.

Buschinger, A., C. Peeters, and R. H. Crozier. 1990. "Life-pattern studies of an Australian *Sphinctomyrmex* (Formicidae: Ponerinae; Cerapachyini): Functional polygyny, brood periodicity and raiding behavior." *Psyche* 96: 287–300.

Buschinger, A., and M. Schreiber. 2002. "Queen polymorphism and queen-morph related facultative polygyny in the ant, *Myrmecina graminicola* (Hymenoptera, Formicidae)." *Insectes Sociaux* 49: 344–353.

Calcaterra, L. A., J. A Briano, D. F. Williams, and D. H. Oi. 2001. "Observations on the sexual castes of the fire ant parasite *Solenopsis daguerrei* (Hymenoptera: Formicidae)." *Florida Entomologist* 84: 446–448.

Clémencet, J., B. Vignier, and C. Doums. 2005. "Hierarchical analysis of population genetic structure in the monogynous ant *Cataglyphis cursor* using microsatellite and mitochondrial DNA markers." *Molecular Ecology* 14: 3735–3744.

Cook, J. M. 1993. "Sex determination in the Hymenoptera: A review of models and evidence." *Heredity* 71: 421–435.

Cook, J. M. and Crozier, R. H. 1995. "Sex determination and population biology in the Hymenoptera." *Trends in Ecology and Evolution* 10: 281–286.

Crozier, R. H., and P. Pamilo. 1996. *Evolution of social insect colonies.* Oxford: Oxford University Press.

Dahbi, A., X. Cerdá, A. Hefetz, and A. Lenoir. 1996. "Social closure, aggressive behavior and cuticular hydrocarbon profiles in the polydomous ant *Cataglyphis iberica* (Hymenoptera Formicidae)." *Journal of Chemical Ecology* 22: 2173–2186.

De Haro, A., and X. Cerdá. 1984. "Communication entre nids à travers le transport d'ouvrières chez *Cataglyphis iberica* Emery 1906 (Hym. Formicidae)." In *Processus d'acquisition précoce. Le communications,* ed. A. de Haro and X. Espadaler, 53–56. Barcelona: Publicaciones de la Universidad Autónoma de Barcelona.

Dobson, S. L., and M. A. Tanouye. 1998. "Evidence for a genomic imprinting sex determination mechanism in *Nasonia vitripennis* (Hymenoptera; Chalcidoidea)." *Genetics* 149: 233–242.

Douglas, A., and W. L. Brown. 1959. "*Myrmecia inquilina* new species: The first parasite among the lower ants." *Insectes Sociaux* 6: 13–19.

Eidmann, H. 1943. "Die Überwinterung der Ameisen." *Zeitschrift für Morphologie und Ökologie der Tiere* 39: 217–225.

Elmes, G. W. 1978. "A morphometric comparison of three closely related species of *Myrmica* (Formicidae), including a new species from England." *Systematic Entomology* 3: 131–145.

Emery, C. 1909. "Über den Ursprung der dulotischen, parasitischen und myrmekophilen Ameisen." *Biologisches Centralblatt* 29: 352–362.

Fersch, R., A. Buschinger, and J. Heinze. 2000. "Queen polymorphism in the Australian ant *Monomorium* sp. 10." *Insectes Sociaux* 47: 280–284.

Fischer-Blass, B., J. Heinze, and S. Foitzik. 2006. "Microsatellite analysis revealed strong, but differential impact of a social parasite on its two host species." *Molecular Ecology* 15: 863–872.

Foitzik, S., V. L. Backus, A. Trindl, and J. M. Herbers. 2004. "Ecology of *Leptothorax* ants: Impact of food, nest sites and social parasites." *Behavioral Ecology and Sociobiology* 55: 484–493.

Foitzik, S., and J. Heinze. 1998. "Nest site limitation and colony take over in the ant, *Leptothorax nylanderi.*" *Behavioral Ecology* 9: 367–375.

Foitzik, S., and J. M. Herbers. 2001. "Colony structure of a slave-making ant: II. Frequency of slave raids and impact on host population." *Evolution* 55: 316–323.

Francoeur, A. 2000. "Liste des espéces de fourmis (Formicides: Hyménoptères)." *Entomofaune du Québec,* online at http://entomofaune.qc.ca/publicat/DF01-formicides.pdf

Goodloe, L., and R. Sanwald. 1985. "Host specificity in colony-Founding by *Polyergus lucidus* queens (Hymenoptera: Formicidae)." *Psyche* 92: 297–302.

Hasegawa, E., A. Tinaut, and F. Ruano. 2002. "Molecular phylogeny of two slave-making ants: *Rossomyrmex* and *Polyergus* (Hymenoptera: Formicidae)." *Annales Zoologici Fennici* 39: 267–271.

Hee, J. J., D. A. Holway, A. V. Suarez, and T. J. Case. 2000. "Role of progagule size in the success of incipient colonies of the invasive Argentine ant." *Conservation Biology* 14: 559–563.

Heinze, J. 1993a. "Queen-queen interactions in polygynous ants." In L. Keller, ed., *Queen number and sociality in insects*, 334–361. Oxford: Oxford University Press.

———. 1993b. "Life history strategies of subarctic ants." *Arctic* 46: 354–358.

———. 1995. "The origin of workerless social parasites in formicoxenine ants." *Psyche* 102: 195–214.

Heinze, J., and A. Buschinger. 1989. "Queen polymorphism in *Leptothorax* sp. A: Its genetic and ecological background (Hymenoptera: Formicidae)." *Insectes Sociaux* 36: 139–155.

Heinze, J., S. Cremer, N. Eckl, and A. Schrempf. 2006. "Stealthy invaders: The biology of *Cardiocondyla* tramp ants." *Insectes Sociaux* 53: 1–17.

Heinze, J., and J. H. C. Delabie. 2005. "Population structure of the male-polymorphic ant *Cardiocondyla obscurior*." *Studies on Neotropical Fauna and Environment* 40: 187–190.

Heinze, J., S. Foitzik, B. Fischer, T. Wanke, and V. E. Kipyatkov. 2003. "The significance of latitudinal variation in body size in a holarctic ant, *Leptothorax acervorum*." *Ecography* 26: 349–355.

Heinze, J., and B. Hölldobler. 1994. "Ants in the cold." *Memorabilia Zoologica* 48: 99–108.

Heinze, J., B. Hölldobler, and S. P. Cover. 1992. "Queen polymorphism in the North American harvester ant, *Ephebomyrmex imberbiculus*." *Insectes Sociaux* 39: 267–273.

Heinze, J., B. Hölldobler, and K. Yamauchi. 1998. "Male competition in *Cardiocondyla* ants." *Behavioral Ecology and Sociobiology* 42: 239–246.

Heinze, J., and L. Keller. 2000. "Alternative reproductive strategies: A queen perspective in ants." *Trends in Ecology and Evolution* 15: 508–512.

Heinze, J., and B. Oberstadt. 2003. "Costs and benefits of subordinate queens in colonies of the ant, *Leptothorax gredleri*." *Naturwissenschaften* 90: 513–516.

Heinze, J., A. Schrempf, B. Seifert, and A. Tinaut. 2002. "Queen morphology and dispersal tactics in the ant, *Cardiocondyla batesii*." *Insectes Sociaux* 49: 129–132.

Heinze, J., M. Stahl, and B. Hölldobler. 1996. "Ecophysiology of hibernation in boreal *Leptothorax* ants (Hymenoptera: Formicidae)." *Écoscience* 3: 429–435.

Heinze, J., A. Trindl, B. Seifert, and K. Yamauchi. 2005. "Evolution of male morphology in the ant genus *Cardiocondyla.*" *Molecular Phylogenetics and Evolution* 37: 278–288.

Heinze, J., and K. Tsuji. 1995. "Ant reproductive strategies." *Researches on Population Ecology* 37: 135–149.

Helms Cahan, S., and J. H. Fewell 2004. "Division of labor and the evolution of task sharing in queen associations of the harvester ant *Pogonomyrmex californicus.*" *Behavioral Ecology and Sociobiology* 56: 9–17.

Herbers, J. M. 1986a. "Nest site limitation and facultative polygyny in the ant *Leptothorax longispinosus.*" *Behavioral Ecology and Sociobiology* 19: 115–122.

———. 1986b. "Effects on ecological parameters on queen number in *Leptothorax longispinosus* (Hymenoptera: Formicidae)." *Journal of the Kansas Entomological Society* 59: 675–686.

Herbers, J. M., and S. Foitzik. 2002. "The ecology of slavemaking ants and their hosts in north temperate forests." *Ecology* 83: 148–163.

Hölldobler, B., and E. O. Wilson. 1977. "The number of queens: An important trait in ant evolution." *Naturwissenschaften* 64: 8–15.

———. 1990. *The ants.* Cambridge: Belknap Press of Harvard University Press.

Johnson, R. A. 1996. "Distribution and natural history of the workerless inquiline ant *Pogonomyrmex anergismus* Cole (Hymenoptera: Formicidae)." *Psyche* 101: 257–262.

———. 2002. "Semi-claustral colony founding in the seed-harvester ant *Pogonomyrmex californicus:* A comparative analysis of colony founding strategies." *Oecologia* 132: 60–67.

Keller, L. 1991. "Queen number, mode of colony founding, and queen reproductive success in ants (Hymenoptera Formicidae)." *Ethology, Ecology and Evolution* 3: 307–316.

———, ed. 1993. *Queen number and sociality in insects.* Oxford: Oxford University Press.

———. 1995. "Social life: The paradox of multiple-queen colonies." *Trends in Ecology and Evolution* 10: 355–360.

———. 1998. "Queen lifespan and colony characteristics in ants and termites." *Insectes Sociaux* 45: 235–246.

Keller, L., and D. Fournier. 2002. "Lack of inbreeding avoidance in the Argentine ant *Linepithema humile.*" *Behavioral Ecology* 13: 28–31.

Keller, L., and M. Genoud. 1997. "Extraordinary lifespans in ants: A test of evolutionary theories of ageing." *Nature* 389: 958–960.

Keller, L., and L. Passera. 1989. "Size and fat content of gynes in relation to the mode of colony founding in ants (Hymenoptera: Formicidae)." *Oecologia* 80: 236–240.

Keller, L., and E. L. Vargo. 1993. "Reproductive structure and reproductive roles in colonies of eusocial insects." In *Queen number and sociality in insects,* ed. L. Keller, 16–44. Oxford: Oxford University Press.

Kinomura, K., and K. Yamauchi. 1987. "Fighting and mating behaviors of dimorphic males in the ant *Cardiocondyla wroughtoni.*" *Journal of Ethology* 5: 75–81.

———. 1992. "A new workerless socially parasitic species of the genus *Vollenhovia* (Hymenoptera, Formicidae) from Japan." *Japanese Journal of Entomology* 60: 203–206.

Kugler, J. 1983. "The males of *Cardiocondyla* Emery (Hymenoptera: Formicidae) with the description of the winged male of *Cardiocondyla wroughtoni* (Forel)." *Israel Journal of Entomology* 17: 1–21.

Leirikh, A. N. 1989. "Sezonnye izmeneniya kholodoustoichivosti murav'ev na verkhnei Kolyme." *Izvestiya Akademii Nauk SSSR. Seriia Biologicheskaia* 5: 752–759

Lenoir, A., L. Querard, N. Pondicq, and F. Berton. 1988. "Reproduction and dispersal in the ant *Cataglyphis cursor* (Hymenoptera, Formicidae)." *Psyche* 95: 21–44.

Lenoir, J.-C., A. Schrempf, A.Lenoir, J. Heinze, and J.-L. Mercier. 2007. "Genetic structure and reproductive strategy of the ant *Cardiocondyla elegans:* Strictly monogynous nests invaded by unrelated sexuals." *Molecular Ecology* 16: 345–354.

Marikovskii, P. I., and V. T. Yakushkin. 1974. "Muraveij *Cardiocondya uljanini* Em., 1889 i sistematicheskoe polozhenie 'parraziticheskogo murav'ya *Xenometra.*'" *Izvestiya Akademii Nauk Kazakhskoi SSR. Seriya Biologicheskikh Nauk* 3: 57–62.

Marlin, J. C. 1968. "Notes on a new method of colony formation employed by *Polyergus lucidus lucidus* Mayr (Hymenoptera: Formicidae)." *Transactions of the Illinois State Academy of Science* 61: 207–209.

Maynard Smith, J., and E. Szathmáry. 1995. *The major transitions in evolution.* Oxford, UK: W.H. Freeman.

Mintzer, A. C. 1987. "Primary polygyny in the ant *Atta texana:* Number and weight of females and colony foundation success in the laboratory." *Insectes Sociaux* 34: 108–117.

Mori, A., D. A. Grasso, R. Visicchio, and F. Le Moli. 2001. "Comparison of reproductive strategies and raiding behaviour in facultative and obligatory slave-making ants: The case of *Formica sanguinea* and *Polyergus rufescens.*" *Insectes Sociaux* 48: 302–314.

Mori, A., and F. Le Moli. 1998. "Mating behavior and colony founding of the slave-making ant *Formica sanguinea* (Hymenoptera: Formicidae)." *Journal of Insect Behavior* 11: 235–245.

Pamilo, P. 1981. "Genetic organization of *Formica sanguinea* populations." *Behavioral Ecology and Sociobiology* 9: 45–50.

Pamilo, P., and P. Seppä. 1994. "Reproductive competition and conflicts in colonies of the ant *Formica sanguinea*." *Animal Behaviour* 48: 1201–1206.

Passera, L. 1994. "Characteristics of tramp species." In *Exotic ants: Biology, impact, and control of introduced species,* ed. D. F. Williams, 23–43. Boulder: Westview Press.

Plateaux, L. 1986. "Comparaison des cycles saisonniers, des durées des sociétés et des productions des trois espèces de fourmis *Leptothorax (Myrafant)* du groupe *nylanderi*." *Actes des Colloques Insectes Sociaux* 3: 221–234.

Reeve, H. K., and F. L. W. Ratnieks. 1993. "Queen-queen conflicts in polygynous societies: Mutual tolerance and reproductive skew." In L. Keller, ed., *Queen number and sociality in insects,* 45–85. Oxford: Oxford University Press.

Rettenmeyer, C. W., and J. F. Watkins. 1978. "Polygyny and monogyny in army ants (Hymenoptera: Formicidae)." *Journal of the Kansas Entomological Society* 51: 581–591.

Rissing, S. W., and G. B. Pollock. 1988. "Pleometrosis and polygyny in ants." In R. L. Jeanne, ed., *Interindividual behavioral variability in social insects,* 179–222. Boulder: Westview Press.

Rissing, S. W., G. B. Pollock, M. R. Higgins, R. H. Hagen, and D. R. Smith. 1989. "Foraging specialization without relatedness or dominance among co-founding ant queens." *Nature* 338: 420–422.

Rosengren, R., and P. Pamilo. 1983. "The evolution of polygyny and polydomy in mound-building *Formica* ants." *Acta Entomologica Fennica* 42: 65–77.

Rosengren, R., L. Sundström, and W. Fortelius. 1993. "Monogyny and polygyny of *Formica* ants: A result of alternative dispersal tactics." In L. Keller, ed., *Queen number and sociality in insects,* 308–333. Oxford: Oxford University Press.

Rüppell, O., and J. Heinze. 1999. "Alternative reproductive tactics in females: The case of size polymorphism in winged ant queens." *Insectes Sociaux* 46: 6–17.

Rüppell, O., J. Heinze, and B. Hölldobler. 2001. "Complex determination of queen body size in the queen size dimorphic ant *Leptothorax rugatulus* (Formicidae: Hymenoptera)." *Heredity* 87: 33–40.

Schlick-Steiner, B. C., F. M. Steiner, M. Sanetra, G. Heller, C. Stauffer, E. Christian, and B. Seifert. 2005. "Queen size dimorphism in the ant *Tetramorium moravicum* (Hymenoptera, Formicidae): Morphometric, molecular genetic and experimental evidence." *Insectes Sociaux* 52: 186–193.

Schrempf, A., S. Aron, and J. Heinze. 2006. "Sex determination and inbreeding depression in an ant with regular sib-mating." *Heredity* 97: 75–80.

Schrempf, A., and J. Heinze. 2007. "Back to one: Consequences of derived monogyny in an ant with polygynous ancestors." *Journal of Evolutionary Biology* 20: 792–799.

Schrempf, A., J. Heinze, and S. Cremer. 2005. "Sexual cooperation: Mating increases longevity in ant queens." *Current Biology* 15: 267–270.

Schrempf, A., C. Reber, A. Tinaut, and J. Heinze. 2005. "Inbreeding and local mate competition in the ant *Cardiocondyla batesii*." *Behavioral Ecology and Sociobiology* 57: 502–510.

Seifert, B., 2003. "The ant genus *Cardiocondyla* (Insecta: Hymenoptera: Formicidae)—A taxonomic revision of the *C. elegans, C. bulgarica, C. batesii, C. nuda, C. shuckardi, C. stambuloffii, C. wroughtonii, C. emeryi,* and *C. minutior* species groups." *Annalen des Naturhistorischen Museums Wien* 104B: 203–338.

Seppä, P., L. Sundström, and P. Punttila. 1995. "Facultative polygyny and habitat succession in boreal ants." *Biological Journal of the Linnean Society* 56: 533–551.

Siddall, M. E., D. S. Martin, D. Bridge, S. Desser, and D. K. Cone. 1995. "The demise of a phylum of protists: Phylogeny of Myxozoa and other parasitic cnidaria." *Journal of Parasitology* 81: 961–967.

Steiner, F. M., B. C. Schlick-Steiner, S. Schödl, X. Espadaler, B. Seifert, E. Christian, and C. Stauffer. 2004. "Phylogeny and bionomics of *Lasius austriacus* (Hymenoptera, Formicidae)." *Insectes Sociaux* 51: 24–29.

Stille, M. 1996. "Queen/worker thorax volume ratios and nest-founding strategies in ants." *Oecologia* 105: 87–93.

Strassmann, J. E. 1989. "Altruism and relatedness at colony foundation in social insects." *Trends in Ecology and Evolution* 4: 371–374.

Strätz, M., C. P. Strehl, and J. Heinze. 2002. "Behavior of usurping queens in colonies of the ant species *Leptothorax nylanderi* (Hymenoptera: Formicidae)." *Entomologia Generalis* 26: 73–84.

Stuart, R. J., and T. M. Alloway. 1983. "The slave-making ant, *Harpagoxenus canadensis* M. R. Smith, and its host-species, *Leptothorax muscorum* (Nylander): Slave raiding and territoriality." *Behaviour* 85: 58–90.

Stuart, R. J., A. Francoeur, and R. Loiselle. 1987. "Lethal fighting among dimorphic males of the ant, *Cardiocondyla wroughtonii*." *Naturwissenschaften* 74: 548–549.

Talbot, M. 1957. "Population studies of the slave-making ant *Leptothorax duloticus* and its slave, *Leptothorax curvispinosus*." *Ecology* 38: 449–456.

Tinaut, A., and J. Heinze. 1992. "Wing reduction in ant queens from arid habitats." *Naturwissenschaften* 79: 84–85.

Tinaut, A., and F. Ruano. 1992. "Braquipterismo y apterismo en formicidos. Morfologia y biometria en las hembras de especies ibericas de vida libre (Hymenoptera: Formicidae)." *Graellsia* 48: 121–131.

Trunzer, B., J. Heinze, and B. Hölldobler. 1998. "Cooperative colony founding and experimental primary polygyny in the ponerine ant *Pachycondyla villosa.*" *Insectes Sociaux* 45: 267–276.

Tsuji, K., and N. Tsuji. 1996. "Evolution of life history strategies in ants: Variation in queen number and mode of colony founding." *Oikos* 76: 83–92.

Wcislo, W. T., and B. N. Danforth. 1997. "Secondarily solitary: The evolutionary loss of social behavior." *Trends in Ecology and Evolution* 12: 468–474.

Wehner, R., S. Wehner, and D. Agosti. 1994. "Patterns of biogeographic distribution within the *bicolor* species group of the North African desert ant, *Cataglyphis* Foerster 1850 (Insecta: Hymenoptera: Formicidae)." *Senckenbergiana Biologica* 74: 163–191.

Wesson, L. G. 1939. "Contributions to the natural history of *Harpagoxenus americanus* Emery (Hymenoptera: Formicidae)." *Transactions of the American Entomological Society* 65: 97–122.

Wilson, E. O. 1975. "*Leptothorax duloticus* and the beginnings of slavery in ants." *Evolution* 29: 108–119.

Wilson, E. O., and W. L. Brown. 1956. "New parasitic ants of the genus *Kyidris,* with notes on ecology and behavior." *Insectes Sociaux* 3: 439–454.

Yamauchi, K., and K. Kinomura. 1993. "Lethal fighting and reproductive strategies of dimorphic males in the ant genus *Cardiocondyla.*" In T. Inoue and S. Yamane, eds., *Evolution of insect societies,* 369–397. Tokyo: Hakuhinsha (in Japanese).

Yamauchi, K., and K. Ogata. 1995. "Social structure and reproductive systems of tramp versus endemic ants (Hymenoptera: Formicidae) of the Ryukyu Islands." *Pacific Science* 49: 55–68.

Yasuno, M. 1964. "The study of the ant population in the grassland at Mt. Hakkōda. III. The effect of the slave making ant, *Polyergus samurai,* upon the nest distribution pattern of the slave ant, *Formica fusca japonica.*" *Science Reports of the Tōhoku University, ser. 4 (biol)* 30: 43–55.

CHAPTER THREE

Aging of Social Insects

OLAV RUEPPELL

AGING IS A UNIVERSAL yet highly variable property of living systems (Finch 1990). Following two decades of rapid progress in understanding aging in a few short-lived nonsocial model organisms, comparative approaches are called for to investigate the biological generality of these findings. Social insects are increasingly noticed as alternative models with great scientific potential. Social evolution involves special adaptations that create unique opportunities for aging research. After reviewing the major aging hypotheses (Hughes and Reynolds 2005) in reference to social insects, this chapter focuses on two particular properties of social insects that entail special opportunities for aging research. First, group living creates many interactions between individuals and emergent properties at the colony level. Social evolution has selected for extreme longevity in some individuals (Keller and Genoud 1997), but longevity may have also contributed to the evolution of sociality (Carey 2001a). Little is known about the relations between social group structure and individual life expectancy, except for human studies. The second property of social insects is the existence of castes exhibiting pronounced intra-specific life expectancy differences. This phenotypic plasticity for aging provides an interesting model for proximate studies of the epigenetic regulation of aging (Omholt and Amdam 2004) which can illuminate the proximate constraints on interdependent life history optimization (Oster and Wilson 1978).

Aging Theory and Social Evolution

There are three established evolutionary (ultimate) hypotheses of aging, and some of the most compelling experimental support for them comes from social insects. The disposable soma theory (Kirkwood 1987; Figure 3.1) proposes that most available resources should be used to maintain the germ line of an organism at the expense of the somatic cells. Quantification of the resource allocation to soma and germ line is difficult within a multicellular, solitary organism, but social insect colonies provide a good model to test this hypothesis. In advanced social insect species, reproduction is monopolized by a morphologically specialized caste. These reproductives can be regarded as the germ line of the colony and their workers are analogous to the somatic cells of multicellular organisms (Wilson and Sober 1989). Hypotheses based on evolutionary theory correctly predict that reproductives live longer than their "somatic" support, the workers. Thus most available resources are used to maintain the reproductives (germ line) at the expense of the workers (soma) (Rueppell and Kirkman 2005).

The two remaining evolutionary hypotheses of aging are based on the observation that selection decreases with age because the number of individuals subjected to selection declines with advancing age (Rose 1991). Thus, mutations with late-acting deleterious effects are not strongly selected against and can accumulate to cause senescence (mutation accumulation hypothesis; Medawar 1952). Such mutations might even be selectively favored if they are beneficial earlier in life (antagonistic pleiotropy hypothesis, Williams 1957). Delayed reproduction increases the duration of selection against harmful mutations and, thus, should lead to a longer life span. Reproduction is delayed in social insects, most prominently in species that found new colonies independently. Consistent with evolutionary theory, independently founding species are generally longer lived than dependent colony founders (Hölldobler and Wilson 1977; Keller and Genoud 1997). Additionally, the delayed onset of reproduction in social insects abolishes the inter-sexual conflict over parental investment. In contrast to many solitary, promiscuous species, social insect males are selected against manipulating their mates to increase current reproduction at the expense of longevity. Mating in social insects can indeed increase queen longevity (Schrempf, Heinze, and Cremer 2005; but see Baer and Schmid-Hempel 2005).

Another prediction by the latter two hypotheses is that the intrinsic rate of aging co-evolves with the extrinsic mortality risk: if the extrinsic mortality is high, fewer individuals survive for selection to act on late-acting genes. To the degree that castes evolve independently, the phenotypic plasticity of social insects (West-Eberhard 2003) provides excellent test cases of this prediction. For example, ant queens and termite reproductives live in the center of their colony, shielded from external mortality factors such as predators or other environmental hazards. Only when first leaving the natal nest to establish a new colony do they experience high mortality (Hölldobler and Wilson 1990). Therefore, it is predicted that the intrinsic mortality (deaths without external insults under environmentally benign conditions) of reproductives is comparatively low, which seems to be the case (Keller and Genoud 1997). In contrast, workers progress from the protected colony environment to tasks associated with higher extrinsic mortality as they age (Tofilski 2002; Cole, this volume). As expected, workers have a relatively low life expectancy, particularly in species with highly divergent worker and reproductive phenotypes (Hölldobler and Wilson 1990; Keller 1998; Carey 2001a; Page and Peng 2001). The predicted correlation between extrinsic and intrinsic mortality was also confirmed within the worker caste of the weaver ant *(Oecophylla smaragdina)* in laboratory tests (Chapuisat and Keller 2002).

Evolutionary life history theory also predicts that life stages associated with high mortality should be minimized. Such an association has been found in the duration of the pupal stage of honey bees. The developmental time of queen pupae, which experience high mortality due to fatal sister-sister conflict (Tarpy, Gilley, and Seeley 2004), lasts only 6 to 7 days. In contrast, worker and drone pupae, which experience low mortality (Sakagami and Fukuda 1968; Fukuda and Ohtani 1977), develop in 11 to 12 days and 14 to 15 days, respectively (Winston 1987). Furthermore, the evolution of age-based division of labor among social insect workers (Beshers and Fewell 2001) probably has been influenced by varied extrinsic mortality associated with different tasks (O'Donnell and Jeanne 1995a; Tofilski 2002). Tasks associated with a higher external mortality are performed primarily by old individuals with a low residual value to the colony (Woyciechowski and Kozlowski 1998; Amdam et al. 2004). In sum, compelling empirical support for the evolutionary theory of aging comes from social insects and recent elaborations of the evolutionary theory of aging (e.g., Lee 2003) have conceptual parallels to the pervasive kin selection theory.

A number of proximate hypotheses of aging (Figure 3.1) do not explicitly invoke or contradict evolutionary arguments. Rather, these hypotheses complement evolutionary explanations by focusing on possible mechanisms at the cellular and molecular level. In some cases, the molecular mechanisms directly corroborate one of the evolutionary theories of aging. For example, insulin/insulin-like growth factor 1 (IGF-1) signaling (see below) antagonistically affects reproduction and life span (Tatar and Yin 2001). However, in some cases unique, divergent predictions arise, particularly in social insects.

A classic mechanistic hypothesis is the "rate-of-living" hypothesis (Pearl 1928), which suggests that organisms with a faster metabolism also age faster. The underlying cause could be a limitation of internal resources or physiological capacity (Neukirch 1982); however, more research has focused on the damaging effects of metabolic by-products, such as oxidative damage through free radicals (Bokov, Chaudhuri, and Richardson 2004).

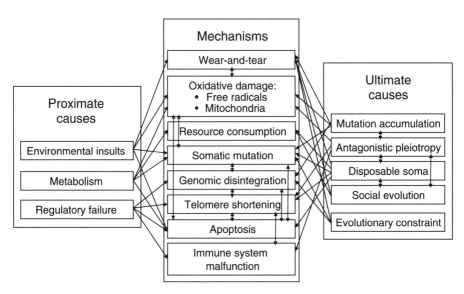

Figure 3.1. Different hypotheses of aging exist that are not mutually exclusive and explain aging at different levels. Social insect research has so far supported mainly ultimate (evolutionary) explanations. Nevertheless, these are each compatible with several mechanistic hypotheses (arrows) and social insects could become important models in aging research to make these connections in the future.

Predominantly, the mitochondria have been implicated, but the "mitochondrial theory of aging" includes additional mechanisms such as mitochondrial signaling pathways (Jacobs 2003). The relative rates of damage accumulation from metabolic activity can be partially offset by cellular repair, resulting in potentially complex relationships between metabolic activity and life expectancy (Corona et al. 2005). Complimentary evidence is provided by the increase of life expectancy through caloric restriction, which reduces metabolic activity (Sohal and Weindruch 1996), and mutations in cellular signaling pathways that slow metabolism (Guarente and Kenyon 2000). Tentative support for the role of metabolic rate as a life span determinant in social insects comes from the fundamental taxonomic life span differences between the long-lived perennial termites and ants and the relatively short-lived annual wasps and bees. The metabolically intense flight (Harrison and Roberts 2000) of wasps and bees may cause them to age much faster than non-flying social insects. However, more refined phylogenetic comparisons between metabolic load measurements and life expectancies are needed. It would also be interesting to compare metabolic rates and molecular damage accumulation between reproductives and workers to investigate whether their different life expectancy correlates with differential damage accumulation, despite higher levels of classic antioxidants in workers (Parker et al. 2004; Corona et al. 2005).

A second set of mechanistic explanations for aging implicates DNA changes (mutations and epigenetic changes) and resulting regulatory malfunctions. The somatic mutation hypothesis (reviewed by Finch 1990) states that somatic cells become less viable with age because they accumulate mutations in their genomic and organelle DNA (e.g., due to oxidative damage, see above). Somatic mutations can cause genomic instability, another mechanism of suggested importance in organism senescence, but no data exist from any social insect so far.

A specific case of genomic instability is the shortening of the chromosome telomeres, which leads to replicative senescence of cells and limits the growth of cell lines in vitro (Hayflick and Moorhead 1961). A correlation between telomere length and life expectancy exists in *Lasius* ants (Jemielity et al. 2007). Telomere maintenance in honey bees presumably employs a mechanism similar to humans (Honeybee Genome Sequencing Consortium 2006), but because most adult insect tissues are post-mitotic (Finch 1990), cellular senescence may be of less general importance. Yet honey bee intestinal epithelium cells (Snodgrass 1956; Figure 3.2) and

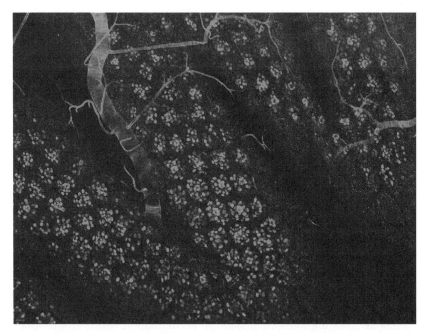

Figure 3.2. Incorporation of the thymidine-analogue 5′-bromo-2′ deoxyuridine (BrdU) demonstrates cell proliferation (bright spots) in the intestine of an adult honey bee queen and can be used to quantify potential replicative senescence of intestinal stem cells.

immune cells (Amdam, et al. 2005) proliferate in adults, raising the possibility of a link between aging and cellular senescence. The intestinal replicative cells of insects have been characterized as bona-fide stem cells (Ohlstein and Spradling 2006) but the association between stem cell replicative senescence and organismal life expectancy is not simple in honey bees (Ward et al. 2008). Similarly, the immune system is essential in a social context (Schmid-Hempel 1998) and its functional decline may impact longevity to a greater extent than degradation of other major organ systems (Doums et al. 2002; Amdam et al. 2004; Effros 2004).

The Social Context of Aging

The emerging interdisciplinary field of biodemography combines biological experiments at different levels with the actuarial concepts and techniques of demography (Carey 2001b), with a (current) focus on aging and life span. The central premise is that biological entities such as an organism, a colony, or a population have an age structure, and biological processes such as reproduction, growth, and mortality risk are age dependent. This implies that the age dynamics of any of these variables conveys more information than static population mean value across different ages. For example, studies of mortality dynamics over the entire life span point to underlying biological processes of mortality (Sakagami and Fukuda 1968) and question the paradigm of a finite, species-specific life span (Vaupel et al. 1998). Few studies of social insects have taken full advantage of biodemographic approaches (Carey 2001b), even though the age dynamics of many variables are particularly important for understanding social insect biology. In addition, social insects provide ideal demographic study systems because they typically occur in large numbers and reliably return to their colony as long as they live. Observation hives for honey bees and bumblebees as well as the open natural nests of many wasp species allow for high-quality, longitudinal survival analyses under natural conditions. The study of aging in an evolutionarily relevant context allows for an understanding of the interactions between individual adaptations and extrinsic mortality factors. In addition to the ecological context that individuals of every biological species experience, social organisms have to integrate with a dynamic social environment that has shaped their life histories. In the following two sections I summarize biodemographic studies of social insects to illustrate the influence of ecological and social variables on aging rates and life span.

Ecological Influences

Social insects have adapted to a large diversity of lifestyles and habitats (Hölldobler and Wilson 1990) and their evolutionary relationships are increasingly resolved at different taxonomic levels. Thus, they offer many possibilities for comparative studies of the evolutionary ecology of aging. At the intraspecific level, many "tramp" species that have successfully established themselves in a variety of habitats (McGlynn 1999) are interesting study cases for micro-evolutionary life span adaptations to new ecological environments (Amdam, Norberg, et al. 2005). Population comparisons would be informative for the formation of broader ecogerontological rules (i.e., which environmental factors favor longevity; Carey 2001b) and for understanding the success of invasive social insects (McGlynn 1999). Studies of clinal variation (Kaspari and Vargo 1995) and the effects of population density (see Cole, this volume) or predator-free habitats should yield especially insightful results.

Seasonal changes affect whole sets of environmental variables and determine the activity period of social insects. For annual species such as bumblebees and many temperate wasps, the life span of individuals and colonies is limited to one growth season. This relative similarity in life expectancy of workers, queens, and colonies increases the cost/benefit ratio of replacement of the old queen by a younger individual (supersedure), and thus has profound consequences for social organization and evolution (Tsuji and Tsuji 2005). Seasonality may explain why life expectancy in temperate species is lower than in tropical species, at least for workers (Rodd, Plowright, and Owen 1980). Worker life expectancy of the perennial honey bee also depends on the season in temperate climates. During the active season, worker life span averages 15 to 38 days (Page and Peng 2001), with a constant decline in life expectancy over the summer (Neukirch 1982). However, their life span can exceed 140 days in the winter (Page and Peng 2001), possibly up to a year (Maurizio, 1950). This increased longevity is presumably due to a combination of factors, including an altered protein metabolism (Omholt and Amdam 2004). In addition, the lowered extrinsic mortality risk for nonforaging winter workers may be significant. Long-lived workers can also be obtained during the summer (Maurizio 1950) under broodless conditions, indicating the importance of the regulation of resource allocation (Amdam and Omholt 2002). The life span of reproductives seems to be largely under social control (Winston 1987; Page and Peng 2001) since they are buffered against predation and fluctuations in the external environment (Rueppell et al. 2004). Seasonal effects on queen mortality have not been documented (except in the Argentine ant *Linepithema humile;* Keller, Passera, and Suzzoni 1989), but male hymenoptera are generally short-lived, and in honey bees it is well-documented that drones are expelled from colonies at the end of the reproductive season (Winston 1987).

Colony growth in social insects mostly follows a logistic pattern (Hölldobler and Wilson 1990; Cole, this volume) because the birth and death rates of workers are dependent on colony size (Oster and Wilson 1978). Growth rates also depend on several environmental variables, such as resource availability, population density, and pathogens/predators. Intraspecifically, large colony size can confer a longevity advantage to individual workers (Fukuda and Sekiguchi 1966; Winston 1979; O'Donnell and Jeanne 1992). This advantage may be due to more effective environmental homeostasis within the nest, improved defensive capability, or increased colony efficiency that decreases the demands on individuals. In contrast, colony size is negatively correlated with worker longevity at the interspecific level. This may be due to the fact that species with large colonies generally have a higher degree of divergence between workers and reproductives (Bourke 1999), which results in relatively low worker life expectancy.

Social Environment

Intraspecific interactions influence the longevity of many animals but the unsurpassed integration of individuals in social insect societies (Wilson 1971) has created an intimate link between individual life span and colony social structure. The most prominent social determinant of life span is the distinction between reproductives and nonreproductive workers. On the one hand, the relatively long life of the reproductive has probably co-evolved with sociality (Carey 2001a), while on the other hand, social cooperation assures fitness returns of reproductive effort beyond an individuals' life span (Gadagkar 1990). Thus, sociality may have arisen due to low adult survival in some taxa (Landi et al. 2003).

Hymenopteran queens live much longer than workers (Hölldobler and Wilson 1990; Keller and Genoud 1997; Carey 2001a; Page and Peng 2001) and males (Page and Peng 2001; Rueppell, Fondrk, and Page 2005; Figure 3.3). In contrast, the workers and kings of termite colonies can be as long-lived as their queens (Thorne, Breisch, and Haverty 2002). Thus, reproduction does not necessarily confer a longevity advantage in social species. However, in social hymenoptera reproductives usually outlive nonreproductives, even in species without physical worker or queen castes. In ants of the genus *Diacamma*, workers facultatively mate and assume the reproductive role in their colony, resulting in a three to five–fold life span extension compared to their nonreproductive nestmates (Tsuji, Nakata, and Heinze 1996; Andre, Peeters, and Doums 2001). Similarly, the colony's queen in the facultatively parthenogenetic ant *Platythyrea punctata* lives approximately four-fold longer than her genetically identical nonreproductive sisters (Hartmann and Heinze 2003).

Aging and mortality patterns have been studied mostly in workers.

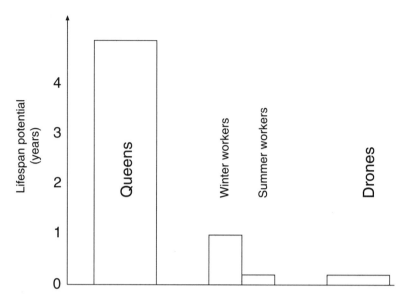

Figure 3.3. The intraspecific variability of life spans in social insects (shown for the honey bee) provides unique opportunities to study naturally-evolved plasticity in aging rates.

Across different taxa, foraging outside of the nest greatly increases mortality. In bumblebees and ants with physical worker castes (Hölldobler and Wilson 1990) individuals specialize on either foraging or inside tasks throughout their lives. Foragers consistently have lower life expectancies than inside workers, even in a protected laboratory environment (Porter and Tschinkel 1985; Chapuisat and Keller 2002). In many social insects without specific forager castes, these duties are consistently performed by older individuals (Beshers and Fewell 2001; Nascimento, Simoes, and Zucchi 2005). The transition from hive activities to foraging is characterized by a dramatic increase in mortality, resulting in a type I survivorship curve (Sakagami and Fukuda 1968). This robust finding (Fukuda and Sekiguchi 1966) is uncommon in short-lived organisms such as insects, and likely results from the protected and provisioned colony environment in which potentially vulnerable offspring are reared. Under these circumstances, the central life history variable for social insect workers becomes the transition from the low-risk colony environment to the high-risk foraging environment (Porter and Jorgensen 1981; Visscher and Dukas 1997; Rueppell, Bachelier, et al. 2007), referred to as the "age of first foraging" (Guzmán-Novoa, Page, and Gary 1994). This transition is accompanied by regulatory changes in physiology (see below) which shorten forager life span (O'Donnell and Jeanne 1995b; Amdam et al. 2004; Amdam, Aase, et al. 2005). These changes can only be understood as adaptations to minimize resource loss at the colony level, reinforcing the notion of workers as disposable soma. Nutrient transfers can reduce individual life span (Amdam et al. 2004) but may promote colony survival, especially under starvation stress (Rueppell and Kirkman 2005).

The concept of adaptive demography suggests that colony caste ratios are optimized (Oster and Wilson 1978), and some empirical support for this exists (Yang, Martin, and Nijhout 2004). However, the typical age structure of colonies in itself is not necessarily adaptive, but can be a necessary consequence of individual mortality schedules. For example, single- and double-cohort colonies of honey bees are not inferior to colonies of natural age composition (Rueppell, Linford, et al. 2008). Demographic manipulations have been used to assess the influence of colony size (Fukuda and Sekiguchi 1966; Winston 1979; Harbo 1986; O'Donnell and Jeanne 1992) and the brood/adult ratio. Despite a large proportion of apparently inactive workers (Winston 1987; Hölldobler and Wilson 1990), the brood/adult ratio is negatively correlated with the life span of the reared brood (Eischen, Rothenbuhler, and Kulincevic 1982) and the adult caregivers (Winston and Fergusson 1985). These few studies provide limited information on the relationship between individual life span and social organization and need to be further elaborated and integrated with mechanistic studies at the social, individual, and molecular level.

Phenotypic Plasticity of Aging

Between Caste Comparisons

The large variation in life span between castes and the exceptional longevity of the reproductives (Keller and Genoud 1997) is the foremost reason for researchers on aging to be interested in social insects (Finch 1990). Reproductives differ from workers in many physiological and behavioral aspects that may or may not be related to their differential aging rates (Page and Peng 2001; Rueppell et al. 2004); therefore, general trait comparisons between workers and reproductives, such as global comparisons of gene expression patterns, are unlikely to reveal the primary causes of their differential life expectancy. However, between-caste comparisons can be useful in more complex designs that address the age dynamics of the studied differences (e.g., sequential analyses of gene expression differences). Additionally, between-caste comparisons provide powerful, naturally-evolved systems to test specific processes that may be associated with differential life expectancy.

One such study focused on the important antioxidant enzyme Cu-Zn superoxide dismutase I (SOD1), which has been shown to increase life expectancy in various aging models (Parker et al. 2004). Yet in the small black garden ant *Lasius niger,* queens consistently expressed SOD1 less than the shorter-lived workers and males in all three body compartments (head, thorax, abdomen) and this expression mirrored the enzyme activity levels. Although no age information was available for the investigated individuals (Parker et al. 2004), the study suggests that SOD1 does not have a major role in determining longevity in this ant.

In a second candidate gene study, the expression of eight antioxidant and five respiration-related genes (including SOD1) were compared between queen and worker honey bees (Corona et al. 2005). Antioxidant gene expression decreased with age in queens in all body compartments, but increased in workers. The respiration-related genes also decreased in queens with age, but remained constant in workers. Overall, antioxidant and respiration-related genes showed higher expression levels in workers than queens. This suggests that reactive oxidant production and protection against it may be higher in workers than in queens, but no clear pattern emerges as the cause for the life span difference between the two castes (Corona et al. 2005).

Both studies together suggest that the natural evolution of longevity may be more complex than suggested by laboratory life span extensions obtained through genetic manipulations. However, numerous genes can potentially increase life span when mutated and are worth studying as possible candidates (Partridge and Gems 2002; Tatar, Bartke, and Antebi 2003). This is particularly true for genes involved in insulin/IGF-1, juvenile hormone, and ecdysone signaling (Simon et al. 2003) because these mechanisms are widely involved in pleiotropic behavioral and life-history regulation, social insect caste differentiation, and physiological caste differences (Hartfelder and Engels 1998; Wheeler, Buck, and Evans 2006; Hunt et al. 2007).

The mechanisms responsible for the different aging rates among social insect castes may be also addressed by a top-down comparative approach. Characterization of the actual cause of death, or even symptoms of senescence of either caste, can provide a starting point for mechanistic hypotheses about why reproductives outlive workers. Many biomarkers of senescence at different biological levels (organismal, cellular, and molecular) have already been identified in humans and classic aging models (e.g., Markowska and Breckler 1999; Martin 2000; Grotewiel et al. 2005), which could be used in comparative social insect studies.

Within-Caste Studies

While queens and males have not been extensively studied, it is clear that the worker caste(s) of most social insects display considerable plasticity in life expectancy. In large part, this is due to task specialization, which entails differences in external mortality risk, work load (metabolic activity), and physiology (Amdam and Omholt 2002). In honey bees, the different mortality schedules arising from behavioral specialization are superimposed on seasonal effects (Fukuda and Sekiguchi 1966) and a pattern of increasing worker mortality with chronological age (Rueppell, Christine, et al. 2007). Worker subcastes with different aging rates can be defined based on season, physiology, or behavioral profile. For example, workers can be extremely long-lived in a diutinus (overwintering) state (Omholt and Amdam 2004) to survive unfavorable periods without brood-rearing or foraging, such as overwintering (Maurizio 1950). It has been suggested that senescence of diutinus bees is independent of age (Omholt and Amdam 2004). More generally, aging seems to be more a function of task exposure than of chronological age; generally, hive workers senescence much slower than foragers (Page and Peng 2001) and molecular damage accumulation in worker brains is linked to foraging experience (Seehuus and Amdam 2006) rather than chronological age. This may explain why sensory and learning performance does not decline with chronological age (Rueppell, Christine, et al. 2007), but learning declines with foraging time (Behrends et al. 2007). In contrast, mortality risk and stress resistance seem to be a function of chronological age (Remolina et al. 2007; Rueppell, Christine, et al. 2007). Workers can also enter reproductive status upon the loss of social inhibition by the brood and queen but mortality consequences are less clear than in some ants where workers regularly assume reproductive status with increased life expectancy (Tsuji, Nakata, and Heinze 1996; Hartmann and Heinze 2003).

Foraging activity is widely associated with high mortality in social insects and foraging life span is generally short (Visscher and Dukas 1997). Foragers can be lost to predation, disease, disorientation, dehydration, and accidents (Wilson 1971; Schmid-Hempel and Schmid-Hempel 1984; Hölldobler and Wilson 1990; Schmid-Hempel 1998). Foraging is also more energetically and biomechanically demanding than colony-internal tasks, particularly in flying species. Abrasion of the wings (Figure 3.4) represents mechanical senescence that is easy to quantify and experimentally manipulate. It may be responsible for declining foraging performance in honey bees (Tofilski 2000), and it affects foraging decisions and mortality (Cartar 1992; Higginson and Barnard 2004). In addition, foraging entails very high metabolic rates (Roces and Lighton 1995) that may translate into faster molecular aging through oxidative damage (Bokov Chaudhuri, and Richardson 2004; Corona et al. 2005; Seehuus and Amdam 2006). However, quantitative

Figure 3.4. Young honey bee worker (A) and old forager (B). The comparison illustrates the external wear-and-tear incurred by the old bee as abrasion of cuticular hairs and wing damage.

reduction of foraging has only a modest effect and regulatory events seem to be just as important (Rueppell, Bachelier, et al. 2007). The vitellogenin levels of foragers are low (Fluri et al. 1982; Amdam and Omholt 2002) and the replenishment of glycogen stores in the flight muscles may also be impaired (Neukirch 1982). Similarly, foraging coincides with a decrease in lipid stores in workers of the wasp *Polybia occidentalis* (O'Donnell and Jeanne 1995b), and a 40 percent reduction of the overall dry weight in harvester ants (Porter and Jorgensen 1981). Low nutrient levels in foragers may lead to exhaustion and prohibit damage repair at the cellular level, accelerating individual senescence but preserving resources at the colony level.

In particular, vitellogenin is associated with longevity in honey bees (Corona et al. 2007). Queens have high levels of vitellogenin; in workers its titer declines sharply with the onset of foraging, but remains high in the diutinus state (Amdam and Omholt 2002). Vitellogenin serves as a major storage mechanism for zinc, which in turn is important for immune function. Down-regulation of vitellogenin in foragers thus compromises their immune system (Amdam et al. 2004; Amdam, Aase, et al. 2005a). Vitellogenin can also increase honey bee survival by serving as an antioxidant to prevent the accumulation of oxidative damage, particularly in foragers (Seehuus et al. 2006). In addition, it delays the onset of worker foraging, resulting in a longer life (Nelson et al. 2007).

The Integrative Future

The recent development of broadly applicable molecular tools and theoretical concepts is greatly enhancing studies of social insects and aging research. Common to both disciplines is a solid theoretical foundation that provides a framework for empirical progress at multiple levels of biological organization. For the immediate future, the completed honey bee genome project (Honeybee Genome Sequencing Consortium 2006) and the integration of biodemographic approaches and mechanistic studies promise many opportunities for significant progress in understanding aging in social insects.

On the one hand, social insects provide excellent model organisms for comparative aging research. Their evolutionary success has been achieved by combining short generation times with modular growth and social networks for resource transfers to build long-lived, perennial colonies. Their extraordinary plasticity and intraspecific variation of aging rates, resulting in large natural differences in life expectancy, directly attest to the evolvability of this trait. Kin selection entails particular selection pressures on colony members that provide many opportunities to test hypotheses on aging. The pronounced epigenetic caste and subcaste differences allow us to systematically study the same genotype in different phenotypic expressions, a field that is still in its infancy. These comparisons provide many opportunities for mechanistic studies of causes, correlates, and consequences of a naturally low versus high life expectancy. In addition, social insect colonies constitute small, experimental microcosms with well-defined and quantifiable social roles and resource flows. They are complex systems that have very successfully adapted to diverse environments. The detailed demographic analysis of these societies, combined with mechanistic studies, will yield unique data for the social/behavioral science of aging, a field that is almost entirely devoid of biological model organisms. The sociobiology of aging thus holds great potential for comparative and experimental studies based on evolutionary theory.

One the other hand, studies on aging are important for understanding social insect biology. Colonies are critically dependent on the birth and death schedules of their members and face resource allocation trade-offs that lead to organismal aging in solitary species. Biological aging research has identified general, central life-history regulators, and it becomes increasingly apparent that these mechanisms have been co-opted as proximate control modules of social evolution. In addition, biodemographic research has demonstrated how focusing on age-specific mortality dynamics can be a powerful approach to understand aging. This approach should lead to theoretical approaches to insect sociality that go beyond basic "adaptive demography" models and could be applied to several problems to understand the biology of the inherently age-structured social insect colonies.

Acknowledgments

I would like to thank the editors for their support and hard work to produce this volume. This chapter was greatly improved by the valuable comments of Colin Brent and an anonymous reviewer. Financial support came from the American Federation for Aging Research and the National Science Foundation (grant #0615502).

Literature Cited

Amdam, G. V., A. L. T. O. Aase, S. C. Seehuus, M. K. Fondrk, K. Norberg, and K. Hartfelder. 2005. "Social reversal of immunosenescence in honeybee workers." *Experimental Gerontology* 40: 939–947.

Amdam, G. V., and S. W. Omholt. 2002. "The regulatory anatomy of honeybee life span." *Journal of Theoretical Biology* 216: 209–228.

Amdam, G. V., K. Norberg, S. W. Omholt, P. Kryger, A. P. Lourenco, M. M. G. Bitondi, and Z. L. P. Simoes. 2005. "Higher vitellogenin concentrations in honey bee workers may be an adaptation to life in temperate climates." *Insectes Sociaux* 52: 316–319.

Amdam, G. V., Z. L. P. Simoes, A. Hagen, K. Norberg, K. Schroder, O. Mikkelsen, T. B. L. Kirkwood, and S. W. Omholt. 2004. "Hormonal control of the yolk precursor vitellogenin regulates immune function and longevity in honeybees." *Experimental Gerontology* 39: 767–773.

Andre, J. B., C. Peeters, and C. Doums. 2001. "Serial polygyny and colony genetic structure in the monogynous queenless ant *Diacamma cyaneiventre*." *Behavioral Ecology and Sociobiology* 50: 72–80.

Baer, B., and P. Schmid-Hempel. 2005. "Sperm influences female hibernation success, survival and fitness in the bumble-bee *Bombus terrestris*." *Proceedings of the Royal Society B—Biological Sciences* 272: 319–323.

Behrends, A., R. Scheiner, N. Baker, and G. V. Amdam. 2007. "Cognitive aging is linked to social role in honey bees (Apis mellifera)." *Experimental Gerontology* 42: 1146–1153.

Beshers, S. N., and J. H. Fewell. 2001. "Models of division of labor in social insects." *Annual Review of Entomology* 46: 413–440.

Bokov, A., A. Chaudhuri, and A. Richardson. 2004. "The role of oxidative damage and stress in aging." *Mechanisms of Ageing and Development* 125: 811–826.

Bourke, A. F. G. 1999. "Colony size, social complexity and reproductive conflict in social insects." *Journal of Evolutionary Biology* 12: 245–257.

Carey, J. R. 2001a. "Demographic mechanisms for the evolution of long life in social insects." *Experimental Gerontology* 36: 713–722.

———. 2001b. "Insect biodemography." *Annual Review of Entomology* 46: 79–110.

Cartar, R. V. 1992. "Morphological senescence and longevity—An experiment relating wing wear and life span in foraging wild bumble bees." *Journal of Animal Ecology* 61: 225–231.

Chapuisat, M., and L. Keller. 2002. "Division of labour influences the rate of ageing in weaver ant workers." *Proceedings of the Royal Society of London Series B-Biological Sciences* 269: 909–913.

Corona, M., K. A. Hughes, D. B. Weaver, and G. E. Robinson. 2005. "Gene expression patterns associated with queen honeybee longevity." *Mechanisms of Ageing and Development* 126: 1230–1238.

Corona, M., R. A. Velarde, S. Remolina, A. Moran-Lauter, Y. Wang, K. A. Hughes, and G. E. Robinson. 2007. "Vitellogenin, juvenile hormone, insulin signaling, and queen honey bee longevity." *Proceedings of the National Academy of Sciences USA* 104: 7128–7133.

Doums, C., Y. Moret, E. Benelli, and P. Schmid-Hempel. 2002. "Senescence of immune defence in Bombus workers." *Ecological Entomology* 27: 138–144.

Effros, R. B. 2004. "From Hayflick to Walford: The role of T cell replicative senescence in human aging." *Experimental Gerontology* 39: 885–890.

Eischen, F. A., W. C. Rothenbuhler, and J. M. Kulincevic. 1982. "Length of life and dry weight of worker honeybees reared in colonies with different worker-larva ratios." *Journal of Apicultural Research* 21: 19–25.

Finch, C. E. 1990. *Longevity, senescence, and the genome.* Chicago: University of Chicago Press.

Fluri, P., M. Luscher, H. Wille, and L. Gerig. 1982. "Changes in weight of the pharyngeal gland and haemolymph titres of juvenile hormone, protein, and vitellogenin in worker honeybees." *Journal of Insect Physiology* 28: 61–68.

Fukuda, H., and T. Ohtani. 1977. "Survival and life span of drone honeybees." *Research in Population Ecology* 10: 31–39.

Fukuda, H., and K. Sekiguchi. 1966. "Seasonal change of the honeybee worker longevity in Sapporo, north Japan, with notes on some factors affecting the life-span." *Japanese Journal of Ecology* 16: 206–212.

Gadagkar, R. 1990. "Evolution of eusociality—The advantage of assured fitness returns." *Philosophical Transactions of the Royal Society of London B* 329: 17–25.

Grotewiel, M. S., I. Martin, P. Bhandari, and E. Cook-Wiens. 2005. "Functional senescence in *Drosophila melanogaster.*" *Ageing Research Reviews* 4: 372–397.

Guarente, L., and C. Kenyon. 2000. "Genetic pathways that regulate ageing in model organisms." *Nature* 408: 255–262.

Guzmán-Novoa, E., R. E. Page, and N. E. Gary. 1994. "Behavioral and life-history components of division of labor in honey bees (*Apis mellifera* L.)." *Behavioral Ecology and Sociobiology* 34: 409–417.

Harbo, J. R. 1986. "Effect of population size on brood production, worker survival and honey gain in colonies of honeybees." *Journal of Apicultural Research* 25: 22–29.

Harrison, J. F., and S. P. Roberts. 2000. "Flight respiration and energetics." *Annual Review of Physiology* 62: 179–205.

Hartfelder, K., and W. Engels. 1998. "Social insect polymorphism: Hormonal regulation of plasticity in development and reproduction in the honeybee." *Current Topics in Developmental Biology* 40: 45–77.

Hartmann, A., and J. Heinze. 2003. "Lay eggs, live longer: Division of labor and life span in a clonal ant species." *Evolution* 57: 2424–2429.

Hayflick, L., and P. S. Moorhead. 1961. "The serial cultivation of human diploid cells strains." *Experimental Cell Research* 25: 585–621.

Higginson, A. D., and C. J. Barnard. 2004. "Accumulating wing damage affects foraging decisions in honeybees (*Apis mellifera* L.)." *Ecological Entomology* 29: 52–59.

Hölldobler, B., and E. O. Wilson. 1977. "The number of queens: An important trait in ant evolution." *Naturwissenschaften* 64: 8–15.

———. 1990. *The ants.* Cambridge: The Belknap Press of Harvard University Press.

The Honeybee Genome Sequencing Consortium. 2006. "Insights into social insects from the genome of the honeybee *Apis mellifera.*" *Nature* 443: 931–949.

Hughes, K. A., and R. M. Reynolds. 2005. "Evolutionary and mechanistic theories of aging." *Annual Review of Entomology* 50: 421–445.

Hunt, James H., Bart J. Kensinger, Jessica A. Kossuth, Michael T. Henshaw, Kari Norberg, Florian Wolschin, and Gro V. Amdam. 2007. "A diapause pathway underlies the gyne phenotype in Polistes wasps, revealing an evolutionary route to caste-containing insect societies." *Proceedings of the National Academy of Sciences, U.S.A.* 104: 14020–14025.

Jacobs, H. T. 2003. "The mitochondrial theory of aging: Dead or alive?" *Aging Cell* 2: 11–17.

Jemielity, S., M. Kimura, K. M. Parker, J. D. Parker, X. J. Cao, A. Aviv, and Keller, L. 2007. "Short telomeres in short-lived males: What are the molecular and evolutionary causes?" *Aging Cell* 6: 225–232.

Kaspari, M., and E. L. Vargo. 1995. "Colony size as a buffer against seasonality: Bergmann's rule in social insects." *American Naturalist* 145: 610–632.

Keller, L. 1998. "Queen life span and colony characteristics in ants and termites." *Insectes Sociaux* 45: 235–246.

Keller, L., and M. Genoud. 1997. "Extraordinary lifespans in ants: A test of evolutionary theories of ageing." *Nature* 389: 958–960.

Keller, L., L. Passera, and J. P. Suzzoni. 1989. "Queen execution in the Argentine Ant, Iridomyrmex humilis." *Physiological Entomology* 14: 157–163.

Kirkwood, T. B. L. 1987. "Immortality of the germ-line versus disposability of the soma." In A. D. Woodhead and K. H. Thompson, eds., *Evolution of longevity of animals: A comparative approach,* 209–218. New York: Plenum Press.

Landi, M., D. C. Queller, S. Turillazzi, and J. E. Strassmann. 2003. "Low relatedness and frequent queen turnover in the stenogastrine wasp *Eustenogaster fraterna* favor the life insurance over the haplodiploid hypothesis for the origin of eusociality." *Insectes Sociaux* 50: 262–267.

Lee, R. D. 2003. "Rethinking the evolutionary theory of aging: Transfers, not births, shape social species." *Proceedings of the National Academy of Sciences USA* 100: 9637–9642.

Markowska, A. L., and S. J. Breckler. 1999. "Behavioral biomarkers of aging: Illustration of a multivariate approach for detecting age-related behavioral changes." *Journals of Gerontology Series a-Biological Sciences and Medical Sciences* 54: B549–B566.

Martin, H. 2000. "On the significance of markers of ageing." *Zeitschrift für Gerontologie und Geriatrie* 33: 1–7.

Maurizio, A. 1950. "The influence of pollen feeding and brood rearing on the length of life and physiological condition of the honeybee." *Bee World* 31: 9–12.

McGlynn, T. P. 1999. "The worldwide transfer of ants: Geographical distribution and ecological invasions." *Journal of Biogeography* 26: 535–548.

Medawar, P. B. 1952. *An unsolved problem of biology.* London: H.K. Lewis.

Nascimento, F. S., D. Simoes, and R. Zucchi. 2005. "Temporal polyethism and survivorship of workers of *Agelaia pallipes* (Hymenoptera, Vespidae, Epiponini)." *Sociobiology* 45: 377–387.

Nelson, C. M., K. Ihle, M. K. Fondrk, R. E. Page, and G. V. Amdam. 2007. "The gene vitellogenin has multiple coordinating effects on social organization." *PLoS Biology* 5: 673–677.

Neukirch, A. 1982. "Dependence of the life span of the honeybee *(Apis mellifica)* upon flight performance and energy consumption." *Journal of Comparative Physiology* 146: 35–40.

O'Donnell, S., and R. L. Jeanne. 1992. "The effects of colony characteristics on life span and foraging behavior of individual wasps (*Polybia occidentalis*, Hymenoptera, Vespidae)." *Insectes Sociaux* 39: 73–80.

O'Donnell, S., and R. L. Jeanne. 1995a. "Implications of senescence patterns for the evolution of age polyethism in eusocial insects." *Behavioral Ecology* 6: 269–273.

———. 1995b. "Worker lipid stores decrease with outside-nest task performance in wasps: Implication for the evolution of age polyethism." *Experientia* 51: 749–752.

Ohlstein, B., and A. Spradling. 2006. "The adult Drosophila posterior midgut is maintained by pluripotent stem cells." *Nature* 439: 470–474.

Omholt, S. W., and G. V. Amdam. 2004. "Epigenic regulation of aging in honeybee workers. *Science of Aging Knowledge Environment*

Oster, G. F., and E. O. Wilson. 1978. *Caste and ecology in the social insects.* Princeton: Princeton University Press.

Page, R. E., and Y.-S. C. Peng. 2001. "Aging and development in social insects with emphasis on the honey bee, *Apis mellifera* L." *Experimental Gerontology* 36: 695–711.

Parker, J. D., K. M. Parker, B. H. Sohal, R. S. Sohal, and L. Keller. 2004. "Decreased expression of Cu-Zn superoxide dismutase 1 in ants with extreme life span. *Proceedings of the National Academy of Sciences USA* 101: 3486–3489.

Partridge, L., and D. Gems. 2002. "Mechanisms of aging: Private or public?" *Nature Reviews Genetics* 3: 165–175.

Pearl, R. 1928. *The Rate of living.* New York: Knopf.

Porter, S. D., and C. D. Jorgensen. 1981. "Foragers of the harvester ant, *Pogonomyrmex owyheei:* A disposable caste?" *Behavioral Ecology and Sociobiology* 9: 247–256.

Porter, S. D., and W. R. Tschinkel. 1985. "Fire ant polymorphism: The ergonomics of brood production." *Behavioral Ecology and Sociobiology* 16: 323–336.

Remolina, S.C., D. Hafez, G.E. Robinson, and K. A. Hughes. 2007. "Senescence in the worker honey bee Apis mellifera." *Journal of Insect Physiology* 53: 1027–1033.

Roces, F., and J. R. B. Lighton. 1995. "Larger bites of leaf-cutting ants." *Nature* 373: 392–393.

Rodd, F. H., R. C. Plowright, and R. E. Owen. 1980. "Mortality rates of adult bumblebee workers (Hymenoptera: Apidae)." *Canadian Journal of Zoology-Revue Canadienne De Zoologie* 58: 1718–1721.

Rose, M. 1991. *Evolutionary biology of aging.* New York: Oxford University Press.

Rueppell, O., G. V. Amdam, R. E. Page, and J. R. Carey. 2004. "From genes to society: Social insects as models for research on aging." *Science Aging Knowledge Environment.: pe5.*

Rueppell, O., C. Bachelier, M. K. Fondrk, and R. E. Page. 2007. Regulation of life history determines life span of worker honeybees (*Apis mellifera* L.)." *Experimental Gerontology* 42: 1020–1032.

Rueppell, O., R. Linford, P. Gardner, J. Coleman, K. Fine. 2008. "Aging and demographic plasticity in response to experimental age structures in honeybees (*Apis mellifera* L)." *Behavioral Ecology and Sociobiology* 62: 1621–1631.

Rueppell, O., S. Christine, C. Mulcrone, and L. Groves. 2007. "Aging without functional senescence in honey bee workers." *Current Biology* 17: R274–R275.

Rueppell, O., M. K. Fondrk, and R. E. Page. 2005. "Biodemographic analysis of male honey bee mortality." *Aging Cell* 4: 13–19.

Rueppell, O., and R. W. Kirkman. 2005. "Extraordinary starvation resistance in *Temnothorax rugatulus* (Hymenoptera: Formicidae) colonies: Demography and adaptive behavior." *Insectes Sociaux* 52: 282–290.

Sakagami, S. F., and H. Fukuda. 1968. "Life tables for worker honeybees." *Research in Population Ecology* 10: 127–139.

Schmid-Hempel, P. 1998. *Parasites in social insects.* Princeton: Princeton University Press.

Schmid-Hempel, P., and R. Schmid-Hempel. 1984. "Life duration and turnover of foragers in the ant *Cataglyphis bicolor* (Hymenoptera, Formicidae)." *Insectes Sociaux* 31: 345–360.

Schrempf, A., J. Heinze, and S. Cremer. 2005. "Sexual cooperation: Mating increases longevity in ant queens." *Current Biology* 15: 267–270.

Seehuus, S. C., and G. V. Amdam. 2006. "Cellular senescence in honeybee brain is largely independent of chronological age." *Experimental Gerontology* 41: 1117–1125.

Seehuus, S. C., K. Norberg, U. Gimsa, T. Krekling, and G. V. Amdam. 2006. "Reproductive protein protects sterile honeybee workers from oxidative stress." *Proceedings of the National Academy of Sciences USA* 103: 962–967.

Simon, A. F., C. Shih, A. Mack, and S. Benzer. 2003. "Steroid control of longevity in *Drosophila melanogaster.*" *Science* 299: 1407–1410.

Snodgrass, R. E. 1956. *Anatomy of the honey bee.* Ithaca: Cornell University Press.

Sohal, R. S., and R. Weindruch. 1996. "Oxidative stress, caloric restriction, and aging." *Science* 273: 59–63.

Tarpy, D. R., D. C. Gilley, and T. D. Seeley. 2004. "Levels of selection in social insects: A review of conflict and cooperation during honey bee *(Apis mellifera)* queen replacement." *Behavioral Ecology and Sociobiology* 55: 513–523.

Tatar, M., A. Bartke, and A. Antebi. 2003. "The endocrine regulation of aging by insulin-like signals." *Science* 299: 1346–1350.

Tatar, M., and C. M. Yin. 2001. "Slow aging during insect reproductive diapause: Why butterflies, grasshoppers and flies are like worms." *Experimental Gerontology* 36: 723–738.

Thorne, B. L., N. L. Breisch, and M. I. Haverty. 2002. "Longevity of kings and queens and first time of production of fertile progeny in dampwood termite (Isoptera; Termopsidae; *Zootermopsis*) colonies with different reproductive structures." *Journal of Animal Ecology* 71: 1030–1041.

Tofilski, A. 2000. "Senescence and learning in honeybee *(Apis mellifera)* workers." *Acta Neurobiologiae Experimentalis* 60: 35–39.

———. 2002. "Influence of age polyethism on longevity of workers in social insects." *Behavioral Ecology and Sociobiology* 51: 234–237.

Tsuji, K., K. Nakata, and J. Heinze. 1996. "Life span and reproduction in a queenless ant." *Naturwissenschaften* 83: 577–578.

Tsuji, K., and N. Tsuji. 2005. "Why is dominance hierarchy age-related in social insects? The relative longevity hypothesis." *Behavioral Ecology and Sociobiology* 58: 517–526.

Vaupel, J. W., J. R. Carey, K. Christensen, T. E. Johnson, A. I. Yashin, N. V. Holm, I. A. Ichine, V. Kannisto, A. A. Khazaeli, P. Liedo, V. D. Longo, Y. Zeng, K. G. Manton, and J. W. Curtsinger. 1998. "Biodemographic trajectories of longevity." *Science* 280: 855–859.

Visscher, P. K., and R. Dukas. 1997. "Survivorship of foraging honeybees." *Insectes Sociaux* 44: 1–5.

Ward, K. N., J. Coleman, K. Clittin, S. Fahrbach, O. Rueppell. 2008. "Age, caste, and behavior determine the replicative activity of intestinal Stem Cells in honeybees (*Apis mellifera* L.)." *Experimental Gerontology* 43: 530–537.

West-Eberhard, M. J. 2003. *Developmental plasticity and evolution.* New York: Oxford University Press.

Wheeler, D. E., N. Buck, and J. D. Evans. 2006. "Expression of insulin pathway genes during the period of caste determination in the honeybee, *Apis mellifera*." *Insect Molecular Biology* 15: 597–602.

Williams, G. C. 1957. "Pleiotropy, natural selection, and the evolution of senescence." *Evolution* 11: 398–411.

Wilson, D. S., and E. Sober. 1989. "Reviving the superorganism." *Journal of Theoretical Biology* 136: 337–356.

Wilson, E. O. 1971. *The insect societies.* Cambridge: The Belknap Press of Harvard University Press.

Winston, M. L. 1979. "Intra-colony demography and reproductive rate of the Africanized honeybee in South America." *Behavioral Ecology and Sociobiology* 4: 279–292.

———. 1987. *The biology of the honey bee.* Cambridge: Harvard University Press.

Winston, M. L., and L. A. Fergusson. 1985. "The effect of worker loss on temporal caste structure in colonies of the honeybee (*Apis mellifera* L.)." *Canadian Journal of Zoology* 63: 777–780.

Woyciechowski, M., and J. Kozlowski. 1998. "Division of labor by division of risk according to worker life expectancy in the honeybee (*Apis mellifera* L.)." *Apidologie* 29: 191–205.

Yang, A. S., C. H. Martin, and H. F. Nijhout. 2004. "Geographic variation of caste structure among ant populations." *Current Biology* 14: 514–519.

CHAPTER FOUR

The Ecological Setting of Social Evolution:
The Demography of Ant Populations

BLAINE J. COLE

THIS CHAPTER IS CENTERED on the demography of ant populations: the growth, reproduction, and mortality of colonies. I have chosen to focus on this rather specific set of topics in the ecology of ants because they are a central feature of life history that is a prerequisite for understanding social evolution. I will first describe the utility of demographic data and its analysis and then review information on the demography of ant populations. More detailed descriptions of some of the demographic methods are given in an online appendix (http://sols.asu.edu/publications/pdf/cole_chapter4.pdf).

Age-specific mortality, age-specific fecundity, and population growth rate interact to influence the fitness of an organism. For ants, when we say that strategies concerning communication, recognition, division of labor, foraging, or sex allocation are selectively advantageous, we are making a statement about the fitness consequences of a behavioral tactic or syndrome. What we often measure, however, are the functional consequences of a trait: how a particular change in behavior influences the efficiency of communication or division of labor. We then use the functional differences to infer the direction of selection by a logical argument. This approach is probably correct in a qualitative sense: a difference in behavior that increases the efficiency of communication or division of labor probably is selectively advantageous. However, it is nearly impossible to say how significant even a major change in behavior will be without considering the evolutionary ecology of the organism. It is even less likely that we can fruitfully argue about the relative importance of one behavioral change (e.g., sex allocation) over another (e.g., division of labor) without understanding the ecological setting of the trait. To make significant progress in understanding the evolution of any trait, including social behavior, we must place our evolutionary, genetic, and behavioral studies in a demographic context. The state of this information in ants is not very satisfactory, but there is some information, and I hope that a review of techniques and data will stimulate further research in this area.

The demographic literature on natural populations of ants is incomplete and often sketchy. There are a number of reasons for the lack of information, but to some degree we can blame the ants. Ant colonies are long-lived, iteroparous organisms that place their nests where workers cannot be easily seen. Colonies may move frequently or have cryptic locations often making them impossible to follow. If one cannot identify individual colonies, determine their ages, follow their survival, measure their reproduction, or be able to carry out the study for long enough, it is difficult to gather demographic data. An ideal study for obtaining demographic data would be a detailed longitudinal study, but even when the study does not have the most desirable properties, it can be possible to extract useful information. For example, when we do not have direct information on the age of colonies, it may still be possible to infer their age-specific mortality and fecundity.

One of the main techniques used in this chapter is manipulation of demographic matrices. The online appendix shows how to extract a variety of information from the age- and stage-based data that can be collected from ant populations. Here I work with essentially two types of data: age-or stage-specific survival and reproduction. The survival data are encapsulated by the transition matrix, **T,** which measures the probability of making a transition from one age or stage to another. The reproductive data are given by the fecundity matrix, **F,** which measures the reproductive contribution of one age or stage to another. When the data are age-specific, then the sum of the two matrices is the familiar Leslie matrix, **L,** and when the data are stage-specific we refer to this matrix as **S.** By manipulating these matrices a variety of information about life expectancy, population growth rate, age distribution, and reproductive value can be obtained. The online appendix relies heavily on Cochran and Ellner (1992) and especially Caswell (2001), the definitive source for demographic analysis of populations using the matrix methods described below. Age within stage distributions are discussed by Boucher (1997).

Demography of Ant Populations

This part of the chapter is divided into components that reflect life cycle stages and the type of data (age- or stage-based). Queens produce colonies; colonies grow, eventually reproduce, and finally they die. While there is some information on the probability that queens will successfully found a colony, on the growth and mortality of colonies, and on the reproduction of colonies, it is fair to say that there has been more of a focus on gathering data on reproduction than on growth and survival.

Colony Growth and Survival

Queen Survival during Colony Founding

The demography of queens, the least conspicuous stage in the life cycle of the colony, is poorly known. Although it is assumed that colony founding is the most dangerous portion of the life cycle, the probability that a queen will survive to found a colony is known for few species. Laboratory studies have shown that under appropriate conditions the probability of successful colony founding can be quite high (Johnson 1998), so the observed high mortality must be due to a combination of parasites, predators, pathogens, competitors, environmental stresses, and the physiological cost of producing the first workers that is expressed in the field.

When queens disappear into the soil or wood during claustral colony founding, it may be impossible to associate a queen with a specific colony. If queens are only observable for a matter of hours, then it is difficult even to measure the density of queens and the density of incipient colonies. This latter measure may provide a possible way to assay the success of colony founding, but estimates of the density of founding queens are rare. Colony founding by non-independent means (e.g., colony fission, temporary social parasitism, or through re-adoption of queens into natal nests) yield higher values for queen survival, but these types of colony founding are not discussed here.

There are published data for colony founding success by queens in *Pogonomyrmex occidentalis*, *Crematogaster ashmeadi*, *Solenopsis invicta*, and *Atta bisphaerica* (Table 4.1). The probabilities range from 0.001–0.076 that a queen produces colonies through the incipient stage. Indirect data would be a welcome addition to these scanty direct measurements. One approach would be to relate the density of queens to the density of

Table 4.1. Queen and population density

Species	Queen density[a]	Colony density[a]	Notes	Reference
Acromyrmex landolti fracticornis	4,500	2,400	Paraguay[b]	Data cited in Fowler et al. 1986
Acromyrmex muticonodis	200	2.5	Sao Paulo, Brazil[b]	Data cited in Fowler et al. 1986
Atta capiguara	300	2.3	Sao Paulo, Brazil[b]	Data cited in Fowler et al. 1986
Atta capiguara	450	8	Paraguay[b]	Data cited in Fowler et al. 1986
Atta capiguara	8,000	.	Sao Paolo, Brazil[b]	Data cited in Fowler et al. 1986
Atta cephalotes	6,000	0.5	Guatemala[b]	Data cited in Fowler et al. 1986
Atta sexdens rubropilosa	500	3	Sao Paulo, Brazil[b]	Data cited in Fowler et al. 1986
Atta vollenweideri	1,800	0.9	Paraguay[b]	Data cited in Fowler et al. 1986
Atta bisphaerica	900		Sao Paolo, Brazil[b]	Fowler 1987. Data cited in Fowler et al. 1986
Pogonomyrmex occidentalis	112–1,104	20–96	Survival of queens 0.0009 20 plots, Colorado, USA	Cole & Wiernasz 2002
Pogonomyrmex occidentalis	671	81	Survival of queens 0.013 Colorado, USA	Cole & Wiernasz, unpubl.
Crematogaster ashmeadi	197	49	Florida, USA	Hahn & Tschinkel 1997
Solenopsis invicta	3,000	40	Survival of queens 0.076 Florida, USA Survival of queens 0.002–0.04	Tschinkel 1992
Lasius flavus	10,000	2,600	England[c]	Elmes
Lasius niger	12,000	100	England[c]	Elmes
Lasius niger	55,173	392	Denmark[c]	Boomsma et al. 1982
Lasius niger	5,699	181	Denmark[c]	Boomsma et al. 1982
Lasius niger	1,092	16	Denmark[c]	Boomsma et al. 1982
Lasius niger	23,360	800	Poland[c]	Petal 1980
Myrmica spp.	67,100	1,100	Poland[c]	Petal 1980

a. Except for *C. ashmeadi* density ha⁻¹. Where multiple values given by source, the midpoint of range.

b. Data for queen density and colony density come from same location, but not necessarily the same population.

c. Data calculated from queen production per colony, colony density, and proportion of colonies, where applicable, that produce queens.

colonies. In a population, the number of queens that land on a particular area should be related to the number of colonies in that area. The form of the relationship is a function of the survival of queens to the colony stage and colony life span. The correlation between the number of queens of a variety of species and the density of colonies is shown in Figure 4.1. As expected, the number of queens is substantially more than the number of colonies. For two species that have been measured repeatedly, *P. occidentalis* and *Lasius flavus*, the density of queens and colonies fall roughly on a

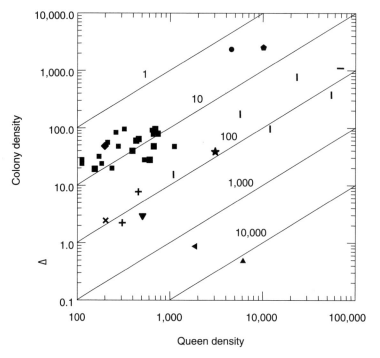

Figure 4.1. Plotting density of colonies and founding queens within a site for multiple species and populations. Densities are given in colonies or queens per hectare (except for *Crematogaster ashmeadi*). Data are given in Table 4.1. Symbols represent different species: ●—*Acromyrmex landolti*; ■—*Pogonomyrmex occidentalis*; ◆—*Crematogaster ashmeadi*; ×—*Acromyrmex multiconodis*; +—*Atta capiguara*; ▲—*Atta cephalotes*; ▼—*Atta sexdens*; ◀—*Atta vollenweideri*; I—*Lasius niger*; ◆—*Lasius flavus*; I—*Myrmica*; ★—*Solenopsis invicta*. The diagonal lines indicate equal ratios of queens to colonies at a location.

line, indicating a relatively constant relation of queen and colony density. Because *P. occidentalis* colonies cluster around the line indicating about ten times as many queens as colonies, it would require at least 90% mortality (10% survival) for the entire population to be replaced in one year. Since it is known that survival is actually closer to 1%, about 10 years are required for population replacement. This rough method gives an estimate of colony survival that is rather accurate. Consistent with direct measures, it estimates approximately 1 year for *S. invicta* and 3.3 years for *C. ashmeadi.* Remembering that these are not longevities (which can be at least 45 years in the case of *P. occidentalis*), but life expectancies at the smallest colony size, estimates are probably consistent to within a factor of two or three.

For species that cluster in the vicinity of the 100 isocline, with a life expectancy (after colony founding) of approximately 10 years, we would predict that queen survival must be approximately 0.001. For *Atta bisphaerica* the estimate of founding success is 0.002, and while there is an estimate of queen density, there is no estimate of colony density at that location (Fowler 1987). In *A. capiguara,* the survival of incipient colonies is again 0.001 over the first three months after colony founding (Fowler, Robinson, and Diehl 1984). Comparing queen and colony density for two other species of *Atta,* the probability of founding a colony should be as low as 0.0001–0.00001. Whether this estimate is off by a factor of two or three it seems clear that survival of queens in many species must be less than 1 per thousand and for certain species may be far less than that.

For two species, *S. invicta* and *P. occidentalis,* there is information about survival at more than one time point during colony founding. The survival of *S. invicta* queens was followed daily during the formation of incipient colonies and high mortality rates of 5 to 6% per day were followed by a period of declining mortality (0.6–2% per day) as incipient colonies formed (Tschinkel 1992). *P. occidentalis* had a similar pattern, with a mortality of 7% per day falling to 5.5% per day at the time when incipient colonies were produced. The concordance of the two measures suggests qualitative generality if not quantitative congruence. These mortality rates mean that the life expectancy of a queen after a mating flight is 14 to 18 days. This is a rather amazing value for an insect that, in the case of *P. occidentalis,* can live for 45 years (Keeler 1988, 1993). It seems likely that the life expectancy of an *Atta* queen must be measured in hours.

Survival of Colonies

For colonies with a single queen, the survival of the queen and the colony coincide. For many of the best studied species, the survival of colonies is another life-cycle stage in the survival of queens. In these cases the demography of colonies cannot be separated from the demography of queens.

Age-Related Mortality

Age-specific mortality of colonies has been measured in several species. One might predict that mortality would be higher among young colonies and lower among older colonies, if for no other reason than the increased size of the colony. For example, in *S. invicta* survival increases with colony size (Adams and Tschinkel 1995a), colony size increases with age (Tschinkel 1993), and survivorship is particularly low in young colonies (Adams and Tschinkel 1995b).

Although age-specific mortality rates decline in *P. occidentalis* (Cole and Wiernas unpubl. data) initially, as expected, they increase in *P. barbatus* (Gordon and Kulig 1998). Both species are seed harvesting ants in arid environments, with relatively large adult colony sizes and single queens. It seems unlikely that the increase in mortality reflects senescence. It is not clear why these two species, with superficially similar ecologies, are different. These data were collected by direct observations of the survival of individual colonies for at least 10 years. The age-specific survival for *P. barbatus* colonies together with *P. occidentalis* is shown in Figure 4.2. There are substantial differences in the pattern of survival between *P. occidentalis* and *P. barbatus*. Although survival is fairly consistent later in life with annual survival being approximately 0.8 in *P. occidentalis* and 0.9 in *P. barbatus*, there is a ten-fold greater mortality among first year *P. occidentalis* young colonies (0.4) compared to *P. barbatus* (0.04).

Survival of queens during colony founding is 1.8% for *P. occidentalis*. This means that when we imagine the colony from 0 to 1 year, the curve would climb to 56 times the starting value shown in Figure 4.2. The survival of *P. barbatus* queens is unknown, but presumed to be similarly low. All evidence suggests that queens have an extreme Type III survivorship curve, with most mortality occurring early in life and a very small fraction of the population reaching reproductive maturity. While many insects are known to have extreme Type III survivorship, most do not combine it with extreme longevity. In this way ant colonies are most similar in their patterns of mortality to trees, codfish, and corals. They do not have many

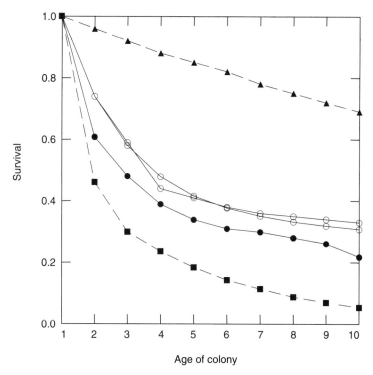

Figure 4.2. Survival curves of colonies from approximately one year of age. The three solid lines are *P. occidentalis* colony survival. The line with solid circles is the survival curve from age-based data. The two curves with hollow circles are the survival curves for colonies based on stage-based data. The upper line with solid triangles is from age-based data for *P. barbatus.* The lower dashed line, with solid squares, is the estimated survival curve based on the stage-specific survival probabilities for *Formica opaciventris* from Scherba's (1963) data.

similarities, apart from longevity, to the survival patterns of the social vertebrates with which they are often compared.

The data points that produce Figure 4.2 constitute the entries in the transition matrix, **T** (see online appendix). From the transition matrix we can obtain an estimate of the amount of time that a colony can be expected to spend in each subsequent age category as a function of age $= (\mathbf{I}\text{-}\mathbf{T})^{-1}$, where **I** is the identity matrix (ones on the diagonal and zeros elsewhere). The sum of these estimates is the colony life expectancy, which for *P. occidentalis* rises to a maximum of about 10 years at the age of 4 to 5 years.

Stage-Related Patterns of Mortality

For ants, it is often far easier to obtain information about stage- or size-related patterns of mortality than about age-related patterns of mortality. It may be easier to obtain survival information about incipient colonies, young colonies, and mature colonies, for example, than to follow individual colonies for long enough to obtain age information. Size information, as one specific type of stage information, is particularly informative. Data must be collected in at least two time intervals and individuals of known stage or size must be classed as surviving (to a size class) or dying. Because individuals are assessed after a time interval, there is temporal information implicit in the data set, and it is possible to extract a considerable amount of age-related information from the size- or stage-related data.

The size of an ant colony is usually given as the number of workers, often a difficult measurement to make. Little is known about the architecture of subterranean colonies (but see Tschinkel 2004, 2005, for recent examples), so it is difficult to know how to excavate a nest. One alternative is to measure the sizes of ant mounds or other colony constructions, but it is not clear whether these external nest measurements are good indicators of colony size. The good news is that in many cases the number of workers correlates well with the size of the nest structure (Table 4.2); the bad news is that most species do not make conspicuous colony constructions.

If we use stage-based data, the basic tool is again the transition matrix. Scherba (1963) presented information on the transition of colonies of *Formica opaciventris* between five colony categories based on mound structure and colony activity levels, and calculated mortality rates for each of his colony categories. This is important information for understanding the dynamics of a population; however, with the category transition data that he also collected, it is possible to infer life expectancy and age-specific survival. The data were from censuses in 1957–1959 and can be used to construct a transition matrix between categories for each year (1957–1958 and 1958–1959). In this case the entries refer to the probability that a colony will make a transition from one stage to another, making it possible to calculate life expectancy (2.7, 3.3, 5.2, 4.2, 7.0 years) for colonies found in each of his five activity categories. Additionally, one can obtain an estimate of age-specific survival based on stage-specific data. One obtains this result by iterating the transition matrix (see online appendix), obtaining the result shown in Figure 4.2. It is not generally appreciated that this sort of information can be extracted from stage-classified data.

Table 4.2. Colony size and nest sizes

Species	Relation of worker population to nest measurements	Sample sizes	Notes	References
Pogonomyrmex occidentalis	$r = 0.88$	31	Log mound volume with log foragers from mark/recapture	Wiernasz & Cole 1995
Pogonomyrmex occidentalis	$r = 0.7$	33	Mound area with worker number	Lavigne 1969
Solenopsis invicta	$r^2 = 0.9$	55	Mound volume with worker biomass/numbers	Tschinkel et al. 1995
Solenopsis invicta	$r^2 = 0.85$	89	Mound volume with worker biomass/numbers	Tschinkel 1993
Pogonomyrmex badius	$r^2 = 0.73 - 0.93$	31	Various measure of (subterranean) nest architecture	Tschinkel 1999
Pogonomyrmex barbatus	Nest mound area directly related to colony number		Mound area with worker number	Gordon 1992
Trachymyrmex septentrionalis	$r = 0.57$	55	Surface area of craters and worker number	Beshers & Traniello 1994
Formica exsecta	$r = 0.77$	59	Nest surface area and mark-recapture of workers	Liautard et al. 2003
Pogonomyrmex salinus	No relation of nest characters to colony size	25	Excavation of nests	Gaglio et al. 1998
Lasius flavus	Strongly correlated, r not given	10	Excavation	Nielsen et al. 1976

A size-based transition matrix for harvester ants (an example of which is shown in the appendix) was obtained by measuring the size of nests in two consecutive years. In this case the transition matrix is between colonies of different sizes. Finding the life expectancy as outlined above yields an estimate of life expectancy for the largest colonies of about 35 years. This is not unexpected given other estimates of longevity, based on 25 years of observation of *P. occidentalis* (Keeler 1988, 1993), of about 45 years. This latter estimate is based on regressing survivorship on age and assuming constant mortality.

Estimates of longevity based on size transitions may be better for this long-lived organism than estimates based on age itself, even when the data set is more than 10 years in duration. The age-specific survival derived from stage-based data for colonies starting in either of the two smallest size categories is shown in Figure 4.2, along with the measured age-specific survival for the first 10 years of this study. The size-based estimates are always somewhat higher and begin to deviate from the age-based data in later years, but the overall agreement between two methods based on completely different data sets lends confidence in the utility of this approach. It is important to emphasize that the stage-specific survival curves based on size are data that were collected over 2 years, while the data from colonies of known age required 10 years to assemble.

For other species (Table 4.3), investigators have provided a simple measure of the proportion of colonies remaining alive after a time interval. In a stage-based transition model, the transition matrix becomes the single probability of survival. The life expectancy is estimated as the reciprocal of the mortality rate. It is natural that investigators are more likely to measure survival of larger colonies and in many cases the authors recognize this bias in their calculation (e.g., Jonkman 1979; Fowler, Robinson, and Diehl 1984 were particularly aware of this because their data are from colonies visible in aerial photographs). Because mortality usually declines with colony size, data from larger colonies will overestimate life expectancy for the population as a whole, but estimates for the subset of colonies observed should be accurate. If there has been a thorough search for all colonies, large and small, and the data are aggregated across those size categories, then the mortality rates represent an average of the population.

The only data set where a single estimate of survival can be compared to the survival spectrum for age- or stage-based data is in *P. occidentalis*. In this case the data present a point estimate that yields a life expectancy of

13.3 years, while the life expectancy ranges from 11 to 35 years for the smallest to largest colonies. These values must agree with one another and therefore reflect the higher size-specific mortality of small colonies as well as the size distribution in the population. From Table 4.3 we see that averaged life expectancy varies over a relatively small range compared to other life history measures. In part, this is probably due to the fact that when one measures life expectancy based on the entire population of colonies, one obtains an aggregate measure reflecting the size/age distribution in a population, colony growth rates, and size-specific survival. All the data from *Atta* seem lower than expected for reasons that are not clear. The survival of *Paraponera clavata* is the outlier among species, although the figures may reflect movement of nests as well as mortality.

Reproduction of Colonies

The second main ingredient of population demography is reproduction. As with mortality, we can look at reproduction as either age-related or size/stage related. Ideally, we want to know the age-specific pattern of reproduction so that we can calculate reproductive values and interpret selection operating at various life-cycle stages. However, if there was little information on age-related patterns of mortality among ants, there is virtually nothing on age-related patterns of reproduction. Data from *P. barbatus* are the only published information that directly touch on this point (Gordon 1995; Wagner and Gordon 1999). Gordon (1995) reported that the number of reproductive colonies increased from 3 to 5 years of age, but it was not clear how many colonies of each age were observed. Wagner and Gordon (1999) found that reproductive output increased slightly with age ($r^2 = 0.03$–0.06), but unfortunately the number of queens produced could not be measured. While the regression of reproductive output on colony age was significant for those colonies that produced reproductives, it was not significant when all colonies, even those that did not reproduce, were included in the analysis. Since colony size presumably increases with age, it is difficult to conclude that there is any direct effect of colony age.

Size-Related Reproduction

Apart from the importance of size to survival, size is most often linked to reproduction. There are at least two ways in which colony size can be

Table 4.3. Annual mortality rates

Species	Gross annual mortality rate (extrapolated life expectancy)	Notes	References
Paraponera clavata	0.36, 0.23 (2.8, 4.3)	Survivorship for 3 three yrs at one site, 2 at another, $N = 217$ colonies	Thurber et al. 1993
Formica exsecta	0.049 (20.4)	Censuses at intervals of 1–3 yrs for 10 years, average over time span, $N = 57$ colonies	Pamilo 1991
Atta cephalotes	0.26 (3.8)	Mortality rates 0.14–0.57, across sites combined, mortality measured over two-yr. interval, $N = 74$ colonies	Perfecto & Vandermeer 1993
Pogonomyrmex owyheei	0.07 (14.3)	Mortality from annual censuses, 3 yrs, 2 sites, $N = 88$ colonies	Porter & Jorgensen 1988
Pogonomyrmex owyheei	0.05 (20)	Mortality from 2 yrs, 3 sites, $N = 201$ colonies	Sharp & Barr 1960, reported in Porter and Jorgensen 1988
Pogonomyrmex occidentalis	0.028 (35.7)	Mortality from 14 yrs, $N = 107$ colonies. Concordant with previous report (Keeler 1988). Longevity est. 45 yr.	Keeler 1993
Atta vollenweideri	0.14 (7.1)	Mortality estimated from aerial photos 30 yrs. apart. Only colonies > 3 yrs	Jonkman 1979

Species	Mortality rate (years)	Method	Reference
Atta capiguara	0.15 (6.7)	Survival estimated from photos 10-yr. interval. Survival of large nests estimated from table as 0.2 for 10 yrs.	Fowler 1984
Atta columbica	0.095 (10.5)	Survival from multiple censuses. $N=92$ colonies, 2 yrs.	Wirth et al. 2003
Myrmecocystus depilis	0.075 (13.3)	Mortality from 5 time intervals. Weighted average by sample size, $N=133$ colonies, 1958–1993	Chew 1995
Myrmecocystus mexicanus	0.053 (18.9)	Mortality from 5 time intervals. 4 cohorts from Chew 1993 one from Chew 1987. Weighted average by sample size, $N=82$ colonies	Chew 1987, 1995
Aphaenogaster cockerelli	0.15 (6.7)	Mortality from 3 time intervals. Weighted average by sample size, $N=32$ colonies, 1958–1976	Chew 1987
Pogonomyrmex barbatus[a]	0.048 (20.8)	Estimate derived from deaths of all ages, $N=250$ colonies, 13 yrs.	Gordon & Kulig 1998
Pogonomyrmex occidentalis[a]	0.075 (13.3)	Estimate for one year derived from deaths of all ages/sizes, $N=1{,}121$ colonies.	Cole & Wiernasz, unpub. data

a. Overall mortality rate included for comparison purposes. Age and/or size based mortality functions are available.

related to reproductive output. The first is that there may be a positive correlation between the number of workers and the number of queens that the colony produces, and the second is that there is a threshold size for reproduction. However, there may be no relationship between colony size and reproductive output per se, yet there is a relationship between colony size and the probability that a colony reproduces. Both patterns have been observed (Table 4.4). In a few species or populations, the number of queens produced and colony size are correlated (e.g., *Myrmica sulcinodis,* Elmes and Wardlaw 1982), but that occurs far less commonly than expected. A much more common pattern is that there is essentially a size threshold for reproduction (e.g., *Camponotus,* Fowler 1986; *Pogonomyrmex,* Cole and Wiernasz 2000). It is nearly universal that the size of a colony affects the probability that a colony will reproduce. The number of queens that are produced is much less strongly (and frequently not) related to colony size. Interestingly, there is usually a stronger relationship between colony size and the production of males (e.g., *Myrmica,* Elmes and Wardlaw 1982). Given the complex interactions among local food abundance, within-colony demography, and differing queen and worker interests (Herbers 1991), perhaps it should not surprise us that new queen production bears a complicated relationship to colony size. However, the long-standing assumption about colony demography that bigger colonies have greater reproduction, which may form the basis of ideas about topics as diverse as population dynamics and reproductive conflict, does not deserve the status of generalization.

The Age/Size Frequency Distribution

In a stable age distribution the proportion of individuals of a given age must decline with age. The few non-invading populations for which we have age distributions show this pattern *(P. barbatus, P. occidentalis).* This is not proof that the populations are at a stable age distribution, but it is a necessary prerequisite. For species that may be invading a new habitat, such as *Diacamma ceylonense* (Karpakakunjaram et al. 2003), there is no expectation that the age distribution will have any particular form. If numerous colonies invade over a short time, then the age distribution will initially contain a few cohorts. Invasion of a new habitat by one or a few colonies will produce an age distribution that is characteristic of an expanding population. The age distribution may indicate more about the progress of the invasion than about demographic processes within the population.

Table 4.4. Size/age and reproductive output

Species	Method	Result	Notes	Reference
Pogonomyrmex occidentalis	Logistic regression, correlation	Threshold for reproduction (N=89–324 colonies, 9 years)	Reproduction does not increase for colonies that reproduce (N=37–61 colonies, 9 years)	Cole and Wiernasz 2000 unpubl. data
Pogonomyrmex barbatus	Correlation	Begin to reproduce at 3 yrs. Apparently increasing probability with age, but size vs. age unclear	Increasing no. of reproductive colonies with age, but unknown total no. of colonies of given age (Gordon 1995). Queen production itself not measured by Wagner & Gordon.	Gordon 1995; Wagner & Gordon 1999
Tetramorium caespitum	Correlation	No correlation	N=49 colonies from 2 years	Brian et al. 1967
Lasius niger	Correlation	1 positive, 1 nonsignificant, 1 significant for total reproduction but not for queens	Three populations	Boomsma et al. 1982
Lasius niger	Correlation	No correlation	2.7% of variation in sexual numbers explained by worker no. Presumed nonsignificant.	Petal 1980

(continued)

Table 4.4 (*continued*)

Species	Method	Result	Notes	Reference
Myrmica spp.	Correlation	See notes	9.9% of variation in total sexual reproduction explained by worker no. Unknown significance.	Petal 1980
Camponotus pennsylvanicus	Correlation	From graph a clear threshold for reproduction. N unknown.	For colonies that reproduce apparent increase of reproduction with size. Some possible lab colonies.	Fowler 1986
Camponotus ferrugineus	Correlation	A clear threshold from graph	No relation to size above threshold for reproduction. Some possible lab colonies.	Fowler 1986
Myrmica sabuleti	Correlation and categorical analysis of reproducing and non-reproducing colonies	Threshold (?) at Site X (reproductive nests bigger, but no correlation of size and gyne positive output); correlation at Stonehill.	Site X, N=64 nests, two years; Stonehill, N=99 nests, two years.	Elmes & Wardlaw 1982
Myrmica sulcinodis	As for *M. sabuleti*	Reproductive colonies are bigger, R²=0.27 for reproductive colonies only	Two sites, 7 years. All nests, N=224; all gyne producing nests N=44.	Elmes & Wardlaw 1982

Species	Method	Results	Notes	Reference
Myrmica sulcinodis	As for *M. sabuleti*	Threshold (?) at site X (reproductive nests bigger, but no correlation of size and gyne output).	N = 49, two years	Elmes & Wardlaw 1982
Solenopsis invicta	Correlation	Small size class with much less reproduction, apparent increase in reproduction with increasing size classes.	Analysis based on size classes and reproduction assayed throughout a season.	Tschinkel 1993
Leptothorax longispinosus	Correlation, path analysis	1. No pattern in path analysis when other variables considered; 2. Smaller colonies more likely to be nonreproductive.	Small correlations of variable directions depending on queens and other factors. N=7 years, 2 popls, 679 colonies.	Herbers 1990
Trachymyrmex septentrionalis	Correlation	Alate biomass correlates with worker popl at one site.	FL correlation with all alate biomass; Long Island no correlation. For large colonies, no relation of size to reproduction, N=55 colonies.	Beshers and Traniello 1994

The size distribution of colonies reflects the convolution of colony growth and survival. Unlike age distributions, with size distribution a variety of patterns are possible. If colony growth and mortality rates are declining functions of colony size, colony size distributions may have a peak. If colonies grow rapidly through small sizes and survive for a long time at large sizes they will accumulate in the larger size classes. The size distribution will thus be affected by the patterns of colony growth. Some have suggested that colony growth patterns should follow a logistic function (Brian 1965; Wilson 1971; Oster and Wilson 1978) by analogy with simple models of population growth. The often-cited example of honey bee colony growth (in *Wilson 1971, p. 431*) is a spectacular fit to a logistic function; however, there are few data from ants to allow us to conclude that logistic colony growth in nature is common (Table 4.5). Tschinkel (1993) showed

Table 4.5. Colony age and colony size

Species	Pattern	Notes	Reference
Oecophylla smaragdina	Linear or logistic increase	No. nests increases; 5 colonies	Gupta 1968
Solenopsis invicta	Logistic growth with variation or logistic growth with overlying cycles. Colony growth declines with size.	Multiple colonies of known age fit to function. Field measurements of growth rate.	Adams & Tschinkel 2001; Tschinkel 1993
Pogonomyrmex barbatus	Colonies increase in size over 4 years.	N = 12 colonies	Gordon 1992
Atta vollenweideri	Logistic growth and/or maximum size reached with decline.	From aerial photographs 15 years apart.	Jonkman 1980
Lasius flavus	Linear increase	Correlate 3 nest dimensions to age for 8 years. N = 8 colonies.	Waloff and Blackith 1962
Pogonomyrmex occidentalis	Growth rate declines with colony size	Can fit linear function to data (i.e., logistic) but huge scatter in data means other declining functions fit equally well.	Wiernasz and Cole 1995, unpubl. data.

that a logistic function with overlaid periodicity fit the colony size data that he and co-workers collected. While logistic colony growth is generated by a negative linear function between growth rate and colony size, virtually any sort of negative relationship between growth rate and colony size will result in an upper limit to colony size, especially when combined with mortality.

There are more data on the size distributions of colonies than age distributions in nature (Figure 4.3). Size information from sufficient numbers of colonies to generate a distribution limits the number of studies that can be included. Additionally, it is important that the data be comprehensive and not restricted to mature or representative colonies, as this biases data to larger colonies. I have standardized the presentation of the data so that the frequency distribution is divided into ten size categories with a maximum relative frequency near one. Species may have a declining distribution (Figure 4.3a), a right-skewed distribution (Figure 4.3b), or a left-skewed distribution (Figure 4.3c).

The differences in size distribution reflect differences in colony growth and survival, but we cannot completely disentangle their relative contributions. Declining distributions are most likely to be associated with very high mortality rates relative to colony growth. A fairly short life span or tremendous differences among colonies in growth rate may also contribute to this pattern. While we do not have information about demographic properties from species with these characteristics, they represent a suite of covarying characters that are required to produce the observed distributions. Distributions that are skewed left may indicate that those colonies that achieve a maximum size have a long life span. There has been no systematic investigation of the life history correlates of colony size distributions, but the fact that there is considerable variation suggests that it may prove profitable to explore them.

It can be useful to determine the age distribution to associate with a particular stage (e.g., colony size). That there is an age distribution rather than a particular age is due to the fact that not all colonies of the same size will be the same. To determine the distribution of ages within stages requires information that can be obtained from the **S** matrix: the population growth rate and the stable stage distribution, stage-based fecundity, and stage-based survival. For *P. occidentalis* the distribution of ages within a stage (see online appendix) is shown in Figure 4.4. Because colonies can increase or decrease in size, the age distribution of larger colonies is flatter,

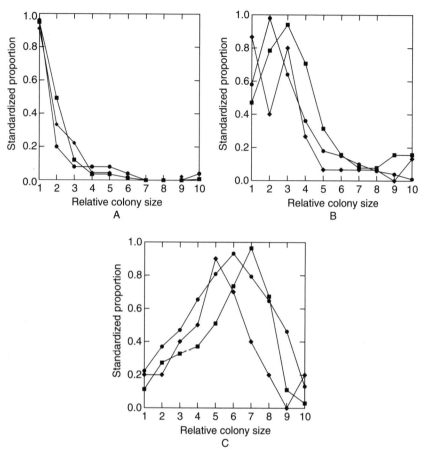

Figure 4.3. Several representative colony-size distributions. Species do not fall into three categories, they are just shown that way for presentation. A: ●—*Polyrhachis dives,* Yamauchi et al. 1987; ■—*Dolichoderus quadripunctatus,* Torossian 1967; ◆—*Aphaenogaster rudis,* Talbot 1951. B: ●—*Myrmica rubra,* Elmes 1973; ■—*Tetramorium caespitum,* Brian et al. 1967; ◆—*Dolichoderus pustulosus,* Kannowski 1967. C: ●—*Odontomachus haematodes,* Colombel 1970; ■—*Pogonomyrmex occidentalis,* Wiernasz & Cole unpubl. data; ◆—*Myrmica schencki,* Talbot 1945. The range of colony sizes is divided into ten categories and the most abundant size class is set to near 1.

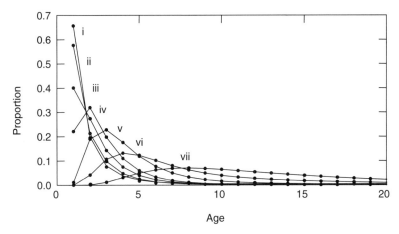

Figure 4.4. The age-within-stage frequency distributions for the seven size classes (i–vii) of *P. occidentalis* used in this chapter. Each line represents the probability that a colony in an observed size class will be a given age. Very small colonies (stage i) are very likely to be young, while size class vi shows much greater variation in size. Some colonies grow directly to a given size while others decline to this size after being much older.

with long tails. The distribution gives an estimate of the ages of colonies that occur in a population with the stage transition matrix **S.**

Life History Evolution

I have been trying to assemble data on the demography of ant species so that it is possible to make inferences about age- or stage-based survival and age- or stage-based reproduction of the species. With this information one can obtain several derived parameters such as life expectancy, population growth rates, and stable age or stage distributions. The goal of obtaining these data is to use the information to make inferences about the strength of selection operating on supposed adaptations. For that a different tool is necessary.

Reproductive Value

A population that has a consistent pattern of age-related mortality and reproduction will eventually attain a stable age distribution. Fisher (1930) introduced the concept of the reproductive value of an individual of age x in a population that has reached a stable age distribution:

$$V(x) = \frac{e^{r(x-1)}}{l(x)} \sum_{i=x}^{\infty} e^{-ri}l(x)m(x)$$

where $l(x)$ is the probability of surviving to age x, $m(x)$ is the reproductive output at age x (i.e., the schedules of mortality and fecundity that we have been discussing), and r is the rate of population increase (Roff 2002). If the population size is not changing ($r = 0$), the reproductive value (of a female) at age x is the expectation of future reproductive success. Additional mortality (i.e., selection) will have a disproportionate effect when it operates on individuals with higher reproductive value. Reproductive value rises to a maximum at about the age of first reproduction and then declines independently of senescence, simply due to the greater cumulative probability of mortality. To understand the operation of natural selection in a population, it is necessary to understand the age-specific schedules of mortality and fecundity and the growth of the population. To determine whether a change in life history is at a selective advantage, we can examine the change in the reproductive value function.

The age-specific reproductive value is the dominant eigenvector of the transposed Leslie matrix (see Figure 4.5 for the *P. occidentalis* data set). By estimating the age-specific survival to be 0.9 after the age of 10 years, this function has been extrapolated beyond the actual duration of this study. There are not, to my knowledge, other comparable data for ants. Reproductive value reaches a peak at around 6 to 7 years of age and then declines, but the value of future reproduction is usually greater than that of current reproduction. For example, if a colony is more than 7 years old, the value of current reproduction is about 0.6 (the probability of reproducing), while that of reproduction next year is 0.8×0.6 (survival $\times$ reproduction) $= 0.48$ and in 2 years is $0.8^2 \times 0.6 = .38$. The value over the next 2 years of reproduction (0.86) outweighs the value of reproduction this year. This means that it is difficult for any strategy that increases current reproduction at the expense of future reproduction to be advantageous. Because survival is rather high from year to year, reproduction to exhaustion is not expected. Bang-bang control strategies, in which colonies invest all resources in colony growth until a critical moment when they switch all resources to

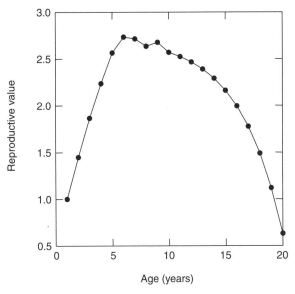

Figure 4.5. The reproductive value function for the harvester ant *P. occidentalis*. Reproductive value is standardized so that it is 1 in the earliest age category—reproductive value is measured with respect to this. Reproduction does not begin until age 5–6, so that the increase in reproductive value before that is a function of survival to greater ages. The decrease in reproductive value after age 10 reflects accumulating mortality rather than a decline in reproduction.

reproduction, have been elegantly analyzed in annually reproducing vespids (Macevicz and Oster 1976). Such a life history is not expected in this long-lived ant species, but until we have information that is obtainable only from the demographic data that allows us to calculate reproductive values can we estimate how valuable future reproduction may be.

With the distribution of ages within stages, it is possible to compute the mean age (and confidence intervals) of colonies of a given size. The reproductive value function calculated for the age-based demography for *P. occidentalis* (circles) is shown in Figure 4.6. The reproductive value of the oldest age category includes the reproductive value of individuals that are ≥ 10 years old. The squares show the function derived from the reproductive value of each size category of the population and the mean age of colonies that are in that size category. Inferences made using measures of reproductive value obtained from age-based and size-based survival and

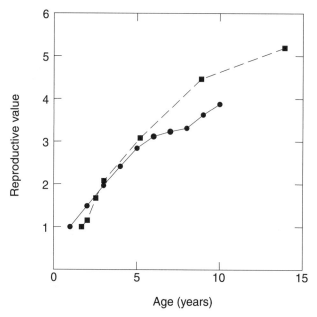

Figure 4.6. The reproductive value function calculated from age-based data (circles) and using size-based data to infer average age (squares). The reproductive value of the oldest colonies includes the summed reproductive value for the rest of their life. The size-based reproductive value calculates the reproductive value of a size category and associates a particular age by calculating the mean of the age-within-stage distribution given in Figure 4.4.

reproduction would probably be the same. It is not possible currently to know whether other species will give similar results and therefore how broadly this can be applied.

Investigating Adaptation

It is a rather simple matter to investigate the selective advantage or disadvantage of a particular change to a life history by altering the terms of the age- or stage-based matrix. For example, suppose one wanted to determine whether a strategy that resulted in increased probability of reproduction in 1 year and decreased probability of reproduction and survival in the following year was advantageous. Within a demographic context, we could solve that quantitatively. For combinations of the size and stage of an effect, we can determine whether the eigenvalue of **S** had increased or decreased; that is, whether an individual with this life history would have a greater or lower intrinsic growth rate and thus whether the trait was selectively advantageous or disadvantageous. The reproductive value function, calculated from the same matrix, is also changed and the magnitude of the change of alternative life history or behavioral strategies could be measured. We have the opportunity to quantify the magnitude of selection on life history differences, but only with information on the age/stage-specific mortality and reproduction. This leads us into a more analytic stage in the behavioral ecology of social insects.

The size distribution of colonies should be the result of the size transition matrix, **S** (contingent on the assumption that the population approximates the stable size distribution). This matrix condenses information on survival, growth, and reproduction of colonies. The conclusion that one is forced into is that the **S** matrix of populations—the patterns of survival, growth, and reproduction—vary enormously (see Figure 4.3). This is another way of emphasizing that understanding the variety of demography in ant populations is guaranteed to generate interesting and surprising results.

Conclusion

Tschinkel (1991) called for the development of a subfield within social insect biology that he called "Sociometry." He made the point that in studying the biology of social insects, we have skipped over the step of gathering basic information on the colonies and life cycles of social insects. The situation in ants has improved somewhat, in no small part due to Tschinkel's work. However, we are still lacking most quantitative information on demography.

In this chapter both the utility of demography in social insect evolutionary biology and the lack of data for most species is emphasized. It is not very likely that the species that have received the most attention are in any way representative of ants in general. *Pogonomyrmex* ants, which have the best demographic information, form large colonies with very long-lived queens. They produce colonies that are genetically diverse, due to multiply-mated queens, and the colonies gather seeds as a main food source. This deviates in every way from more typical ants. Leaf-cutter ants, which also have substantial demographic data, have the same list of oddities only perhaps more extreme–they also have multiply-mated queens, huge colonies, and the most intricate caste system specialized for processing vegetation and cultivating fungi. The demography of fire ants is known primarily from its introduced range (Tschinkel 2006). The list of species that are well known are those that have advantages for collecting demographic data, so it is important to remember that we are not in a position to make generalizations about ant demography. However, the acquisition of these data in more species is necessary for understanding the dynamics of evolutionary change.

Literature Cited

Adams, E. S., and W. R. Tschinkel. 1995a. "Density-dependent competition in fire ants: Effects on colony survivorship and size variation." *Journal of Animal Ecology* 64: 315–324.

———. 1995b. "Spatial dynamics of colony interactions in young populations of the fire ant *Solenopsis invicta.*" *Oecologia* 102: 156–163.

Adler, F. R. and D. M. Gordon. 1992. "Information collection and spread by networks of patrolling ants." *American Naturalist* 40: 373–400.

Baroni-Urbani, C. 1978. "Adult populations in ant colonies." *Production ecology of ants and termites*, M. V. Brian, ed. New York: Cambridge University Press.

Baroni-Urbani C., G. Josens, and G. J. Peakin. 1978. "Empirical data and demographic parameters." *Production ecology of ants and termites,* M. V. Brian, ed. New York: Cambridge University Press.

Beshers, S.N. and J. F A. Traniello. 1994. "The adaptiveness of worker demography in the attine ant *Trachymyrmex septentrionalis.*" *Ecology* 75: 763–775.

Boomsma, J. J., G. A. van der Lee, and T. M. van der Have. 1982. "On the production ecology of *Lasius niger (Hymenoptera: Formicidae)* in successive coastal dune valleys." *Journal of Animal Ecology* 51: 975–991.

Boucher, D. H. 1997. "General patterns of age-by-stage distributions." *Journal of Animal Ecology* 85: 235–240.

Brian, M. V. 1965. *Social insect populations.* New York: Academic Press.

Brian, M.V., G. Elmes, and A. F. Kelly. 1967. "Populations of the ant *Tetramorium caespitum* Latrielle." *Journal of Animal Ecology* 36: 337–342.

Caswell, H. 2001. *Matrix population models.* Sunderland: Sinauer Associates, Inc.

Chew, R. M. 1987. "Population dynamics of colonies of three species of ants in desertified grassland, southeastern Arizona, 1958–1981." *American Midland Naturalist* 118: 177–188.

Chew, R. M. 1995. "Aspects of the ecology of three species of ants (*Myrmecocystus* spp., *Aphaenogaster* sp.) in desertified grassland in southeastern Arizona, 1958–1993." *American Midland Naturalist* 134: 75–83.

Cochran, M. E., and S. Ellner. 1992. "Simple methods for calculating age-based life-history parameters for stage-structured populations." *Ecological Monographs* 62: 345–364.

Cole, B. J., and D. C. Wiernasz. 2000. "Colony size and reproduction in the western harvester ant, *Pogonomyrmex occidentalis*." *Insectes Sociaux* 47: 249–255.

Cole, B. J. and D. C. Wiernasz. 2002. "Recruitment limitation and population density in the harvester ant, *Pogonomyrmex occidentalis*." *Ecology* 83: 1433–1442.

Colombel, P. 1970. "Recherches sur la biologie et l'ethologie d'*Odontomachus haematodes* L." *Insectes Sociaux* 17: 183–198.

Elmes, G. W. 1973. "Observations on the density of queens in natural colonies of *Myrmica rubra* L. (Hymenoptera: Formicidae)." *Journal of Animal Ecology* 42: 761–771.

Elmes, G. W. 1987. "Temporal variation in colony populations of the ant *Myrmica sulcinodis*. II. Sexual production and sex ratios." *Journal of Animal Ecology* 56: 573–583.

Elmes, G. W., and J. C. Wardlaw. 1982. "A population study of the ants *Myrmica sabuleti* and *Myrmica scabrinodis*, living at two sites in the south of England. I. A comparison of colony populations." *Journal of Animal Ecology* 51: 651–664.

Fisher, R. A. 1930. *The Genetical Theory of Natural Selection.* Oxford: Oxford University Press.

Fowler, H. G. 1984. "Population dynamics of the leaf-cutting ant, *Atta capiguara*, in Paraguay." *Ciência e Cultura* 36: 628–632.

Fowler, H. G. 1986. "Polymorphism and colony ontogeny in North American carpenter ants (Hymenoptera: Formicidae: *Camponotus pennsylvanicus* and *Camponotus ferrugineus*)."

———. 1987. "Colonization patterns of the leaf-cutting ant, *Atta bisphaerica* Forel: Evidence for population regulation." *Journal of Applied Entomology* 104: 102–105.

Fowler, H. G. and B. L. Haines. 1983. "Diversidad de especies de hormigas cordoras y termitas de tumulo en cuanto a la sucesion vegetal en praderas paraguayas." In *Social Insects in the Tropics, vol. II*, P. Jaisson, ed. Paris: Presses de l'Université Paris XIII.

Fowler, H. G., V. Pereira-daSilva, L. C. Forti, N. B. Saes. 1986. "Population dynamics of leaf-cutting ants: a brief review." In *Fire Ants and Leaf-Cutting Ants*, C. S. Lofgren and R. K. Vander Meer, eds. Boulder, CO: Westview Press.

Fowler, H. G., S. W. Robinson, and J. Diehl. 1984. "Effect of mature colony density on colonization and initial colony survivorship in *Atta capiguara*, a leaf cutting ant." *Biotropica* 16: 51–54.

Gaglio, M. D., W. P. MacKay, E. A. Osorio, and I. Iniguez. 1998. "Nest populations of *Pogonomyrmex salinus* harvester ants (Hymenoptera: Formicidae)." *Sociobiology* 32: 459–463.

Gordon, D. M. 1991. "Behavioral flexibility and the foraging ecology of seed-eating ants." *American Naturalist* 138: 379–411.

Gordon, D. M. 1992. "How colony growth affects forager intrusion between neighboring harvest ant colonies." *Behavioral Ecology and Sociobiology* 31: 417–427.

Gordon, D. M. 1995. "The development of ant colony's foraging range." *Animal Behaviour* 49: 649–659.

Gordon, D. M., and A. W. Kulig. 1996. "Founding, foraging and fighting: Colony size and the spatial distribution of harvester ant nests." *Ecology* 77: 2393–2409.

———. 1998. "The effect of neighboring colonies on mortality in harvester ants." *Journal of Animal Ecology* 67: 141–148.

Gordon, D. M. and D. Wagner. 1997. "Neighborhood density and reproductive potential in harvester ants." *Oecologia* 109: 556–560.

Greenslade, P. J. M. 1971. "Interspecific competition and frequency changes among ants in Solomon Islands coconut plantations." *Journal of Applied Ecology* 8: 323–352.

Gupta, C. S. 1968. "Studies on the population structure of the nest of the Indian red ant—*Oecophylla smaragdina*." Symposium on recent advances in tropical ecology, International Society for Tropical Ecology. Pp. 187–198.

Hahn, D. A. and W. R. Tschinkel. 1997. "Settlement and distribution of colony-founding queens of the arboreal ant, *Crematogaster ashmeadi*, in a longleaf pine forest." *Insectes Sociaux* 44: 323–336.

Herbers, J. M. 1991. "The population biology of *Tapinoma minutum* (Hymenoptera: Formicidae) in Australia." *Insectes Sociaux* 38: 195–204.

Hölldobler, B. and E. O. Wilson. 1990. *The Ants.* Cambridge, MA: Belknap Press.

Johnson, R. A. 1998. "Foundress survival and brood production in the desert seed-harvester ants *Pogonomyrmex rugosus* and *P. barbatus* (Hymenoptera: Formicidae)." *Insectes Sociaux* 45: 255–266.

Jonkman, J. C. M. 1979. "Population dynamics of leaf-cutting ant nests in a Paraguayan pasture." *Zeitschrift für Angewande Entomologie* 87: 281–293.

Jonkman, J. C. M. 1980. "The external and internal structure and growth of nests of the leaf-cutting ant *Atta vollenweideri* Forel, 1893 (Hym: Formicidae), Part I." *Zeitschrift für Angewandte Entomologie* 89: 158–172.

Karpakakunjaram, V., P. Nair, T. Varghese, G. Royappa, M. Kolatkar, and

R. Gadagkar. 2003. "Contributions to the biology of the queenless ponerine ant *Diacamma ceylonense* Emery (Formicidae)." *Journal of the Bombay Natural Historical Society* 100: 533–543.

Keeler, K. H. 1988. "Colony survivorship in *Pogonomyrmex occidentalis*, the western harvester ant, in western Nebraska." *Southwestern Naturalist* 33: 480–482.

———. 1993. "Fifteen years of colony dynamics in *Pogonomyrmex occidentalis*, the western harvester ant, in western Nebraska." *Southwestern Naturalist* 38: 286–289.

Lavigne, R. J. 1969. "Bionomics and net structure of *Pogonomyrmex occidentalis* (Hymenoptera: Formicidae)." *Annals of the Entomological Society of America* 62: 1166–1175.

Macevicz, S., and G. Oster. 1976. "Modeling social insect populations II: Optimal reproductive strategies in annual eusocial insect colonies." *Behavioral Ecology and Sociobiology* 1: 265–282.

Oster, G. F., and E. O. Wilson. 1978. *Caste and ecology in the social insects.* Princeton: Princeton University Press.

Perfecto, I. and J. Vanderemeer. 1993. "Distribution and turnover rate of a population of *Atta cephalotes* in a tropical rain forest in Costa Rica." *Biotropica* 25: 316–321.

Petal, J. 1980. "Ant populations, their regulation and effect on soil in meadows." *Ekologia Polska* 28: 297–326.

Porter, S. D. and C. D. Jorgensen. 1988. "Longevity of harvester ant colonies in southern Idaho." *Journal of Range Management* 41: 104–107.

Roff, D. 2002. *The evolution of life histories.* New York: Chapman-Hall.

Scherba, G. 1961. "Nest structure and reproduction in the mound-building ant *Formica opaciventris* Emery in Wyoming." *Annals of the New York Entomological Society* 59: 71–87.

Scherba, G. 1963. "Population characteristics among colonies of the ant *Formica opaciventris* Emery (Hymenoptera: Formicidae)." *Annals of the New York Academy of Sciences* 71: 219–232.

Talbot, M. 1945. "Population studies of the ant *Myrmica schencki* ssp. *emeryana* Forel." *Annals of the Entomological Society of America* 38: 365–372.

Talbot, M. 1951. "Populations and hibernating conditions of the ant *Aphaenogaster (Attomyrma) rudis* Emery." *Annals of the Entomological Society of America* 44: 302–307.

Talbot, M. 1961. "Mounds of the *Formica ulkei* at the Edwin S. George Reserve, Livingston County Michigan." *Ecology* 42: 202–205.

Thurber, D. K., M. C. Belk, H. L. Black, C. D. Jorgensen, S. P. Hubbell, and R. B. Foster. 1993. "Dispersion and mortality of colonies of the tropical ant *Paraponera clavata*." *Biotropica* 25: 215–221.

Torossian, C. 1967. "Recerches usr la biologie et l'ethologie de *Dolichoderus quadripunctatus* (L.) Hym Formicoidea Dolichoderidae." *Insectes Sociaux* 14: 105–122.

Tschinkel, W. R. 1991. "Insect sociometry, a field in search of data." *Insectes Sociaux* 38: 77–82.

———. 1992. "Brood raiding and the population dynamics of founding and incipient colonies of the fire ant *Solenopsis invicta*." *Ecological Entomology* 17: 179–188.

———. 1993. "Sociometry and sociogenesis of colonies of the fire ant *Solenopsis invicta* during one annual cycle." *Ecological Monographs* 63: 425–457.

Tschinkel, W. R. 1998. "Sociometry and sociogenesis of colonies of the harvester ant, *Pogonomyrmex badius*: worker characteristics in relation to colony size and season." *Insectes Sociaux* 45: 385–410.

Tschinkel, W. R. 1999. "Sociometry and sociogenesis of colony-level attributes of the Florida harvester ant (Hymenoptera: Formicidae)." *Annals of the Entomological Society of America* 92: 80–89.

———. 2004. "The nest architecture of the Florida harvester ant, *Pogonomyrmex badius*." *Journal of Insect Science* 4: 21.

———. 2005. "The nest architecture of the ant, *Camponotus socius*." *Journal of Insect Science* 5: 9.

———. 2006. *The fire ants.* Cambridge: The Belknap Press of Harvard University Press.

Tschinkel, W. R., E. S. Adams, and T. Macom. 1995. "Territory area and colony size in the fire ant *Solenopsis invicta*." *Journal of Animal Ecology* 64: 473–480.

Wagner, D., and D. M. Gordon. 1999. "Colony age, neighborhood density and reproductive potential in harvester ants." *Oecologia* 119: 175–182.

Waloff, N. and R. E. Blackith. 1962. "The growth and distribution of the mounds of *Lasius flavus* (Fabricius) (Hym: Formicidae) in Silwood Park, Berkshire." *Journal of Animal Ecology* 31: 421–437.

Wiernasz, D. C. and B. J. Cole. 1995. "Spatial distribution of *Pogonomyrmex occidentalis*: recruitment, mortality and overdispersion." *Journal of Animal Ecology* 64: 519–527.

Wiernasz, D. C. and B. J. Cole. 2003. "Queen size mediates queen survival and colony fitness in harvester ants." *Evolution* 57: 2179–2183.

Wilson, E. O. 1971. *The insect societies.* Cambridge: The Belknap Press of Harvard University Press.

Wirth, R., H. Herz, R. J. Ryel, W. Beyschlag, and B. Hölldobler. 2003. "Herbivory of Leaf-Cutting Ants. A case study on *Atta colombica* in the tropical rainforest of Panama." New York: Springer.

Yamauchi, K., Y. Ito, K. Kinomura, and H. Takamine. 1987. "Polycalic colonies of the weaver ant *Polyrhachis dives*." *Kontyu* 55: 410–420.

CHAPTER FIVE

Control of Termite Caste Differentiation

COLIN S. BRENT

ONE OF THE DEFINING features of eusociality is a reproductive division of labor, with the offspring of one or a few fecund individuals being cared for by functionally sterile nestmates (Wilson 1971). In more derived social insects, these nonreproductive individuals have evolved into specialized castes exhibiting different behavioral repertoires and sometimes profound morphological modifications. Specialized castes are expected to enhance colony efficiency, but this requires adequate mechanisms to optimize colony composition (Oster and Wilson 1978). During their evolution, these control mechanisms faced both phylogenetic constraints and selective pressures from the competing strategies of different castes. As a result, rigorous comparative studies have the potential to provide significant insights about what drives and shapes social complexity. Unfortunately, acquiring information about caste mechanisms has faced substantial technological constraints and for the vast majority of species there is a paucity of substantive data.

During the last decade, as new techniques and improved instruments have become available, there have been some important advances in our understanding of the exogenous and endogenous controls regulating caste. The majority of these studies have focused on genetic, hormonal, and pheromonal caste regulation in the eusocial Hymenoptera. Oddly, termites (Order Isoptera), one of the most socially complex, evolutionarily interesting, and economically important groups, have been comparatively neglected. This is especially surprising given that some of the earliest studies of colony dynamics, and the roles of hormones and pheromones, were conducted with termites (e.g., Grassi and Sandias 1896; Heath 1903; Imms 1919; Lüscher 1952, 1960). Although this focus on the Hymenoptera has increased our understanding of social evolution and caste composition in this order, it has done little to enhance our knowledge of the termites given their fundamentally different origins. Unlike the Hymenoptera, termites have hemimetabolous development, bisexual societies, and diplo-diploid sex determination, all of which may have profoundly influenced adaptations favoring sociality (Noirot 1989). Fortunately, a new generation of researchers is again looking to the termites with better tools and a host of long-standing hypotheses to be tested.

Social Control of Development

Our current understanding of the mechanisms controlling termite caste differentiation is fragmentary and focused on just a few species. For some termites caste differentiation is a relatively flexible process (Kalotermitidae, Termopsidae, some Rhinotermitidae), while in others the process is more restricted and occurs earlier (Hodotermitidae, Mastotermitidae, Termitidae, and some Rhinotermitidae). Although there is a division between reproductive and neuter castes in both groups, only in the latter group is a true worker morph found (Figure 5.1; Noirot 1985a, 1985b). It is likely that the earliest regulatory mechanisms to evolve were those limiting reproduction to just one or a few members of a colony. The most parsimonious means to generate this mechanism would be through modification of pre-existing controls used to guide normal maturation in the subsocial ancestors of termites, the prototermites. A simple delay in the imaginal molt might have been sufficient to retain colony members that could then serve as helpers. In the closely-related roaches (Kambhampati 1995; Nalepa and Bandi 1999), whose development may be representative of the ancestral condition, there are numerous examples of how stimuli from conspecifics can affect the timing and extent of reproductive activity (Engelmann 1970; Gadot, Burns, and Schal 1989; Nalepa 1994; Holbrook, Bachmann, and Schal 2000). As social complexity grew in prototermites and termites, the ancestral regulatory mechanisms may have undergone considerable elaboration to accommodate caste controls, or completely novel mechanisms may have evolved.

Within the few extensively studied eusocial insects, there is tremendous variability in the stimuli used to influence the development of nestmates.

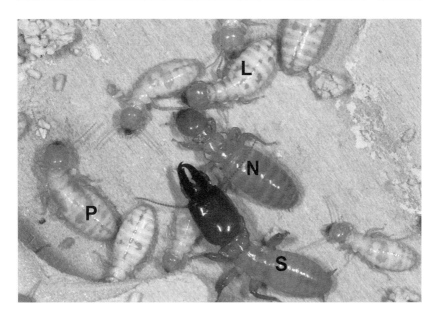

Figure 5.1. Several castes found in colonies of *Zootermopsis nevadensis*.
Shown are: (L) a larval helper (not a true worker caste) with a relatively
undifferentiated body plan; (N) a neotenic reproductive with a larval form
but a pigmented cuticle; (P) a prealate with wing buds; and (S) a soldier with
enlarged mandibles and legs, and a pigmented body. Photograph courtesy of
J. Liebig.

Some species utilize aggressive interactions to assert reproductive dominance (West-Eberhard 1978; Cole 1981; Michener 1990; Packer 1993),
while others rely on pheromones, often in combination with behavioral interactions (Lüscher 1972; Zimmerman 1983; Röseler and van Honk 1990;
Spradbery 1991; Keller and Nonacs 1993; Vargo 1998; Slessor, Foster, and
Winston 1998). These cues can directly influence conspecifics such as primer
pheromones, which can modulate endocrine activity (Wilson and Bossert
1963). Whether these are honest signals promulgating a self-regulatory response that is beneficial to nestmates or manipulative signals that benefit
only the reproductive may vary considerably between species (Keller and
Nonacs 1993).

In termites, the social control of reproduction is complicated by the existence of two reproductive forms, primary and secondary, that have their

development arrested at different stages. Primary reproductives are individuals that leave their natal nest as winged alates to found new colonies (Thorne 1996; Roisin 2000). Maturation into an alate is limited to a subset of individuals and occurs after successive molts that include a nymph stage where the individual possesses wing buds (Noirot 1990). Social stimuli may regulate which individuals pass through each stage, but the evidence to date is sparse (Korb and Schmidinger 2004; Korb, this volume). Zimmerman (1983) suggested that workers may limit the number of nymphs developing into alates by destroying their wing buds; such evidence of mutilation has been found in several species (Roisin 1994) but there is scant support for such manipulation (Korb 2005), and other factors such as the age distribution of larvae, the availability of colony resources, and seasonal environmental effects may be more important (Noirot 1985a; Noirot 1985b; Korb and Katrantzis 2004). The strongest evidence for social control of maturing primaries is observed after their adult molt, when they remain reproductively inactive until leaving their natal nest on their nuptial flight (Noirot 1969). The inhibition was assumed to be an effect of suppressive stimuli from functional reproductives, which may be the case for those termites in which removal of the king or queen can lead to one or more alates becoming reproductively active within their natal nest (Noirot 1985b, 1990). However, for some primitive termites there is mounting evidence that their reproductive development may be controlled by inhibitory stimuli from other alates (Hewitt et al. 1972; Greenberg and Stuart 1979; Brent and Traniello 2001a; Brent, Schal, and Vargo 2005) and promoting stimuli from mates (Hewitt et al. 1972; Vieau 1990; Brent and Traniello 2001a; Brent, Schal, and Vargo 2005).

The development of secondary reproductives, which arise as replacements to ill or absent reproductives within their natal nest (Thorne 1996; Roisin 2000), is normally arrested before they can molt away from a neuter caste. The origins of secondary reproductives can vary widely depending on the species studied, but in species lacking a true worker they are usually derived from either older larvae or pre-alates (Miller 1969; Noirot 1985a, 1990; Myles 1999). The pattern of their differentiation may be the result of an incomplete expression of the alate phenotype (reviewed in Myles 1999), the result of selection for heterochronous changes in developmental timing (Nalepa and Bandi 2000). In termites with a true worker caste, replacement reproductives are often derived from alates remaining in their natal nest (Myles 1999), although worker- and nymph-derived forms can occur (Noirot 1990). The principal control over sexual maturation of secondaries appears to be exerted by functional reproductives, particularly in primitive species as demonstrated by numerous orphaning experiments (Lüscher 1956; Lenz 1985; Noirot 1985a, 1990). Reproductives might also promote maturation of opposite-sexed replacements (Lüscher 1975; Brent and Traniello 2001b).

The specific stimuli responsible for inducing reproductive suppression are unknown even for the best studied termites. Although inhibitory primer pheromones have been long suspected (Light 1944; Lüscher 1974, 1976; Bordereau 1985), there is no definitive proof of their existence. Sex-specific pheromones used for mate recognition (Pasteels and Bordereau 1998; Bordereau et al. 2002; Peppuy, Robert, and Bordereau 2004) and caste-specific cuticular hydrocarbon profiles (Sevala et al. 2000; Uva, Clement, and Bagneres 2004) both occur in termites, but their effects on gonadal development have yet to be tested. Agonistic interactions are another possible mechanism for limiting sexuals, and reproductives have been observed to engage in direct combat to displace or eliminate potential competitors (Ruppli 1969; Myles and Chang 1984; Lenz 1985; Lenz et al. 1985; Shellman-Reeve 1994; Myles 1999). Functional reproductives might also indirectly influence nestmate development by manipulating the behavior of workers. Workers aggressively purge supernumerary reproductives in some species (Mensa-Bonsu 1976; Myles and Chang 1984; Myles 1999; Miyata, Furuichi, and Kitade 2004) and, given the dependence of most reproductives on workers for task assistance and nutrients, they may also eliminate others through neglect (Light and Illg 1945; Buchli 1958; Lenz 1976, 1994; Mensa-Bonsu 1976; Greenberg and Stuart 1979; Grandi, Barbieri, and Colombo 1988; Vieau 1990; Myles 1999; Brent and Traniello 2001c).

Soldier development may be inhibited by social cues similar in nature to those regulating reproductive development. This would be especially probable if the soldier caste was a direct evolutionary offshoot of the reproductive development program, as has been suggested (Myles 1986, 1988; Thorne, Breisch, and Muscedere 2003; but see Roisin 2000). The cost of maintaining this dependent defensive caste would have favored the evolution of a mechanism to optimize their number (Oster and Wilson 1978). In incipient colonies only one or a few soldiers will develop, but as the colony matures their number increases in proportion with worker availability (Noirot 1969). Although nutritional dependence leaves soldiers susceptible to loss through worker neglect, there is ample evidence that soldier-produced pheromones inhibit nestmate differentiation (Castle 1934; Springhetti 1970; Nagin 1972; Renoux 1975; Lefeuve and Bordereau 1984; Hrdy 1976, 1985; Korb, Roux, and Lenz 2003; Park and Raina 2003, 2004, 2005). Although no specific pheromones have been identified, they appear to be associated with secretory products from the soldiers' heads (Lefeuve and Bordereau 1984; Korb, Roux, and Lenz 2003; Roux and Korb 2004). There is also some indication that stimuli from functional reproductives may promote soldier differentiation (Miller 1942; Bordereau and Han 1986), which could work in conjunction with inhibitory stimuli to regulate the proportion of soldiers developing. External factors, such as the prevalence of competitors or predators, distance to suitable foraging locations, and physical characteristics of the nesting site are also likely to influence soldier number.

The social control of worker development is one of the most problematic, neglected, yet critical areas of termite caste differentiation research. A great deal of variability in what constitutes a worker, coupled with varying degrees of developmental flexibility across the Isoptera, creates a complicated picture. In termites without true workers, there is a dependence on relatively undifferentiated larval helpers to perform colony tasks. These can be derived larvae following a course of prolonged static development, which may be a default condition given that most will be inhibited from developing into secondary reproductives or soldiers. They can also be derived from nymphs or adult forms which have undergone a regressive molt. Some individuals eventually differentiate into alates, but this may be delayed pending accumulation of sufficient endogenous resources or colony conditions that support alate development (LaFage and Nutting 1978; Noirot 1985a; Lenz 1994). By retaining a larval form, these individuals maintain all their developmental options and can readily respond to changing colony conditions. Some of this flexibility has been lost in termite species with a true worker caste. Unless exposed to a permissive social environment during a critical window in their early development, larvae in these colonies differentiate into a distinct neuter caste and lose the option to mature into alates (Noirot 1985b; Noirot and Bordereau 1990). Despite this loss, true workers often maintain the ability to become soldiers, and in some species can become replacement reproductives (Noirot 1969; Noirot 1985b).

For both larval helpers and true workers, current evidence suggests mechanisms preventing individuals from differentiating away from the worker caste rather than promoting differentiation toward it. But until such inhibitory cues have been clearly identified and shown to be sufficient for maintaining the caste ratios observed in healthy termite colonies, we cannot eliminate the possibility of promoting stimuli produced by nestmates. It is also possible that caste fate may be determined by factors outside of the social environment. In termites with true workers, queens may influence the differentiation of their offspring by sequestering regulatory hormones in their eggs (reviewed in Lanzrein, Gentinetta, and Fehr 1985). Genetic differences between individuals might also cause a predisposition toward worker differentiation (Goodisman and Crozier 2003). And finally, the nutritional and physiological state of an individual may be crucial in determining which caste options are available (reviewed in Lenz 1994).

Endocrine Control of Development

Although the specific social stimuli influencing caste differentiation have not been identified, there is progress elucidating the transduction of these social signals into physiological responses. Caste changes occur during molting (Hartfelder and Emlen 2005) when variations in endocrine activity determine gene expression (reviewed in Hartfelder and Emlen 2005; Zitnan and Adams 2005). Social stimuli may influence development by directly or indirectly manipulating endocrine activity. The primary endocrine regulators of development and reproduction for insects are ecdysteroids and juvenile hormones (JH) (Goodman and Granger 2005; Raikhel, Brown, and Belles 2005; Zitnan and Adams 2005). Ecdysteroids play a prominent role in activating gene transcription via the ecdysone receptor (EcR) to regulate insect development and molting. In termites, the roles of various ecdysteroids have not been identified, but 20-hydroxyecdysone and ecdysone are the known predominant forms (Bordereau et al. 1976; Delbeque et al. 1978; Okot-Kotber 1983). The exact sources of these ecdysteroids have not been identified for termites, but they are probably produced by the prothoracic glands in neuter castes and by the gonads in reproductives (Noirot and Bordereau 1990; Brent, Schal, and Vargo 2005). Despite the regulatory importance of these steroidal hormones and the ready availability of methods to both monitor and manipulate its titer, few studies have been conducted to understand its role in termite morphogenesis and reproduction.

Juvenile hormones have received much more attention than ecdysteroids, although understanding of their role in termite caste differentiation is still limited. The only homologue identified in the few termite species examined is JH III (Meyer et al. 1976; Greenberg and Tobe 1985; Park and Raina 2004; Yagi et al. 2005), but other forms or metabolic products, such as JH-acid, may be active. JH is produced by the paired corpora allata glands, whose activity is reversibly suppressed by allatostatins (Goodman and Granger 2005). These neuropeptides are produced by the brain and have only recently been identified in termites (Yagi et al. 2005), creating new opportunities to identify the role of JH in morphogenesis. JH may influence caste differentiation by modulating the gene transcription activity of the ecdysone receptor (Barchuk, Maleszka, and Simões 2004; Henrich 2005). JH also promotes the production of vitellogenins and hexamerins in termites (Scharf, Wu-Scharf, et al. 2005; Zhou, Tarver, et al. 2006), and these storage proteins can affect hormone titers and metabolic processes crucial to development (Sappington and Raikhel, 1998; Burmester, 1999; Zhou, Oi, and Scharf 2006c; Zhou, Tarver, et al. 2006).

As with all other aspects of termite caste differentiation, current information on hormonal regulation of normal development is fragmentary and from just a few species. Information on one of the most fascinating processes occurring in termites, the regressive molt, is nonexistent. The best defined mechanism is for the inducement of soldier development. A high concentration of JH in the hemolymph has been shown to promote differentiation into a presoldier or soldier in numerous species (Lüscher 1972; Wanyonyi 1974; Lenz 1976; Howard and Haverty 1979; Okot-Kotber 1982; Okot-Kotber et al. 1991; Korb, Roux, and Lenz 2003; Scharf, Wu-Scharf, et al. 2003; Park and Raina 2004, 2005; Mao et al. 2005). This increase occurs either during or after peak concentrations of ecdysteroids that coincide with the initiation of the molt, and appears to enhance mandible development and cuticle deposition (Okot-Kotber 1983). Termites cease molting after becoming soldiers (Noirot 1985a, 1985b), which may be the result of becoming hormonally locked into their caste. Recent advances in termite genomics are showing how these hormonal changes promote soldier development by inducing differential gene expression (Miura et al. 1999; Miura 2001; Scharf, Ratliff, et al. 2003, Scharf, Wu-Scharf et al. 2003; Scharf, Wu-Scharf, et al. 2005; Koshikawa et al. 2005).

The control of reproductive differentiation is less well understood and the available data is not always in agreement. Current evidence indicates that molting individuals with low JH titers are competent to differentiate toward a reproductive caste (Lüscher 1972, 1974; Wanyonyi 1974; Yin and Gillot 1975; Korb, Roux, and Lenz 2003; but see Greenberg and Tobe 1985; Hartfelder and Emlen 2005), while those with moderate JH titers have stationary molts preventing gonad development (Lüscher 1972, 1976). Most of the focus of this research has been on the role of JH, but interactive effects with ecdysteroids and the timing for the release of both hormones are doubtless crucial in determining the course of development. Even after the molt to a reproductive form, egg production is still under hormonally-mediated social control. In alates of *Zootermopsis angusticollis*, reproductive inhibition appears to be correlated with a high JH titer and low ecdysteroid titer. The shift to a reproductive state appears to require an increase in ecdysteroids and a decrease in JH during a brief organizational period following disinhibition (Brent, Schal, and Vargo 2005). Stimuli from other immature primary queens delay these hormonal shifts, while stimuli from a male reproductive quicken the changes (Brent Schal, and Vargo 2007). Once the ovaries are ready, increasing JH appears to drive vitellogenesis (Vieau and Lebrun 1981; Greenberg, Kunkel, and Stuart 1978; Greenberg and Tobe 1985; Brent, Schal, and Vargo 2005; Scharf, Ratliff, et al. 2005) and probably a host of other processes related to gamete production (Raikhel, Brown, and Belles 2005), but extensive research is still needed to delineate its exact role as well as that of the ecdysteroids.

Conclusions and Future Directions

As this brief review suggests, there are numerous and large gaps in our understanding of the process of termite caste differentiation. The first priority is to better understand the proximate mechanisms involved. The most efficient approach will be to focus on a single species of termite rather than trying to construct a cohesive picture from fragmentary studies of numerous species. There is such tremendous variety even between closely related termites that findings in one species may not hold true for others, particularly with regard to influential social stimuli. The species chosen should be drawn from a group of termites that comes closest to approximating the earliest termites. Basal species with small colonies nesting in discrete nests, such as found in the Termopsidae, would be the easiest to work with, although understanding more cosmopolitan and economically important pest species (*Reticulitermes, Coptotermes,* etc.) should be given priority. Once a cohesive model has been constructed, studies can be expanded to look at other species with varying nesting habits and patterns of caste determination to see how the basic program might have been modified.

Previously it had been a challenge to determine the nature and source of the stimuli to which maturing termites respond because of the normally protracted periods between exposure and an observable morphological response. Now that it is possible to monitor rapid changes in hormone titers and gene expression in response to social conditions, more effective and sensitive bioassays can be designed to determine which stimuli are relevant and how they act. Stimuli that may be of particular interest are the epicuticular hydrocarbons. The expression of these lipids varies significantly between species and between castes (Bagnères et al. 1991; Haverty et al. 1996; Vauchot et al. 1996; Clément and Bagnères 1998; Sevala et al. 2000; Klochkov, Kozlovski, and Belyaeva 2005). Hydrocarbons may encode individual-specific information that can influence the development and behavior of nestmates. Currently, they are the best candidate for termite primer pheromones and there are well-developed techniques and bioassays to determine their composition and function. But other elements of the termite's social and physical environment should also be re-examined using these more sensitive assays, for these stimuli may act cumulatively to regulate caste determination.

Determining how exogenous stimuli are transduced into developmental responses that influence caste has also become significantly easier with the development of effective means for manipulating hormonal titers. A variety of agonists, antagonists, and regulatory peptides (i.e., allatostatins) and proteins (i.e., carriers, enzymes, etc.) are now available. These can be used to fully elucidate the control and role of the endocrine system in mediating responses to hydrocarbons and other candidate stimuli. Identifying interactive effects between the endocrine system and individual developmental state is of particular importance. In addition, using antibodies to the hormones and their receptors can reveal the sites of synthesis, storage, and activity of the hormones, and may also reveal which parts of the brain are involved.

Recent advances in caste-specific gene expression (Miura et al. 1999; Miura 2001; Scharf, Wu-Scharf, et al. 2003, Scharf, Wu-Scharf, et al. 2005, Scharf, Ratliff, et al. 2005 Cornette et al. 2006; Liénard et al. 2006; Zhou, Tarver, et al. 2006) also hold tremendous promise. They provide the opportunity to understand how the endocrine responses to social conditions are translated into the observed morphological differences. They can also tell us about some of the functions of the various castes which might not be readily apparent from just their behavior. Gene expression can be manipulated by amplification and RNA interference to look for feedback mechanisms, interactive effects, and individual and cumulative contributions to morphological and behavioral changes associated with specific castes. Current research is primarily focused on identifying gene expression differences between fully differentiated castes; however, as we characterize the influence of exogenous stimuli and the endocrine system on caste differentiation, it will become increasingly easier to control the timing and direction of development so that gene expression can be monitored at the most crucial periods of caste determination. These new tools should greatly speed the process of discovery and will permit the creation of a fully detailed model of the caste differentiation process.

An enhanced understanding of the proximate mechanisms regulating differentiation might also aid in better understanding the evolutionary origins of termite castes. By extending these studies to include closely related subsocial roaches, additional basal termites, and more derived termites with true workers, it may be possible to determine which primitive mechanistic elements for maturation have been conserved and how they have been modified to create distinct castes during the course of evolution. The endocrine processes involved in regulating reproductive maturation and activity in roaches and termites appear to share many common elements, and the types of social stimuli to which individuals respond could also be conserved. For instance, it is quite likely that the evolution of an apterous reproductive form involved duplicating the controls governing reproductive development in primaries with a simple omission of the steps promoting wing development. Similar modification of existing developmental programs has been suggested for caste evolution in the eusocial Hymenoptera (West-Eberhard 1996; Andam et al. 2004). Other mechanistic elements may prove to be novel. One such possibility might be the inhibitory pheromones produced by unique organs in the heads of soldiers (Lefeuve and Bordereau 1984), especially if they utilize novel chemicals or means of delivery. Another might be the proximate mechanisms regulating the sexual division of labor in more advanced termites, which has received very little recent attention (Matsuura 2006). A clearer understanding of the origins of these various elements could highlight the phylogenetic constraints faced by termites as they evolved and the means by which they overcame these hurdles.

Comparative studies of caste differentiation within the continuum of termite species can enhance our understanding of the environmental constraints that might have shaped termite evolution. Termites employ several different nesting and foraging strategies (reviewed in Noirot 1970). Some species, particularly the primitive Termopsids, create relatively short-term nests in the logs they are consuming for nutrients and normally only dispersing alates will ever leave. Other species nest in centralized mounds out of which foragers are regularly sent to retrieve food and water. And still other species are subterranean and can have decentralized nests encompassing a broad area. Temperature, humidity, nutrient availability, the pervasiveness of predators and competitors, and other environmental factors could contribute to the nesting habits and caste composition of each species, just as they influence the process of caste differentiation (e.g., Korb and Schmidinger 2004; Liu et al. 2005, 2005b). Examining caste control mechanisms in these differing species would reveal which environmental factors are the most influential and would highlight the interplay between phylogeny and ecology during termite evolution.

I anticipate that should a concerted effort be made over the next decade, the social and physiological mechanisms controlling termite caste differentiation can be substantially clarified. While it will never be possible to know all of the evolutionary steps that ancestral termites took toward increasing social complexity, a better understanding of some of the current adaptations to a eusocial lifestyle should illuminate part of this history and generate new questions. There are, of course, additional research directions that could be explored than were addressed in this brief review, but there is very little now that we cannot undertake due to lack of technical innovation. In order to best utilize the rapid advances in our understanding of the Hymenoptera, we need to ensure a broad perspective which is best achieved through comparison with unrelated groups exhibiting similar social complexity. While studies of social aphids and thrips are without a doubt of great interest, the key to a fuller understanding of the evolution of eusociality may lie in the hands of the Isopterists.

Literature Cited

Amdam, G. V., Z. L. P. Simoes, A. Hagen, K. Norberg, K. Schroder, O. Mikkelsen, T. B. L. Kirkwood, and S. W. Omholt. 2004. "Hormonal control of the yolk precursor vitellogenin regulates immune function and longevity in honeybees." *Experimental Gerontology* 39: 767–773.

Bagnères, A. G., A. Killian, J. L. Clement, and C. Lange. 1991. "Interspecific recognition among termites of the genus *Reticulitermes*—Evidence for a role for the cuticular hydrocarbons." *Journal of Chemical Ecology* 17: 2397–2420.

Barchuk, A. R., R. Maleszka, and Z. L. P. Simões. 2004. "Apis mellifera ultraspiracle: cDNA sequence and rapid upregulation by juvenile hormone." *Insect Molecular Biology* 13: 459–467.

Bordereau, C. 1985. "The role of pheromones in termite caste differentiation." In J. A. L. Watson, B. M. Okot-Kotber, and C. Noirot, eds., *Caste differentiation in social insects*, 221–226. New York: Pergamon Press.

Bordereau, C., E. M. Cancello, E. Sémon, A. Courrent, and B. Quennedey. 2002. "Sex pheromone identified after solid phase microextraction from tergal glands of female alates in *Cornitermes bequaerti* (Isoptera, Nasutitermitinae)." *Insectes Sociaux* 49: 209–215.

Bordereau, C., and S. H. Han. 1986. "Stimulatory influence of the queen and king on soldier differentiation in the higher termites *Nasutitermes lujae* and *Cubitermes fungifaber*." *Insectes Sociaux* 33: 296–305.

Bordereau, C., M. Hirn, J.-P. Delbecque, and M. de Reggi. 1976. "Presence d'ecdysones chez un insecte aulte: La reine de termite." Comptes Rendus de l'Académie des Sciences D 282: 885–888.

Brent, C. S., C. Schal, and E. L. Vargo. 2005. "Endocrine changes in maturing primary queens of *Zootermopsis angusticollis*." *Journal of Insect Physiology* 51: 1200–1209.

———. 2007. "Effects of social stimuli on the endocrine systems of maturing queens of the dampwood termite *Zootermopsis angusticollis*." *Physiological Entomology* 32: 26–33.

Brent, C. S., and J. F. A. Traniello. 2001a. "Influence of sex-specific stimuli on ovarian maturation in both primary and secondary reproductives of the dampwood termite *Zootermopsis angusticollis*." *Physiological Entomology* 26: 239–247.

———. 2001b. "Social regulation of testicular development in primary and secondary males of the dampwood termite *Zootermopsis angusticollis* Hagen." *Insectes Sociaux* 48: 384–391.

———. 2001c. "Social influence of larvae on ovarian maturation in both primary and secondary reproductives of *Zootermopsis angusticollis*." *Physiological Entomology* 26: 78–85.

Buchli, H. H. R. 1958. "L'orgine des castes et les potentialites ontogéniques des termites européens du genre *Reticulitermes* Holmgren." *Annales des Sciences Naturelles* 20: 263–429.

Burmester, T. 1999. "Evolution and function of the insect hexamerins." *European Journal of Entomology* 96: 213–225.

Castle, G. B. 1934. "The damp-wood termites of western United States, genus *Zootermopsis*." In C. Kofoid, ed., *Termites and termite control*, 264–282. Berkeley: University of California Press.

Clément, J.-L., and A.-G. Bagnères. 1998. "Nestmate recognition in termites." In R. K. Vander Meer, M. Breed, M. Winston, and K. Espelie, eds., *Pheromone communication in social insects: Ants, wasps, bees, and termites*, 126–155. Boulder, CO: Westview Press.

Cole, B. J. 1981. "Dominance hierarchies in Leptothorax ants." *Science* 212: 83–84.

Cornette, R., S. Koshikawa, M. Hojo, T. Matsumoto, and T. Miura. (2006). "Caste-specific cytochrome P450 in the damp-wood termite *Hodotermopsis sjostedti* (Isoptera, Termopsidae)." *Insect Molecular Biology* 15: 235–244.

Cuvillier-Hot, V., A. Lenoir, R. Crewe, C. Malosse, and C. Peeters. 2004. "Fertility signaling and reproductive skew in queenless ants." *Animal Behavior* 68: 1209–1219.

Delbeque, J. P., B. Lanzrein, C. Bordereau, H. Imboden, M. Hirn, J. D. O'Connor, C. Noirot, and M. Lüscher. 1978. "Ecdysone and ecdysterone in physogastric termite queens and eggs of *Macrotermes bellicosus* and *Macrotermes subhyalinus*." *General and Comparative Endocrinology* 36: 40–47.

Engelmann, F. 1970. *The physiology of insect reproduction*. New York: Pergamon Press.

Gadot, M., E. Burns, and C. Schal. 1989. "Juvenile hormone biosynthesis and oocyte development in adult female Blattella germanica: Effects of grouping and mating." *Archives of Insect Biochemistry and Physiology* 11; 189–200.

Goodisman, M. A. D., and R. H. Crozier. 2003. "Association between caste and genotype in the termite *Mastotermes darwiniensis* Froggatt (Isoptera: Mastotermitidae)." *Australian Journal of Entomology* 42: 1–5.

Goodman, W. G., and N. A. Granger. 2005. "The juvenile hormones." In L. I. Gilbert, K. Iatrou, and S. S. Gill, eds., *Comprehensive molecular insect science* 3: 320–408. Boston: Elsevier.

Grandi, G., R. Barbieri, and G. Colombo. 1988. Oogenesis in *Kalotermes flavicollis* Fabr. (Isoptera, Kalotermitidae) I. Differentiation and maturation of oocytes in female supplementary reproductives." *Bolletino di Zoologia* 55: 279–292.

Grassi, B., and A. Sandias. 1896. "The constitution and development of the society of termites: Observations on their habits; with appendices on the parasitic Protozoa of Termitidae, and on the Embiidae." *Quarterly Journal of Microscopical Science* 39: 245–322.

Greenberg, S. L. W., J. G. Kunkel, and A. M. Stuart. 1978. "Vitellogenesis in a primitive termite, *Zootermopsis angusticollis* Hagen (Hodotermitidae)." *Biological Bulletin* 155: 336–346.

Greenberg, S. L. W., and A. M. Stuart. 1979. "The influence of group size on ovarian development in adult and neotenic reproductives of the termite *Zootermopsis angusticollis* Hagen (Hododtermitidae)." *International Journal of Invertebrate Reproduction* 1: 99–108.

Greenberg, S. L. W., and S. S. Tobe. 1985. "Adaptation of a radiochemical assay for juvenile hormone biosynthesis to study caste differentiation in a primitive termite." *Journal of Insect Physiology* 31: 347–352.

Hartfelder, K., and D. J. Emlen. 2005. "Endocrine control of insect polyphenism." In L. I. Gilbert, K. Iatrou, and S. S. Gill, eds., *Comprehensive molecular insect science*, 3: 651–703. Boston: Elsevier.

Haverty, M. I., J. K. Grace, L. J. Nelson, R. T. Yamamoto. 1996. "Intercaste, intercolony, and temporal variation in cuticular hydrocarbons of Coptotermes formosanus Shiraki (Isoptera: Rhinotermitidae)." *Journal of Chemical Ecology* 22: 1813–1834.

Heath, H. 1903. "The habits of California termites." *Biological Bulletin* 4: 7–63.

Henrich, V. C. 2005. "The ecdysteroid receptor." In L. I. Gilbert, K. Iatrou and S. S. Gill, eds., *Comprehensive molecular insect science*, 3: 651–703. Boston: Elsevier.

Hewitt, P. H., J. A. L. Watson, J. J. C. Nel, and I. Schoeman. 1972. "Control of the change from group to pair behavior by Hodotermes mossambicus reproductives." *Journal of Insect Physiology* 18: 143–150.

Holbrook, G. L., J. A. S. Bachmann, and C. Schal. 2000. "Effects of ovariectomy and mating on the activity of the corpora allata in adult female *Blattella germanica* (L.) (Dictyoptera: Blattellidae). *Physiological Entomology* 25: 27–34.

Howard, R. W., and M. I. Haverty. 1979. "Termites and juvenile hormone analogs: A review of methodology and observed effects." *Sociobiology*, 4: 269–278.

Hrdý, I. 1976. "The influence of juvenile hormone analogues on caste development in termites." In M. Lüscher, ed., *Phase and caste determination in insects: Endocrine aspects*, 71–72. Oxford: Pergamon Press.

———. 1985. "The role of juvenile hormones and juvenoids in soldier formation in Rhinotermitidae." In J. A. L. Watson, B. M. Okot-Kotber, and C. Noirot, eds., *Caste differentiation in social insects*, 245–250. Oxford: Pergamon Press.

Imms, M. A. 1919. "On the structure and biology of Archotermopsis, together with descriptions of new species of intestinal protozoa, and general observations in the Isoptera." *Philosophical Transcripts of the Royal Society, London B* 209: 75–180.

Kambhampati, S. 1995. "A phylogeny of cockroaches and related insects based on DNA sequence of mitochondrial ribosomal RNA genes." *Proceedings of the National Academy of Science* 92: 2017–2020.

Keller, L., and P. Nonacs. 1993. "The role of queen pheromones in social insects: Queen control or queen signal?" *Animal Behavior* 45: 787–794.

Keller, L., and E. L. Vargo. 1993. "Reproductive structure and reproductive roles in eusocial insect colonies." In L. Keller, ed., *Queen number and sociality in insects*, 16–44. New York: Oxford University Press.

Klochkov, S. G., V. I. Kozlovskii, and N. V. Belyaeva. 2005. "Caste and population specificity of termite cuticular hydrocarbons." *Chemistry of Natural Compounds* 41: 1–6.

Korb, J. 2005. "Regulation of sexual development in the basal termite *Cryptotermes secundus*: Mutilation, pheromonal manipulation or honest signal?" *Naturwissenschaften* 92: 45–49.

Korb, J., E. A. Roux, and M. Lenz. 2003. "Proximate factors influencing soldier development in the basal termite *Cryptotermes secundus* (Hill)." *Insectes Sociaux* 50: 299–303.

Korb, J., and S. Schmidinger. 2004. "Help or disperse? Cooperation in termites influenced by food conditions." *Behavioral Ecology and Sociobiology* 56: 89–95.

Koshikawa, S., R. Cornette, M. Hojo, K. Maekawa, T. Matsumoto, and T. Miura. 2005. "Screening of genes expressed in developing mandibles during soldier differentiation in the termite *Hodotermopsis sjostedti*." *FEBS Letters* 579: 1365–1370.

LaFage, J. P., and W. L. Nutting. 1978. "Nutrient dynamics of termites." In M. V. Brian, ed., *Production ecology of ants and termites*, 165–232. Cambridge: Cambridge University Press.

Lanzrein, B., V. Gentinetta, and R. Fehr. 1985. "Titres of juvenile hormone and ecdysteroids in reproducion and eggs of *Macrotermes michaelseni*: Relation to caste determination." In J. A. L. Watson, B. M. Okot-Kotber, C. Noirot, eds., *Caste differentiation in social insects*, 307–327. Oxford: Pergamon Press.

Lefeuve, P., and C. Bordereau. 1984. "Soldier formation regulated by a primer pheromone from the soldier frontal gland in a higher termite, *Nasutitermes lujae*." *Proceedings of the National Academy of Sciences* 81: 7665–7668.

Lenz, M. 1976. "The dependence of hormone effects in termite caste determination on external factors." In M. Lüscher, ed., *Phase and caste determination in insects*, 73–89. Oxford: Pergamon Press.

———. 1985. "Is inter- and intraspecific variability of lower termite neotenic number due to adaptive thresholds for neotenic elimination? Considerations from studies on *Porotermes adamsoni* (Froggatt) (Isoptera: Termopsidae)." In C. Noirot, ed., *Caste differentiation in social insects*, 125–146. Oxford: Pergamon Press.

———. 1994. "Food resources, colony growth and caste development in wood-feeding termites." In J. H. Hunt and C. A. Nalepa, eds.,

Nourishment and evolution in insect societies, 159–210. San Francisco: Westview Press.

Liénard, M. A., J.-M. X. S. Lassance, I. Paulmier, J.-F. Picimbon, and C. Löfstedt. 2006. "Differential expression of cytochrome c oxidase subunit III gene in castes of the termite *Reticulitermes santonensis*." *Journal of Insect Physiology* 52: 551–557.

Light, S. F. 1944. "Parthenogenesis in termites of the genus *Zootermopsis*." *University of California Publications in Zoology* 43: 405–412.

Light, S. F., and P. Illg. 1945. "Rate and extent of development of neotenic reproductives in groups of nymphs of the termite genus *Zootermopsis* (formerly *Termopsis*)." *University of California Publications in Zoology* 53: 1–40.

Liu, Y. X., G. Henderson, L. X. Mao, and R. A. Laine. 2005a. "Seasonal variation of juvenile hormone titers of the formosan subterranean termite, *Coptotermes formosanus* (Rhinotermitidae)." *Environmental Entomology* 34: 557–562.

———. 2005b. "Effects of temperature and nutrition on juvenile hormone titers of *Coptotermes formosanus* (Isoptera: Rhinotermitidae). *Annals of the Entomological Society of America* 98: 732–737.

Lüscher, M. 1952. "New evidence for an ectohormonal control of caste determination in termites." *Transactions of the Ninth International Congress of Entomology, Amsterdam* 1: 289–294.

———. 1956. "Die Entstehung von Ersatzgeschlechtstieren bei der Termite *Kalotermes flavicollis* (Fabr.)." *Insectes Sociaux* 3: 119–128.

———. 1960. "Hormonal control of caste differentiation in termites." *Annals of the New York Academy of Sciences* 89: 549–563.

———. 1972. "Environmental control of juvenile hormone (JH) secretion and caste differentiation in termites." *General and Comparative Endocrinology* 3 (suppl.): 509–514.

———. 1974. "Kasten und Kasten differenzierung bei neideren Termiten." In G. H. Schmidt, ed., *Sozialpolymorphismus bei Insekten*, 694–739. Stuttgart: Wissenschaftl. Verlagsges.

———. 1975. "Pheromones and polymorphism in bees and termites." In C. Noirot, P. E. Howse, and G. Le Manse, eds., *Pheromones and defensive secretions in social insects*, 123–141. University of Dijon.

———. 1976. "Evidence for an endocrine control of caste determination in higher termites." In M. Lüscher, ed., *Phase and caste determination in insects, endocrine aspects*, 91–103. New York: Pergamon Press.

Mao, L. X., G. Henderson, Y. X. Liu, and R. A. Laine. 2005. "Formosan subterranean termite (Isoptera: Rhinotermitidae) soldiers regulate juvenile hormone levels and caste differentiation in workers." *Annals of the Entomological Society of America* 98: 340–345.

Matsuura, K. 2006. "A novel hypothesis for the origin of the sexual division of labor in termites: Which sex should be soldiers?" *Evolutionary Ecology* 20: 565–574.

Mensa-Bonsu, A. 1976. "The production and elimination of supplementary reproductives in *Porotermes adamsoni* (Froggatt)." *Insectes Sociaux* 23: 133–154.

Meyer D. R., B. Lanzrein, M. Lüscher, and K. Nakanishi. 1976. "Isolation and identification of a juvenile hormone in termites." *Experientia* 32: 773.

Michener, C. D. 1990. "Reproduction and castes in the social halictine bees." In W. Engels, ed., *Social insects: An evolutionary approach to castes and reproduction*, 123–146. New York: Springer Verlag.

Miller, E. M. 1942. "The problem of castes and caste differentiation in *Prorhinotermes simplex* (Hagen)." *Bulletin of the University of Miami* 15: 3–27.

———. 1969. "Caste differentiation in the lower termites." In K. Krishna and F. M. Weesner, eds., *Biology of termites*, 1: 283–310. New York: Academic Press.

Miura, T. 2001. "Morphogenesis and gene expression in the soldier-caste differentiation of termites." *Insectes Sociaux* 48: 216–223.

Miura, T., A. Kamikouchi, M. Sawata, H. Takeuchi, S. Natori, T. Kubo, and T. Matsumoto. 1999. "Soldier caste-specific gene expression in the mandibular glands of *Hodotermopsis japonica* (Isoptera: Termopsidae)." *Proceedings of the National Academy of Sciences* 96: 13874–13879.

Miyata, H., H. Furuichi, and O. Kitade. 2004. "Patterns of neotenic differentiation in a subterranean termite, *Reticulitermes speratus* (Isoptera: Rhinotermitidae)." *Entomological Science* 7: 309–314.

Myles, T. G. 1986. "Reproductive soldiers in the Termopsidae (Isoptera)." *Pan-Pacific Entomologist* 62: 293–299.

———. 1988. "Resource inheritance in social evolution from termites to man." In C. Slobodchikoff, ed., *The ecology of social behavior*, 379–423. San Diego, CA: Academic Press.

———. 1999. "Review of secondary reproduction in termites (Insecta: Isoptera) with comments on its role in termite ecology and social evolution." *Sociobiology* 33: 1–91.

Myles, T. G., and F. Chang. 1984. "The caste system and caste mechanisms of *Neotermes connexus* (Isoptera: Kalotermitidae)." *Sociobiology* 9: 163–319.

Nagin, R. 1972. "Caste determination in *Neotermes jouteli* (Banks)." *Insectes Sociaux* 19: 39–61.

Nalepa, C. A. 1994. "Nourishment and the origin of termite eusociality." In J. H. Hunt and C. A. Nalepa. eds., *Nourishment and evolution in insect societies*, 57–104. San Francisco: Westview Press.

Nalepa, C. A., and C. Bandi. 1999. "Phylogenetic status, distribution, and bio-geography of *Cryptocercus* (Dictyoptera: Cryptocercidae)." *Annals of the Entomological Society of America* 92: 292–302.

———. 2000. "Characterizing the ancestors: Paedomorphosis and termite evolution." In T. Abe, D. E. Bignell, and M. Higashi, eds., *Termites: Evolution, sociality, symbioses, ecology*, 53–76. Boston: Kluwer Academic Publishers.

Noirot, C. 1969. "Formation of castes in higher termites." In K. Krishna and F. M. Weesner, eds., *Biology of termites*, 1: 311–350. New York: Academic Press.

———. 1970. "The nests of termites." In K. Krishna and M. Weesner, eds., *Biology of termites*, 2: 73–125. New York: Academic Press.

———. 1985a. "Pathways of caste development in the lower termites." In J. A. L. Watson, B. M. Okot-Kotber, and C. Noirot, eds., *Caste differentiation in social insects*, 41–58. New York: Pergamon Press.

———. 1985b. "The caste system in higher termites." In J. A. L. Watson, B. M. Okot-Kotber, and C. Noirot, eds., *Caste differentiation in social insects*, 75–86. New York: Pergamon Press.

———. 1989. "Social structure in termite societies." *Ethology, Ecology and Evolution* 1: 1–17.

———. 1990. "Sexual castes and reproductive strategies in termites." In W. Engels, ed., *Social insects: An evolutionary approach to castes and repro-duction*, 5–35. New York: Springer-Verlag.

Noirot, C., and C. Bordereau. 1990. "Termite polymorphism and morphogene-tic hormones." In A. P. Gupta, ed., *Morphogenetic hormones and Arthropods: Roles in histogenesis, organogenesis and morphogenesis*, 3: 293–324. Rutgers University Press; New Brunswick (New Jersey)

Okot-Kotber, B. M. 1982. "Correlation between larval weights, endocrine gland activities and competence period during differentiation of workers and soldiers in *Macrotermes michaelseni* (Isoptera: Termitidae)." *Journal of Insect Physiology* 28: 905–910.

———. 1983. "Ecdysteroid levels associated with epidermal events during worker and soldier differentiation in *Macrotermes michaelseni* (Isoptera: Macrotermitinae)." *General and Comparative Endocrinology* 52: 409–417.

Okot-Kotber, B. M., I. Ujváry, R. Mollaaghababa, F. Szurdoki, G. Matolcsy, and G. D. Prestwich. 1991. "Physiological influence of fenoxycarb proinsecticides and soldier head extracts of various termite species on soldier differentiation in *Reticulitermes flavipes* (Isoptera)." *Sociobiology* 19: 77–90.

Oster, G. F., and E. O. Wilson. 1978. *Caste and ecology in the social insects*. Princeton: Princeton University Press.

Packer, L. 1993. "Multiple-foundress associations in sweat bees." In L. Keller, ed., *Queen number and sociality in insects*, 215–233. New York: Oxford University Press.

Park, Y. I., and A. K. Raina. 2003. "Factors regulating caste differentiation in the Formosan subterranean termite with emphasis on soldier formation." *Sociobiology* 41: 49–60.

———. 2004. Juvenile hormone III titers and regulation of soldier caste in *Coptotermes formosanus* (Isoptera: Rhinotermitidae)." *Journal of Insect Physiology* 50: 561–566.

———. 2005. "Regulation of juvenile hormone titers by soldiers in the Formosan subterranean termite, *Coptotermes formosanus*." *Journal of Insect Physiology* 51: 385–391.

Pasteels, J. M., and C. Bordereau. 1998. "Releaser pheromones in termites." In R. K. Vander Meer, M. D. Breed, M. L. Winston, and K. E. Espelie, eds., *Pheromone communication in social insects*, 193–215. Boulder: Westview Press.

Peppuy, A., A. Robert, and C. Bordereau. 2004. "Species-specific sex pheromones secreted from new sexual glands in two sympatric fungus-growing termites from northern Vietnam, *Macrotermes annandalei* and *M. barneyi*." *Insectes Sociaux* 51: 91–98.

Raikhel, A. S., M. R. Brown, and X. Belles. 2005. "Hormonal control of reproductive processes." In L. I. Gilbert, K. Iatrou, and S. S. Gill, eds., *Comprehensive molecular insect science*, 3: 433–491. Boston: Elsevier.

Renoux, J. 1975. "Le polymorphisme de *Schedorhinotermes lamanianus* (Sjöstedt) (Isoptera: Rhinotermitidae). Essai d'interprétation." *Insectes Sociaux* 23: 279–494.

Roisin, Y. 1994. "Intragroup conflicts and the evolution of sterile castes in termites." *American Naturalist* 143: 751–765.

———. 2000. "Diversity and evolution of caste patterns." In T. Abe, D. E. Bignell, and M. Higashi, eds., *Termites: Evolution, sociality, symbioses, ecology*, 95–119. Dordrecht, The Netherlands: Kluwer Academic Publishers.

Röseler, P.-F., and C. G. J. van Honk. 1990. "Castes and reproduction in bumblebees." In W. Engels, ed., *Social insects: An evolutionary approach to castes and reproduction*, 147–166. Heidelberg: Springer-Verlag.

Roux, E. A., and J. Korb. 2004. "Evolution of eusociality and the soldier caste in termites: A validation of the intrinsic benefit hypothesis." *Journal of Evolutionary Biology* 17: 869–875.

Ruppli, E. 1969. "Die elimination uberzahliger ersatzgeschlechtstiere bei der termite *Kalotermes flavicollis* (Fabr.)." *Insectes Sociaux* 16: 235–248.

Sappington, T. W., and A. S. Raikhel. 1998. "Molecular characteristics of insect vitellogenins and vitellogenin receptors." *Insect Biochemistry and Molecular Biology* 28: 277–300.

Scharf, M. E., C. R. Ratliff, J. T. Hoteling, B. R. Pittendrigh, and G. W. Bennett. 2003. "Caste differentiation responses of two sympatric *Reticulitermes* termite species to juvenile hormone homologs and synthetic juvenoids in two laboratory assays." *Insectes Sociaux* 50: 346–354.

Scharf, M. E., C. R. Ratliff, D. Wu-Scharf, X. Zhou, B. R. Pittendrigh, and G. W. Bennett. 2005. "Effects of juvenile hormone III on *Reticulitermes flavipes*: Changes in hemolymph protein composition and gene expression." *Insect Biochemistry and Molecular Bioliogy* 35: 207–215.

Scharf, M. E., D. Wu-Scharf, B. R. Pittendrigh, and G. W. Bennett. 2003. "Caste- and development-associated gene expression in a lower termite." *Genome Biology* 4: R62.

Scharf, M. E., D. Wu-Scharf, X. Zhou, B. R. Pittendrigh, and G. W. Bennett. 2005. "Gene expression profiles among immature and adult reproductive castes of the termite *Reticulitermes flavipes*." *Insect Molecular Biology* 14: 31–44.

Sevala, V. L., A. G. Bagneres, M. Kuenzli, G. J. Blomquist, and C. Schal. 2000. "Cuticular hydrocarbons of the dampwood termite, *Zootermopsis nevadensis*: Caste differences and role of lipophorin in transport of hydrocarbons and hydrocarbon metabolites." *Journal of Chemical Ecology* 26: 765–789.

Shellman-Reeve, J. 1994. "Limited nutrients in a dampwood termite: Nest preference, competition and cooperative nest defense." *Journal of Animal Ecology* 63: 921–932.

———. 1997. "The spectrum of eusociality in termites." In B. J. Crespi, ed., *The evolution of social behavior in insects and arachnids*, 52–93. Cambridge: Cambridge University Press.

Slessor, K. N., L. J. Foster, and M. L. Winston. 1998. "Royal flavors: honey bee queen pheromones." In R. K. Vander Meer, M. D. Breed, K. E. Espelie, and M. L. Winston, eds., *Pheromone communication in social insects*, 331–344. Boulder, CO: Westview Press.

Spradbery, J. P. 1991. "Evolution of queen number and queen control." In K. G. Ross and R. W. Matthews, eds., *The social biology of wasps*, 336–388. Ithaca, NY: Comstock Publishing Associates.

Springhetti, A. 1970. "Influence of the king and queen on the differentiation of soldiers in *Kalotermes flavicollis* Fabr. (Isoptera)." *Monitore Zoologia Italia* 4: 99–105.

Thorne, B. L. 1996. "Termite terminology." *Sociobiology* 28: 253–263.

———. 1997. "Evolution of eusociality in termites." *Annual Review of Ecological Systematics* 28: 27–54.

Thorne, B. L., N. L. Breisch, and M. L. Muscedere. 2003. "Evolution of eusociality and the soldier caste in termites: Influence of intraspecific competition and accelerated inheritance." *Proceedings of the National Academy of Science* 100: 12808–12813.

Uva, P., J. L. Clement, and A. G. Bagneres. 2004. "Colonial and geographic variations in agonistic behaviour, cuticular hydrocarbons and mtDNA of Italian populations of *Reticulitermes lucifugus* (Isoptera, Rhinotermitidae)." *Insectes Sociaux* 51: 163–170.

Vargo, E. L. 1998. "Primer pheromones in ants." In R. K. Vander Meer, M. D. Breed, M. L. Winston, and K. E. Espelie, eds., *Pheromone communication in social insects: Ants, wasps, bees and termites*, 293–313. Boulder, CO: Westview Press.

Vargo, E. L., and M. Laurel. 1994. "Studies on the mode of action of a queen primer pheromone of the fire ant *Solenopsis invicta*." *Journal of Insect Physiology* 40: 601–610.

Vauchot, B., E. Provost, A.-G. Bagnères, and J.-L.Clement. 1996. "Regulation of the chemical signatures of two termite species, *Reticulitermes santonensis* and *Reticulitermes lucifugus grassei*, living in mixed experimental colonies." *Journal of Insect Pysiology* 42: 309–321.

Vieau, F. 1990. "The male effect upon the female reproductive potency in the incipient laboratory colonies of *Kalotermes flavicollis* Fabr." *Insectes Sociaux* 37: 169–180.

Vieau, F., and D. Lebrun. 1981. "Juvenile hormone vitellogenesis and egg laying in the termite *Kalotermes flavicollis*." *Comptes Rendus des Seances de l'Academie des Sciences Serie III-Sciences de la Vie* 293: 399–402.

Wanyonyi, K. 1974. "The influence of the juvenile hormone analogue ZR512 (Zoecon) on caste development in *Zootermopsis Nevadensis* (Hagen) (Isoptera)." *Insectes Sociaux* 21: 35–44.

West-Eberhard, M. J. 1978. "Polygyny and the evolution of social behavior in wasps." *Journal of the Kansas Entomological Society* 51: 832–856.

———. 1996. "Wasp societies as microcosms for the study of development and evolution." In S. Turillazzi and M. J. West-Eberhard, eds., *Natural history and evolution of paper wasp*, 290–317. New York: Oxford Univsity Press.

Wilson, E. O. 1971. "The insect societies." Cambridge: Belknap Press of Harvard University.

Wilson, E. O., and W. H. Bossert. 1963. "Chemical communication among animals." *Recent Progress in Hormone Research* 19: 673–716.

Yagi, K. J., R. Kwok, K. K. Chan, R. R. Setter, T. G. Myles, S. S. Tobe, and B. Stay. 2005. "Phe-Gly-Leu-amide allatostatin in the termite *Reticulitermes flavipes*: Content in brain and corpus allatum and effect on juvenile hormone synthesis." *Journal of Insect Physiology* 51: 357–365.

Yin, C.-M., and C. Gillot. 1975. "Endocrine activity during caste differentiation in *Zootermopsis angusticollis* Hagen (Isoptera): A morphometric and autoradiographic study." *Canadian Journal of Zoology* 53: 1690–1700.

Zhou X., F. M. Oi, and M. E. Scharf. 2006. "Social exploitation of hexamerin: RNAi reveals a major caste-regulatory factor in termites." *Proceedings of the National Academy of Sciences USA* 103: 4499–4504.

Zhou X., C. Song, T. L. Grzymala, F. M. Oi, and M. E. Scharf. 2006. "Juvenile hormone and colony conditions differentially influence cytochrome P450 gene expression in the termite *Reticulitermes flavipes*." *Insect Molecular Biology* 15: 749–761.

Zhou X., M. R. Tarver, G. W. Bennett, F. M. Oi, and M. E. Scharf. 2006. "Two hexamerin genes from the termite *Reticulitermes flavipes*: Sequence, expression, and proposed functions in caste regulation." *Gene* 376: 47–58.

Zimmerman, B. 1983. "Sibling manipulation and indirect fitness in termites." *Behavioral Ecology and Sociobiology* 12: 143–145.

Zitnan, D., and M. E. Adams. 2005. "Neuroendocrine regulation of insect ecdysis." In L. I. Gilbert, K. Iatrou, and S. S. Gill, eds., *Comprehensive molecular insect science* 1: 87–155. Boston: Elsevier.

CHAPTER SIX

Termites: An Alternative Road to Eusociality and the Importance of Group Benefits in Social Insects

JUDITH KORB

SOCIAL HYMENOPTERA are favorite model organisms of kin selection. In contrast, termites have received much less attention even though they are the oldest social insects. In the Hymenoptera the evolution of eusociality has been linked with a genetic predisposition brought about by their haplodiploid sex determination. The diplodiploid termites lack this predisposition, and several hypotheses have been put forward to account for their independent origin of eusociality. However, quantitative data testing these hypotheses have been rare. One question raised by recent studies of totipotent workers in a termite with an ancestral life-type is whether the workers really stay in order to gain indirect benefits by helping to raise siblings. This strongly contrasts with the role of workers in social Hymenoptera or higher termites. Instead, the potential for significant direct benefits through nest inheritance seem to be the driving force for these lower termite workers to remain in the natal nest. The safe nest environment, which offers a reasonable chance to become a philopatric breeder, must be seen alongside high mortality risks during dispersal to found a new colony. Such considerations show the importance of ecological and demographic factors in explaining complex insect societies (see also Cole, this volume).

Furthermore, the advantages of living in a group, like emergent properties, that arise through the interaction of many individuals (e.g., division of labor) might be fundamental forces that have been generally overlooked (see also Jeanson and Deneubourg, this volume). In the future, investigating the size-dependent properties of these groups and quantifying their benefits in natural systems may produce important insights about the causes for the evolutionary and ecological success of social insects.

Background

Kin selection, defined as the propagation of altruistic traits via relatives, is central to current theories explaining the evolution of social insects (Bourke and Franks 1995; Crozier and Pamilo 1996; Hamilton 1964). Insect societies, exemplified by the termites, ants, and some bees and wasps, are typically characterized by a reproductive division of labor where individuals forgo personal reproduction to help others reproduce (see also Peeters and Liebig, this volume). This behavior is usually directed toward close relatives. According to the kin selection theory, such altruistic behavior can be evolutionary favored when the benefit given to the relative, weighed by the relatedness between altruist and beneficiary, (indirect fitness) is larger than the costs in direct reproduction for the altruist (direct fitness) (Hamilton 1964; Maynard Smith 1964). The importance of indirect fitness benefits, and especially of relatedness, was intensively studied in social Hymenoptera, which became the key organisms used to test kin selection theory (Bourke and Franks 1995; Crozier and Pamilo 1996; see also Boomsma, Kronauer, and Pedersen, this volume; Heinze and Foitzik, this volume).

Termites (Isoptera) have received much less attention than the Hymenoptera (Crozier and Pamilo 1996). (For a review on the proximate mechanisms underlying caste differentiation in termites, see Chapter 5). Their resemblance to ants, which is so striking that they are commonly referred to as "white ants," lead to the assumption of convergent evolution driven by the same common factors. Indeed, more than 75% of all termite species (all higher termites, Termitidae) are characterized by altruistic castes (workers and soldiers) that help to raise offspring and have a greatly reduced capability to reproduce (Kambhampati and Eggleton 2000; Roisin 2000). However, the Termitidae are a highly derived group and cannot be used to assess how eusociality may have first evolved within the termites. In fact, recent quantitative studies of species from families with a life type that is thought to be ancestral in termite's evolution suggest that this does not apply to them (for a recent discussion of termite phylogeny, see Inward, Vogler, and Eggleton 2007; Korb 2007a). The direct benefits of inheriting the natal nest rather than altruistic helping appear to be the main selective forces for the occurrence of defensive reproductive morphs in dampwood termites (Thorne, Breisch, and Haverty 2003) and for workers in drywood termites (Korb 2007b). This contrasts sharply with social Hymenoptera, where altruistic helping is generally thought to be the major driving force (probably with the exception of some wasps).

Although it is generally difficult to deduce the ancient evolutionary history from studies on extant species, research on wood-dwelling termites may allow one to draw important conclusions. As will be explained in detail below, the wood-dwelling life type has idiosyncratic properties (e.g., poor nutritive quality of food, bonanza type food resource) which *per se* set the selective environment for the evolution of cooperation and altruism in termites. Therefore, it seems reasonable to extrapolate the recent results to the evolutionary history and conclude that in termites costly altruistic helping probably only evolved after living in extended family groups. This implies that benefits from group living, such as protection against predators, increased parasite resistance, or division of labor, might have been essential for the transition toward truly altruistic helping and the evolution of eusociality. For extant termite species in which workers retain their reproductive options (all Termopsidae, Kalotermitidae, and *Prorhinotermes*), only sterile soldiers that occur after the colony has reached a threshold size exhibit true altruism (reviewed in Roisin 2000).

The known principle factors that have influenced the social evolution of termites are their nesting and foraging habits. The basal termites (Termopsidae, Kalotermitidae, and *Prorhinotermes*), which all have a flexible development, generally reside in and feed on a single piece of wood (so-called one-piece life type termites, Abe 1987; for simplicity I will call them "wood-dwelling termites" hereafter). Because they never leave the wood to exploit a new resource, group size is limited by resource availability within the natal nest. The switch to exploit new resources ("multiple-pieces life type" termites according to Abe 1987), which is seen in all current termites with true altruistic workers (Mastotermitidae, most Rhinotermitidae, and Termitidae; reviewed in Roisin 2000), is associated with costly foraging outside the nest and the necessity for brood care, which increases the indirect benefits of helping.

In this chapter, I first summarize the state of the art in research on termite social evolution, concentrating on workers of termites with an ancestral wood-dwelling life type as their developmental flexibility offers unique opportunities to study factors influencing reproductive decision making. Soldiers will be treated separately as worker and soldier castes have independent origins in termites (Noirot and Pasteels 1987), contrasting sharply with social Hymenoptera where soldiers are a behaviorally specialized form of worker. Second, I explain why helping and indirect benefits are probably of low importance in these basal termites. Third, I discuss consequences of individuals staying at the nest, which should be a focus for future research. Fourth, I summarize the various factors that, during colony development, seem to influence the reproductive decision making of workers in wood-dwelling termites. Finally, by outlining why it might be justified to extrapolate these results for extant termites to their evolutionary history, I conclude with a potential explanation of why termites are a special case and why a similar transition to eusociality rarely happened in other diploid species, especially cooperatively breeding vertebrates to which wood-dwelling termites are in many aspects more similar than to social Hymenoptera.

Evolution of Workers in Termites: The Importance of Nest Inheritance

As with other social insects, explanations for the evolution of termite sociality have for a long time focused on the role of genetics and kinship (e.g., see Boomsma, Kronauer, and Pedersen, this volume; Cole, this volume; Heinze and Foitzik, this volume). In contrast to the haplodiploid Hymenoptera, the diplodiploid termites do not have a system of sex determination providing an apparently easy genetic explanation for the evolution of eusociality (Queller and Strassmann 1998). The discovery of chromosomal translocations, in which a tight linkage of genes to the sex chromosomes occurs, seemed to provide a welcomed haplodiploidy analogy (Lacy 1980; Luykx and Syren 1979); however, chromosomal translocations are not common in the clades thought to be closest to the noneusocial ancestors of termites, and the species in which such translocations occur do not show the predicted sex-discriminative behavior (Crozier and Luykx 1985; Hahn and Stuart 1987; Leinaas 1983; Roisin 2001). An alternative genetic explanation evoked cycles of inbreeding and outbreeding to create conditions favoring altruism. Bartz (1979) suggested that unrelated winged reproductives (alates) found new colonies but are later replaced as king and queen by their own offspring, which remain in the colony and breed together. After several cycles of such replacements, highly inbred and therefore homozygous alate offspring are produced. Assuming that such inbred alates mate with unrelated, but equally inbred partners, relatedness asymmetries comparable to those of the social Hymenoptera could arise: offspring from the colony founders would be more related to each other than they would be to their own offspring, favoring altruism. Although inbreeding does occur in termites, the number of cycles is insufficient to produce highly inbred offspring and outbred reproductives often produce alates (e.g., Husseneder et al. 1999; Myles and Nutting 1988). Thus, exclusively genetic explanations have proven inadequate and ecological conditions appear to be of paramount importance for the evolution of eusociality in termites (Korb 2008; Shellman-Reeve 1997; Thorne 1997).

Philopatric benefits have been suggested to explain the evolution of termites staying at the nest, preceding helping behavior (Myles 1988; Shellman-Reeve 1997; Thorne 1997); however, this hypothesis had never been explicitly tested. Similar to cooperatively breeding vertebrates, where some individuals (helpers) stay at the nest to gain direct benefits by an increased probability of future inheritance of a breeding position (Heinsohn and Legge 1999; Koenig and Dickinson 2004), termite workers could stay in their natal nest in the hope of replacing their parents (see also reproductive skew theory: Johnstone 2000). Replacement or supplementary reproductives are common, especially in basal termites, where workers can develop via a single molt into neotenic reproductives when the same sex reproductive of the colony dies or shows signs of senescence (reviewed by Roisin 2000; Shellman-Reeve 1997; Thorne 1997). Roisin (1999) dismissed the philopatric explanation because founding reproductives are long-lived (Shellman-Reeve 1997; Thorne, Breisch, and Haverty 2002), therefore reducing opportunities for the first brood to inherit the nest and providing them little incentive to stay. However, field data were then lacking and nest inheritance may be more common than once thought (see below). Thorne, Breisch, and Muscedere (2003) noted that during intercolonial encounters for the dampwood termite *Zootermopsis nevadensis*, reproductives were the primary focus of attacks and were frequently killed, creating opportunities for inheritance. At the same time the slow development of wood-nesting termites, resulting from their nitrogen-poor diet, necessitates a relatively lengthy stay of several years in the natal nest (Lenz 1994; Nalepa 1994; Shellman-Reeve 1997) during which the reproductives might die. As the individuals also face few competitors for the breeding position compared to older, larger colonies, remaining in the nest can be selected.

Another factor that increases the importance of nest inheritance as a reproductive option for an individual would be a low probability of successfully founding a nest (Johnstone 2000). The first available quantitative field data for reproductive survivorship in termites show that in the drywood termite *Cryptotermes secundus* the probability of inheriting the natal nest is in the same order of magnitude as that for founding a new colony. In both cases more than 99% of all individuals die without a chance to reproduce (Korb and Schneider 2007). In this species, like in all wood-dwelling termites (Roisin 2000; Thorne and Traniello 2003), the workers are immatures (undifferentiated older larvae) that have the option to either develop into winged sexuals that leave the nest to found their own colony or to remain in the nest with a chance to become a replacement reproductive when the same-sex reproductive dies (Korb and Katrantzis 2004). A model based on long-term field data revealed that in *C. secundus* the number of individuals staying at the nest can be explained by the probability of inheriting the nest versus founding a new colony (Korb 2008). Depending on colony size, the age of the reproductives, and the potential longevity of the nest, the probability of nest inheritance changes and these three variables can accurately predict the number of individuals developing into dispersing sexuals in field colonies. This provides strong evidence that the workers adjust their "dispersal versus staying decision" according to reproductive opportunities. In contrast to cooperative breeding vertebrates that can search their environment for vacant breeding sites and adjust their decision to stay according to the chance to become a breeder elsewhere (Emlen 1997), wood-dwelling termites have no opportunity to check the availability of breeding vacancies as they remain in their natal nest until dispersal (Roisin 1999). Because numerous founding sites are available to *C. secundus* alates (Korb and Lenz 2004), the probability of successfully establishing a colony is relatively constant, and the decision to leave the colony only depends on the likelihood of inheritance.

The cost-benefit ratio for staying at the nest also depends on the costs directly associated with staying, such as delayed maturity, the possibility of local resource competition with relatives, or costs of helping, which may all reduce the residual reproductive value of individuals (Stearns 1992). These costs, however, seem to be negligible in *C. secundus* and may be offset by several benefits. First, workers of this and other wood-dwelling species are immatures that dwell in an environment sheltered from external conditions and predators (Roux 2004). They experience low extrinsic juvenile mortality and their prolonged development allows them to accumulate the nitrogen reserves necessary for improved reproduction. This selection for delayed maturity (Stearns 1992) comes with relatively little costs on future survival and longevity. Second, local resource competition at the nest is absent, especially within newly founded colonies where the food supply is ample. Food availability begins to decline only after several years and under these conditions workers develop into alates that leave the nest (Korb and Schmidinger 2004). Third, workers do not seem to engage in costly help. There is no brood care behavior and interactions among workers, like proctodeal feeding and allogrooming, are reciprocal (Korb 2007b). Reproductives also feed for themselves and the only individuals that needed to be fed are soldiers. This presents a minor cost as less than 5% of all individuals (median: 2%) are soldiers. On a per capita basis each worker feeds a soldier less frequently than once per day (Korb 2007b). In accordance with the lack of brood care, the development of workers into dispersing winged sexuals is not influenced by the number of young offspring present at the nest: workers in colonies where the number of eggs and first and second instar larvae was experimentally increased were not more likely to stay in the nest than workers that had less "dependent" siblings (Korb 2007b). Thus, contrary to general perception, these supposed termite workers apparently do not remain in their natal nest to gain indirect benefits by brood rearing, but rather in the hope of nest inheritance. Re-introducing the old-fashioned term "false workers" (Roisin 2000) for the developmentally flexible workers of wood-dwelling termites would nicely reflect their functional distinction from those of the "true workers" of the multiple-pieces nesting termites.

The potential importance of nest inheritance in termites' social evolution was also recently shown by Thorne, Breisch, and Muscedere (2003). In the dampwood termite, *Zootermopsis nevadensis,* the development of soldier-like replacement reproductives is linked to intercolonial fusion events after which these soldier-like reproductives have a higher chance to inherit the colony as replacement reproductives. This implies that soldier traits of reproductives have an adaptive function in Termopsidae, but it is unlikely to explain the evolution of soldiers as, from a developmental point of view, they are reproductives rather than soldiers (Roisin 2000) and it is improbable for soldiers to become sterile if soldiers were originally reproductives (Roisin 1994, 1999; Jeon and Choe 2003).

To summarize the state of the art in termite social evolution, there are only two detailed studies on reproductive survival and nest inheritance in wood-dwelling termites reflecting an ancestral life type and both suggest that indirect benefits of raising siblings are of minor importance relative to potential direct fitness gains.

Why Workers Do Not Help

There are two non-mutually exclusive explanations for why workers of wood-dwelling termites do not help to raise siblings. First, the development and the life type of the wood-dwelling termites present a case where care for offspring is less/not needed and thus provides little benefit. In contrast to Hymenoptera, termites are hemimetabolous insects and the workers are in fact larval instars (see also Brent, this volume), although they can be morphologically specialized. This major difference between the two orders has long been stressed, but the full consequences were hardly ever considered (notable exception Nalepa 1994). The need for care in termites is limited as the young become independent very early. Generally, individuals from the third larval instar onward can feed and care for themselves, and in young colonies even second instar individuals can be autonomous (e.g., reviewed by Shellman-Reeve 1997). Results for *C. secundus* and *Cryptotermes domesticus* show that second instar larvae regularly feed themselves regardless of the colony's age. This is possible as the termites live inside their food. The eggs and first instar larvae are not cared for (Korb 2007b), which suggests that they utilize body reserves, but certainly more research is necessary to reveal details. The lack of allogrooming of eggs and first instar larvae, a basic care behavior provided by parents or helpers to young in many species ranging from insects to mammals (e.g., Rosengaus et al. 1999; Solomon and French 1997), can be explained by the parasite-free environment these termites live in. Allogrooming is generally directed to remove ectoparasites and to prevent fungal infections, often by applying fungicides; however, very few microbes occur in the dry wood that these termites inhabit (Rosengaus et al. 2003). During rare outbreaks of mite infestations *C. secundus* and *C. domesticus* lack effective means to ward off mites, but young instars are never infected with mites probably due to their small size (Korb and Fuchs 2006). It is not yet clear how the first instars acquire their gut symbionts when proctodeal feeding is absent, but feeding the shed skin from newly molted individuals can often be observed, even when the termite is still molting. Though it needs to be studied whether gut symbionts will survive this transfer, together with proctodeal feeding among older instars, this might be the mechanism to receive symbionts for the first inoculation of larvae and the re-inoculation after each subsequent molt.

A second reason why workers may not help is that they are not forced to do so by the reproductives. According to the "pay to stay" model (Kokko, Johnstone, and Wright 2002), individuals have to help if their remaining at the nest imposes costs on the parents by causing local resource competition. In many species this competition causes either selection for offspring dispersal (Hamilton and May 1977) or for helping as a payment (Kokko, Johnstone, and Wright 2002). In wood-dwelling termites, food is a bonanza resource, meaning that it is abundant for a long time after founding a colony and that it generally outlasts by far the lifetime of the founding primary reproductives. Thus local resource competition between parent and offspring or among siblings is absent. This creates a unique opportunity for many siblings to remain safely within their natal nest, protected from environmental hazards such as predators, pathogens, or unfavorable climates. Because the benefit for their helping behavior would be low, there is little selection for helping to raise siblings.

Consequences of Staying at Home: Benefits of Group Living

The slow development toward maturation in wood-dwelling termites allows for the formation of an extended group. This offers additional benefits to those that stay as can be seen even among associations of nonrelatives (Krause and Ruxton 2002; Wilson and Hölldobler 2005; see also Jeanson and Deneubourg, this volume; Pratt, this volume). Grouping can improve defense against predators; for example, through cooperative attacks as observed in colonial ground-nesting birds (reviewed in Caro 2005). Dilution and predator saturation effects can also provide protection through numbers, even in nonsocial species (e.g., Caro 2005). Similar benefits probably also accrue to wood-dwelling termites. In *C. secundus* the main threat consists of intraspecific encounters where large colonies are more likely to win these encounters (Roux 2004; Roux and Korb 2004). Such encounters occur when two colonies which were founded in the same tree meet. This meeting can result in aggressive interactions where the same-sex reproductives of both colonies fight against each other (workers and soldiers are not directly involved in fights) and where the reproductives of the larger colony are more likely to kill the reproductives of the smaller colony. An additional general benefit to group living in termites may be an enhanced ability to resist disease (e.g., Traniello, Rosengaus, and Savoie 2002; see also Fefferman and Traniello, this volume). Grouping can buffer the effects of perturbations in both the social (e.g., Oster and Wilson 1978) and physical environment (e.g., Korb 2003), providing a greater resilience to the colony as a whole.

Another major advantage of extended grouping is that it facilitates the development of individual specialization on subsets of colony tasks (e.g. Alexander, Noonan, and Crespi 1991; Beshers and Fewell 2001; Fewell and Taylor, this volume). Such task specializations increase colony efficiency (Oster and Wilson 1978), thereby reducing the costs to remain in the colony. Division of labor resulting from such task specializations generally seems to be a major incentive for the evolution of cooperation among separate units because it provides net direct benefits for each partner. Thus, similar to social insects, division of labor is regarded to be decisive for the evolution of cooperation among genes in chromosomes, of nuclear and mitochondrial genomes in eukaryotic cells, and among differentiated cells in multicellular organisms (Keller 1999; Maynard Smith and Szathmáry 1995). In *C. secundus,* colony tasks are divided between soldiers and the rest of the colony with soldiers performing some specialized defensive functions like the plugging of galleries, while all other tasks, including generalized defense, are done by all colony members (Roux 2004). This organization may be similar for many basal termites (Rosengaus and Traniello 1993), although exceptions exist in species exhibiting temporal polyethism (Crosland 1997). In more derived termites division of labor among workers is common (e.g., Noirot 1989) up to very elaborate caste polymorphisms in fungus-growing termites (Macrotermitinae) where age- as well as caste-polyethism occurs (e.g., Traniello and Leuthold 2000).

Colony size appears to be a major determining factor in both the timing and degree of caste specialization. There is a strong association between colony size and the occurrence of the first soldiers in *C. secundus* (Roux and Korb 2004) and other termites (Haverty 1977; Haverty and Howard 1981), and between maximum colony size and the caste complexity among termite families (reviewed in Shellman-Reeve 1997). These associations indicate that mutualistic benefits of division of labor only arise after the colonies reach a threshold size. As colony size increases, self-organizational processes add further attributes to colonies and differentiates them from simple groups (Camazine et al. 2001; Jeanson and Deneubourg, this volume; Queller and Strassmann 1998). For example, pheromone trails become efficient recruiting mechanisms after a certain threshold group size is reached (Beekman, Sumpter, and Ratnieks 2001). Also, the construction of elaborate nests is a result of self-organization that provides effective protection against predators and improved thermoregulation (Heinrich 1993; Korb 2003; Noirot and Darlington 2000).

This list of group advantages is far from complete, and although many were first defined by Michener in 1964, few have been systematically studied in either the social Hymenoptera or the termites (e.g., Karsai and Wenzel 1998; Shreeves and Field 2002). Nonlinear growth rates observed during the first years after colony foundation in many species (ergonomic phase; Oster and Wilson 1978) might reflect these advantages. As colony size increases so do the advantages of grouping, resulting in a faster growth rate. Once a colony reaches its maximal growth rate, which in *C. secundus* seems to be limited by the fecundity of the queen, the group starts to produce dispersing sexuals. The nonlinear effects during the ergonomic phase provide direct benefits to the individuals staying at the nest as well as indirect benefits as related individuals also gain through these advantages. This illustrates the importance group size advantages might play in social evolution (Karsai and Wenzel 1998; Korb and Heinze 2004; Wilson and Hölldobler 2005). To quantify these benefits in the future, trait-group selection models (multilevel selection) that partition selection into a within-group and between-group selection component might be helpful. Although these models are known to be equivalent to kin selection models (Frank 1998; Hamilton 1975; Wilson 1997), they can sometimes highlight important colony-level effects that might otherwise go unnoticed (Foster, Wenseleers, and Ratnieks 2006; Korb and Heinze 2004).

Factors Influencing Reproductive Decision Making during Colony Development in Wood-Dwelling Termites

According to the before mentioned results, the factors most important for reproductive decision making during colony development in wood-dwelling termites can be summarized as follows (Figure 6.1): wood-dwelling termites have a very slow development of several years due to their low quality food; this, together with the continued egg-laying of the queen, results in multi-age family groups (Figure 6.1b); and death of the reproductives during this period favors the development of neotenic reproductives. Thus, when the first individuals approach maturity, they: (1) live in a well protected nest, (2) gain benefits from group living, (3) do not experience competition for food, and (4) have the chance to inherit the natal breeding position without strong competition among relatives. This favors individuals staying at the natal nest (Figure 6.1c).

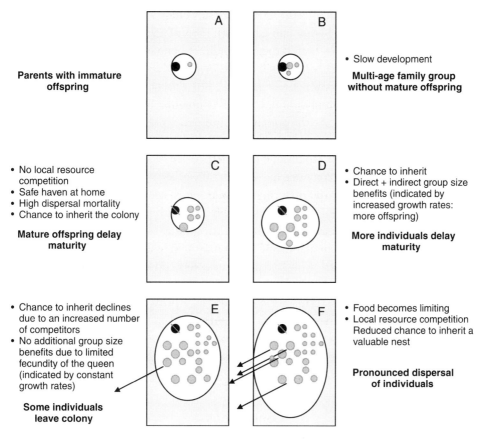

Figure 6.1. Crucial factors for reproductive decision making during colony development. Shown are colonies (circle) within wood blocks (rectangle), at different stages of development. Smaller circles inside the colonies present the castes: black = reproductives; gray = immature individuals (workers), where the size signifies the instar. Offspring sufficiently mature to leave the nest are indicated by the larger gray circles. An increased mortality probability of the reproductives is indicated by a gray bar through the circle. Different factors primarily influence the decision of mature offspring to stay or leave (indicated by arrows) during each stage of colony development: (C) direct benefits of nest inheritance; (D) direct benefits of nest inheritance plus direct and indirect benefits through group size advantages; (E) competition among relatives for breeding; (F) competition among relatives for food plus decreased benefits of nest inheritance.

With increasing colony size, group size advantages become more pronounced resulting in higher annual growth rates of colonies (Figure 6.1d), which in turn provides both direct and indirect benefits to individuals delaying dispersal. Along with this increase in colony age the intrinsic mortality rate of the reproductives increase, but this becomes increasingly offset by a growing number of individuals competing to inherit the breeding position. Eventually, the colony will grow to a point when group size benefits no longer increase, and the probability per individual for inheritance of the nest begins to decline (Figure 6.1e). In *C. secundus* this seems to occur when the colony reaches about 200 workers; it then has a steady growth rate of about 100 offspring per year which seems to be set by the fecundity of the queen. After reaching this size, a rather constant proportion of the workers leave the colony in *C. secundus* (Korb and Schneider 2007). This leads to comparatively stable maximum colony sizes of around 200 to 300 individuals. The last phase of colony development in wood-dwelling termites starts when the wood becomes limited. Then competition among relatives occurs which results in a pronounced dispersal of almost all workers that are capable of doing so (Figure 6.1f) (Lenz 1994; Korb and Schmidinger 2004). Thereafter, the maximum colony size seems to be reduced to account for the decreasing value of the nest as an inherited resource.

Conclusions

Although it is generally problematic to deduce the evolution of sociality, which happened in the past, from extant species, the case of wood-dwelling termites allows us to draw some important conclusions. The termites' wood-dwelling life type is idiosyncratically linked with a unique combination of traits that favors offspring staying at their nest (Table 6.1). First, the nest provides a safe haven while dispersal involves considerable risk. These factors, coupled with the poor nutritive quality of their food, selects for a delayed maturation. Second, the bonanza-type food resource allows staying at the nest without appreciable costs to relatives. Third, hemimetabolous development and the ready availability of food allow even the youngest individuals to remain relatively independent of care from others, minimizing the costs to siblings that choose to stay in the natal nest. Fourth, the prolonged development of immature termites enhances the likelihood that they will be present in the natal nest when the reproductives

Table 6.1. Occurrence of traits favoring the evolution of delayed dispersal in wood-dwelling termites and cooperative breeding vertebrates. ✓: presence; —: absence.

Traits	Wood-dwelling termites	Helpers in:		
		Fish	Birds	Mammals
Safe nest, costly dispersal	✓	—/✓	✓	✓
No local resource competition at nest	✓	—/✓	—	—/✓
Independent young	✓	—	—	—

die, increasing the opportunity for nest inheritance. Thus, if the wood-dwelling life type is the ancestral state in termite evolution—and more and more evidence supports this general assumption (for recent reviews, see Inward, Vogler, and Eggleton 2007; Korb 2007a)—this idiosyncratic combination of factors follows as a by-product of their life type and selects for delayed dispersal with a chance to inherit the natal breeding position. Accordingly, it seems likely that the first step in termites' social evolution was large family groups without costly altruistic sibrearing. The emerging advantages of living in a large group may then have further increased the incentives for staying in the nest and lead to group sizes where the costs of altruistic helping via defense are offset by the indirect benefits of saving many relatives' life.

This stepwise evolutionary hypothesis with the occurrence of large family groups first also overcomes problems associated with the evolution of altruism in diploid organisms (Teyssèdre et al., 2006) and it might explain why eusociality is rare elsewhere in diploids. Teyssèdre, Couvet, and Nunney (2006) showed that the spread of altruism in diploids, but not in haplodiploids, requires a link between altruism and enhanced reproductive efficiency. This link is unlikely when a mutant-causing altruism occurs in a solitary organism. However, if such an altruistic mutant evolves in an already established group of coexisting relatives then it can automatically increase group productivity, as is exemplified by a defensive soldier morph. The uniqueness of the combination of traits found in wood-dwelling termites might explain why cooperatively breeding vertebrates, although facing similar ecological conditions, rarely evolved eusociality (Alexander, Noonan, and Crespi 1991) (Table 6.1).

Generally, family groups in vertebrates are too small for the costs of altruism, for example through defense, to be offset by the indirect benefits of saving many relatives' life. At least one reason for this limitation is the occurrence of local resource competition which selects for dispersal of mature offspring (Hamilton and May 1977) (Table 6.1). Interestingly, the eusocial mole rats, like the wood-dwelling termites, utilize Bonanza-type resources (Sherman, Jarvis, and Alexander 1991), thus avoiding local resource competition and selection for dispersal of mature offspring.

Literature Cited

Abe, T. 1987. "Evolution of life types in termites." In S. Kawano, J. H. Connell, and T. Hidaka, eds., *Evolution and coadaptation in biotic communities,* 125–148. Tokyo: University of Tokyo Press.

Alexander, R. D., K. M. Noonan, and B. J. Crespi. 1991. "The evolution of eusociality." In P. Sherman, J. Jarvis, and R. Alexander, eds., *The biology of the naked mole-rat,* 3–44. Princeton: Princeton University Press.

Bartz, S. 1979. "Evolution of eusociality in termites." *Proceedings of the National Academy of Sciences USA* 76: 5764–5768.

Beekman, M., D. J. T. Sumpter, and F. L. W. Ratnieks. 2001. "Phase transition between disordered and ordered foraging in Pharaoh's ants." *Proceedings of the National Academy of Sciences USA* 98: 9703–9706.

Beshers, S. N., and J. H. Fewell. 2001. "Models of division of labor in social insects." *Annual Reviews of Entomology* 46: 413–440.

Bourke, A. F. G., and N. R. Franks. 1995. *Social evolution in ants.* Princeton: Princeton University Press.

Camazine, S., J. L. Deneubourg, N. R. Franks, J. Sneyd, G. Theraulaz, and E. Bonabeau. 2001. *Self-organization in biological systems.* Princeton: Princeton University Press.

Caro, T. 2005. *Antipredator defenses in birds and mammals.* Chicago: University of Chicago Press.

Crosland, M. W. J., C. M. Lok, T. C. Wong, M. Shakarad, and J. F. A. Traniello. 1997. "Division of labor in the lower termite: the majority of tasks are performed by older workers." *Animal Behavior* 54: 999–1012.

Crozier, R. H., and P. Luykx. 1985. "The evolution of termite eusociality is unlikely to have been based on a male-haploid analogy." *American Naturalist* 126: 867–869.

Crozier, R. H., and P. Pamilo. 1996. *Evolution of social insect colonies: Sex allocation and kin selection.* Oxford: Oxford University Press.

Emlen, S. T. 1997. "Predicting family dynamics in social vertebrates." In N. B. Davies and J. R. Krebs, eds., *Behavioural ecology* (4th ed.), 228–253. Oxford: Blackwell Science.

Foster, K. R., T. Wenseleers, and F. L. W. Ratnieks. 2006. "Kin selection is the key to altruism." *Trends in Ecology and Evolution* 21: 57–60.

Frank, S. A. 1998. *Foundations of social evolution.* Princeton: Princeton University Press.

Hahn, P. D., and A. M. Stuart. 1987. "Sibling interactions in two species of termites: A test of the haplodiploid analogy (Isoptera; Kalotermitidae; Rhinotermitidae)." *Sociobiology* 13: 83–92.

Hamilton, W. D. 1964. "The genetical evolution of social behaviour. I, II." *Journal of Theoretical Biology* 7: 1–52.

———. 1975. "Innate social aptitudes in man: An approach from evolutionary genetics." In R. Fox, ed., *Biosocial anthropology,* 133–155. London: Malaby.

Hamilton, W. D., and R. M. May. 1977. "Dispersal in stable habitats." *Nature* 269: 578–581.

Haverty, M. I. 1977. "The proportion of soldiers in termite colonies: A list and a bibliography (Isoptera)." *Sociobiology* 2: 199–216.

Haverty, M. I., and R. W. Howard. 1981. "Production of soldiers and maintenance of soldier proportions by laboratory experimental groups of *Reticulitermes flavipes* (Kollar) and *Reticulitermes virginicus* (Banks) (Isoptera: Rhinotermitidae)." *Insectes Sociaux* 28: 32–39.

Heinrich, B. 1993. *The hot-blooded insect.* Cambridge: Harvard University Press.

Heinsohn, R., and S. Legge. 1999. "The cost of helping." *Trends in Ecology and Evolution* 14: 53–57.

Husseneder, C., R. Brandl, J. T. Epplen, and M. Kaib. 1999. "Within colony relatedness in a termite species: Genetic roads to eusociality?" *Behaviour* 136: 1045–1063.

Inward, D. J. G., A. P. Vogler, and P. Eggleton. 2007. "A comprehensive phylogenetic analysis of termites (Isoptera) illuminates key aspects of their evolutionary biology." *Molecular Phylogenetics and Evolution* 44: 953–967.

Jeon, J., and J. C. Choe. 2003. "Reproductive skew and the origin of sterile castes." *American Naturalist* 161: 206–224.

Johnstone, R. A. 2000. "Models of reproductive skew: A review and synthesis." *Ethology,* 106: 5–26.

Kambhampati, S., and P. Eggleton. 2000. "Taxonomy and phylogeny of termites." In T. Abe, D. E. Bignell, and M. Higashi, eds., *Termites: Evolution, sociality, symbioses, ecology,* 1–24. Dordrecht: Kluwer Academic Publishers.

Karsai, I., and J. W. Wenzel. 1998. "Productivity, individual-level and colony-level flexibility, and organization of work as consequences of colony size." *Proceedings of the National Academy of Sciences USA* 96: 8665–8669.

Keller, L. 1999. *Levels of selection in evolution.* Princeton: Princeton University Press.

Koenig, W., and J. Dickinson. 2004. *Ecology and evolution of cooperative breeding in birds.* Cambridge: Cambridge University Press.

Kokko, H., R. A. Johnstone, and J. Wright. 2002. "The evolution of parental and alloparental effort in cooperatively breeding groups: When should helpers pay to stay." *Behavioral Ecology* 13: 291–300.

Korb, J. 2003. "Thermoregulation and ventilation of termite mounds." *Naturwissenschaften* 90: 212–219.

Korb, J. 2007a. "Termites." *Current Biology* 17:R995–R999

Korb, J. 2007b. "Workers of a drywood termite do not work." *Frontiers in Zoology* 4: e7.

Korb, J. 2008. "The ecology of social evolution in termites." In J. Korb and J. Heinze, eds., *The ecology of social evolution,* 151–174 Heidelberg: Springer.

Korb, J., and A. Fuchs. 2006. "Termites and mites—Adaptive behavioural responses to infestation." *Behaviour* 143: 891–907.

Korb, J., and J. Heinze. 2004. "Multilevel selection and social evolution of insect societies." *Naturwissenschaften* 91: 291–304.

Korb, J., and S. Katrantzis. 2004. "Influence of environmental conditions on the expression of the sexual dispersal phenotype in a lower termite: Implications for the evolution of workers in termites." *Evolution & Development* 6: 342–352.

Korb, J., and M. Lenz. 2004. "Reproductive decision-making in the termite *Cryptotermes secundus* (Kalotermitidae) under variable food conditions." *Behavioral Ecology* 15: 390–395.

Korb, J., and S. Schmidinger. 2004. "Help or disperse? Cooperation in termites influenced by food conditions." *Behavioral Ecology and Sociobiology* 56: 89–95.

Korb, J., and K. Schneider. 2007. "Does kin structure explain the occurrence of workers in a lower termite." *Evolutionary Ecology* 21: 817–828.

Krause, J., and G. D. Ruxton. 2002. *Living in groups.* Oxford: Oxford University Press.

Lacy, R. C. 1980. "The evolution of eusociality in termites: A haplodiploid analogy?" *American Naturalist* 116: 449–451.

Leinaas, H. P. 1983. "A haplodiploid analogy in the evolution of termite eusociality? Reply to Lacy." *American Naturalist* 121: 302–304.

Lenz, M. 1994. "Food resources, colony growth and caste development in wood-feeding termites." In J. Hunt and C. A. Nalepa, eds., *Nourishment and evolution in insect societies,* 159–210. Boulder: Westview Press.

Luykx, P., and R. M. Syren. 1979. "The cytogenetics of *Incisitermes schwarzi* and other Florida termites." *Sociobiology* 4: 91–210.

Maynard Smith, J. 1964. "Group selection and kin selection." *Nature* 201: 1145–1146.

Maynard Smith, J., and E. Szathmáry. 1995. *The major transitions in evolution.* Oxford: W.H. Freeman.

Michener, C. D. 1964. "Reproductive efficiency in relation to colony size in hymenopterous societies." *Insectes Sociaux* 11: 317–342.

Myles, T. G. 1988. "Resource inheritance in social evolution from termite to man." In C. N. Slobodchikoff, ed., *The ecology of social behavior,* 379–423. New York: Academic Press.

Myles, T. G., and W. L. Nutting. 1988. "Termite eusocial evolution: A re-examination of Bartz's hypothesis and assumptions." *Quarterly Reviews of Biology* 63: 1–24.

Nalepa, C. A. 1994. "Nourishment and the origin of termite eusociality." In J. Hunt and C. A. Nalepa, eds., *Nourishment and evolution in insect societies,* 57–104. Boulder: Westview Press.

Noirot, C. 1989. "Social structure in termite societies." *Ethology, Ecology & Evolution* 1: 1–17.

Noirot, C., and J. P. E. C. Darlington. 2000. "Termite nests: Architecture, regulation and defence." In T. Abe, D. E. Bignell, and M. Higashi, eds., *Termites: Evolution, sociality, symbioses, ecology,* 121–140. Dordrecht: Kluwer Academic Publishers.

Noirot, C., and J. M. Pasteels. 1987. "Ontogenetic development and evolution of the worker caste in termites." *Experientia* 43: 851–860.

Oster, G. F., and E. O. Wilson. 1978. *Caste and ecology of social insects.* Princeton: Princeton University Press.

Queller, D. C., and J. E. Strassmann. 1998. "Kin selection and social insects." *BioScience* 48: 165–178.

Roisin, Y. 1994. "Intragroup conflict and the evolution of sterile castes in termites." *American Naturalist* 143: 751–765.

———. 1999. "Philopatric reproduction, a prime mover in the evolution of eusociality?" *Insectes Sociaux* 46: 297–305.

———. 2000. "Diversity and evolution of caste patterns." In T. Abe, D. E. Bignell, and M. Higashi, eds., *Termites: Evolution, sociality, symbioses, ecology,* 95–120. Dordrecht: Kluwer Academic Publishers.

———. 2001. "Caste sex ratios, sex linkage, and reproductive strategies in termites." *Insectes Sociaux* 48: 224–230.

Rosengaus, R. B., J. E. Moustakas, D. V. Calleri, and J. F. A.Traniello. 2003. "Nesting ecology and cuticular microbial loads in dampwood (*Zootermopsis angusticollis*) and drywood termites (*Incisitermes minor, I. schwarzi, Cryptotermes cavifrons*)." *Journal of Insect Science* 3: 31.

Rosengaus, R. B., and J. F. A. Traniello. 1993. "Temporal polyethism in incipient colonies of the primitive termite *Zootermopsis angusticollis:* A single multiage caste." *Journal of Insect Behavior* 6: 237–252.

Rosengaus, R. B., J. F. A. Traniello, T. Chen, T. Lefebrve and Karp R.D. 1999. "Immunity in a social insect." *Naturwissenschaften* 86: 588–591.

Roux, E. A. 2004. "Evolution of eusociality and the soldier caste: A case study in a drywood termite." Ph.D. thesis, University of Regensburg.

Roux, E. A., and J. Korb. 2004. "Evolution of eusociality and the soldier caste in termites: A validation of the intrinsic benefit hypothesis." *Journal of Evolutionary Biology* 6: 342–352.

Shellman-Reeve, J. S. 1997. "The spectrum of eusociality in termites." In J. C. Choe and B. J. Crespi, eds., *The evolution of social behaviour in insects and arachnids,* 52–93. Cambridge: Cambridge University Press.

Sherman, P. W., J. U. M. Jarvis, and R. D. Alexander. 1991. *The biology of the naked mole-rat.* Princeton: Princeton University Press.

Shreeves, G., and J. Field. 2002. "Group size and direct fitness in social queens." *American Naturalist* 159: 81–95.

Solomon, N. G., and J. A. French. 1997. *Cooperative breeding in mammals.* Cambridge: Cambridge University Press.

Stearns, S. C. 1992. *The evolution of life histories.* Oxford: University Press.

Teyssèdre, A., D. Couvet, and L. Nunney. 2006. "Lower group productivity under kin-selected reproductive altruism." *Evolution* 60: 2023–2031.

Thorne, B. L. 1997. "Evolution of eusociality in termites." *Annual Review of Ecology and Systematics* 28: 27–54.

Thorne, B. L., N. L. Breisch, and M. I. Haverty. 2002. "Longevity of kings and queens and the first time of production of fertile progeny in dampwood termite (Isoptera; Termopsidae; *Zootermopsis*) colonies with different reproductive structures." *Journal of Animal Ecology* 71: 1030–1041.

Thorne, B. L., N. L. Breisch, and M. L. Muscedere. 2003. "Evolution of eusociality and the soldier caste in termites: Influence of intraspecific competition and accelerated inheritance." *Proceedings of the National Academy of Sciences USA* 100: 12808–12813.

Thorne, B. L., and J. F. A. Traniello. 2003. "Comparative social biology of basal taxa of ants and termites." *Annual Reviews of Entomology* 48: 283–306.

Traniello, J. F. A., and R. H. Leuthold. 2000. "Behavior and ecology of foraging termites." In T. Abe, D. E. Bignell, and M. Higashi, eds., *Termites: Evolution, sociality, symbioses, ecology,* 141–168. Dordrecht: Kluwer Academic Publishers.

Traniello, J. F. A., R. B. Rosengaus, and K. Savoie. 2002. "The development of immunity in a social insect: Evidence for the group facilitation of disease resistance." *Proceedings of the National Academy of Sciences USA* 99: 6838–6842.

Wilson, D. S. 1997. "Altruism and organism: Disentangling the themes of multilevel selection theory." *American Naturalist* 150: S122–134.

Wilson, E. O., and B. Hölldobler. 2005. "Eusociality: Origin and consequences." *Proceedings of the National Academy of Sciences USA* 102: 13367–13371.

CHAPTER SEVEN

The Evolution of Communal Behavior in Bees and Wasps: An Alternative to Eusociality

WILLIAM T. WCISLO

SIMON M. TIERNEY

BEGINNING WITH DARWIN (1859), a fundamental question for research on social insects has concerned the evolution of divergent phenotypes among already-sterile individuals (Linksvayer and Wade 2005). As reviewed in this volume, the caste-based (eusocial) societies of termites (Isoptera), and ants, wasps, and bees (Hymenoptera), have been extensively studied. A second question concerns the evolutionary *origins* of group life, for which appropriate foci are societies in which group members share a nest but work is not organized by caste differences (i.e., cooperative breeders and communal nesters). Unlike cooperatively breeding vertebrates (see Brockmann 1997; Hayes 2000; de Waal and Tyack 2003; Ekman and Ericson 2006) and eusocial invertebrates (see Cole, this volume), much less is known about the behavior and evolution of communal, casteless societies and the ecological contexts in which they occur.

Despite arguments for and against revising social terminology (see references in Costa and Fitzgerald 2005; Wcislo 2005), we use Michener's (1974) definition of *communal behavior*: cohabiting females that share a nest but build, provision, and oviposit in their own cells. Theoretically it is important to distinguish such associations from quasisocial ones in which reproductively competent females jointly build and provision cells. In practice, however, it is difficult to distinguish between communal and quasisocial associations because of the difficulties in observing behavior within nests. As Michener (1974, 2007) and others have pointed out, for many taxa these social terms apply to individual colonies or nests, but not to entire species, unless there is no intraspecific variation. Thus, although we sometimes refer to a particular species as having communal behavior, this is shorthand and we intend to refer to the particular nests studied in a given population. Conversely, many species are regarded as "solitary" even though some nests occasionally contain communal females.

A body of largely anecdotal evidence shows that communal behavior is widespread within the aculeate Hymenoptera. It evolved repeatedly and is stable over evolutionary time. Communal social groups are often described as egalitarian, or being comprised of reproductive equals, but in most cases this is an untested assertion. As West-Eberhard (1978) argued, in most group-living bees and wasps it is difficult to imagine that there are *no* reproductive inequalities, due to differences in age, genetic make-up, and nutritional status. It is more likely that female reproductive opportunities and output vary considerably, and that rudimentary dominant and subordinate behaviors are widespread (West-Eberhard 1978; Shimizu 2004; Jeanson, Kukuk, and Fewell 2005).

Perspectives on the evolution of communal behavior have shifted since early work by Wheeler (1923) and others. Communal societies of insects were thought to represent an intermediate stage in an evolutionary transition from solitary to eusocial behavior, with a step-wise increment in social complexity, such that solitary species first evolved communal behavior and then, via subsequent steps, the communal species gave rise to eusocial ones (Wheeler 1923; Evans 1958; Wilson 1971; West-Eberhard 1978). Studies of intraspecific variation in social behavior, coupled with a renewed emphasis on the importance of taking historical (phylogenetic) patterns into account, have suggested that this step-wise view of behavioral evolution is usually not supported by available data (Michener 1985; Carpenter 1989; Schwarz et al. 2003; Schwarz et al. 2007). Michener (1985, 2007) and others (e.g., West-Eberhard 2003) have argued that in some lineages behavior is so flexible that social organization may evolve without a series of intervening species, and that social behavior may be gained and lost so frequently that the phylogenetic signal in the behavioral data is unreliable. Indeed, Michener (2007, p. 15) suggests that the perennial question of the number of times eusociality has arisen during bee evolution is "both unknowable and useless. It is the wrong question." As discussed below, a phylogenetic question in need of more attention is why communal societies and caste-based ones appear to be evolutionary alternatives in different lineages.

In this chapter, we briefly summarize the occurrence of communal behavior in bees and wasps, and review studies showing that nestmates are
often not kin, and that there is little evidence that females discriminate between familiar and unfamiliar individuals. Consequently, any potential
benefits from indirect (inclusive) fitness benefits (*sensu* Hamilton 1964)
are likely to be minimal. We discuss environmental conditions that might
favor communal organization *vis-à-vis* solitary or eusocial behavior due to
direct fitness benefits. Recognizing that successful organisms "solve" environmental "problems" (Wcislo 1989; Odling-Smee, Laland, and Eldman
2003), we consider whether communal social organization is an alternative
solution to tackle the same environmental problems that were solved by
the repeated evolution of caste-based social organizations. We also draw
attention to critical questions for which data are almost totally lacking, and
thus emphasize at the outset that any conclusions are necessarily tentative.

Phyletic Distribution of Communal Behavior

We restrict our discussion to the nest-building Aculeata (Hymenoptera),
excluding ants (i.e., Apoidea including both Apiformes and Spheciformes
[Michener 2007]; Vespidae; and Pompilidae). We do not intend to provide
an exhaustive survey, but cite reviews and representative examples. In
some taxa, communal social organization occurs as an ontogenetic phase
(e.g., co-founding gynes of many eusocial ants, see Heinze and Foitzik, this
volume; Fefferman and Traniello, this volume) or as a temporary condition, while in others communal organization is maintained throughout the
life of a colony. In many "solitary" bees and wasps, some nests are occupied by two or rarely more females. Such species are critical for understanding the evolution of tolerance (Reeve 1989; Moynihan 1998), which is
a fundamental prerequisite for any kind of social organization. Opportunistic communal associations are frequently reported for species in which
nests are spatially clustered in large aggregations (Eickwort 1981; West-
Eberhard 1978; Cowan 1991; Matthews 1991; O'Neill 2001); however,
these species are inherently more likely to attract the attention of interested biologists, and little is known of the relative frequency of opportunistic associations in species that nest in isolation relative to those that
aggregate nests. In other species, females frequently switch nests and transient communal associations occur when two females overlap in the same
nest (Wcislo, Low, and Karr 1985; O'Neill 2001).

Regular communal associations are known in many of the major lineages of bees, including Colletidae (Sakagami and Zucchi 1978; Spessa, Schwarz, and Adams, 2000; *contra* Michener 2007: p. 133, who stated that "all colletids are solitary"); Andrenidae (Paxton, Kukuk, and Tengo 1999; Paxton et al. 1996; Danforth 1989, 1991); Halictinae, especially the agapostemon sweat bees (Roberts 1969; Abrams and Eickwort 1981; Kukuk 1992; Richards, von Wettberg, and Rogers 2003); Nomiinae (Wcislo and Engel 1996); Megachilinae (Garófalo et al. 1992); Xylocopinae (Camillo and Garófalo 1989); and Apinae (Rozen 1984; Soucy, Giray, and Roubik 2003; Cameron 2004). Among wasps, these associations are known for Pompilidae (Wcislo, West-Eberhard, and Eberhard 1988; Evans and Shimizu 1996), Sphecidae (Evans and Hook 1986), and Eumeninae (Cowan 1991). For general reviews, see Michener (1974), Wilson (1975), Iwata (1976), West-Eberhard (1978), Eickwort (1981), Cowan (1991), Matthews (1991), and O'Neill (2001).

A striking fact about the phyletic distribution of communal behavior in aculeate Hymenoptera is that it typically occurs in clades in which there are no examples of caste-based societies (Eickwort 1981; Kukuk and Eickwort 1987; Danforth, Neff, and Baretto-Ko 1996). For example, most (28 of 39) species of nest-sharing wasps discussed by West-Eberhard (1978) are from clades where worker castes have never evolved. One exception to this phyletic pattern involves an intraspecific polyphenism in the sweat bee *Halictus sexcinctus* (Richards et al. 2003), in which both communal and eusocial nests are known from a single locality; an analysis of mitochondrial DNA sequences suggests that the two forms do not represent cryptic species. A second exception is known in a spheciforme wasp taxon (Pemphredonini), in which communal species and the eusocial *Microstigmus* co-occur (see references in Matthews 1991; Wcislo 1992).

Possible Evolutionary Transitions Involving Communal Behavior

West-Eberhard (1978) proposed that nest-sharing, casteless social groups would give rise quickly to polygynous family groups because of inclusive fitness benefits derived from associating with kin. She also hypothesized that once family groups evolve, then they are more likely to persist because kin selection will dampen the disruptive effects of intraspecific parasitic behaviors (e.g., cell and prey usurpation: Eberhard 1972; Eickwort 1975;

Ward and Kukuk 1998). Although she did not discuss them, two phylogenetic predictions follow from West-Eberhard's hypothesis. First, in lineages in which females do not live in family groups, such as communal *Perdita* or *Andrena* bees (e.g., Danforth, Neff, and Barretto-Ko 1996; Paxton et al. 1996), group-living should be evolutionarily labile and species should repeatedly switch from solitary to communal life histories. Phylogenetic studies show that some lineages with caste-based social family groups contain species with secondarily solitary behavior (Wcislo and Danforth 1997; Danforth, Conway, and Ji 2003), though there may be a "point of no return" where it is impossible to lose social behavior (Wcislo and Danforth 1997; Wilson and Hölldobler 2005) without going extinct or becoming an obligate social parasite (but see Chenoweth et al. 2007). In contrast, for lineages of bees and wasps with communal behavior there is no evidence for secondary reversions to solitary behavior, unlike birds in which communal roosting has been lost repeatedly (Beauchamp 1999; Ekman and Ericson 2006).

A second prediction following West-Eberhard (1978) is that obligate brood and social parasites that attack related heterospecifics should evolve more frequently in lineages with females living in nest-sharing, nonfamily groups. The behavioral antecedents to obligate brood parasitism (i.e., opportunistic cheating or robbing) are widespread and occur in species with communal nesting behavior (Wcislo 1987; Field 1992). Lineages that contain many communal species (e.g., andrenid and nomiine bees, *Cerceris* wasps) have not generated any known obligate brood parasites, although communal species serve as hosts for some obligate brood parasites (reviewed in Wcislo 1987).

Costs of Communal Social Organization

The relative costs and benefits of communal living have been discussed repeatedly (Lin and Michener 1972; Eickwort 1981; Cowan 1991; Kukuk 1992), although the empirical data are scant (see summary in Table 7.1). Potential costs fall into two general classes: one associated with increased competition for resources, and one with increased risk of losing those resources to conspecific cheaters.

Resource Competition

As with all social groups, there is likely to be increased competition for resources because conspecifics by definition are the closest competitors for the same resources (Alexander 1974; Dittus 1988). Different females in a communal nest use the same set of floral resources, or the same set of prey resources (Wcislo et al. 1988, Wcislo 1993), so there is potential for competition. There are no studies, however, that assess the impact of resource limitations on the foraging efficiency of communal females, nor are there studies that indicate that any information transfer takes place among them. Indeed, a study of the facultatively communal bee, *Perdita coreopsidis* (Andrenidae), showed that there are no differences between solitary and communal females with respect to duration of foraging trips; time spent within nests between trips; time of day spent foraging; number of trips needed to provision a cell; or the number of cells provisioned per day (Danforth 1989).

Usurpation and Intraspecific Parasitism

The second potential cost for communal females is associated with increased probabilities of nest (or cell) usurpation and intraspecific brood parasitism. Although it is routinely claimed that group-living is associated with such increased risks (e.g., Eickwort 1981; Cowan 1991), few empirical studies have demonstrated that *intra*specific parasitism and usurpation are more likely among communal nesters than among solitary ones (e.g., Eberhard 1972; Eickwort 1975; Wcislo 1987; Field 1992). Nevertheless, anecdotal evidence suggests that the claim is valid. The spider wasp *Auplopus semialatus* (Pompilidae), for example, regularly lives in small communal groups with fewer than eight females; cohabiting females vigorously contest the use of empty brood cells and captured spiders, and repeatedly usurp prey from nestmates (Wcislo, West-Eberhard, and Eberhard 1988). In most species of communal bees and wasps, however, it is extremely difficult to assess true rates of intraspecific parasitism for two reasons. First, many species nest in the soil, which makes it impractical to record behavioral observations without the use of observation nests (e.g., Danforth 1991), and thus it is difficult to observe usurpation. Secondly, genetic markers are not particularly informative in assessing whether cells have been usurped, unless they are coupled with behavioral observations, since they will otherwise show that multiple females are reproducing within a given nest.

Table 7.1. Overview of proposed costs and benefits associated with communal societies, in comparison with an assessment of those costs and benefits associated with caste-based societies. An assessment of available empirical evidence for communal nesting is also given.

Cost/benefit to communal organization	Comparison with caste-based societies	Empirical evidence from communal bees and wasps
Proposed costs		
Competition		
Intense competition within nests for *resources* needed for brood rearing	Less intense competition among principal egg layers	Anecdotal
Conspecific cheating		
Increased *nest parasitism* and cell usurpation	Equivalent risk of intraspecific parasitism among foundresses; some worker reproduction but not phylogenetically widespread	Largely anecdotal
Temporal costs due to extra time spent protecting nests and cells from intraspecific parasites	Lower costs associated with guarding against intraspecific parasites	Largely anecdotal
Proposed benefits		
Improved defense		
Passive *defense* against interspecific parasites and predators	Equivalent	Few studies inconclusive as to whether conspecific solitary nests are worse off than communal ones

Active defense (nest guarding)	Equivalent	Few studies inconclusive as to whether conspecific solitary nests are worse off than communal ones
Energetic savings		
Nest site limitations make it advantageous to join rather than initiate a nest	Equivalent savings	None
Shared cost of general *nest construction*	Equivalent for cofounders and swarm founders	True for some arid zone taxa but not widespread among all communal vs. solitary comparisons
Increased foraging efficiency via *information transfer* amongst nestmates	Equivalent for species with communication among foragers	None

Temporal Constraints

Another potential cost associated with intraspecific parasitism is the loss of time that could be spent on foraging and other tasks rather than guarding cells or nests. A communal spider wasp *Auplopus semialatus,* for example, required up to 120 minutes to seal a brood cell before it was safe from marauding nestmates, suggesting there is a cost to defending a cell (but see Soucy, Giray and Roubik 2003 for a counterexample). In general, however, we lack detailed time budgets for solitary and communal bees and wasps, which are needed to assess how much time is actually lost in guarding a nest or brood cells.

Benefits of Communal Social Organization

Potential benefits associated with group-living in bees and wasps also fall into two general classes: one associated with improved defense, and one associated with energetic savings from shared nest construction and maintenance (see Table 7.1). Benefits associated with improved defense may be passive or active.

Passive Defense

Enhanced passive defenses arise from a dilution effect associated with increasing group size, analogous to Hamilton's (1971) "selfish-herd" arguments. If a parasite or predator attacks a brood cell at random, then the probability that any given cell is attacked is $1/N$, where N is the number of cohabiting females. Establishing nests in soil versus twigs may also shape relative rates of parasitism, if the relative complexity of environmental space (roughly two- versus three-dimensional, respectively) influences the success rate for searching parasites (Matthews 1991; Wcislo 1996). Comparisons involving pairs of related sister taxa showed that ground-nesting species more frequently had higher rates of parasitism than did twig-nesting species of bees and wasps; however, confounding factors suggest that this conclusion should be accepted with caution (Wcislo 1996). If this conclusion is sustained, then group-living should evolve more frequently among soil-dwelling lineages rather than among stick- or mud-nesters (Michener 1985).

Improved defense may also arise as an incidental by-product of increased activity at the nest entrance. In many cases it is difficult to ascertain whether individuals are behaving as functional guards or are merely standing near the nest entrance, which effectively guards it as a by-product of some other behavior. Both solitary and group-nesting females of an orchid bee spent approximately the same amount of time in their nests, but as a result of staggered foraging trips multi-female nests were unoccupied only 1.7% of the time, while solitary nests were unoccupied 30.1% of the time (Soucy, Giray, and Roubik 2003). Satellite flies *(Leucophora)* that entered a bee host nest *(Andrena agilissima)* remained within the nest for shorter periods of time if another host female returned to the nest, suggesting activity *per se* effectively guards a nest (Polidori et al. 2005). In contrast, anecdotal evidence for the facultatively communal bee, *Perdita coreopsidis,* showed that the most populous nest also had the highest rate of cell parasitism (Danforth 1989), implying that more individuals do not translate into better defense, and in fact may render the nest more attractive to parasites and predators.

Active Defense and Nest Guarding

Wilson and Hölldobler (2005) argued that a prime driving force behind the evolution of eusociality are advantages associated with improved defense (also, e.g., Lin and Michener 1972; West-Eberhard 1978). They argued that small groups are better defenders than are solitary individuals, and larger groups are better than smaller groups. The same argument applies to communal nests (Lin and Michener 1972; West-Eberhard 1978; Forbes et al. 2002). Interspecific comparisons are confounded by the fact that the suite of natural enemies which attack solitary or social forms are not always the same; macroparasites and predators may have greater impact on solitary females, while microparasites (e.g., bacteria) may have greater impact on social ones (Wcislo 1997; Schmid-Hempel 1998). Conclusive tests of the idea that groups are more effective defenders than singletons are relatively scarce because they require comparisons of solitary and group-living individuals of the same species in the same location. Furthermore, from a defensive perspective, there is no reason to expect fundamental differences between communal and eusocial groups unless the latter have specialized defender morphs ("soldiers"), which are unknown in bees and wasps.

Active defenses may be associated with increased guarding behavior.

Most nests of the sweat bee *Agapostemon virescens*, for example, were occupied by communally nesting females, and one bee was continuously present at the nest entrance when bees were foraging (Abrams and Eickwort 1981). Communal nests were not attacked by cleptoparasitic bees *(Nomada articulata)*, whereas the parasites entered the only solitary nest in the study when the occupant was out foraging. Likewise, a nest of a communal spider wasp, *Auplopus semialatus*, was unattended for less than a minute during more than 42 hours of observation, and in that brief time an obligate cleptoparasitic wasp, *Irenangelus eberhardi*, successfully oviposited in an open brood cell (Wcislo, West-Eberhard, and Eberhard 1988; for other examples, McCorquodale 1989a; Garófalo et al. 1992; Spessa, Schwarz, and Adams 2000). In contrast, females of the obligately communal bee, *Perdita portalis* (Andrenidae), were never observed guarding nests, suggesting that improved nest defense is unlikely to be associated with the maintenance of communal behavior in this species (Danforth 1991).

Nest Site Limitations

It may be advantageous to share a nest with others if nest sites are rare or nests are difficult to establish (Michener 1974; McCorquodale 1989a, 1989b). The availability of nesting substrata helps shape the community-level composition of Mediterranean bee communities (Potts et al. 2005), suggesting that nest site availability may be a limiting resource (Schwarz, Bull, and Hogendoorn 1998; Langer, Hogendoorn, and Keller 2004 for eusocial bees). It is not clear, however, whether suitable nesting substrata are more limiting for communal versus solitary species.

Sharing the Costs of Nest Construction

Various researchers have noted that communal nesting seems to be especially common in Australian halictine bees and sphecid wasps (e.g., Knerer and Schwarz 1976; Evans and Hook 1986), and andrenid bees in the southwest deserts of North America and Mexico (e.g., Danforth 1991), where they frequently nest in very hard and compact soil. In such regions nests tend to be initiated following rains when the soil is soft (McCorquodale 1989b). Although there are no studies to ascertain the relative costs associated with establishing nests in wood or soil, or in relatively hard versus soft soil, causal links between substrate hardness and nest sharing have been proposed repeatedly (Evans and Hook 1986; McCorquodale 1989a, 1989b; Wcislo, Fernandez-Marin, and di Trani 2004). In communal *Perdita* bees, for example, Danforth (1991) suggested that nest sharing may be advantageous due to the energetic and temporal costs associated with solitary nest excavation. Danforth also noted that if the period of resource availability is relatively brief, there may be temporal factors that select against solitary nesting, assuming that it takes a solitary individual more time to dig a nest relative to joining an already established one. In sphecid wasps, however, spatial patterns of soil hardness were not associated consistently with patterns of nest provisioning behavior (see Wcislo, Low, and Karr 1985; McCorquodale 1989b, and references therein).

Information Transfer

In birds, increased foraging efficiency is hypothesized to be one of the prime advantages associated with communal roosting (Ward and Zahavi 1973; Beauchamp 1999), assuming that roosts act as centers for information transfer whereby unsuccessful foragers can follow successful ones to a feeding site. To date there are no behavioral studies showing that communal bees and wasps use cues from foraging nestmates to reduce search time, as is well-known for numerous eusocial insects.

Nest Productivity and Communal Behavior

Measures of per capita productivity imply that there are either benefits or costs to group nesting, depending on whether productivity is an increasing or decreasing function of group size (Michener 1974). In many taxa with behavioral castes there is an overall decrease in per capita productivity with increasing group size (Michener 1974; Karsai and Wenzel 1998), while in other taxa, especially in eusocial allodapine bees, there is increased productivity with initial increases in group size (Schwarz, Bull, and Hogendoorn 1998; Tierney, Schwarz, and Adams 1997; Tierney et al. 2002). In a communal colletid bee, *Amphylaeus morosus,* per capita productivity was not significantly different for communal or solitary nests (Spessa et al. 2000; also e.g., Danforth 1989). Published estimates of productivity may be biased, however, if solitary nests or those with fewer females suffer higher rates of nest failure than those with more females, or if brood development is not followed through to the adult stage.

Kinship and Recognition Systems in Communal Societies

An ability to recognize and discriminate between nestmates and non-nestmates (frequently kin versus non-kin) is widespread in caste-based social insect groups (Fletcher and Michener 1987). West-Eberhard (1978) reviewed early literature on nest-sharing wasps and argued that about half of the 29 species she tabulated lived in groups that were comprised of relatives, based on behavioral inferences (e.g., frequent re-use of cells, low rates of dispersal) and on theoretical grounds: if group living is advantageous, then family-based group living will be favored because of additional indirect fitness benefits that accrue via kin selection, and genes that enable associations with relatives will be shared by those relatives (Hamilton 1964; Wilson and Hölldobler 2005). More recent studies using genetic markers are inconsistent with her arguments, and instead have shown that nest-sharing, casteless groups often are comprised of nonrelatives (Kukuk and Sage 1994; Danforth, Neff, and Barretto-Ko 1996; Paxton et al. 1996; Spessa, Schwarz, and Adams 2000; Kukuk, Bitney, and Forbes 2005; but see McCorquodale 1988 and Pfennig and Reeve 1993 for examples of kin associations).

Few studies are available that assess recognition capabilities in communal species for comparison with caste-based societies, or in solitary species that occasionally share nests (e.g., Pfennig and Reeve 1989). Kukuk and co-workers studied patterns of food exchange in a communal Australian bee (*Lasioglossum hemichalceum;* Halictidae) and found no tendency to preferentially direct food toward familiar individuals or nestmates (Kukuk and Crozier 1990; Kukuk 1992). Furthermore, females from distant nests can be introduced into another nest without evidence of fighting, and they begin to provision cells in the new nest (Ward and Kukuk 1998; Wcislo, personal observation), again suggesting the lack of any discrimination. A lack of discrimination between familiar and unfamiliar females was also observed during staged encounters for two species of communal andrenid bees (*Andrena* and *Panurgus*) (Paxton, Kukuk, and Tengo 1999). Females have been reported to join nests in other communal bees and wasps with no signs of aggression or guarding by the resident females (Danforth 1991; Abrams and Eickwort 1981), implying that communal nests generally are relatively open societies.

The occurrence of open societies raises the question of whether group members are incapable of recognizing familiar individuals. There is a biased phyletic distribution of caste-based societies among bees and sphe-ciforme wasps (Apoidea) (Wcislo 1992), in that eusociality has evolved re-peatedly in the former, while it is extremely rare in the latter. Wcislo (1992) hypothesized that this pattern was associated with the phyletic distribu-tion of chemically-mediated individual nest recognition as an evolutionary antecedent for kin recognition mechanisms. If valid, this hypothesis in turn suggests that studies are needed to assess such nest recognition capabili-ties (or lack thereof) in communal lineages of bees and wasps, which might help explain why lineages with communal behavior rarely generate euso-cial species.

Conclusions and Future Directions

Theory and limited empirical evidence suggest that the main advantage derived from group living is improved defense against predators and para-sites. The evolution of communal nesting as a stable state is therefore par-adoxical in that if individuals benefit from group nesting, they would then further benefit from doing so with relatives (West-Eberhard 1978; Wilson and Hölldobler 2005). Yet, as mentioned above, studies using genetic markers indicate that communal nest-sharing females often are not rela-tives. On the other hand, if individuals benefit from cooperating with non-relatives, then they would further benefit by cheating and exploiting the cooperative behavior of those non-relatives (Eberhard 1972; Eickwort 1975; Axelrod 1984). A major unresolved empirical question, therefore, is what limits or constrains cheating in such communal societies of non-relatives?

Avilles (2002) developed a model in which tendencies to cooperate and to form groups dynamically co-evolve, and she showed that the problem of cheaters ("freeloaders") is resolved if there are significant group-size ef-fects on fitness. That is, cheaters will increase in frequency when rare, but then decrease in frequency when they are common because groups that contain excessive cheaters will have lower per capita productivity. Avilles' model shows that per capita productivity increases up to a certain group size (~8), and then decreases. Empirical data on per capita productivity for different size social groups varies considerably (see section Nest Produc-tivity and Communal Behavior). The effect of freeloading on per capita productivity is empirically unknown. In a social sphecid wasp, *Trigonopsis cameronii,* females occasionally robbed prey from nestmates, especially when their own hunting success was poor, which raises the possibility that freeloading in these wasps may actually enhance productivity (Eberhard 1972), but additional studies are needed.

If group living is beneficial, then what constellation of genetic and environmental factors helps explain tendencies to evolve communal- versus caste-based societies? Many examples of caste-based societies involve mother-daughter associations (i.e., matrifilial eusocial groups). These associations form during a window of opportunity determined by egg-to-adult developmental times, adult longevity, and the length of the local growing season (Wcislo and Danforth 1997). For example, if larval development is too slow relative to adult longevity, or relative to the length of the growing season, then overlap of generations will be precluded. We speculate that communal behavior is an alternative form of social organization especially suited to environments with short growing seasons or where the length of the growing season is relatively unpredictable. This hypothesis is consistent with the observation that arid regions tend to be especially rich in species with communal behavior, though quantitative data are lacking.

Nearly 50 years ago Evans (1958) synthesized available information and theory concerning the evolution of group living in wasps, and he pointed out that "speculation on the origin of social life seems to have outstripped the observational data (p. 457)." Twenty years later, in a like-minded pursuit, West-Eberhard (1978, p. 853) wrote that data were still so scarce that Evans' sentence "is now a model of understatement." More recently, in reviews of primitively social wasps, Cowan (1991, p. 73) lamented how "much of the information about these insects consists of the barest anecdotes," while Matthews (1991, p. 601) listed critical factors for understanding social evolution and he noted that data "are virtually nonexistent." Similar concerns hold true for bees (Wcislo and Engel 1996). In a book dedicated to Bert Hölldobler, who has done so much to advance our understanding of the origins and evolution of social behavior in insects, one would like to end in a positive manner. Unfortunately, our review ends on a note that echoes Evans, West-Eberhard, Cowan, and Matthews, because empirical studies of solitary and communal nest-sharing bees and wasps continue to lag behind the rest of the field, despite their critical position for understanding the origins of social behavior.

Ironically, the most fitting tribute to one who has so eloquently argued for making a "journey to the ants" is to veer away from ants and their highly eusocial counterparts among the bees and wasps, and turn instead to the bees and wasps that have made but a short journey in the realm of sociality.

Literature Cited

Abrams, J., and G. C. Eickwort. 1981. "Nest switching and guarding by the communal sweat bee *Agapostemon virescens* (Hymenoptera, Halictidae)." *Insectes Sociaux* 28: 105–116.

Alexander, R. D. 1974. "The evolution of social behavior." *Annual Review of Ecology and Systematics* 5: 325–383.

Avilles, L. 2002. "Solving the freeloaders paradox: Genetic associations and frequency-dependent selection in the evolution of cooperation among non-relatives." *Proceedings of the National Academy of Sciences USA* 99: 14268–14273.

Axelrod, R. 1984. *The evolution of cooperation.* New York: Basic Books.

Beauchamp, G. 1999. "The evolution of communal roosting in birds: Origin and secondary losses." *Behavioral Ecology* 10: 675–687.

Brockmann, H. J. 1997. "Cooperative breeding in wasps and vertebrates: The role of ecological costraints." In J. C. Choe and B. J. Crespi, eds., *The evolution of social behavior in insects and arachnids,* 347–371. Cambridge: Cambridge University Press.

Cameron, S. A. 2004. "Phylogeny and biology of neotropical orchid bees (Euglossini)." *Annual Review of Entomology* 49: 377–404.

Camillo, E., and C. A. Garófalo. 1989. "Social organization in reactivated nests of three species of *Xylocopa* (Hymenoptera, Anthophoridae) in southeastern Brasil." *Insectes Sociaux* 36: 92–105.

Carpenter, J. M. 1989. "Testing scenarios: Wasp social behavior." *Cladistics* 5: 131–144

Chenoweth, L. B., S. M. Tierney, J. A. Smith, S. J. B. Cooper, and M. P. Schwarz. 2007. "Social complexity in bees is not sufficient to explain lack of reversions to solitary living over long time scales." *BMC Evolutionary Biology* 7: 246.

Costa J. T., and T. D. Fitzgerald. 2005. "Social terminology revisited: Where are we ten years later?" *Annales Zoologica Fennici* 42: 559–564.

Cowan, D. P. 1991. "The solitary and presocial Vespidae." In K. G. Ross and R. W. Matthews, eds., *The social biology of wasps,* 33–73. Ithaca: Cornell University Press.

Danforth, B. N. 1989. "Nesting behavior of four species of *Perdita* (Hymenoptera, Andrenidae)." *Journal of the Kansas Entomological Society* 62: 59–79.

———. 1991. "Female foraging and intranest behavior of a communal bee, *Perdita portalis* (Hymenoptera: Andrenidae)." *Annals of the Entomological Society of America* 84: 537–548.

Danforth, B. N., L. Conway, and S. Ji. 2003. "Phylogeny of eusocial *Lasioglossum* reveals multiple losses of eusociality within a primitively eusocial clade of bees (Hymenoptera: Halictidae)." *Systematic Biology* 52: 23–36.

Danforth, B. N., J. L. Neff, and P. Barretto-Ko. 1996. "Nestmate relatedness in a communal bee, *Perdita texana* (Hymenoptera: Andrenidae), based on DNA fingerprinting." *Evolution* 50: 276–284.

Darwin, C. [1859] 1964. *On the origin of species.* Reprint. Cambridge: Harvard University Press.

de Waal, F., and P. Tyack, eds. 2003. *Animal social complexity.* Cambridge: Harvard University Press.

Dittus, W. P. J. 1988. "Group fission among wild toque macaques as a consequence of female resource competition and environmental stress." *Animal Behaviour* 36: 1626–1645.

Eberhard, W. 1972. "Altruistic behavior in a sphecid wasp: Support for kin-selection theory." *Science* 172: 1390–1391.

Eickwort, G. C. 1975. "Gregarious nesting of the mason bee *Hoplitis anthocopoides* and the evolution of parasitism and sociality among megachilid bees." *Evolution* 29: 142–150.

Eickwort, G. C. 1981. "Presocial insects." In H. R. Hermann, ed., *Social insects,* 2: 199–280. New York: Academic Press.

Ekman, J., and P. G. P. Ericson. 2006. "Out of Gondwanaland: The evolutionary history of cooperative breeding and social behaviour among crows, magpies, jays and allies." *Proceedings of the Royal Society B* 273: 1117–1125.

Evans, H. E. 1958. "The evolution of social life in wasps." *Proceedings of the 10th International Congress of Entomology* 2: 449–457.

Evans, H. E., and A. W. Hook. 1986. "Nesting behavior of Australian *Cerceris* digger wasps, with special reference to nest reutilization and nest sharing (Hymenoptera, Sphecidae)." *Sociobiology* 11: 275–302.

Evans, H. E., and A. Shimizu. 1996. "The evolution of nest building and communal nesting in Ageniellini (Insecta: Hymenoptera: Pompilidae)." *Journal of Natural History* 30: 1633–1648.

Field, J. 1992. "Intraspecific parasitism as an alternative reproductive tactic in nest-building wasps and bees." *Biological Reviews* 67: 79–126.

Fletcher, D. J. C., and C. D. Michener, eds. 1987. *Kin recognition in animals.* Chichester: John Wiley and Sons.

Forbes, S. H., R. M. M. Adams, C. Bitney, and P. F. Kukuk. 2002. "Extended parental care in communal social groups." *Journal of Insect Science* 2: 22–28.

Garófalo, C. A., E. Camillo, M. J. O. Campos, and J. C. Serrano. 1992. "Nest re-use and communal nesting in *Microthurge corumbae* (Hymenoptera, Megachilidae), with special reference to nest defense." *Insectes Sociaux* 39: 301–311.

Hamilton, W. D. 1964. "The genetical evolution of social behaviour, *Part I and II*." *Journal of Theoretical Biology* 7: 1–16; 17–52.

———. 1971. "Geometry for the selfish herd." *Journal of Theoretical Biology* 31: 295–311.

Hayes, L. D. 2000. "To nest communally or not to nest communally: A review of rodent communal nesting and nursing." *Animal Behaviour* 59: 677–688.

Iwata, K. 1976. *Evolution of instinct: Comparative ethology of Hymenoptera.* New Delhi: Amerind Publishing Co.

Jeanson, R. L., P. F. Kukuk, and J. H. Fewell. 2005. "Emergence of division of labour in halictine bees: Contributions of social interactions and behavioural variance." *Animal Behaviour* 70: 1183–1193.

Karsai, I., and J. W. Wenzel. 1998. "Productivity, individual-level and colony-level flexibility, and organizations of work as consequences of colony size." *Proceedings of the National Academy of Sciences USA* 95: 8665–8669.

Knerer, G., and M. P. Schwarz. 1976. "Halictine social evolution: The Australian enigma." *Science* 194: 445–448.

Kukuk, P. F. 1992. "Social interactions and familiarity in a communal halictine bee *Lasioglossum hemichalceum*." *Ethology* 91: 291–300.

Kukuk P. F., C. Bitney, and S. H. Forbes. 2005. "Maintaining low intragroup relatedness: Evolutionary stability of non-kin social groups." *Animal Behaviour* 70: 1305–1311.

Kukuk, P. F., and R. H. Crozier. 1990. "Trophallaxis in a communal halictine bee *Lasioglossum (Chilalictus) erythrurum*." *Proceedings of the National Academy of Science USA* 87: 5402–5404.

Kukuk, P. F., and G. C. Eickwort. 1987. "Alternative social structures in halictine bees. In J. Elder and H. Rembold, eds., *Chemistry and biology of social insects, Proceedings of the X International Congress IUSSI (1986, Munich)*, 555–556. Munich: Verlag J. Peperny.

Kukuk, P. F., and G. K. Sage. 1994. "Reproductivity and relatedness in a communal halictine bee *Lasioglossum (Chilalictus) hemichalceum*." *Insectes Sociaux* 41: 443–455.

Langer, P., K. Hogendoorn, and L. Keller. 2004. "Tug-of-war over reproduction in a social bee." *Nature* 428: 844–847.

Lin, N., and C. D. Michener. 1972. "Evolution of sociality in insects." *Quarterly Review of Biology* 47: 131–159.

Linksvayer, T. A., and M. J. Wade. 2005. "The evolutionary origin and elaboration of sociality in the aculeate Hymenoptera: Maternal effects, sib-social effects, and heterochrony." *Quarterly Review of Biology* 80: 317–336.

Matthews, R. W. 1991. "Evolution of social behavior in sphecid wasps." In
    K. G. Ross and R. W. Matthews, eds., *The social biology of wasps*, 570–602.
    Ithaca: Cornell University Press.

McCorquodale, D. B. 1988. "Relatedness among nestmates in a primitively
    social wasp, *Cerceris antipodes* (Hymenoptera: Sphecidae)." *Behavioral
    Ecology and Sociobiology* 23: 401–406.

McCorquodale, D. B. 1989a. "Nest defense in single and multifemale nests of
    Cerceris antipodes (Hymenoptera: Sphecidae)." *Journal of Insect Behavior*
    2: 267–276.

———. 1989b. "Soil softness, nest initiation and nest sharing in the wasp
    Cerceris antipodes (Hymenoptera: Sphecidae)." *Ecological Entomology* 14:
    191–196.

Michener, C. D. 1974. *The social behavior of the bees.* Cambridge: Harvard
    University Press.

———. 1985. "From solitary to eusocial: Need there be a series of intervening
    species?" *Fortschrifte der Zoologie* 31: 293–306.

———. 2007. *The bees of the world* (second edition). Baltimore: The Johns
    Hopkins University Press.

Moynihan, M. H. 1998. *The social regulation of competition and aggression in
    animals.* Washington, DC: Smithsonian Institution Press.

Odling-Smee, F. J., K. N. Laland, and M. W. Feldman. 2003. *Niche construc-
    tion: The neglected process in evolution.* Princeton: Princeton University
    Press.

O'Neill, K. M. 2001. *Solitary wasps: Behavior and natural history.* Ithaca:
    Cornell University Press.

Paxton, R. J., P. F. Kukuk, and J. Tengö. 1999. "Effects of familiarity and nest-
    mate number on social interactions in two communal bees, *Andrena scotica*
    and *Panurgus calcaratus* (Hymenoptera, Andrenidae)." *Insectes Sociaux* 46:
    109–118.

Paxton, R. J., P. A. Thoren, J. Tengö, A. Estoup, and P. Pamilo. 1996. "Mating
    structure and nestmate relatedness in a communal bee, *Andrena jacobi*
    (Hymenoptera, Andrenidae), using microsatellites." *Molecular Ecology* 5:
    511–519.

Pfennig, D. W., and H. K. Reeve. 1989. "Neighbor recognition and context-
    dependent aggression in a solitary wasp, *Sphecius speciosus* (Hymenoptera:
    Sphecidae)." *Ethology* 80: 1–18.

———. 1993. "Nepotism in a solitary wasp as revealed by DNA fingerprint-
    ing." *Evolution* 47: 700–704.

Polidori, C., B. Scanni, E. Scamoni, M. Giovanetti, F. Andrietti, and R. J.
    Paxton. 2005. "Satellite flies (*Leucophora personata*, Diptera:

Anthomyiidae) and other dipteran parasites of the communal bee *Andrena agilissima* (Hymenoptera: Andrenidae) on the island of Elba, Italy." *Journal of Natural History* 39: 2745–2758.

Potts, S. G., B. Vulliamy, S. Roberts, C. O'Toole, A. Dafni, G. Ne'eman, and P. Willmer. 2005. "Role of nesting resources in organising diverse bee communities in a Mediterranean landscape." *Ecological Entomology* 30: 78–85.

Reeve, H. K. 1989. "The evolution of conspecific acceptance thresholds." *American Naturalist* 133: 407–435.

Richards, M. H., E. J. von Wettberg, and A. C. Rutgers. 2003. "A novel social polymorphism in a primitively eusocial bee." *Proceedings of the National Academy of Sciences USA* 100: 7175–7180.

Roberts, R. B. 1969. "Biology of the bee genus *Agapostemon* (Hymenoptera, Halictidae)." *University of Kansas Science Bulletin* 48: 689–719.

Rozen, J. G., Jr. 1984. "Comparative nesting biology of the Exomalopsini." *American Museum Novitates* 2798: 1–37.

Sakagami, S. F., and R. Zucchi. 1978. "Nests of *Hylaeus (Hylaeopsis) tricolor*: The first record of non-solitary life in colletid bees, with notes on communal and quasi-social colonies (Hymenoptera, Colletidae)." *Journal of the Kansas Entomological Society* 51: 597–614.

Schmid-Hempel, P. 1998. *Parasites and social insects.* Princeton: Princeton University Press.

Schwarz, M. P., N. J. Bull, and S. J. B. Cooper. 2003. "The molecular phylogenetics of allodapine bees, with implications for the evolution of sociality and progressive rearing." *Systematic Biology* 52: 1–14.

Schwarz, M. P., N. J. Bull, and K. Hogendoorn. 1998. "Evolution of sociality in the allodapine bees: A review of sex allocation, ecology and evolution." *Insectes Sociaux* 45: 349–368.

Schwarz, M. P., M. H. Richards, and B. N. Danforth. 2007. "Changing paradigms in insect social evolution: Insights from halictine and allodapine bees." *Annual Review of Entomology* 52: 127–150.

Shimizu, A. 2004. "Natural history and behavior of a Japanese parasocial spider wasp, *Machaerithrix tsushimensis* (Hymenoptera: Pompilidae)." *Journal of the Kansas Entomological Society* 77: 383–401.

Soucy, S. L., T. Giray, and D. W. Roubik. 2003. "Solitary and group nesting in the orchid bee *Euglossa hyacinthina* (Hymenoptera, Apidae)." *Insectes Sociaux* 50: 248–255.

Spessa, A., M. P. Schwarz, and M. Adams. 2000. "Sociality in *Amphylaeus morosus* (Hymenoptera: Colletidae: Hylaeinae)." *Annals of the Entomological Society of America* 93: 684–692.

Tierney, S. M., M. P. Schwarz, and M. Adams. 1997. "Social behaviour in an

Australian allodapine bee *Exoneura (Brevineura) xanthoclypeata* (Hymenoptera: Apidae)." *Australian Journal of Zoology* 45: 384–398.

Tierney, S. M., M. P. Schwarz, T. Neville, and P. M. Schwarz. 2002. "Sociality in the phylogenetically basal allodapine bee genus *Macrogalea* (Apidae, Xylocopinae): Implications for social evolution in the tribe Allodapini." *Biological Journal of the Linnean Society* 76: 211–224.

Ward, S. A., and P. F. Kukuk. 1998. "Context-dependent behavior and the benefits of communal nesting." *American Naturalist* 152: 249–263.

Ward, P., and A. Zahavi. 1973. "The importance of certain assemblages of birds as information centers for food finding." *Ibis* 115: 517–534.

Wcislo, W. T. 1987. "The roles of seasonality, host synchrony, and behaviour in the evolutions and distributions of nest parasites in Hymenoptera (Insecta), with special reference to bees (Apoidea)." *Biological Reviews of the Cambridge Philosophical Society* 62: 415–443.

———. 1989. "Behavioral environments and evolutionary change." *Annual Review of Ecology and Systematics* 20: 137–169.

———. 1992. "Nest localization and recognition in a solitary bee, *Lasioglossum figueresi* Wcislo (Hymenoptera: Halictidae), in relation to sociality." *Ethology* 92: 108–123.

———. 1993. "Communal nesting in a North American pearly-banded-bee, *Nomia tetrazonata*, with notes on nesting behavior of *Dieunomia heteropoda* (Hymenoptera: Halictidae: Nomiinae)." *Annals of the Entomological Society of America* 86: 813–821.

———. 1996. "Rates of parasitism in relation to nest site in bees and wasps (Hymenoptera: Apoidea)." *Journal of Insect Behavior* 9: 643–656.

———. 1997. "Behavioral environments of sweat bees (Halictinae) in relation to variability." In J. C. Choe and B. J. Crespi, eds., *The evolution of social behavior in insects and arachnids*, 316–332. Cambridge: Cambridge University Press.

———. 2005. "Social labels: We should emphasize biology over terminology and not *vice versa*." *Annales Zoologici Fennici* 42: 565–568.

Wcislo, W. T., and B. N. Danforth. 1997. "Secondarily solitary: The evolutionary loss of social behavior." *Trends in Ecology and Evolution* 12: 468–474.

Wcislo, W. T., and M. S. Engel. 1996. "Social behavior and nest architecture of nomiine bees (Hymenoptera: Halictidae: Nomiinae)." *Journal of the Kansas Entomological Society* 69(Suppl.): 158–167.

Wcislo, W. T., H. Fernandez-Marin, and J. C. di Trani. 2004. "Communal use of nests by male and female *Trachypus petiolatus* (Hymenoptera: Sphecidae)." *Journal of the Kansas Entomological Society* 77 (Suppl.): 323–331.

Wcislo, W. T., B. S. Low, and C. J. Karr. 1985. "Parasite pressure and repeated burrow use by different individuals of *Crabro* (Hymenoptera: Sphecidae; Diptera: Sarcophagidae)." *Sociobiology* 11: 115–126.

Wcislo, W. T., M. J. West-Eberhard, and W. G. Eberhard. 1988. "Natural history and behavior of a primitively social wasp, *Auplopus semialatus,* and its parasite, *Irenangelus eberhardi* (Hymenoptera: Pompilidae)." *Journal of Insect Behavior* 1: 247–260.

West-Eberhard, M. J. 1978. "Polygyny and the evolution of social behavior in wasps." *Journal of the Kansas Entomological Society* 51: 832–856.

West-Eberhard, M. J. 2003. *Developmental plasticity and evolution.* Oxford: Oxford University Press.

Wheeler, W. M. 1923. *The social insects: Their origin and evolution.* New York: Kegan, Paul, Trench, Trubner and Co.

Wilson, E. O. 1971. *The insect societies.* Cambridge: Harvard University Press.

———. 1975. *Sociobiology.* Cambridge: Harvard University Press.

Wilson, E. O., and B. Hölldobler. 2005. "Eusociality: Origin and consequences." *Proceedings of the National Academy of Sciences USA* 102: 13367–13371.

PART TWO

Communication

THOMAS D. SEELEY

ONE KEY TO UNDERSTANDING how an insect society is organized—how it functions as an integrated whole—is elucidating its communication processes. Communication is critical because each individual within a society needs information to decide, moment-by-moment, how to behave so as to contribute to the common good. In particular, an individual needs to be sufficiently well informed to decide correctly and repeatedly what task to perform and how to do this task. While it is true that much of this information is obtained without social intercourse—for example, a worker honey bee can sense by herself her particular location in the nest (i.e., brood nest, honeycombs, dance floor), the particular time (i.e., time of day, season of the year), the particular behavioral context (i.e., defending the nest, tending the brood, resting), and her particular social identity (i.e., age, experience, physiological state)—it is also true that much of the needed information is obtained through social interactions, especially by receiving signals produced by nestmates. Within most insect societies there is an extensive overlap of the reproductive interests of individuals, so it is not surprising that social insects have evolved many special means for sharing information: signals for communicating.

It is appropriate that communication is a prominent part of this monograph, both because communication is a central feature of insect social life and because the analysis of social insect communication is a strong theme in the scientific work of Bert Hölldobler, who is honored by this book. Indeed, more than 75% of the several hundred papers published by Hölldobler are investigations of communication. More importantly, his papers have shown us how to conduct a beautiful study of a communication system in an insect society: first provide a careful description of the (often multimodal) signals involved, and then perform an incisive experimental analysis of how the signaling system contributes to colony functioning. Who is not awed by Hölldobler's gorgeously stained sagittal sections of ants showing their exocrine glands, by his vivid illustrations of the mechanical components of ants' recruitment displays, or by his ingenious unraveling of the five distinct recruitment systems of weaver ants?

Chapters 8 through 13 highlight six important directions in current studies of social insect communication. Michael Breed and Robert Buchwald report on communication for nestmate recognition, a process that is critically important in keeping acts of altruism aimed at relatives. They concentrate on the intriguing puzzles of how diversity among recognition cues is generated and is maintained over evolutionary time. Christoph Kleineidam and Wolfgang Röseler review the functional organization of olfactory systems in social Hymenoptera, and show how the neuroanatomy of this sensory system has evolved to complement the evolution of physical castes in certain ant species. Christian Peeters and Jürgen Liebig summarize current work on queen pheromones and look closely at the evidence that these pheromones provide workers with useful information about their queen's fertility, as opposed to chemically manipulating the workers' reproductive activities. Robert Jeanne provides a much-needed synthesis of what is known about mechanical signals in the societies of social wasps, whose carton nests are particularly favorable for the transmission of substrate vibrations. James Nieh makes a novel comparison of the mechanisms of recruitment to food sources in bees and wasps, carefully comparing the various mechanisms that have evolved for activating foragers in the nest and then guiding them to the food outside the nest. And finally, Flavio Roces reviews the organization of foraging by ant colonies, reminding us that foragers bring home information as well as food, and that forager behavior is likely adapted to optimize the intake of both commodities.

CHAPTER EIGHT

Cue Diversity and Social Recognition

MICHAEL D. BREED

ROBERT BUCHWALD

ONE OF THE MOST INTRIGUING problems in evolution is understanding how some processes favor high phenotypic diversity. Two arenas in which phenotypes are particularly diverse are in cellular recognition in immune systems (Penn and Potts 1999) and cues used in social recognition (Downs and Ratnieks 2000). Our particular interest is in the phenotypic diversity of cues used in social recognition in eusocial insects, and in this chapter we address three questions: How many cue phenotypes are necessary for social recognition, how is this phenotypic diversity achieved, and what impact do mistakes or limitations in social recognition have on social insect biology (Ratnieks 1991)? We focus on nestmate recognition in social insects (Vander Meer and Morel 1998), but the principles we derive are broadly applicable across animal taxa. Ultimately, we argue that phenotypic diversity in social recognition is the result of a selective balance between needs for accuracy in identification (Downs and Ratnieks 2000) and the limitations of the sensory and information processing systems. Following this reasoning, we employ probabilistic models (Millor et al. 2006) to predict the minimum number of phenotypes needed for effective social recognition.

Recognition is one of the few evolutionary contexts in which selection favors phenotypic diversity (Kelley, Walter, and Trowsdale 2005; Aertsen and Michiels 2005). Persistence of diverse phenotypes in evolutionary time is important in both intrinsic recognition systems such as the immune system and in social recognition among animals. In fact, the association of the major histocompatiblity complex (MHC) with social recognition (Penn and Potts 1999) intertwines, at a fundamental level, the immune system and social recognition across a large segment of Animalia. In immune systems, genetic mechanisms that foster the generation of diversity are important (Kelley, Walter, and Trowsdale 2005; Aertsen and Michiels 2005); immune recognition moves beyond neutral substitution to active amplification of phenotypic diversity. In this chapter we consider how diversity in recognition phenotypes might be generated in social insects, and we develop a model that contrasts with models that invoke active generation of phenotypic diversity that are used to explain immunological diversity.

Social Recognition

Social recognition shapes behavioral interactions among animals in a fundamental way. It is a key ingredient in diverse social contexts, such as nestmate recognition in eusocial insects, identification of herd or flock members in mammals and birds, mating-partner recognition, sibling recognition, and parent-offspring recognition (Tsutsui 2004). In the following discussion, we focus on identification of nestmates by social insects (Vander Meer and Morel 1998); this type of social recognition takes place at the colony level, typically with all members of a colony carrying the same identifying cues, regardless of genetic similarity. The principles we develop, however, apply broadly in social recognition and can operate at the individual level as well as at the colony level.

Social recognition can be viewed as a series of gradations from relatively coarse-grained discriminations to very finely tuned processes. Social recognition begins with species recognition, then gender recognition, social group or family recognition, and finally individual recognition. Species and gender recognition normally rely on a small number of features that are relatively invariant among members of the species or gender; these features are often thought of as recognition badges (e.g., Goth and Hauber 2004). In species recognition, variability among species is paramount, and selection typically favors differentiation of species (thereby preventing mating errors). Interspecific elaboration is more likely when phylogenetically associated species live sympatrically. Species within a genus may be restricted to the same overall morphology but elaborate differences in coloration allow obvious specific identifications. Gender discrimination is similar to species recognition in that male-female differences in a few key features are usually adequate. In some, perhaps many, species, specific and gender identification is mediated by a single badge, such as a species- and gender-unique pheromonal blend or a color signal (Sinervot et al. 2006).

In contrast, social group, family, or individual recognition can only function if phenotypes vary among groups, families, or individuals within a population. It is at this level of recognition that phenotypic variation is critical.

Cue Diversity

How much information does a eusocial insect need to identify its social companions? On the surface, this is a remarkably easy question to answer. There must be enough unique combinations of cues to give each colony in the population a unique identifier. The simplest system would operate like telephone numbers, assigning an identifier to each group of animals. In such a circumstance, the number of identifiers available need only match the number of groups in the population.

There are a number of reasons, though, that cues used in social recognition are not fully analogous to identifiers like telephone numbers. Foremost, there is no executive function, at the population level, to assign identifiers. Instead, identifiers are drawn from a preexisting resource pool and may therefore be duplicated within the pool; this means that although there are a finite number of identifiers to choose from, there must be enough so that the probability of two or more individuals in the population carrying the same identifier is tolerably low.

What defines "tolerably low?" To answer this question we must first look at the kinds of mistakes that can occur in nestmate recognition. A related animal can be erroneously treated as non-kin, or an unrelated animal can be erroneously treated as kin. An overly sensitive recognition system is likely to lead to the first type of error, and a recognition system that is not sensitive enough is likely to lead to the second type. Agrawal (2001) likened the balance between these two types of errors to Type I and Type II statistical errors. This approach to thinking about the costs and benefits of nestmate recognition has considerable merit, and we return to the issue of error rates below in the section on modeling cue diversity. A particularly important point is that the costs and benefits affect both residents and intruders. Both types of mistakes have costs, which are balanced by the cost of producing and processing the information needed to avoid errors in discrimination.

Decisions made by foragers entering colonies are a function of nest or colony recognition, which may rely on the same cues as nestmate recognition by guards (Breed 1998; Vander Meer and Morel 1998). Foragers may make unintentional errors in entering the wrong nest; in western honey bees, *Apis mellifera,* this "worker drift" (movement of adult workers from their natal colony to neighboring colonies) is thought to be unintentional, but can result in real costs to colonies that lose workers and corresponding benefits to colonies that gain workers (Pfeiffer and Crailsheim 1998). For social insect species that aggregate their colonies (nests are located in close proximity; this is reasonably common in eusocial bees and wasps, but rare or absent in termites and ants), colony identification and potential unintentional errors in colony identification are, at least in theory, quite important. In addition to drift, intentional attempts to enter other colonies, with the purpose of robbing food or brood, social parasitism, or of displacing the resident queen, are also common. This behavior is found in *A. mellifera* (Breed, Diaz, and Lucero 2004) and in a variety of ants (Lenoir et al. 2001) and may be found in other taxa as well (Ratnieks et al. 2006). If incursions into other nests never occurred, there would be no need for nestmate recognition cues, or for guards to refuse entrance to non-nestmates; these intrusions are the primary force driving the evolution of nestmate recognition. Any mistake in colony identification or purposeful entry into the wrong colony by a forager leads to possible exclusion by a guard. As the stakes increase due to higher potential losses, the expression of recognition should become more intense (Downs and Ratnieks 2000). As Agrawal (2001) pointed out, the expression of nestmate recognition by guards is the evolutionary result of the balance between the costs and benefits associated with making correct decisions or mistakes of either type.

Perhaps an overlooked point that emerges from this consideration is the costs of allowing a diseased or parasite-carrying worker to enter a colony and the corresponding benefits from excluding such a worker. To our knowledge, the reactions of guards to diseased or parasite-ladened workers have not been sampled; this is an interesting point for future inquiry.

A hidden, or less well understood, cost-benefit relationship in nestmate recognition comes from the costs of producing and analyzing cues of the complexity required to make perfect nestmate discriminations. We argue that guards will not be perfect in making discriminations because the cost or complexity of perfect information processing is too high and would require an advanced, and therefore costly, physiological framework and neurological fine-tuning. This reasoning leads us to probabilistic models in which we examine how the number of recognition cues, discriminative levels, and colonies in a population can predict the number of identifying signatures needed for a social insect.

Tolerance for Mismatching

In honey bees, guards discriminate nestmates from non-nestmates by comparing the cue profile of a potential intruder with a mental template that is learned by the guard (Breed, Diaz, and Lucero 2004; Dani 2006). Similar patterns of nestmate recognition are seen in many other social insects, including stingless bees (Buchwald and Breed 2005), *Polistes* wasps (Dani 2006), and many species of ant (Dani 2006). Selection may favor some tolerance for mismatches with a guard's template that will accommodate incidental exposure and absorption of cues outside the nest environment; for example, if one cue in the profile does not match the expected concentration, but the others do, a worker at the nest entrance would be treated as a nestmate. The value of tolerance for mismatches lies in reducing the number of rejection errors of the sort involving mistaken exclusion of a nestmate. The loss of cues due to volatilization is potentially a very important issue, as the thorax of flying bees and wasps may reach temperatures approaching 40°C (Stabentheiner et al. 2003); this is hot enough to significantly increase the volatility of many recognition compounds found on the insect cuticle.

What happens when tolerance for mismatches is driven by the need to accommodate environmental influences on cue profiles, but selection also favors highly exclusive behavior? In this case the effect is to increase the necessary number of possible cues in the cue profile, thereby increasing the degree of redundancy in the cue. This allows the guard to crosscheck information within the cue profile in order to tolerate minor deviations from the expected and still have enough information to make accurate discriminations.

If recognition phenotypes are genetically constrained, then the likelihood of identical or partial matches in recognition phenotypes will be correlated with the relatedness between the animals. Monozygotic twins are an extreme example of this; in humans strong resemblance in appearance between monozygotic twins suggests nearly complete genetic constraint on physical appearance, with little phenotypic variation due to the environment. If, on the one hand, selection favors preferential treatment of close relatives, then the most parsimonious system is one in which close relatives all have very similar cue profiles (and these profiles are similar to the animal in question). In this case, we predict that recognition phenotypes should be genetically constrained. If, on the other hand, the critical recognition discrimination is between members of a social insect colony and members of other colonies, then selection should favor recognition phenotypes for which the underlying variation is largely environmental. Environmental variation reduces the likelihood of mistaking a worker from a closely related colony as a colony member. It also allows the formation of a colony-level recognition phenotype, even if there is considerable genetic diversity in the colony, as may occur if the queen has mated more than once (polyandry) or if there is more than one queen in the colony (polygyny).

There are other routes for dealing with changed cue profiles due to environmental influences while a worker is away from the colony. Dimensional information, such as the ratio between the concentrations of two compounds, should be relatively robust against the effects of dilution due to evaporation, as long as volatilities of the cue compounds are similar. For example, the use of dimensional olfactory information ratios (relative concentrations) are probably more important than absolute concentrations for honey bees making discriminations between nestmates and non-nestmates (Breed, Diaz, and Lucero 2004).

In sum, tolerance for template mismatches is an important aspect of nestmate recognition systems, as it reduces the potential for recognition errors in which nestmates are incorrectly excluded. Balancing this tolerance is an increased likelihood that social parasites and robbers might enter the nest. Increased complexity of the recognition cue profile can ameliorate the effect of tolerance for mismatches of individual cues, as can the use of dimensionality and redundancy.

Generating Variation: Neutral Substitution

What mechanisms might allow for an adequate number of available cues to make discriminations? In other words, how does cue diversity arise in evolution? There are two possible routes, selective and neutral. The selective route would involve a specific mechanism that has evolved to add cue diversity over time, analogous to the generation of diversity mechanisms found in immune systems (Kelley, Walter, and Trowsdale 2005; Aertsen and Michiels 2005). It is possible, as in the case of the immune system of vertebrates, that allelic diversity at certain loci is actually fostered or enhanced by special mechanisms that generate variation at far more rapid rates than in other portions of the genome. This variation allows organisms to generate novel immune recognition phenotypes that accumulate and help to anticipate new infections. For example, social recognition in many vertebrates is mediated by odors that are correlated with MHC phenotypes, meaning that social recognition piggybacks on the phenotypic recognition generated by the immune system (Penn and Potts 1999).

In contrast to the selective route, we favor the neutral route as an explanation for the evolution of diversity in social recognition phenotypes. Natural selection generally trims variation that arises due to mutation or novel recombinations; we hypothesize that in recognition systems we are more likely to find variation that is generated by accumulated mutational changes under relaxed selective conditions than by novel genetic processes that create new variation. This leads to consideration of how such variation might develop in a way that evades the evolutionary processes that normally trim such variation.

Neutral substitutability of recognition phenotypes lies at the core of our understanding about phenotypic diversity in recognition cues (Buchwald et al. in prep.). Neutral substitution is the ability of evolutionary processes to substitute one phenotype for another, in a recognition context, without impairing the functionality of the phenotype in its use in another context. For example, nestmate recognition cues are most likely features (generally chemical compounds) that are already present as phenotypes in other contexts and that are co-opted for use as recognition cues. However, in order to function as recognition cues, these features must vary phenotypically among individuals and/or colonies.

Human facial phenotypes provide a useful analogy to clarify this point (Loffler et al. 2005). Human faces vary substantially in appearance, yet as long as the basic functions of chewing, breathing, smelling, and seeing are not impaired, this variability is easily accommodated. Human mate choice does not dictate a single facial phenotype, as there is variation among humans in their preferences for facial phenotypes of their mates (Grammer et al. 2003). Therefore, the variation in human facial phenotypes that is essential for individual recognition is neutrally substitutable as long as that variation does not effect motor and sensory function, and to a certain extent, mate choice.

In neutral models there is an accumulation of genetic variation over time that is not subject to the pruning or canalization that typically results from selection. In the case of nestmate recognition by social insects, epicuticular compounds important in waterproofing the exoskeleton have been co-opted as recognition cues. The plethora of possible epicuticular hydrocarbons including alkanes, alkenes, methylalkanes, and in the case of honey bees, fatty acids (Breed 1998), provides a possible basis for exactly the sort of variation required for nestmate identifications. If many different compounds can serve the same waterproofing purpose—this seems to be the case, although this has not been directly tested—then over time, variation in the genes for enzymes underlying the manufacture of these compounds should, in fact, accumulate.

Under these conditions, how does the recognition phenotype reflect the underlying genotype of the animal? We do not expect a 1:1 correspondence between gene and recognition cue, as each of these types of compounds is the result of a multi-step metabolic pathway, with each step catalyzed by one or more enzymes. Variation between individuals in cuticular compounds is the cumulative result of variation in the kinetics of many enzymes. Thus, studies of the genetics of social recognition cues usually find strong genotype/phenotype correlations (Breed 1998). Neutral substitution can then be seen as permitting alleles for a broader range of enzyme kinetics than might be expected if the concentration of each compound on the cuticle of the insect was critical to waterproofing the animal.

We have tested this concept of neutral substitutability in A. mellifera. This model differs somewhat from the generalized system discussed above in that the key recognition compounds have a structural role in beeswax, rather than serving as barriers to water loss through the cuticle. In this species, cuticular fatty acids are key recognition compounds, but these compounds are also prominent components of beeswax (Breed 1998). All colonies have some of the same fatty acids in their wax and on the bees, but the acids differ in relative concentration among colonies. Adult bees emerge from the pupa with a blank recognition profile. Only after physical contact between worker bees and the wax comb in their colony do the bees acquire their chemical identity (Breed et al. 1995). Because it is acquired from nest materials, the workers in a colony all carry a similar fatty acid profile regardless of genetic relatedness.

These same fatty acids also play a critical role in determining the physical characteristics of beeswax. This raises the question of how variation in wax composition, and hence variation in recognition cues among colonies, is tolerated. We found that, among saturated or unsaturated fatty acids, substitution of one fatty acid for another results in no significant change in the mechanical characteristics of the beeswax. The same substitution, however, would create a completely different recognition signature. This model of neutral substitution also works well when considering phenotypes used in visual and auditory recognition systems. Thus neutral substitution should be viewed as sharing importance with generation of diversity mechanisms in explaining diversity of recognition phenotypes.

Discriminatory Ability

The idea of neutral substitution is complemented by models that focus on the information processing abilities of the guards in a social insect colony. To maintain cue diversity, selection operates on the animal using a template to make more accurate discriminations and on the animal presenting a cue profile to produce the needed information accurately. An obvious route for improved discrimination ability in social insects is the evolution of compound-specific chemoreceptors, which offer finer resolution than receptors that respond to general classes of compounds. Another route is to move beyond a simple presence/absence approach in olfactory information processing toward the ability to discriminate concentrations of compounds. From an information processing point of view, using relative concentrations of compounds in a recognition signature, rather than processing each compound separately, allows animals making social discriminations to glean more information from a smaller number of cues. That, however, requires a more chemosensitive and possibly more expensive recognition system. For olfactory cues, information based on relative concentrations of more than one compound are reasonably robust against the effects of dilution; an animal perceived from a distance will be incorrectly identified if the concentration of each compound is assessed separately in a template, but will likely be correctly identified if the concentrations of the compounds relative to each other are considered (Breed, Diaz, and Lucero 2004). If a compound stimulates a generalized receptor that responds to many compounds of the same functional group, then the information from a set of compounds with similar structures is limited at best. This leads to selection for receptors with high specificity, like those that are typically involved in pheromone perception.

Selection should act on the animal making discriminations in other ways, as well. Humans can make very fine distinctions based on details of facial features, identifying gender, assigning age, and recognizing individuals with great accuracy. It is not surprising that human social behavior has led to selection on the receiver of recognition visual cues to be capable of finely dividing the variation in cues (Breed, Diaz, and Lucero 2004). The same principle should apply to receivers of olfactory information, and there are three possible ways in which this may play out at the receptor level. First, as pointed out above, compound-specific receptors, or at least compound-specific coding from groups of broad-spectrum receptors, would increase olfactory acuity. Second, ability to make fine discriminations based on concentrations would be an advantage, as long as dilution was not an issue. Finally, sensitivity to very low concentrations of compounds might add to the number of possible recognition cues.

Returning to the human model, there is no doubt that our perceptual and information processing abilities are finely tuned for making social distinctions. Humans can recognize hundreds, or even thousands, of other individuals, and can do so in both visual (facial recognition) and auditory (voice recognition) modes. For social insects, the general problem is simpler; typically, one cue profile that represents their colony is learned and is used in what amounts to self versus non-self discriminations (Ratnieks 1991). Nevertheless, distinctions between relatively small differences in cue profiles may be important in intracolony discriminations. Many studies have postulated within-colony discriminations in social insects (e.g., Dani 2006), but there is little empirical evidence to support the expression of such discriminations (Breed, Diaz, and Lucero 1994). Within-colony discriminations between patrilines or matrilines would require very highly tuned discriminatory abilities, and presumably genetic and environmental variation among these groups would be very low.

Redundancy

The critical nature of social recognition in the survival and reproduction of social animals favors the evolution of redundant systems to double-check social identifications. Two or more distinct modes of recognition, each with moderate or even low accuracy, can yield more accurate overall identifications than a single recognition mode, even if that phenotype can support discriminations with a low error rate.

Cue redundancy does insure against mistakes, but it also carries the costs associated with producing, perceiving, and processing additional cues. Recognition of individuals by humans is a good example of recognition redundancy; we can recognize others by facial appearance, voice, gait, posture, and perhaps odor. Any one of these information sets is typically adequate for recognition; the multiple recognition modes allow for confirmation of identifications and for identification in a variety of environmental contexts. Redundancy is achieved by dividing the overall cue profile into subsets. When identifying another animal, an individual may first assess whether there is a match within one learned recognition subset and then assess among other subsets for verification.

On initial consideration, it may seem that redundancy would require production of parallel cue sets, each with the same amount of information that would be needed if the animal was relying on one cue set. Suppose that the error rate in recognition from a single cue set is high, perhaps 20% misidentifications. If the animal also uses a second, independent, cue set with an equally high rate of misidentifications, the overall error rate is the product of the two error rates, or 4%. Thus, use of redundant cue sets can significantly reduce the number of cues needed for discrimination.

It is easy to see, in the example of human recognition, how facial appearance and voice can be used independently, and how humans can exploit this redundancy to make more accurate social identifications. In the end, it may be easier to evolve the ability to make discriminations based on two or three relatively less informative cue sets that, when used redundantly, give accurate discriminations, than to evolve a single highly informative cue set.

Modeling Cue Diversity

How many cues are necessary for a nestmate recognition system to function? Using principles derived from our answers to the three questions posed at the beginning of the chapter, we now outline simple models for calculating the amount of information needed to recognize individuals in a population, and for predicting the number of distinct recognition phenotypes in a population. While these models are powerful predictors, they quickly run onto the rocky shoals of biological reality—they are here provided as a starting point in developing an understanding of recognition systems, rather than a conclusion.

We predict that the number of available recognition phenotypes in a population will far exceed the number of individuals; the number of phenotypes needed to support discriminations is driven by the probability of phenotype matches when phenotypes are drawn randomly from the pool of available phenotypes. Colonies that are closely related may reside close to one another; therefore, variability in cue profiles must be reflective of genetic variance with a resolution that is high enough to override population viscosity. Alternately, there could be an environmental component to the variation that drives dissimilarities in cue profiles even though colonies share considerable genetic identity by descent. Interestingly, most studies suggest that variation in cue profiles lies in endogenously produced compounds, rather than compounds acquired from the environment (Dani 2006), and that additive genetic factors, rather than environmental influences, affect the metabolic variation among colonies in producing cue compounds (Dani 2006).

Our approach stands in strong contrast to previous attempts to analyze this question. Ratnieks (1991, and studies cited therein) built models of phenotypic diversity using an allelic approach. Our argument is that the number of recognition phenotypes defies allelic modeling because the perceived number of phenotypes is a result of the sensory and information processing limitations in the animal attempting to perceive the phenotype. The ecological context in which a cue is presented also has a dramatic effect on the perceptible number of phenotypes from that cue. Added to this, we must consider that most phenotypes that could be used in social recognition are the result of interactions among several genes.

Basic information theory tells us that one bit of information is adequate to allow recognition of two individuals, two bits combine to allow recognition of four distinct interactants, and so on. This approach can be used to calculate the number of character states needed to provide individually unique labels to each animal in an N-member population. This model is excellent, for example, in determining how many telephone numbers are necessary to uniquely identify all nodes in a telephone network. If there were a way of regulating the assignment of numbers (analogous to phenotypes) so that no number is used more than once, there would be no need to have numbers in excess of the individuals requiring them. However, if numbers are at risk for being re-drawn from the pool, confusion will reign.

The only solution is for the number of distinct identifiers to far exceed the number of individuals to be identified. That way probabilistic processes take over, ensuring a low rate of phenotypic matches in the population.

Recognition phenotypes should express a colony-level identity rather than individual identities in social insects. In addition to overcoming possible difficulties due to high intracolonial genetic variance among workers, a colony-level phenotype simplifies the discrimination task for guards, which need to only master one recognition template if there is a single cue profile that represents the entire colony. While Breed and Bennett (1987), in an early analysis of how cue templates and profiles might be handled, postulated that guards might use multiple templates, subsequent work has found mechanisms for establishing colony-level cue profiles (and therefore templates) in a variety of social insect taxa. In *A. mellifera* the comb wax serves as a repository and source for cue compounds (Breed 1998), while the nest of *Polistes* wasps (Dani 2006) and the hypopharyngeal gland of many ants (Lenoir et al. 2001) serve analogous functions in those taxa.

Based on the assumption that social recognition operates at the colony level in social insects, we can now model cue diversity at the appropriate scale for this system. Figure 8.1 illustrates the probability of a pair of colonies possessing matching cue profiles as a function of the number of available profiles and the size of the population. Unfortunately, few data are available to tell us how many colonies typically form a potentially interacting subpopulation. Logic and casual observation suggests that 10 is too few and 1,000 is too many. Assuming subpopulations contain 50 to 100 potentially interacting colonies, these curves suggest that between 2,000 and 10,000 unique signatures would be adequate to achieve a low—perhaps 1% or less—probability that two colonies would possess matching signatures. The asymptotic nature of the curves suggests that little additional benefit, in terms of preventing matches of this sort, is gained by adding additional cue profiles.

Extending this line of reasoning, we can ask how a certain number of cue profiles might be obtained. The simplest olfactory distinction is presence versus absence (i.e., compound concentration is above or below a sensory threshold). The next step would involve three discernable levels; from this point, finer olfactory discriminations would result in increasing numbers of possible signatures. In Figure 8.2 we use genetic loci and alleles as a basis for modeling. For our purposes, a locus is analogous to a single chemical compound in a cue profile, and an allele is analogous to the

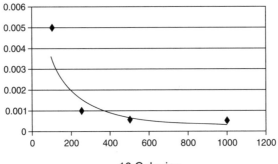

10 Colonies

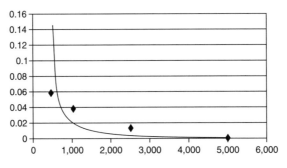

50 Colonies

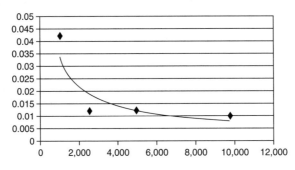

100 Colonies

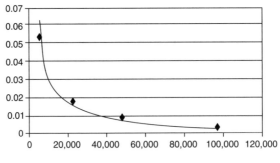

1,000 Colonies

ability of the insect to discriminate among concentrations of the compound. Two "alleles" allows for three discriminable concentrations (aa, ab, bb), and so on. In Figure 8.2 we show that, in terms of generating large numbers of cue profiles, it is far more effective to add loci and have relatively few alleles per locus than it is to add alleles at relatively few loci. By analogy, this leads to the prediction that cue profiles should contain many compounds, although fine distinctions on the basis of concentration are not expected.

Finally, it is necessary to explore, for chemical cues, how both the number of compounds in the cue profile and the ability of the insect to make distinctions among concentrations of those compounds affect the number of potential cue profiles. In Figure 8.3 we assume that population size and the need to avoid matching profiles combine to establish the need for a minimum of 5,000 and a maximum of 100,000 profiles. Based on our prediction that cue profile diversity will depend on larger numbers of compounds rather than fine olfactory distinctions among concentrations, and the curve in Figure 8.3, we predict that cue profiles will, generally, be composed of 13 to 16 compounds for which high and low concentrations can be discriminated, or 8 to 10 compounds with somewhat finer olfactory distinctions.

This argument hinges in part on the assumption that there is a cost to having a larger number of cue profiles available. If there were no cost, then there would be no limit to cue profile complexity and social discriminations would be unfailingly accurate. These costs can take two forms, inter-individual variability in compound concentrations and the chemosensory and information processing abilities, as discussed previously in this chapter. These costs are hypothetical, but we think it is reasonable to

Figure 8.1. Here we illustrate the probability of a pair of colonies possessing matching cue profiles as a function of the number of available profiles and the size of the population. The x-axis is the number of distinct cue profiles available in the population, and the y-axis is the probability, by random draw, of two colonies possessing the same profiles. In small populations, illustrated in upper left and lower left panels, 100 to 1000 cue combinations are adequate to discriminate among colonies with greater than 95% accuracy. The same level of accuracy requires 10000 or more cue combinations of large populations (more than 1000 colonies).

A

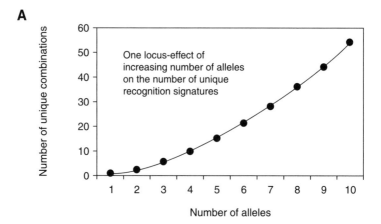

Number of unique combinations (y-axis, 0–60)

One locus-effect of increasing number of alleles on the number of unique recognition signatures

Number of alleles (x-axis, 1–10)

B

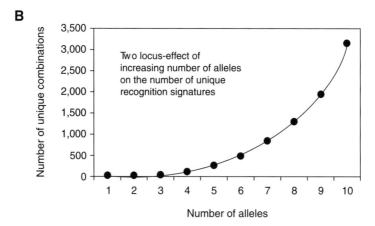

Number of unique combinations (y-axis, 0–3,500)

Two locus-effect of increasing number of alleles on the number of unique recognition signatures

Number of alleles (x-axis, 1–10)

C

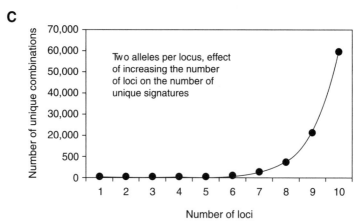

Number of unique combinations (y-axis, 0–70,000)

Two alleles per locus, effect of increasing the number of loci on the number of unique signatures

Number of loci (x-axis, 1–10)

Figure 8.2. How many cue combinations can be generated if there is a one-to-one correspondence between genotype and cues? The top panel, 2a, illustrates the effect of adding alleles at a single locus and the middle panel, 2b, the effect of adding alleles at two loci. The bottom panel, 2c, shows the effect of increasing the number of loci while the alleles are kept constant at two. Far more combinations are generated by increasing the number of loci, rather than increasing the number of alleles at a limited number of loci.

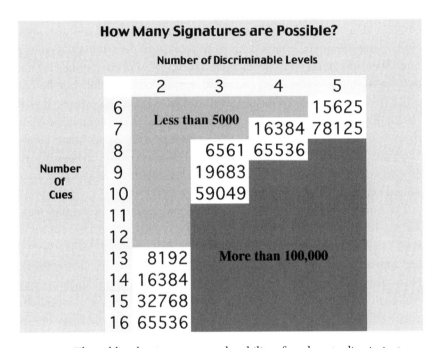

Figure 8.3. This table takes into account the ability of workers to discriminate among cue concentrations or intensities. For convenience these can be thought of a chemical compounds along a concentration gradient. The x-axis of the table is the number of distinct concentrations along a gradient that might be discriminated, the y-axis is the number of cue compounds in the mixture. Unshaded numeric values in the table indicate the number of distinct signatures; these are in the range required for accurate discrimination in large populations (see Figure 8.1). The light gray shaded region yields too few combinations and the dark gray shaded region excess combinations for accurate discrimination. From this table we can predict how many cues and levels of discrimination might be in a cue profile.

assume that they pose a selective counterbalance to factors that favor high cue diversity.

Empirical Studies: How Many Cues Are Out There?

How many cue compounds are actually used by social insects in nestmate identifications? To answer this question we must first ask how we know which compounds are used as cues. Most studies have begun with an examination of insects' surface chemistry or nest environment as a source of candidate cue compounds. Uncritical acceptance of the hypothesis that all compounds are used as cues has resulted in the unfortunate mistake of including many non-cue compounds in an analysis of discrimination among honey bee subfamilies within colonies. Yet bioassays of compounds for activity in nestmate recognition cue profiles need to be carefully applied to avoid misinterpretations due to false positives or false negatives. Breed (1998) discussed the criteria that should be applied in careful bioassays for the role of chemicals in nestmate recognition profiles. Appropriate bioassays have been conducted in only a handful of species, but these serve as an excellent starting point for testing the models for cue diversity presented above.

The number of compounds on the surface of an insect is incredibly vast, but the limited evidence available suggests that only a restricted subset is actually used in making nestmate/non-nestmate discriminations. The honey bee appears to use approximately 14 compounds in its profile (Breed 1998). Six of these are free fatty acids and the remainder are approximately eight odd-chained alkenes (c-21 to c-35). Of other available cue compounds, honey bees do not seem to use floral odors (oils), which could number in the hundreds or thousands of different compounds in the colonial environment (Bowden, Williamson, and Breed 1998). Honey bees also seem not to use alkanes in discriminations even though these are the dominant compounds on the surface of honey bees and in beeswax (Breed 1998). In recognition bioassays, honey bees respond to hexadecane and octadecane as if they were cue profile compounds; however, these compounds are present in only trace amounts in beeswax and on the epicuticle of bees, suggesting that perhaps they are not present in adequate quantities to serve as cues. The predominant alkanes of beeswax and the honey bee epicuticle are higher molecular weight odd-chained alkanes (c-21 to c-35) which do not function as cue compounds in bioassays (Breed et al. 1998).

Polistes sp. uses c-21 to c-31 alkenes and methylalkanes in nestmate recognition, but not straight-chain alkanes; based on Dani et al.'s (2001) data, *Polistes dominulus* may use 12 compounds in its recognition profile. The ant *Myrmecia gulosa* may use 13 compounds in its recognition profile (Dietemann et al. 2003), while another ant, *Camponotus fellah,* which uses c-25 to c-31 methylalkanes as the basis of its recognition profile, may also employ 13 such compounds (Boulay et al. 2000).

Thus there is an apparent convergence on the use of 12 to 14 compounds in cue profiles, although these data are admittedly based on a small sample size and on the assumption that all cue compounds have been identified in these species. If this assumption holds, and the number of cue compounds is, in fact, in this range, then this corresponds well to the prediction made on the basis of Figure 8.3. If each compound is discriminated as being above or below a threshold concentration, or in three concentration steps, then the range of cue profiles available in a population would be from somewhat more than 5,000 to over 100,000, an ample number to provide unique identification of a large number of colonies (Figure 8.1).

Species exceptional to this argument are those that use environmentally derived odors in their cue profile. Notably, the stingless bee *Trigona fulviventris* has shown the ability to use at least some floral oils in its cue profile, in addition to fatty acids, alkenes, and alkanes (Buchwald and Breed 2005). While we know less about the potential number of cue compounds in this species, the addition of floral oils to a cue profile that is otherwise similar to the honey bee's surface chemistry greatly increases the possible number of cue profiles in a population.

One factor that is not considered in this model is similarity in cue profile due to genetic similarity, which may be particularly important in colonies with strong genealogical linkage (such as queens that are sisters) or in populations with inbreeding. As little data is available on the viscosity of social insect populations, there is no way of knowing for most species how common it is for closely related gynes to settle near one another, nor how such conditions might, in an evolutionary time span, affect the recognition system of the species. For example, in honey bees comb wax from closely related colonies is similar in its composition and exposure of workers to similar comb wax can result in recognition errors (Breed et al. 1995). Our analysis applies equally well to environmentally derived or genetically correlated cues.

Conclusions and Future Directions

Social discrimination is a driving force in the evolution of socially interacting species. In answering the questions posed at the beginning of this chapter we have made five key predictions: (1) cues used in social discrimination will be co-opted from other contexts, but that neutral substitution among an array of possible phenotypes will support generation and maintenance of the phenotypic diversity needed for distinct labels at the individual or colony level in a population; (2) there is no special mechanism for the generation of diversity, such as is found in immune systems, that operates in creating enough phenotypic diversity for social recognition; (3) discrimination using chemical compounds will be based on concentration ratios among those compounds, rather than fine distinctions based on individual compound concentrations; (4) cue profiles containing 13 to 16 compounds will be adequate for discrimination at the colony level with a tolerably low error rate in most populations; and, (5) future studies will reveal redundant mechanisms of social identification in eusocial insects, and these redundancies will serve to increase the accuracy of identifications. These predictions could be used to test hypotheses concerning social recognition outside of the eusocial insects, including vertebrate taxa that use visual or auditory cues.

Literature Cited

Aertsen, A., and C. W. Michiels. 2005. "Diversify or die: Generation of diversity in response to stress." *Critical Reviews in Microbiology* 31: 69–78.

Agrawal, A. F. 2001. "Kin recognition and the evolution of altruism." *Proceedings of the Royal Society of London Series B Biological Sciences* 268: 1099–1104.

Bowden, R. M., S. Williamson, and M. D. Breed. 1998. "Floral oils: Their effect on nestmate recognition in the honey bee, *Apis mellifera.*" *Insectes Sociaux* 45: 209–214.

Breed, M. D. 1998. "Chemical cues in kin recognition: Criteria for identification, experimental approaches, and the honey bee as an example." In R. K. Vander Meer, M. L. Winston, K. E. Espelie, and M. D. Breed, eds., *Chemical communication in social insects*, 57–78. Boulder: Westview Press.

Breed, M. D., and B. Bennett. 1987. "Kin recognition in highly eusocial insects." In D. J. C. Fletcher and C. D. Michener, eds., *Kin recognition*, 243–285. New York: John Wiley.

Breed, M. D., P. H. Diaz, and K. D. Lucero. 2004. "Olfactory information processing in honeybee, *Apis mellifera,* nestmate recognition." *Animal Behaviour* 68: 921–928.

Breed, M. D., M. F. Garry, A. N. Pearce, L. Bjostad, B. Hibbard, and R. E. Page. 1995. "The role of wax comb in honey bee nestmate recognition: Genetic effects on comb discrimination, acquisition of comb cues by bees, and passage of cues among individuals." *Animal Behaviour* 50: 489–496.

Breed, M. D., E. A. Leger, A. N. Pearce, and Y. J. Wang. 1998. "Comb wax effects on the ontogeny of honey bee nestmate recognition." *Animal Behaviour* 55: 13–20.

Buchwald, R., and M. D. Breed. 2005. "Nestmate recognition cues in a stingless bee, *Trigona fulviventris.*" *Animal Behaviour.* 70: 1331–1337.

Dani, F. R. 2006. "Cuticular lipids as semiochemicals in paper wasps and other social insects." *Annales Zoologici Fennici* 43: 500–514.

Dani, F. R., G. R. Jones, S. Destri, S. H. Spencer, and S. Turillazzi. 2001. "Deciphering the recognition signature within the cuticular chemical profile of paper wasps." *Animal Behaviour* 62: 165–171.

Dietemann, V., C. Peeters, J. Liebig, V. Thivet, and B. Hölldobler. 2003. "Cuticular hydrocarbons mediate discrimination of reproductives and non-reproductives in the ant *Myrmecia gulosa.*" *Proceedings of the National Academy of Sciences USA* 100: 10341–10346.

Downs, S. G., and F. L. W. Ratnieks. 2000. "Adaptive shifts in honey bee (*Apis mellifera* L.) guarding behavior support predictions of the acceptance threshold model." *Behavioral Ecology* 11: 326–333.

Goth, A., and M. E. Hauber. 2004. "Ecological approaches to species recognition in birds through studies of model and non-model species." *Annales Zoologici Fennici* 41: 823–842.

Kelley, J., L. Walter, and J. Trowsdale. 2005. "Comparative genomics of major histocompatibility complexes." *Immunogenetics* 56: 683–695.

Lenoir, A., P. D'Ettorre, C. Errard, and A. Hefetz. 2001. "Chemical ecology and social parasitism in ants." *Annual Review of Entomology* 46: 573–599.

Loffler, G., G. Yourganov, F. Wilkinson, and H. R. Wilson. 2005. "fMRI evidence for the neural representation of faces." *Nature Neuroscience* 8: 1386–1390.

Millor J., J. M. Ame, J. Halloy, and J. L. Deneubourg. 2006. "Individual discrimination capability and collective decision-making." *Journal of Theoretical Biology* 239: 313–323.

Penn, D. J., and W. K. Potts. 1999. "The evolution of mating preferences and major histocompatibility complex genes." *American Naturalist* 153: 145–164.

Pfeiffer, K. J., and K. Crailsheim. 1998. "Drifting of honeybees." *Insectes Sociaux* 45: 151–167.

Ratnieks, F. L. W. 1991. "The evolution of genetic odor-cue diversity in social Hymenoptera." *American Naturalist* 137: 202–226.

Sinervot, B., A. Chaine, J. Clobert, R. Calsbeek, L. Hazard, L. Lancaster, A. G. McAdam, S. Alonzo, G. Corrigan, and M. E. Hochberg. 2006. "Self-recognition, color signals, and cycles of greenbeard mutualism and altruism." *Proceedings of the National Academy of Sciences of the USA* 103: 7372–7377.

Stabentheiner, A., J. Vollmann, H. Kovac, and K. Crailsheim. 2003. "Oxygen consumption and body temperature of active and resting honeybees." *Journal of Insect Physiology* 49: 881–889.

Tsutsui, N. D. 2004. "Scents of self: The expression component of self/nonself recognition systems." *Annales Zoologici Fennici* 41: 713–727.

Vander Meer, R. M., and L. Morel. 1998. "Nestmate recognition in ants." In R. K. Vander Meer, M. L. Winston, K. E. Espelie, and M. D. Breed, eds., *Chemical communication in social insects,* 79–103. Boulder: Westview Press.

CHAPTER NINE

Adaptations in the Olfactory System of Social Hymenoptera

CHRISTOPH J. KLEINEIDAM

WOLFGANG RÖSSLER

THE SENSE OF SMELL is a very important sensory modality in social insects; numerous pheromones and chemical cues are utilized for fine-tuned interactions between colony members. Trail-following behavior of ants is one prime example illustrating the challenges the olfactory system of social insects had to meet during evolution. Trail following requires the detection of minute amounts of the trail pheromone. Individual workers have to distinguish between conspecific and heterospecific trails. Moreover, a pheromone trail changes over time, has variable concentrations, and can be modified by the ants with additional components leading to distinct recruitment. Besides trail-pheromone communication in ants, the olfactory system is of paramount importance for nestmate recognition and regulation of reproduction. Specificity is often attained by a mixture of several components which differ in their ratios.

To add to the intricacy of pheromone communication, the behavioral responses to pheromones are often context dependent: individuals change their response to the same stimuli in the course of their life and depending on the task they perform (e.g., nursing versus foraging). In order to achieve all these tasks, the olfactory system has to be very flexible with high discrimination power and context sensitivity.

In social Hymenoptera, we find enormous diversity in life history and social structures, allowing for ample comparative studies across species. Within species, we find phenotypic plasticity at different levels of olfactory brain centers and at different life stages. This phenotypic plasticity is important in promoting division of labor and social organization of the colony.

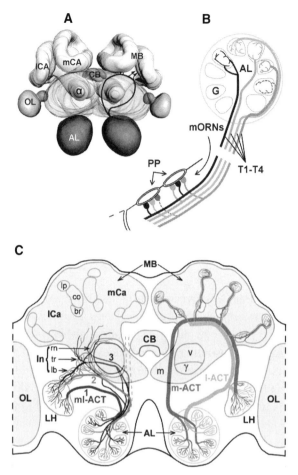

Figure 9.1. The olfactory pathway of Hymenoptera. A: 3-D reconstruction of a whole ant brain *(Atta sexdens)* showing the major neuropils and schematically two prominent antenno-cerebral tracts (L. Kübler). B: Multiple olfactory receptor neurons (only 3 of about 20 mORNs are shown) of two poreplates (PP) send their axons via different sensory tracts (T1–T4) to the glomeruli (G) of the antennal lobe (AL) (C. Kelber). C: Schematic overview of the central olfactory pathway in the honey bee. The left side shows the projection of the multiglomerular projection neurons (ml-ACT) which receive their input from many glomeruli across the antennal lobe and target the lateral horn (LH). The right side shows the projection of the uniglomerular projection neurons via the medial (m-ACT) and the medio-lateral antenno-cerebral tract (mlACT), receiving input from glomeruli in two separated hemispheres of the antennal lobe, and target the mushroom body calyces and the LH. AL, antennal lobe; OL, optic lobes; CB, central body; MB, mushroom body; lCA, lateral calyx; mCA medial calyx; α, alpha lobe of the MB. Modified from Kirschner et al. 2006.

In this chapter we focus on the neuroanatomy and neurophysiology of the olfactory system (Figure 9.1A) and pose the following questions: What are the specific traits within the olfactory system of Hymenoptera compared to other insects? How do these traits relate to the evolution of sociality?

The Insect Nose

The olfactory organ of an insect is the antenna, which is densely packed with various types of sensilla. Most of the sensilla serve odor reception and, together with sensilla for taste and mechano-, thermo-, and hygroreception, the antenna functions as a multimodal sensory organ. Analogous to the compound eye, the term *compound nose* was introduced for the insect antenna to account for the organizational principle of multiple and similar functional units (Hekmat-Scafe, Steinbrecht, and Carlson 1998). The sensilla are the functional units of the antenna and each sensillum houses from one up to numerous receptor neurons (Figure 9.1B). In contrast to only a few associated receptor neurons per sensillum in most insects, olfactory sensilla in Hymenoptera have multiple olfactory receptor neurons (for review, see Keil 1999), ranging from 5 to about 50 neurons in pore plate sensilla (Slifer and Sekhorn 1961; Barlin and Vinson 1981) to more than 100 neurons in basiconic sensilla of some solitary Hymenoptera (Martini 1986; Isidoro, Romani, and Bin 2001). Besides Hymenoptera, multiple olfactory receptor neurons have so far been found only in grasshoppers (Saltatoria) which also have an exceptional olfactory system in many respects (Blaney 1977). The functional significance of multiple olfactory receptor neurons in Hymenoptera is unknown and is subject to one of our case studies presented below.

The odor specificity of an olfactory receptor neuron is given by its receptor molecules, located in the membrane of the dendrites, and by binding proteins which are thought to transport odor molecules through the sensillar lymph to the dendritic membrane. The olfactory receptor neurons of the honey bee for general odors respond to various chemical substances (broadly tuned), and different neurons have overlapping receptive fields (Lacher and Schneider 1963). Thus, different types of olfactory receptor neurons are activated in response to stimulation with a single odor component. General odors in nature are commonly mixtures of many different components. In response to a natural odor, a multitude of the 65,000 olfactory receptor neurons of a honey bee antenna are activated (Esslen and Kaissling 1975). In contrast to olfactory receptor neurons for general odors, pheromone receptor neurons are often very specifically tuned to a single component. This is well documented for sex-pheromone receptor neurons in moths and also seems to be true for some pheromone receptor neurons in ants (Dumpert 1972). The olfactory receptor neurons are primary, bipolar receptor neurons, sending their axons via the double-strand antennal nerve to the antennal lobe where mainly olfactory and gustatory information are processed.

Organization of the Antennal Lobe

The olfactory receptor neurons terminate in spheroid processing units of the antennal lobe, known as *glomeruli* (Kelber, Rössler, and Kleineidam 2006; Mobbs 1982). At the entrance of the antennal lobe, the olfactory receptor neurons are sorted according to their odor specificity. For example, in *Drosophila* it was shown that the sorting into glomeruli relates to the expression of receptor molecules (Vosshall, Wong, and Axel 2000). This organization results in a spatial representation of odors in the antennal lobe which can be visualized with functional imaging techniques (calcium-imaging) first employed in honey bees (Joerges et al. 1997), and later in other insects as well (Galizia, Menzel, and Hölldobler 1999; Carlsson, Galizia, and Hansson 2002; Fiala et al. 2002; Hansson, Carlsson, and Kalinova 2003). Odors evoke neural activity in several glomeruli, and each glomerulus participates in the evoked response of several odors, resulting in an odor specific spatio-temporal activity pattern (Galizia et al. 1997; Sachse, Rappert, and Galizia 1999). The glomerular activity is concentration dependent, but the glomerular patterns are concentration invariant over a wider range of concentrations (Sachse and Galizia 2003). Pheromones (non sex pheromones) in honey bee workers seem to be represented in a similar way as general odors and preliminary data show the same for ants (Zube et al. 2008; Galizia, Menzel, and Hölldobler 1999). The spatial representation of odors in the antennal lobe correlates well with perceptual quality of the same odors tested in behavioral experiments (Guerrieri et al. 2005).

In honey bees, about 160 glomeruli are located in a peripheral layer of the antennal lobe around a nonglomerular central core (Arnold, Masson, and Budharugsa 1985). All of the investigated ants have even more glomeruli, which are arranged in piles or clusters. (Goll 1967; Kleineidam et al. 2005; Zube et al. 2008). Similar arrangements of glomeruli were also found in solitary wasps (Smid et al. 2003). As in other insects, the antennal lobe of Hymenoptera is a site with high neuronal convergence; for example, the ratio of terminating olfactory receptor neurons to neurons leaving the antennal lobe is about 80:1 in the honey bee (65,000 olfactory receptor neurons). It appears that the arrangement of glomeruli in the antennal lobe is highly organized and that functionally corresponding glomeruli are almost invariant in their relative position across individuals (Galizia, Sachse, et al. 1999). Individually identifiable glomeruli (primary glomeruli; Rospars 1988) can be used as landmarks to build an atlas of the entire antennal lobe. In Hymenoptera, we currently have digital atlases available for the honey bee and a parasitoid wasp which facilitate detailed comparative studies (Galizia, McIlwrath, and Menzel 1999; Smid et al. 2003; Brandt et al. 2005). Intrinsic neurons (local interneurons) interconnect the glomeruli. In honey bees, about 4,000 such neurons can be subdivided by their morphology into groups of homogeneous and heterogeneous local neurons (Flanagan and Mercer 1989). All local neurons are multiglomerular, innervating many different glomeruli, and have their somata in a lateral cell cluster of the antennal lobe. Homogeneous local neurons arborize in most glomeruli throughout the antennal lobe, while heterogeneous local neurons innervate a single glomerulus densely and many glomeruli on the opposite side of the antennal lobe sparsely (Fonta, Sun, and Masson 1993; Sun, Fonta, and Masson 1993; Abel, Rybak, and Menzel 2001).

At the output level of the antennal lobe, odor information is relayed via a combinatorial activation of projection neurons (combinatorial code) to higher brain regions. The odor code carried by olfactory receptor neurons is reformatted by the neuronal network of the antennal lobe to the projection neurons. The temporal structure of odor-induced activity in projection neurons is often complex (Laurent 2002; Lei, Christensen, and Hildebrand 2004), and the question of which parameters code for odor quality and intensity still remains unanswered (Menzel et al. 2005).

Besides the sensory input from the olfactory receptor neurons, the antennal lobe is innervated by various descending neuromodulatory systems (Schürmann and Klemm 1984; Bicker 1999; Schachtner, Schmidt, and Homberg 2005; Scheiner and Erber, this volume). A recent phylogenetic study of the serotonergic system within the antennal lobe across several insect orders revealed a disparate innervation in Hymenoptera (Dacks, Christensen, and Hildebrand 2006). Our own work on ants confirm a cluster-specific serotonergic innervation of the antennal lobe, and similar features were found for other neuromodulatory systems as well (Rössler, unpublished data; Hoyer, Liebig, and Rössler 2005; Ziegler et al. 2007). Besides Hymenoptera, however, little is known about the role of serotonin in insect odor coding (Mercer, Kloppenburg, and Hildebrand 1996; Kloppenburg, Ferns, and Mercer 1999).

Researchers are just beginning to exploit such distinct features of the highly evolved olfactory system of social hymenoptera to unveil how, for example, pheromone information is handled in contrast to general odor information and context (Zube et al. 2008).

Higher Brain Regions for Odor Processing

The projection neurons of the antennal lobe leave the first olfactory neuropil toward higher association centers in the mushroom bodies (the structural organization and function of the mushroom bodies is described in detail by Gronenberg and Riveros, this volume) and the lateral horn of the protocerebrum (Mobbs 1982; Abel et al. 2001). Based on their receptive field, two types of projection neurons can be distinguished: uniglomerular and multiglomerular (Figure 9.1C). All projection neurons are bundled in several antenno-cerebral tracts, suggesting multiple pathways for the transmission of olfactory information from the antennal lobe to higher association centers (Abel et al. 2001; Müller et al. 2002; Menzel et al. 2005). In the honey bee, an additional antenno-cerebral tract comprising only uniglomerular projection neurons was found. This tract (lateral antenno-cerebral tract; Figure 9.1C) is not present in Diptera and Lepidoptera and indicates an evolutionary derived or new pathway, further supporting the idea of a highly evolved system in Hymenoptera (see our case study below) (Kirschner et al. 2006). Uniglomerular projection neurons form characteristic synaptic complexes (microglomeruli) in the olfactory input regions of the mushroom bodies where odor information is synaptically relayed to mushroom body intrinsic neurons termed *Kenyon cells* (Mobbs 1982; Ganeshina and Menzel 2001; Frambach et al. 2004; Groh, Ahrens, and Rössler 2004). Interestingly, these large synaptic complexes appear to be very similar across insect species, and our current studies indicate that microglomeruli exhibit a high degree of structural synaptic plasticity.

Odor information is transformed (reformatted) from the combinatorial code of projection neurons to the population code of Kenyon cells (Perez-Orive et al. 2002; Szyszka et al. 2005). In contrast to the projection neurons, odor-activated Kenyon cells are highly odor specific with a brief phasic response (sparse activation). The sparse odor code in Kenyon cells is perhaps supported by synchronized projection neurons (Lei, Christensen, and Hildebrand 2004).

Neuronal Plasticity in Primary and
Secondary Olfactory Centers

Neuronal plasticity in olfactory centers may occur at several levels during the life of insects. In general, we can distinguish postembryonic developmental plasticity from adult plasticity caused by experience, learning, and memory (Meinertzhagen 2001). Adult plasticity and learning and memory have been investigated primarily in the social hymenoptera. Environmental effects during postembryonic development and their influence on the adult nervous systems have not been systematically investigated until recently (see our case studies below). Studies of the honey bee have been mostly focused on volume changes in the antennal lobes and the mushroom bodies associated with the transition from nurse bees to foragers during adult life (e.g., Withers, Fahrbach, and Robinson 1993; Durst, Eichmüller, and Menzel 1994; Sigg, Thompson, and Mercer 1997; Farris, Robinson, and Fahrbach 2001; Fahrbach 2006). The studies show that in the adult antennal lobe, distinct olfactory glomeruli may increase in volume, and this volume change was reversible and task dependent. In a recent electron microscopy study, volume increase in a particular glomerulus was correlated with an increase in synapses, whereas in other cases the density of synapses was not significantly different (Brown, Napper, and Mercer 2004). The most pronounced changes in volume in honey bee and ant adult brain neuropils were found in the mushroom bodies and were shown to be associated with age, experience, and behavioral transitions (Gronenberg and Riveros, this volume). Olfactory learning and memory has been extensively investigated in the honey bee and in *Drosophila*, and has been reviewed in detail elsewhere (Menzel and Giurfa 2001; Heisenberg 2003).

To date, none of the traits described for the olfactory system of social hymenoptera (multiple olfactory receptor neurons, compartmentalization of the antennal lobe, large number of heterogeneous local neurons, additional antenno-cerebral tract, and distinct neuromodulatory systems) has been demonstrated to be a specific adaptation to social life. Comparative data of solitary versus social Hymenoptera are missing with respect to the antennal lobe network (proportion of heterogeneous local neurons), and only little is known about the evolution of the additional antenno-cerebral tract found in honey bees and ants. Likewise, we lack comparative information about the neuromodulatory systems innervating the antennal lobe. In search of adaptations for social life, we have to be aware that traits we find in Hymenoptera might indeed be a necessary prerequisite for an advanced social odor communication; however, solitary hymenoptera might possess and exploit the same traits in a different context.

Case Studies

The Multiple Receptor Neurons of Olfactory Sensilla

In Hymenoptera, the olfactory sensilla house many olfactory receptor neurons. The adaptive value of this organization principle is unknown. Even the odor specificity of the olfactory receptor neurons aggregated within a single sensillum is unknown. We addressed this question by investigating single sensilla in the honey bee. The poreplate sensilla are equipped with 5 to 35 olfactory receptor neurons (Slifer and Sekhorn 1961). Since functionally corresponding glomeruli are almost invariant in relative position to each other across individuals (Galizia, Sachse, et al. 1999), we were able to study the axonal projection pattern in glomeruli of the antennal lobe (glomerular pattern) of olfactory receptor neurons of individual poreplate sensilla and to draw conclusions about the equipment of poreplate sensilla with different olfactory receptor neuron types (Kelber et al. 2006). As a typical feature in Hymenoptera, olfactory receptor neurons innervate the antennal lobe via different tracts. In the honey bee, the olfactory receptor neurons of single poreplate sensilla are found in all four sensory tracts (see Figure 9.1B, T1-T4), which indicates that sensory tracts are not separating information of different modalities (e.g., gustatory and olfactory information). Each olfactory receptor neuron innervates a single glomerulus (uniglomerular) and all olfactory receptor neurons of one poreplate sensillum project to different glomeruli. We conclude that poreplate sensilla are equipped with different olfactory receptor neuron types, but the same olfactory receptor neuron types can be found in different poreplate sensilla.

What is the functional significance of this organization? Is it a matter of miniaturization? The surface area on the honey bee antenna occupied by one poreplate sensillum (diameter ~9 μm) is slightly larger than the area occupied by the major olfactory sensilla in *Drosophila* (large basiconic sensilla; diameter ~7 μm), which are equipped with two or four olfactory receptor neurons (Shanbhag, Müller, and Steinbrecht 1999). We assume that the olfactory sense organs of both *Drosophila* and the honey bee show a maximum extent of miniaturization. The lower density of sensilla in honey bees seems to be overcompensated by the multiple olfactory receptor neurons of poreplate sensilla. With a ~10 times larger surface area of the flagellum, honey bees have ~50 times more olfactory receptor neurons than *Drosophila* on its funiculus. Thus, an increase in the number of olfactory receptor neurons might be one possible reason for the evolution of multiple olfactory receptor neuron sensilla. A larger number of olfactory receptor neurons (with same specificity) increase the sensitivity of the compound nose. The equipment of poreplate sensilla with different types of olfactory receptor neurons does not compromise this advantage as long as olfactory receptor neurons are responding to odors independently of each other. Indeed, this might not be the case. Interactions between olfactory receptor neurons have been shown for moths (O'Connell, Beuchamp, and Grant 1986; Ochieng, Park, and Baker 2002) and honey bees (Akers and Getz 1992; Getz and Breed 1993). It remains to be investigated to what extent odor information processing already takes place at the level of single sensilla and whether multiple olfactory receptor neurons are adapted to allow fine-tuned odor discrimination or context sensitivity.

Ozaki et al. (2005) proposed very advanced processing of chemical information at a single gustatory sensillum of an ant. The authors suggested that the many (estimate of about 200) receptor neurons of one sensillum basiconicum have an almost all-or-none sensitivity to non-nestmate versus nestmate cuticle extracts. Furthermore, the neurons respond only to the non-nestmate cuticle extracts and thus may have a peripheral recognition mechanism implemented. The Ozaki et al. study received notable attention, and was discussed controversially, because of its significant impact on our understanding of how chemical information (odor or taste) is processed in insects (Leonhardt, Brandstaetter, and Kleineidam 2007). At the moment, there is no convincing hypothesis about the mechanisms underlying the possible interactions between sensory neurons to allow such a fine-tuned desensitization.

Dual Pathway of Olfactory Processing in Social Hymenoptera

Olfactory information from the antennal lobes is relayed to higher centers via several tracts (Mobbs 1982; Abel, Rybak, and Menzel 2001; Müller et al. 2002). In a recent study, we revealed important anatomical details of the input and output connections of antennal lobe glomeruli in the honey bee (Kirschner et al. 2006); our comparative work in the carpenter ant *(Camponotus floridanus)* showed a very similar general organization despite a much higher number of input tracts and olfactory glomeruli (Zube et al. 2008). The most important result of these studies is that both uniglomerular projection–neuron output pathways receive a defined input from olfactory glomeruli in two distinct antennal lobe hemispheres, and the output via the two tracts remains, to a great extent, spatially segregated in the higher centers in the mushroom calyx lip and in the lateral horn (Figure 9.1C). The antennal lobe can be divided into two hemispheres according to glomeruli that feed the lateral and medial output tract. This organization appears to be specific for Hymenoptera (Kirschner et al. 2006). Future comparative work is needed to study its functional significance and whether this organization is restricted to Hymenoptera. The two hemispheres may have evolved as an important prerequisite for elaborated chemical communication in social hymenoptera.

Our studies also revealed various novel features of the projections of the mediolateral tracts of multiglomerular projection neurons in a lateral network within the lateral protocerebrum. Future work combining tracing, functional imaging, and electrophysiology techniques should be focused on revealing important insights into the still very unclear principles of olfactory coding. The specific adaptations of the olfactory pathway in Hymenoptera may be one important step to understanding the general principles of odor coding.

Trail Following and Developmental Plasticity in Leaf-Cutting Ants

Leaf-cutting ants express a striking size polymorphism, and size is related to the behavioral repertoire of workers (alloethism). We investigated the antennal lobes of leaf-cutting ants (*Atta vollenweideri* and *A. sexdens*) and discovered that large workers contain a substantially enlarged glomerulus (macroglomerulus) located at the root of the antennal nerve (Figure 9.2A). This was the first description of a macroglomerulus in insects that was not sex-specific.

The macroglomerulus is about 10 times larger than the median glomerulus volume and presumably the result of a large number of innervating olfactory receptor neurons. In the late pupal stage, large workers already have a macroglomerulus, indicating developmental rather than experience-dependent plasticity. In comparison, experience- and age-dependent changes of glomerular volume in honey bees are in the range of only 30% (Winnington, Napper, and Mercer 1996; Sigg, Thompson, and Mercer 1997). Comparative neuroanatomy of the two closely related *Atta* species and measures of the sensory input from the antenna support the interpretation that the macroglomeruli of workers might be involved in the detection and processing of trail pheromone components, in particular of the releaser component (Kleineidam et al. 2005).

Correlation of body size and relative volume of the largest glomerulus at the root of the antennal nerve revealed two distinct neuroanatomical sub-castes of workers, with and without macroglomerulus (Kuebler et al. 2007; Figure 9.2B). The worker polymorphism of the exoskeleton as well as the neuroanatomical traits of the antennal lobe mainly result from environmental conditions during larval development. The phenotypic accommodation at the colony level then leads to inter-individual variance, which promotes division of labor in leaf-cutting ant colonies (West-Eberhard 2005).

What is the significance of an enlarged glomerulus in the antennal lobe for odor mediated behavior? We addressed this question by testing whether small workers differ from large workers in their trail-following behavior, and we found perceptual differences related to body size (Kleineidam et al. 2007). For small and large workers, we measured responsiveness and preference to artificial conspecific and heterospecific pheromone trails made from poison gland extracts of *A. vollenweideri* and *A. sexdens*. Workers followed heterospecific trails, and these trails (after normalizing their concentration) were as effective as conspecific trails. Small workers were less likely to follow a trail of a given concentration than large workers. In the discrimination test, small workers preferred the conspecific trail over the heterospecific trail, whereas large workers showed no significant preference (Figure 9.2C).

Why is it that small foragers are attracted more by the conspecific trails than by the heterospecific trails, while large foragers are equally attracted by both trails? Based on our results from the behavioral experiments, we conclude that small and large foragers evaluate the two competing trails differently. We suggest that large workers primarily respond to the releaser component present in both trails, whereas small workers focus more on the conspecific traits provided by the blend of components contained in the trail pheromone.

How does this hypothesis relate to our neuroanatomical findings? A macroglomerulus may contribute more to the information processing in the antennal lobe than small glomeruli do and may also possess a larger number of projection neurons. If this is the case, and if the releaser component of the trail pheromone is indeed represented in the macroglomerulus, then the releaser component is probably overrepresented in the perception of the trail pheromone in large workers, but not in small workers. Future physiological studies are necessary to confirm these ideas.

A

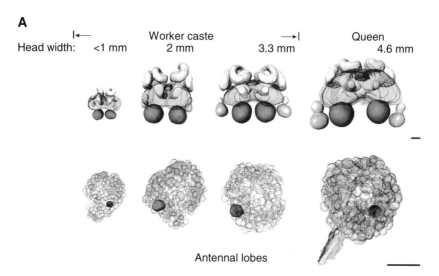

Head width: |←—— Worker caste ——→| Queen
<1 mm   2 mm   3.3 mm   4.6 mm

Antennal lobes

B

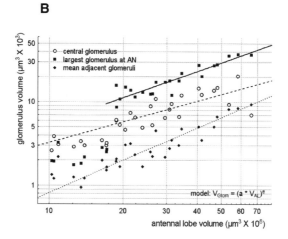

○ central glomerulus
■ largest glomerulus at AN
◆ mean adjacent glomeruli

model: $V_{Glom} = (a \cdot V_{AL})^e$

glomerulus volume ($\mu m^3 \times 10^3$)

antennal lobe volume ($\mu m^3 \times 10^6$)

C

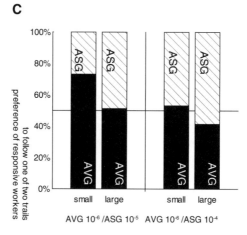

preference of responsive workers to follow one of two trails

ASG   ASG   ASG   ASG

AVG   AVG   AVG   AVG

small   large   small   large

AVG 10^{-6} /ASG 10^{-5} AVG 10^{-6} /ASG 10^{-4}

Brood Care and Developmental Plasticity in Honey Bees

Hymenoptera undergo complete metamorphosis during postembryonic larval-adult development, and the larval nervous system becomes completely remodeled to form the adult nervous system (Weeks and Levine 1990). In recent honey bee studies the effect of brood incubation temperature on the performance of adult bees was investigated. Honey bees incubate their pupae at an extremely narrow temperature range of 35°C (± 0.5), and only in the peripheral brood area will fluctuations range between 33–36°C. Two independent behavior studies showed that the olfactory-learning performance of adult honey bees is influenced by the temperature experienced during their pupal development (Tautz et al. 2003; Jones et al. 2005), and even slight variations in brood incubation temperatures within the physiological range affected adult learning behavior.

Figure 9.2. Neuroanatomical subcastes in leaf-cutting ants and olfactory guided behavior. A: Reconstruction of the major neuropils of three different sized workers and one queen. Brain size is related to body size. Note the large first olfactory neuropil, the antennal lobe. All glomeruli within the antennal lobe were reconstructed. In the two large workers (one soldier-worker), the enlarged glomerulus (dark) is much larger than the median glomerulus volume. In contrast, the largest glomeruli in the small worker and the queen do not differ significantly from the median size of other glomeruli. Scale bar: 100 μm. B: Allometry of antennal lobe volume and the volume of selected glomeruli within the antennal lobe of differently sized workers ($n=31$; head width ranged from 0.5–4.3 mm). The presence of a macroglomerulus in large workers (head width > 1mm) but not in small workers separates the worker caste into two distinct neuroanatomical subcastes (diphasic allometry). ■: largest glomerulus at the entrance of the AN, e = 1.06 for large workers; o: central glomerulus, e = 0.82; ◆: mean of adjacent glomeruli, e = 1.29. Modified from Kuebler et. al 2007. C: Workers' preference on competing trails. The percentages of small and large workers following the conspecific trail (AVG trail, black bars) were compared in two experiments (C1 and C2). The remaining responsive workers (hatched bars) followed the heterospecific trail (ASG trail). In experiment C1 (AVG 10-6 vs. ASG 10-5), the responsive small workers preferred the conspecific trail over the heterospecific trail compared to large workers. In experiment C2 (AVG 10-6 vs. ASG 10-4), the responsive small workers followed both trails with equal probability, and the large workers tended to prefer the heterospecific trail. The responsive small workers and the responsive large workers were compared using the Fishers exact p test and significant differences ($p < 0.05$) are indicated by an asterisk. Modified from Kleineidam et al. 2007.

Groh, Tautz, and Rössler (2004) explored whether there are temperature-mediated effects on the olfactory pathway in the brain. To test this, they raised pupae under different temperatures. The brains of these bees were compared after adult emergence using fluorescent labeling of synaptic proteins and confocal microscopy. The results showed persistent changes in the synaptic organization in the olfactory input regions of the mushroom bodies in adult workers dependent on temperature during metamorphosis. These structural changes potentially may be involved in the observed variation in adult olfactory behavior. Even slight differences (1°C) in pupal-rearing temperature caused changes in the number of synaptic units, the so-called microglomeruli, in the mushroom-body calyx lip (Figure 9.3A, B). Microglomerular numbers were highest in bees raised at the temperature normally maintained in brood cells (34.5°C; Figure 9.3C, D). Interestingly, in the neighboring visual-input region (collar region), microglomerular numbers were less affected by temperature. We conclude that even slight variations in thermoregulatory control of brood rearing may generate modality-specific effects on synaptic neuropils in the adult brain, and propose that resulting differences in the synaptic circuitry may affect neuronal function and may underlie temperature-mediated effects observed in olfactory learning.

Using similar techniques, Groh, Ahrens, and Rössler (2006) investigated environment- and age-dependent effects on the synaptic organization in the brain of honey bee queens. The results revealed that in queens the numbers of microglomeruli in the olfactory and visual input regions of the mushroom-body calyx were significantly lower than in workers. In queens raised in incubators, microglomeruli were also affected by differences in pupal-rearing temperature within the range of naturally occurring temperatures (32–36°C). Interestingly, the highest numbers of microglomeruli developed at a lower temperature in queens compared to workers (33.5°C vs. 34.5°C, respectively).

In addition to this developmental plasticity of mushroom-body calyx input synapses, Groh, Ahrens, and Rössler (2006) found a striking adult plasticity of microglomeruli numbers throughout the extended life span of queens. Whereas the number of microglomeruli in the olfactory lip substantially increased with age (~55%), those in the visual collar significantly decreased (~35%). In summary, these results show that developmental and adult plasticity of the synaptic circuitry in the mushroom-body calyx might underlie caste- and age-specific adaptations in olfactory behavior in the two female honey bee castes.

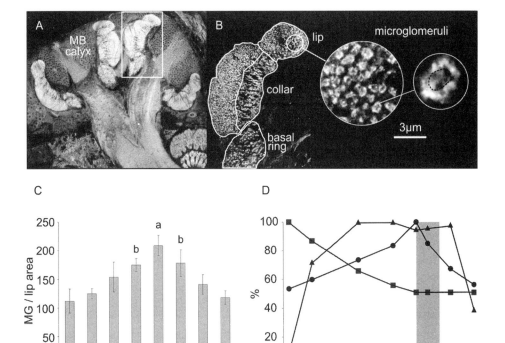

Figure 9.3. A: Confocal images of the mushroom-body calyx in the honey bee brain. B: Differential immunofluorescent labeling of presynaptic synapsin (central region, dashed line) and postsynaptic f-actin (peripheral light staining) reveal the microstructure of synaptic complexes, so-called microglomeruli, in the MB calyx. C. Changes in microglomeruli with different rearing temperatures (1-day-old bees). The number of MG profiles in the olfactory lip of 1-day-old bees critically depends on the pupal rearing temperature. D: Thermoregulation matches optimal developmental time, emergence rate, and postembryonic development of microglomeruli in the MB-calyx lip. Shortest development (rectangles), highest emergence rate (triangles), and highest number of MG in the lip (circles) overlap in the narrow range normally maintained by thermoregulation in central brood cells (gray area). Modified from Groh, Tautz, and Rössler 2004.

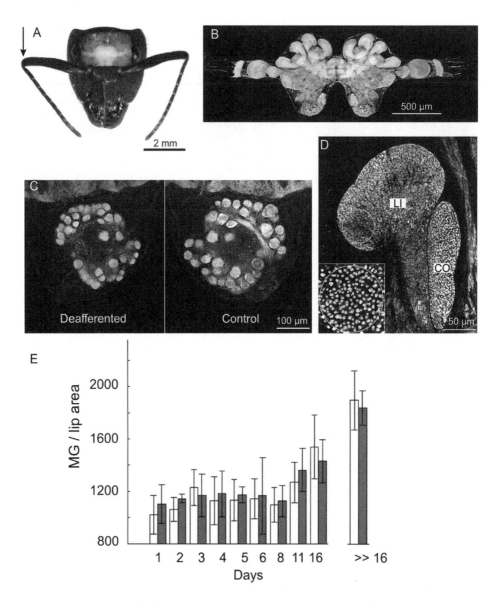

Figure 9.4. A: Head of *Camponotus rufipes* with the frontal cuticle removed to expose the brain. The arrow shows where the right antenna was cut on the first day after emergence to unilaterally remove olfactory sensory input to the antennal lobe. B: Immunofluorescently labeled brain of *Camponotus rufipes*. C: Antennal lobes with glomeruli labeled with anti-synapsin antibody exhibit a reduced size of glomeruli on the deafferented side 15 days after removal of the antenna at day 1 after adult emergence. D: Synaptic boutons in the mushroom

Sensory Experience and Adult Olfactory Plasticity in Carpenter Ants

Compared to the honey bee, many ant species go through a long period of behavioral maturation (up to several months). We chose *Camponotus* ants for these studies because work by Gronenberg, Heeren, and Hölldobler (1996) had already shown that behavioral maturation is correlated with volume changes in brain regions (Gronenberg and Riveros, this volume). The neuronal basis of these volume changes remained unclear. We investigated changes in the synaptic organization of antennal lobe glomeruli and mushroom-body microglomeruli during the first 2 weeks of adult life in the carpenter ant *Camponotus rufipes*. Immunofluorescent techniques were used to label synaptic proteins in order to detect structural changes in the synaptic organization (Figure 9.4A, B). We found that in the olfactory input region of the mushroom-body calyx (lip), the number and density of synaptic complexes (microglomeruli) showed a substantial increase during the first 2 weeks of adult life, especially in the course of the second week, thus demonstrating that volume changes in the mushroom-body calyx may be caused by an increase in synaptic complexes. A recent electron microscopic study by Seid, Harris, and Traniello (2005) on the ant *Phediole dentata* reported that the number of synapses of projection neuron boutons in the mushroom-body calyx lip changed during this period. These results indicate that the mushroom-body calyx undergoes a massive synaptic reorganization during the first weeks or even months of adult life in ants.

body lip were quantified in a circular area in the frontal region of the olfactory lip region. E: The estimated number of microglomeruli (MG) per lip cross-sectional area increased after the first week of adult life ($N=5$–6 ants for each day) and was higher in older ants of variable age (>16 days, $N=8$ ants) compared to young ants, indicating an increase in synaptic capacity of the olfactory input region in the mushroom bodies. White bars correspond to the control sides, gray bars to the deafferented sides. Interestingly, there was no significant difference in the increase on the deafferented sides compared to the control sides indicating crosstalk among the sensory input region of the mushroom bodies on both sides, or via compensation by spontaneous activity. significant difference in the increase on the deafferented sides compared to the control sides indicating crosstalk among the sensory input region of the mushroom bodies on both sides, or via compensation by spontaneous activity.

The extended period of synaptic plasticity in social insects during adulthood offers unique opportunities to look more closely into the parameters that affect plasticity. We began to explore the influence of olfactory sensory input during this period, which we thought likely to be associated with olfactory imprinting to, for example, colony odors. As a first crude experiment, ants were unilaterally deafferented by removal of the antennal flagellum. After a recovery period, unilaterally deafferented ants quickly resumed normal activities inside the nest. After 15 days of unilateral removal of olfactory input, the synaptic neuropil in olfactory glomeruli within the deafferented antennal lobe was reduced by 30% to 50%, indicating that sensory afferents and synapses had degenerated (Figure 9.4C). Interestingly, however, at the level of the mushroom-body calyx lip we observed a similar increase of microglomeruli on both the deafferented and control sides (Figure 9.4D, E), which indicates that compensatory effects are not only able to maintain the synaptic circuitry within the mushroom bodies of the deafferented side, but are also to promote a "normal" increase of synaptic complexes. The bilateral activity of the mushroom bodies may be crucial to perform higher olfactory functions such as learning and memory. Future studies should dissect age- and activity-dependent effects on structural synaptic plasticity in both neuropils. Furthermore, studies should aim to reveal the cellular and molecular bases of synaptic changes and their consequences for olfactory behavior.

Conclusions and Perspectives

We propose three main directions for future studies. First, detailed comparative descriptions of the neuroanatomy of the olfactory pathway within Hymenoptera, social and solitary, are needed. This will provide us with information about traits that distinguish Hymenoptera from other insect orders and possibly about traits specific for social species. To date, we know hardly anything about the olfactory pathway in solitary hymenoptera. Derived traits of the olfactory system may then lead us to study specific adaptations for social life. Second, any of the already known and future discoveries of distinct traits should be explored with respect to its significance for odor information processing and behavior. Like genetic tools in *Drosophila*, the evolution of the Hymenopteran olfactory system provides us with modifications of a system and this unique opportunity can be used to study proximate mechanisms in olfaction.

This new research avenue has just begun to raise fundamental questions, such as:

What does a large number of glomeruli mean and how or why did they evolve?

What does a change in glomeruli volume mean?

Are pheromonal and general odors processed differently?

What causes context specificity of odor responses in behavior?

What is the functional significance of the dual pathway?

What is the role of neuromodulatory systems for processing different odor information?

These questions go beyond social insects, but social insects, due to their specific life history and social organization, are excellent organisms in which to study these questions.

Third, a very powerful arena for study is developing that focuses on phenotypic plasticity and its role for adaptive behavior. What is the neuronal basis for caste, life-stage specific, and experience- and age-dependent changes in the brain and behavior? The pronounced phenotypic plasticity makes ants and bees excellent model systems for the study of neuronal plasticity in the olfactory pathway and the consequences of this plasticity for olfactory-driven behavior. The analyses of the underlying neuronal mechanisms will lead us to a better understanding of the neuronal basis of polyethism and the basis of division of labor.

The technical potential for the different approaches described above is rapidly increasing. Advances in molecular techniques (e.g., tracing of molecules and cell specific markers; differential gene expression studies, intervention with RNAi) will allow us to address more specific questions and will open new approaches. For example, the published honey bee genome project promotes the use of genetic tools to study the molecular basis of neuronal mechanisms in honey bees (gene expression and the discovery of new plasticity relevant molecules). Furthermore, new activity-dependent markers and microscopic techniques (confocal microscopy and life-imaging techniques) combined with three dimensional analyses will further promote progress in studying and quantifying neuronal changes in small brains such as those of social Hymenoptera. Using an integrative approach, therefore, has a very promising future perspective linking physical changes of the olfactory system to changes in olfactory-guided behavior.

Literature Cited

Abel, R., J. Rybak, and R. Menzel. 2001. "Structure and response patterns of olfactory interneurons in the honeybee, *Apis mellifera.*" *Journal of Comparative Neurology* 437: 363–383.

Akers, R. P., and W. M. Getz. 1992. "A test of identified response classes among olfactory receptor neurons in the honey bee worker." *Chemical Senses* 17: 191–209.

Arnold, G., C. Masson, and S. Budharugsa. 1985. "Comparative study of the antennal lobes and their afferent pathways in the worker bee and the drone (*Apis mellifera*)." *Cell and Tissue Research* 242: 593–605.

Barlin, M. R., and S. B. Vinson. 1981. "Multiporous plate sensilla in antennae of the Chalcidoidea (Hymenoptera)." *International Journal of Insect Morphology and Embryology* 10: 29–42.

Bicker, G. 1999. "Histochemistry of classical neurotransmitters in antennal lobes and mushroom bodies of the honeybee." *Microscopy Research and Technique* 45: 174–183.

Blaney, W. M. 1977. "Ultrastructure of an olfactory sensillum on maxillary palps of *Locusta migratoria* (L)." *Cell and Tissue Research* 184: 397–409.

Brandt, R., T. Rohlfing, J. Rybak, S. Krofczik, A. Maye, M. Westerhoff, H. C. Hege, and R. Menzel. 2005. "Three-dimensional average-shape atlas of the honeybee brain and its applications." *Journal of Comparative Neurology* 492: 1–19.

Bromley, A. K., J. A. Dunn, and M. Anderson. 1979. "Ultrastructure of the antennal sensilla of aphids. 1. Coeloconic and placoid sensilla." *Cell and Tissue Research* 203: 427–442.

Brown, S. M., R. M. Napper, and A. R. Mercer. 2004. "Foraging experience, glomerulus volume, and synapse number: A stereological study of the honey bee antennal lobe." *Journal of Neurobiology* 60: 40–50.

Carlsson, M. A., C. G. Galizia, and B. S. Hansson. 2002. "Spatial representation of odours in the antennal lobe of the moth *Spodoptera littoralis* (Lepidoptera: Noctuidae)." *Chemical Senses* 27: 231–244.

Dacks, A. M., T. A. Christensen, and J. G. Hildebrand. 2006. "Phylogeny of a serotonin-immunoreactive neuron in the primary olfactory center of the insect brain." *Journal of Comparative Neurology* 498: 727–746.

Dumpert, K. 1972. "Alarmstoffrezeptoren auf der Antenne von *Lasius fuliginosus* (Latr.) (Hymenoptera, Formicidae)." *Zeitschrift für vergleichende Physioogie der Tiere* 76: 403–425.

Durst, C., S. Eichmüller, and R. Menzel. 1994. "Development and experience lead to increased volume of subcompartments of the honeybee mushroom body." *Behavioral and Neural Biology* 62: 259–263.

Esslen, J., and K.-E. Kaissling. 1975. "Zahl und verteilung antennaler sensillen bei der honigbiene (*Apis mellifera* L.)." *Zoomorphology* 83: 227–251.

Fahrbach, S. E. 2006. "Structure of the mushroom bodies of the insect brain." *Annual Review of Entomology* 51: 209–232.

Farris, S. M., G. E. Robinson, and S. E. Fahrbach. 2001. "Experience- and age-related outgrowth of intrinsic neurons in the mushroom bodies of the adult worker honeybee." *Journal of Neuroscience* 21: 6395–6404.

Fiala, A., T. Spall, S. Diegelmann, B. Eisermann, S. Sachse, J. M. Devaud, E. Buchner, and C. G. Galizia. 2002. "Genetically expressed chameleon in *Drosophila melanogaster* is used to visualize olfactory information in projection neurons." *Current Biology* 12: 1877–1884.

Flanagan, D., and A. R. Mercer. 1989. "Morphology and response characteristics of neurons in the deutocerebrum of the brain in the honeybee *Apis mellifera*." *Journal of Comparative Physiology A, Sensory Neural and Behavioral Physiology* 164: 483–494.

Fonta, C., X. J. Sun, and C. Masson. 1993. "Morphology and spatial-distribution of bee antennal lobe interneurons responsive to odors." *Chemical Senses* 18: 101–119.

Frambach, I., W. Rössler, M. Winkler, and F. W. Schürmann. 2004. "F-actin at identified synapses in the mushroom body neuropil of the insect brain." *Journal of Comparative Neurology* 475: 303–314.

Galizia, C. G., R. Menzel, and B. Hölldobler. 1999. "Optical imaging of odor-evoked glomerular activity patterns in the antennal lobes of the ant *Camponotus rufipes*." *Naturwissenschafte* 86: 533–537.

Galizia, C. G., S. Sachse, A. Rappert, and R. Menzel. 1999. "The glomerular code for odor representation is species specific in the honeybee *Apis mellifera*." *Nature Neuroscience* 2, 473–478.

Galizia, G. C., J. Joerges, A. Küttner, T. Faber, and R. Menzel. 1997. "A semi-in-vivo preparation for optical recording of the insect brain." *Journal of Neuroscience Methods* 76: 61–69.

Galizia, G. C., S. L. McIlwrath, and R. Menzel. 1999. "A digital three-dimensional atlas of the honeybee antennal lobe based on optical sections acquired by confocal microscopy." *Cell and Tissue Research* 295: 383–394.

Ganeshina, O., and R. Menzel. 2001. "GABA-immunoreactive neurons in the mushroom bodies of the honeybee: An electron microscopic study." *Journal of Comparative Neurology* 437: 335–349.

Getz, W. M., and M. D. Breed. 1993. "Odour detection in bees." *Nature* 362: 119–120.

Goll, W. 1967. "Strukturuntersuchungen am Gehirn von Formica." *Zeitschrift für Morphologie und Ökologie der Tiere* 59: 143–210.

Groh, C., D. Ahrens, and W. Rössler. 2006. "Environment- and age-dependent plasticity of synaptic complexes in the mushroom bodies of honeybee queens." *Brain Behavior and Evolution* 68: 1–14.

Groh, C., J. Tautz, and W. Rössler. 2004. "Synaptic organization in the adult honey

bee brain is influenced by brood-temperature control during pupal development." *Proceedings of the National Academy of Sciences USA* 101: 4268–4273.

Gronenberg, W., S. Heeren, and B. Hölldobler. 1996. "Age-dependent and task-
related morphological changes in the brain and the mushroom bodies of the
ant *Camponotus floridanus.*" *Journal of Experimental Biology* 199: 2011–2019.

Guerrieri, F., M. Schubert, J. C. Sandoz, and M. Giurfa. 2005. "Perceptual and
neural olfactory similarity in honeybees." *PLoS Biology* 3: e60.

Hansson, B. S., M. A. Carlsson, and B. Kalinova. 2003. "Olfactory activation
patterns in the antennal lobe of the sphinx moth, *Manduca sexta.*" *Journal
of Comparative Physiology A, Neuroethology Sensory, Neural and
Behavioral Physiology* 189: 301–308.

Heisenberg, M. 2003. "Mushroom body memoir: From maps to models."
Nature Reviews Neuroscience 4: 266.

Hekmat-Scafe, D. S., R. A. Steinbrecht, and J. R. Carlson. 1998. "Olfactory
coding in a compound nose—Coexpression of odorant-binding proteins in
Drosophila." *Olfaction and Taste* Xii, 31–315.

Hoyer, S. C., J. Liebig, and W. Rössler. 2005. "Biogenic amines in the ponerine
ant *Harpegnathos saltator:* Serotonin and dopamine immunoreactivity in
the brain." *Arthopod Structure and Development* 34: 429–440.

Isidoro, N., R. Romani, and F. Bin. 2001. "Antennal multiporous sensilla: Their
gustatory features for host recognition in female parasitic wasps (Insecta,
Hymenoptera: Platygastroidea)." *Microscopy Research and Technique* 55:
350–358.

Joerges, J., A. Küttner, G. C. Galizia, and R. Menzel. 1997. "Representation of
odours and odour mixtures visualized in the honeybee brain." *Nature* 387:
285–288.

Jones, J. C., P. Helliwell, M. Beekman, R. Maleszka, and B. P. Oldroyd. 2005.
"The effects of rearing temperature on developmental stability and learning and memory in the honey bee, *Apis mellifera.*" *Journal of Comparative
Physiology A, Neuroethology Sensory, Neural and Behavioral Physiology*
191: 1121–1129.

Keil, T. 1999. "Morphology and development of the peripheral olfactory
organs. In B. S. Hansson, ed., *Insect olfaction*, 5–48. Heidelberg: Springer.

Kelber, C., W. Rössler, and C. J. Kleineidam. 2006. "Multiple olfactory receptor neurons and their axonal projections in the antennal lobe of the honeybee *Apis mellifera.*" *Journal of Comparative Neurology* 496: 395–405.

Kirschner, S., C. J. Kleineidam, C. Zube, J. Rybak, B. Grünewald, and W.
Rössler. 2006. "Dual olfactory pathway in the honeybee, Apis mellifera."
The Journal of Comparative Neurology 499: 933–952.

Kleineidam, C. J., M. Obermayer, W. Halbich, and W. Rössler. 2005. "A
macroglomerulus in the antennal lobe of leaf-cutting ant workers and its
possible functional significance." *Chemical Senses* 30: 383–392.

Kleineidam, C. J., W. Rössler, B. Hölldobler, and F. Roces. 2007. "Perceptual differences in trail-following leaf-cutting ants relate to body size." *Journal of Insect Physiology* 53: 1233–1241.

Kloppenburg, P., D. Ferns, and A. R. Mercer. 1999. "Serotonin enhances central olfactory neuron responses to female sex pheromone in the male sphinx moth *Manduca sexta.*" *Journal of Neuroscience* 19: 8172–8181.

Kuebler, L. S., C. Kelber, W. Rössler, and C. J. Kleineidam. 2007. "Neuroanatomical sub-castes in leaf-cutting ants: Differences in antennal lobe design correlate with olfactory guided behavior." Paper presented at the 37th Göttingen Meeting of the German Neuroscience Society, TS8–16B. Göttingen: Germany.

Lacher, V., and D. Schneider. 1963. "Elektrophysiologischer Nachweis der Riechfunktion von Porenplatten (Sensilla placodea) auf den Antennen der Drohne und der Arbeitsbiene *(Apis mellifera* L.)." *Zeitschrift für vergleichende Physioogie der Tiere* 47: 274–278.

Laurent, G. 2002. "Olfactory network dynamics and the coding of multidimensional signals." *Nature Reviews Neuroscience* 3: 884–895.

Lei, H., T. A. Christensen, and J. G. Hildebrand. 2004. "Spatial and temporal organization of ensemble representations for different odor classes in the moth antennal lobe." *Journal of Neuroscience* 24: 11108–11119.

Leonhardt, S. D., A. S. Brandstaetter, and C. J. Kleineidam. 2007. "Reformation process of the neuronal template for nestmate recognition cues in the carpenter ant *(Camponotus floridanus).*" *Journal of Comparative Physiology A, Neuroethology Sensory, Neural, and Behavioral Physiology* 193: 993–1000.

Martini, R. 1986. "Ultrastructure and development of single-walled sensilla placodea and basiconica on the antennae of the Sphecoidea (Hymenoptera, Aculeata)." *International Journal of Insect Morphology and Embryology* 15: 183–200.

Meinertzhagen, I. A. 2001. "Plasticity in the insect nervous system." *Advances in Insect Physiology* 28: 84–167.

Menzel, R., G. Galizia, D. Müller, and P. Szyszka. 2005. "Odor coding in projection neurons of the honeybee brain." *Chemical Senses* 30: i301–302.

Menzel, R., and M. Giurfa. 2001. "Cognitive architecture of a mini-brain: The honeybee." *Trends in Cognitive Sciences* 5: 62–71.

Mercer, A. R., P. Kloppenburg, and J. G. Hildebrand. 1996. "Serotonin induced changes in the excitability of cultured antennal lobe neurons of the sphinx moth *Manduca sexta.*" *Journal of Comparative Physiology A, Sensory, Neural, and Behavioral Physiology* 178: 21–31.

Mobbs, P. G. 1982. "The brain of the honeybee *Apis mellifera*. 1. The connections and spatial-organization of the mushroom bodies." *Philosophical*

Transactions of the Royal Society of London Series B-Biological Sciences 298: 309–354.

Müller, D., R. Abel, R. Brandt, M. Zöckler, and R. Menzel. 2002. "Differential parallel processing of olfactory information in the honeybee, *Apis mellifera* L." *Journal of Comparative Physiology A, Neuroethology Sensory, Neural and Behavioral Physiology* 188: 359–370.

Ochieng, S. A., K. C. Park, and T. C. Baker. 2002. "Host plant volatiles synergize responses of sex pheromone-specific olfactory receptor neurons in male *Helicoverpa zea.*" *Journal of Comparative Physiology A, Neuroethology Sensory, Neural and Behavioral Physiology* 188: 325–333.

O'Connell, R. J., J. T. Beuchamp, and A. J. Grant. 1986. "Insect olfactory receptor responses to components of pheromone blends." *Journal of Chemical Ecology* 12: 451–467.

Ozaki, M., A. Wada-Katsumata, K. Fujikawa, M. Iwasaki, F. Yokohari, Y. Satoji, T. Nisimura, and R. Yamaoka. 2005. "Ant nestmate and non-nestmate discrimination by a chemosensory sensillum." *Science* 309: 311–314.

Perez-Orive, J., O. Mazor, G. C. Turner, S. Cassenaer, R. I. Wilson, and G. Laurent. 2002. "Oscillations and sparsening of odor representations in the mushroom body." *Science* 297: 359–365.

Rospars, J. P. 1988. "Structure and development of the insect antennodeutocerebral system." *International Journal of Insect Morphology and Embryology* 17: 243–294.

Sachse, S., and C. G. Galizia. 2003. "The coding of odour-intensity in the honeybee antennal lobe: Local computation optimizes odour representation." *European Journal of Neuroscience* 18: 2119–2132.

Sachse, S., A. Rappert, and C. G. Galizia. 1999. "The spatial representation of chemical structures in the antennal lobe of honeybees: Step towards the olfactory code." *European Journal of Neuroscience* 11: 3970–3982.

Schachtner, J., M. Schmidt, and U. Homberg. 2005. "Organization and evolutionary trends of primary olfactory brain centers in Tetraconata (Crustacea + Hexapoda)." *Arthropod Structure and Development* 34: 257–299.

Schürmann, F. W., and N. Klemm. 1984. "Serotonin-immunoreactive neurons in the brain of the honeybee." *Journal of Comparative Neurology* 225: 570–580.

Seid, M. A., K. M. Harris, and J. F. A. Traniello. 2005. "Age-related changes in the number and structure of synapses in the lip region of the mushroom bodies in the ant *Pheidole dentata.*" *Journal of Comparative Neurology* 488: 269–277.

Shanbhag, S. R., B. Müller, and R. A. Steinbrecht. 1999. "Atlas of olfactory organs of *Drosophila melanogaster* 1. Types, external organization, innervation and distribution of olfactory sensilla." *International Journal of Insect Morphology and Embryology* 28: 377–397.

Sigg, D., C. M. Thompson, and A. R. Mercer. 1997. "Activity-dependent changes to the brain and behavior of the honey bee, *Apis mellifera* (L.)." *Journal of Neuroscience*, 17: 7148–7156.

Slifer, E. H., and S. S. Sekhorn. 1961. "Fine structure of the sense organs on the antennal flagellum of the honey bee, *Apis mellifera* Linnaeus." *Journal of Morphology* 109: 351–381.

Smid, H. M., M. A. K. Bleeker, J. J. A. van Loon, and L. E. M. Vet. 2003. "Three-dimensional organization of the glomeruli in the antennal lobe of the parasitoid wasps *Cotesia glomerata* and *C. rubecula.*" *Cell and Tissue Research* 312: 237–248.

Sun, X. J., C. Fonta, and C. Masson. 1993. "Odor quality processing by bee antennal lobe interneurons." *Chemical Senses* 18: 355–377.

Szyszka, P., M. Ditzen, A. Galkin, C. G. Galizia, and R. Menzel. 2005. "Sparsening and temporal sharpening of olfactory representations in the honeybee mushroom bodies." *Journal of Neurophysiology* 94: 3303–3313.

Tautz, J., S. Maier, C. Groh, W. Rössler, and A. Brockmann. 2003. "Behavioral performance in adult honey bees is influenced by the temperature experienced during their pupal development." *Proceedings of the National Academy of Sciences USA* 100: 7343–7347.

Vosshall, L. B., A. M. Wong, and R. Axel. 2000. "An olfactory sensory map in the fly brain." *Cell* 102: 147–159.

Weeks, J. C., and R. B. Levine. 1990. "Postembryonic neuronal plasticity and its hormonal-control during insect metamorphosis." *Annual Review of Neuroscience* 13: 183–194.

West-Eberhard, M. J. 2005. "Phenotypic accommodation: Adaptive innovation due to developmental plasticity." *Journal of Experimental Zoology Part B-Molecular and Developmental Evolution* 304B: 610–618.

Winnington, A. P., R. M. Napper, and A. R. Mercer. 1996. "Structural plasticity of identified glomeruli in the antennal lobes of the adult worker honey bee." *Journal of Comparative Neurology* 365: 479–490.

Withers, G. S., S. E. Fahrbach, and G. E. Robinson. 1993. "Selective neuroanatomical plasticity and division-of-labor in the honeybee." *Nature* 364: 238–240.

Ziegler, C., N. K. Starke, C. Zube, S. Kirschner, and W. Rössler. 2007. "Neuromodulation and synaptic plasticity within olfactory centers in the brain of the carpenter ant, *Camponotus floridanus.*" Paper presented at the 37th Göttingen Meeting of the German Neuroscience Society, TS8–14B. Göttingen: Germany.

Zube, C., C. J. Kleineidam, S. Kirschner, J. Neef, and W. Rössler. 2008. "Organization of the olfactory pathway and odor processing in the antennal lobe of the ant *Camponotus floridanus.*" *Journal of Comparative Neurology* 506: 425–441.

CHAPTER TEN

Fertility Signaling as a General Mechanism of Regulating Reproductive Division of Labor in Ants

CHRISTIAN PEETERS

JÜRGEN LIEBIG

DIVISION OF LABOR is the essence of sociality in insects and its most striking manifestation is the sterility of almost all colony members. Unequal reproduction among relatives is associated with many conflicts, and recent studies have reconciled these with kin selection (Bourke and Franks 1995). In this chapter, we examine the essential role of communication in the resolution of conflicts over reproduction. Information exchange is indeed entangled with both cooperation and conflicts.

Social Hymenoptera exhibit a broad variety of colony patterns, ranging from small societies with nestmates having equivalent reproductive potentials (e.g., queenless ants, *Polistes* wasps) to huge societies with morphologically highly specialized queens and workers. Although similar conflicts transcend this heterogeneity, the mechanisms that regulate monogyny or queen supersedure, for example, are not the same at either ends of this spectrum. We will argue that seemingly dissimilar regulatory mechanisms rely all on just one category of olfactory information.

All insects are covered with cuticular hydrocarbons which act as a desiccation barrier. There are many distinct molecule types in each species (chain lengths from C23 to C37, with differing numbers and positions of double bonds or methyl-branches), and blends of cuticular hydrocarbons reflecting variations in age, reproductive physiology, and genotype among individuals. In social Hymenoptera, these cuticular differences encode information that seems instrumental in conflict resolution.

From the Past to the Present

Queen Pheromones: A Paradigm Shift

In this chapter we restrict the terms *queen* and *worker* to morphologically specialized adult females (Peeters and Crozier 1988), in contrast to morphologically undifferentiated *breeders* and *helpers* as in *Polistes*. Pheromones have long been known to be involved in the reproductive division of labor in ants and honey bees; unlike vertebrate societies, physical aggression is usually not needed to regulate reproductive skew. Experimental studies in many social insects have established that secretions from the queen are sufficient to mimic her presence (Passera 1984; Hölldobler and Bartz 1985; Fletcher and Ross 1985; Vargo 1998). It remains controversial whether these queen pheromones have a coercive effect on the receivers (i.e., direct inhibition of ovarian physiology) or whether they simply carry information used by receivers to behave adaptively (Seeley 1985; Keller and Nonacs 1993). Several lines of evidence make chemical inhibition unlikely. Its physiological basis has never been made explicit; for example, is the queen pheromone absorbed by the corpora allata or the ovaries, instead of being detected by the antennae? In addition, there has been no demonstration that receivers ever behave against their own interest, which would support inhibition. Moreover, pheromonal manipulation would be evolutionarily unstable because of a likely "arms race." Instead, self-restraint by receivers brings benefits as long as the reproductive is related and sufficiently productive (Keller and Nonacs 1993). Accordingly, queen pheromones may simply carry information about current fertility.

The sources of queen pheromones have largely remained an enigma, with the exception of *Apis* in which the mandibular, tergal, and Dufour's glands are known to be implicated (Winston and Slessor 1998). In ant queens, the efforts to investigate various exocrine glands have had little progress. In *Solenopsis invicta,* poison gland secretions of queens inhibit alate gynes from shedding their wings and developing their ovaries, but queens without poison sacs caused a similar primer effect. Vargo and Hulsey (2000) hinted that another relevant pheromone could be distributed over the body. In the ponerine ant *Megaponera foetens*, a queen is highly attractive to the workers, and her pheromone may originate from a thick glandular epithelium lining her entire body (Hölldobler, Peeters, and Obermayer 1994). In *Bombus terrestris*, where there is no caste dimorphism, a cuticular wash from reproductives caused inhibition of the helpers' ovaries, unlike the extracts from five exocrine glands (Bloch and Hefetz 1999). In hindsight, experimental results from different social insects are not incompatible with the queen pheromone being located on the cuticle, but researchers tend to explain such results as contamination from exocrine glands.

Dominance Behaviors and Recognition

A large number of Hymenopteran societies consist of female adults with equivalent reproductive potentials. In the absence of morphological castes, aggressive interactions lead to the formation of dominance hierarchies and only the alpha individual (or several high-ranking individuals in polygynous species) has active ovaries and reproduces. These dominance interactions are highly directed, showing that hierarchy members can be recognized. In *Polistes* wasps, aggression and the chemical communication of status were shown to be separate phenomena (West-Eberhard 1977; Downing and Jeanne 1985). Similarly, in *Leptothorax* ants competing queens can recognize individual ovarian status and this affects their pattern of aggression and acceptance (Ortius and Heinze 1999). Chemical signaling of ovarian activity had indeed been inferred by several authors (e.g., Wilson 1971; Ratnieks 1988; Visscher and Dukas 1995), but there was no direct evidence for it. Behavioral observations of dominance interactions in *Harpegnathos saltator* hinted that subordinates detect the onset of ovarian activity in gamergates (mated reproductive workers) by antennating their cuticle (Liebig 1998). Researchers of the wasp *Polistes dominulus* and the bumblebee *Bombus hypnorum* found differences in cuticular hydrocarbons that were associated with ovarian activity (Bonavita-Cougourdan et al. 1991; Ayasse et al. 1995)

Such considerations aside, an investigation of cuticular hydrocarbons and their function started in the queenless ant *Dinoponera quadriceps* because of a conspicuous aggressive behavior. During hierarchy formation, the alpha worker bites an antenna of a subordinate and rubs it against the dorsal region of her gaster (Monnin and Peeters 1999). At that time, solid-phase microextraction (SPME) was being developed for the analysis of pesticide residues in water, and following a suggestion by Robin Crewe, Monnin, Malosse, and Peeters (1998) rubbed an SPME fiber against *Dinoponera*'s gaster in the same way that the subordinates' antennae are

Figure 10.1. Use of SPME to sample cuticular hydrocarbons from a *Harpegnathos* worker. Nylon strings are used to immobilize the ant, and the fiber is gently rubbed over the gaster.

rubbed. Gas chromatography revealed that the alpha worker's blend of cuticular hydrocarbons differed from that of infertile nestmates, with an alkene of 31 carbon atoms present in high proportions. If the beta worker (number two in hierarchy) replaces alpha, her ovarian activity increases; SPME allowed the analysis of live beta workers before as well as after such replacement and the proportion of the alkene increased concomitantly (Peeters, Monnin, and Malosse 1999). The same method (Figure 10.1) used in a study of *Harpegnathos saltator* showed that workers simultaneously shift their cuticular profile when they differentiate into gamergates (Liebig et al. 2000). In this species the whole profile changes; longer molecules gradually increase in proportion while shorter molecules decrease over a period of 118 days. Prior to SPME, destructive extraction with solvents had prevented monitoring individual ants at multiple times during their adult life.

Cuticular Hydrocarbons Are Reliable Markers of Reproductive Physiology

A link between oogenesis and the synthesis of cuticular hydrocarbons is found in many solitary insects (e.g., cockroaches, Diptera) and probably involves gonadotropic hormones (Dillwith, Adams, and Blomquist 1983; Trabalon et al. 1990; Wicker and Jallon 1995). In both *Drosophila* and *Calliphora* some of these hydrocarbons function as sex pheromones (e.g., Ferveur, Cobb, and Jallon 1989). Cuticular hydrocarbons also represent a sex pheromone in the cricket *Gryllus bimaculatus* (Tregenza and Weddell 1997), while in cockroaches they may reveal dominance status of males (Roux et al. 2002). The display of information about fertility makes cuticular hydrocarbons appropriate for communication during courtship behavior in solitary insects. During the evolution of social species this information appears to have been co-opted for new functions relating to cooperation and the regulation of reproductive conflict. To date, intracolonial variations in cuticular hydrocarbons have been studied in at least 26 genera of ants, wasps, and bees (Monnin 2006; Liebig and Peeters, in prep.). In the following paragraphs the results of these studies are summarized for various reproductive contexts in which communication is needed about either the onset or decline in egg-laying activity.

Behavioral Regulation and Recognition of Hierarchy Members

In queenless ants, several high-ranking workers compete aggressively to become the gamergate(s). Once behavioral differentiation has happened and ovarian activity begins, however, aggression declines and is replaced by chemical signaling (e.g., Monnin and Peeters 1999; Sledge, Boscaro, and Turillazzi 2001; Cuvillier-Hot et al. 2002; Cuvillier-Hot et al. 2004). Fertility has a striking effect on hydrocarbon profiles in all species investigated so far. In *Diacamma ceylonense*, C25 and C27 monomethyls occur in high proportions in egg-layers (Cuvillier-Hot et al. 2001). In *Streblognathus peetersi*, the cuticular profiles of gamergates are very distinct compared to that of infertile nestmates (callows and foragers); high-ranking workers are intermediate between the two, while newly differentiated alphas (i.e., oogenesis has just begun) resemble gamergates (Cuvillier-Hot et al. 2004). Such alphas are recognized by nestmates within 1 to 2 days of differentiation, even though they can only begin to lay eggs 30 days later (Cuvillier-Hot, Renault, and Peeters 2005). In *Gnamptogenys striatula*, fertile workers have longer-chained hydrocarbons on their cuticle compared to infertile nestmates (Lommelen et al. 2006). In multiple-foundress colonies of the wasp *Polistes dominulus*, alpha individuals (who monopolize oviposition) and subordinates initially have the same proportions of cuticular hydrocarbons, but these proportions become distinct by the time the first adults start to emerge in a colony (Sledge, Boscaro, and Turillazzi 2001).

Queen Recognition and Worker Sterility

Ants exhibit a large spectrum in the degree of dimorphism (e.g., body size, number of ovarioles) between the castes. In *Harpegnathos saltator,* queens and workers are morphologically similar (except for wings) and gamergates reproduce once the founding queen has died. In these two categories of egg-layers, the proportions of cuticular hydrocarbons change in a similar way with the onset of ovarian activity, while young virgin queens resemble infertile workers (Liebig et al. 2000). Thus the hydrocarbons are not related to morphological caste but to reproductive physiology. Caste dimorphism is more marked in the bulldog ant *Myrmecia gulosa* and workers cannot mate. Fertile queens exhibit high proportions of a C25 alkene (Figure 10.2), while virgin workers lay reproductive eggs once orphaned and their cuticular profiles shift to resemble that of queens (Dietemann et al. 2003). Similarly in *Pachycondyla* cf. *inversa,* the cuticular profile of egg-laying workers differs from that of sterile nestmates but resembles the queen's (Heinze, Stengl, and Sledge 2002).

Variations in cuticular hydrocarbons are also found in phylogenetically derived ants with large caste dimorphism. In *Camponotus floridanus* (Formicinae), 10 out of 34 compounds in the cuticular profile of a highly fertile queen are absent on the workers' cuticle (Endler et al. 2004). These ten compounds represent more than 60% of the total amount of cuticular hydrocarbons in some queens (Endler, Liebig, and Hölldobler 2006). *Linepithema humile* (Dolichoderinae) presents another example of strong differences between workers and mated egg-laying queens. More than 50% of the total amount of the profile is specific to the fertile queens (de Biseau et al. 2004); 9 out of 33 identified compounds underlie this difference. In contrast, the cuticular hydrocarbon pattern of young virgin nonlaying queens overlaps almost completely with that of workers. In this species workers lack ovaries and are thus completely sterile—yet they need information about the fertility of their queens to regulate polygyny. In the weakly polygynous *Formica fusca,* reproductive output is reflected by cuticular blends, and the more eggs a queen produces the more attention she receives from workers (Hannonen et al. 2002). Forsyth (1980) had

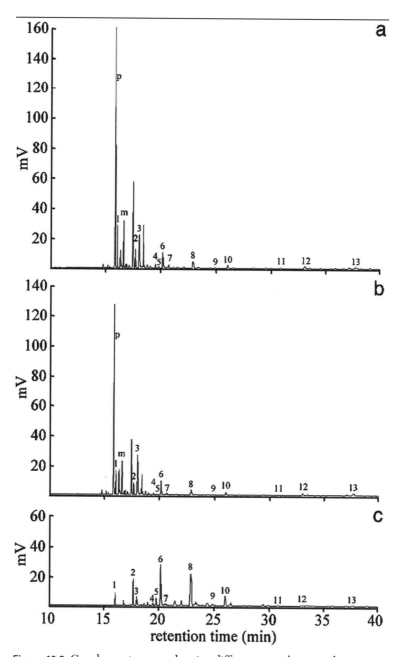

Figure 10.2. Gas chromatograms showing differences in the cuticular hydrocarbons of (a) fertile queen, (b) orphaned worker laying reproductive eggs, and (c) infertile worker of *Myrmecia gulosa* (from Dietemann et al. 2003). Numbers indicate the same compounds across categories. Compounds p and m are specific to reproductive individuals.

evoked the ability of Hymenopteran workers to assess differences in queen productivity during the execution of supernumerary queens. Cuticular differences among queens could serve as a proximate mechanism for the workers to determine which queens are less fertile and can be executed. Workers of *Lasius niger* (Sommer and Hölldobler 1995) and *Solenopsis invicta* (Fletcher and Blum 1983) favor the most fertile queen following pleometrosis (colony founding by multiple queens).

Worker Policing and Recognition of Selfish Egglayers

Several studies showed that ant workers can detect differences in levels of ovarian activity by olfaction. In *Gnamptogenys menadensis*, infertile workers discriminated between virgin workers laying male eggs and gamergates with more active ovaries; other workers laying trophic eggs were ignored (Gobin, Billen, and Peeters 1999). In *Harpegnathos saltator*, infertile workers attacked newly ovipositing gamergates in the presence of gamergates with more active ovaries (Liebig, Peeters, and Hölldobler 1999). The profiles of these newly ovipositing workers were already distinct at the time when they were attacked (Liebig et al. 2000). For *Myrmecia gulosa*, an experimental situation was created where infertile workers could interact with both their queen and workers that started to develop their ovaries; some infertile workers immobilized many of the latter by holding on to their antennae, legs, or body (Dietemann, Liebig, et al. 2005). SPME was used to show a correlation between the change in the victims' cuticular hydrocarbons and the likelihood of immobilization. In *Platythyrea punctata*, worker policing was also correlated with changes in cuticular profiles of new reproductives (Hartmann et al. 2005).

In queenless ants, worker policing has an additional crucial function during the selection of gamergates (Monnin and Ratnieks 2001). High-ranking workers can attempt to replace a gamergate prematurely, while low-ranking workers benefit from preventing the overthrow of a gamergate (usually their mother) as long as she is sufficiently productive. Reliable assessment of fertility is needed to establish the optimal time to replace a senescent gamergate and to identify premature challengers. In *Streblognathus peetersi*, policing workers played a crucial role during the replacement of gamergates with experimentally reduced fertility; they immobilized the manipulated gamergates, thus allowing supersedure by a high-ranking worker. Direct aggression was not observed between the high-ranking individuals and the gamergate (Cuvillier-Hot, Lenoir, and Peeters 2004). Reduced fertility was accompanied by a predictable shift in the cuticular hydrocarbon profile of the gamergate, and an opposite shift occurred in the challenging high-ranking worker due to the onset of oogenesis.

Policing of Worker Eggs

Besides attacking a potential egg-layer in the presence of established reproductives, workers may also destroy the eggs of subordinates by simply eating them. Workers often do so in ants (e.g., Kikuta and Tsuji 1999; D'Ettorre, Heinze, and Ratnieks 2004), in bees (e.g., Oldroyd et al. 2001), and in wasps (e.g., Foster and Ratnieks 2001). In *Dinoponera quadriceps,* the hydrocarbon profile of the egg surfaces resembles the cuticular profile of the egg-layers (Monnin and Peeters 1997). Thus the gamergate can recognize the rare eggs laid by the beta worker and eat them. Long-chained hydrocarbons are synthesized by the oenocytes (cells located in the hemolymph) and then transported to the cuticle and to the ovaries (Schal et al. 1998; Fan et al. 2003). This explains the relative similarity of cuticular profiles and surface profiles of eggs.

Further evidence that information on eggs can be used in the modulation of adult behavior comes from the ant *Camponotus floridanus.* Highly fertile queens show a cuticular hydrocarbon profile that is distinct from that of egg-laying workers; a similar difference is found in the hydrocarbon profiles of their respective eggs (Endler et al. 2004; Endler, Liebig, and Hölldobler 2006). Eggs laid by workers in an orphaned group are usually eaten by workers originating from a colony with a highly fertile queen. When, however, worker eggs are treated with the fractionated hydrocarbon extract of the cuticle of a highly fertile queen, a significantly higher percentage of manipulated eggs survive in comparison to worker eggs treated with the worker profile (Endler et al. 2004). This strongly suggests that the specific hydrocarbon profile of the queen protects eggs from destruction and is used as information by the workers for their decision to police an egg or not.

Evidence for Fertility Signals beyond Correlations

Behavioral and physiological manipulations are excellent tools to identify the function of a signal. Hormonal treatment of alpha workers in *Streblognathus peetersi* decreased fertility and simultaneously changed their cuticular hydrocarbon profile. This treatment triggered two different forms of aggression in nontreated colony members: dominance interactions by high-ranking workers who attempted to become the new alpha and worker policing by low-ranking workers who immobilized the experimentally deficient alpha (Cuvillier-Hot, Lenoir, and Peeters 2004).

In *Myrmecia gulosa,* a bioassay using fractionated cuticular extracts showed that workers can differentiate between the cuticular hydrocarbon profiles of queens and infertile workers (Dietemann et al. 2003). The hydrocarbon fractions of the queen's cuticle were significantly more attractive than extracts from infertile workers. This is not simply a consequence of caste differences because the hydrocarbon profiles of queens and fertile workers are similar (Figure 10.2). This study provided the first evidence that hydrocarbon profiles are differentiated based on fertility differences, but it did not directly show the function of the signal.

The direct involvement of hydrocarbon profiles of the queen in the regulation of reproduction has been demonstrated in *Camponotus floridanus.* As explained earlier, the hydrocarbon profile of the egg surface closely matches the cuticular hydrocarbon profile of the mother (Endler, Liebig, and Hölldobler 2006). Workers are prevented from activating their ovaries by the mere presence of eggs from highly fertile queens, even though the queen is absent (Endler et al. 2004). This result suggests that workers recognize queen presence via the presence of her eggs. Since workers recognize queen eggs on the basis of their specific hydrocarbon profile, it is most likely that these hydrocarbons are also responsible for the induction of self-restraint in workers. Furthermore, since the profile of queen eggs induces self-restraint, the similar cuticular profile of the queen should have the same effect. Thus, hydrocarbon profiles on the surface of queen eggs as well as on the queen cuticle represent a queen signal that regulates reproduction in *C. floridanus.*

Smelling Fertile: Cuticular Information Is Not All or None

Unlike an eventual caste signal that would be all or none (i.e., queen or worker), cuticular hydrocarbons are correlated with degrees of fertility in either of the castes. In *Formica fusca* queens, continuous variation in egg-laying capacity is manifest as gradual changes in cuticular profiles (Hannonen et al. 2002). In *Diacamma ceylonense,* virgin egg-layers and gamergates have different levels of ovarian activity, and this is associated with distinct hydrocarbon blends (Cuvillier-Hot et al. 2001). In queens of *Camponotus floridanus,* the hydrocarbon profile is correlated with colony size as an indicator of the queen's average fertility (Endler, Liebig, and Hölldobler 2006). Young founding queens with low egg-laying rates and colony sizes of less than 10 workers show a profile that is similar to that of workers; however, this profile changes continuously and becomes very distinct in highly fertile queens from colonies with more than 1,000 workers.

A fertility signal can convey graded information to label not only egg-layers but also high-ranking workers with intermediate reproductive potential. Rather than just ovarian activity, cuticular hydrocarbons reveal the individual hormonal state that underlies dominance and oogenesis. Low levels of juvenile hormone (JH) are characteristic of alpha workers in *S. peetersi* (Brent et al. 2006), and topical applications of a JH analog led to both decreased fertility (as measured by levels of vitellogenin in the hemolymph) and shift in cuticular profile (Cuvillier-Hot, Lenoir, and Peeters 2004). In *Polistes dominulus,* the size of the corpora allata (the principal site of JH production) was associated with variation in cuticular hydrocarbon proportions (Sledge et al. 2004). Each species investigated so far had about 50 different hydrocarbons on the cuticle, which provides sufficient variability for differences in sex, colony membership, and age to be detected, in addition to reproductive status. *Drosophila* flies have many long-chain cuticular hydrocarbons (C29–C35) within hours of emergence; as they get older, these are replaced by shorter chain hydrocarbons (C23–29) (Wicker and Jallon 1995). In the ant *Diacamma ceylonense,* workers up to 4 days in age have different hydrocarbon blends compared to older infertile workers (Cuvillier-Hot et al. 2001). In *Camponotus vagus, Harpegnathos saltator, Pogonomyrmex barbatus*, and *Myrmicaria eumenoides,* nurses and foragers can be distinguished by their cuticular profiles (Bonavita-Cougourdan, Clément, and Large 1993; Liebig et al. 2000; Greene and Gordon 2003; Kaib et al. 2000), suggesting age-related hormonal changes. Generally, a close connection among hormone activity (e.g., ecdysone), oenocyte activity, and the formation of the cuticular hydrocarbon profile is assumed (Howard and Blomquist 2005) (Figure 10.3). Application of JH III to *M. eumenoides* nurse workers resulted in a cuticular profile typical of foragers, even though they did not behave as foragers (Lengyel, Westerlund, and Kaib 2007). In *Harpegnathos saltator,* virgin queens develop 2 distinct alkadiene peaks prior to the mating flight (Liebig et al. 2000), which suggests that it is involved in sexual communication. These peaks are not present in founding queens that already lay eggs, or any of the workers.

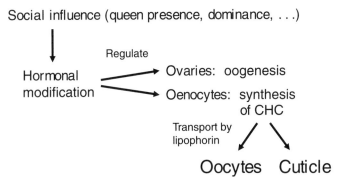

Figure 10.3. Physiological links between endocrine activity, oogenesis, and the synthesis of cuticular hydrocarbons in individual insects. Gonadotropic hormones affect both oogenesis and the activity of oenocytes. Hydrocarbons are then transported in the hemolymph to the ovaries and cuticle.

Workers laying trophic eggs (i.e., unviable yolk sacs) in *M. gulosa* have a distinct cuticular profile from that of workers producing males (Dietemann et al. 2003); that is, when a worker shifts from trophic to fertile eggs its cuticular profile changes also. This is further evidence of the influence of hormonal levels on the activity of both ovaries and oenocytes. In *S. peetersi*, cyclical egg-laying activity does not affect the profile of gamergates, which is appropriate since a temporary halt should not be perceived by colony members as a stimulus to replace their mother. In *C. floridanus*, variation in egg-laying rates among queens is high as well. Nevertheless, their profiles are very similar which indicates discontinuous egg-laying activity with a stable profile (Endler, Liebig, and Hölldobler 2006). Even though eggs may be laid in cycles, oogenesis continues and is linked to hydrocarbon synthesis.

Honest Information, Aggression, and Colony Size

Several authors (e.g., Wilson 1971; Keller and Nonacs 1993; Heinze 2004) have suggested that behavioral regulation of reproduction cannot be effective in larger colonies because a queen or dominant individual cannot interact frequently with all its nestmates. Thus, a shift from aggressive behavior to chemical signaling has been assumed to follow an increase in colony size. Yet signaling also underlies behavioral regulation since olfactory recognition is needed among contenders. Instead of invoking colony size per se, convergence or divergence of genetic interests seems more crucial to understanding which mechanism (behavioral or pheromonal) is used. Behavioral regulation (Figure 10.4) is needed in species where all nestmates are able to mate and reproduce, and such species usually have small colonies. In contrast, in species with similarly small colonies but with dimorphic queens and workers, the former monopolize reproduction without aggression. Increasing caste dimorphism, which is always a feature of species with larger colonies, seems to eliminate the need for aggressive interactions because the interests of egg-layers and helpers converge as a result of the latter being unable to mate (Bourke 1999; Foster 2004; Dietemann, Peeters, and Hölldobler 2005, Endler, Hölldobler, and Liebig 2007).

Recognition pheromones are nonvolatile and require physical contact for transmission. Gamergates and queens separated from orphaned workers by a mesh that allowed airflow but prevented physical contact could not inhibit ovarian activation in these workers (Liebig, Peeters, and Hölldobler 1999; Liebig et al. 2000; Tsuji, Egashira, and Hölldobler 1999). In *S. peetersi*, distance of detection was measured to be 1.3mm, which is compatible with the

Figure 10.4. Two workers of *Harpegnathos saltator* fighting over queen succession. All workers are potentially capable of replacing a senescent queen.

involvement of long-chained hydrocarbons (Cuvillier-Hot, Renault, and Peeters 2005). In *D. quadriceps* and *S. peetersi*, the alpha worker exhibited conspicuous gaster behaviors which may help make its cuticular hydrocarbons more readily detected by competing high-ranking workers. It is commonly observed in colonies of queenless ants that high-rankers remain in close proximity to the gamergate(s); this is further evidence that colony size is unlikely to be a constraint on physical control.

In larger colonies, information about queen presence can be transmitted indirectly via hydrocarbon profiles on eggs. Workers recognize queen eggs in the ant *C. floridanus* based on their surface hydrocarbons (Endler et al. 2004). These eggs are sufficient to inhibit ovarian activity in workers and thus function as an indirect signal of queen presence. This mechanism may also prevail in polydomous colonies such as *Oecophylla*.

Conclusions and Future Directions

Fertility and Reproductive Regulation

Reproductive conflicts in insect societies differ in both proximate and ultimate characteristics: participants may belong to the same or different morphological castes and conflict resolution may manifest as either self-restraint (e.g., workers remain infertile) or aggressive behaviors. One thing in common is that all participants share the need for information about the presence of an active egg-layer. Although oogenesis also affects various exocrine secretions (e.g., Dufour's gland in *Apis*, Dor, Katzaf-Gozansky, and Hefetz 2005), cuticular hydrocarbons carry reliable information about fertility and thus play an essential role in regulating reproductive division of labor.

In species having queen and worker castes, conflicts are often restricted to male production. Senescence of the queen is followed by colony extinction in many monogynous species, so it is adaptive for workers to begin male production (e.g., Dietemann and Peeters 2000). An ageing queen is expected to have reduced fertility, which is useful information for the workers. They can either start activating their ovaries to produce males or rear exclusively sexual offspring as new workers are no longer useful.

Even in the presence of a fertile queen, some workers may behave selfishly and produce male eggs. In various species, policing workers will attack these egg-layers or destroy their eggs. In either case, this is a self-interested response to the queen's fertility signal that they can perceive either directly or indirectly. In the absence of a fertile queen, policing workers modulate their behavior and allow some workers to produce male adults.

In *Polistes* wasps or queenless ants (i.e., all colony members have equivalent reproductive potentials), information about fertility is equally crucial. During hierarchy formation, high-ranking individuals need to recognize each other. Once an alpha begins oogenesis, there is a shift from aggression to chemical signaling. High-ranking subordinates benefit from this shift as it may stop additional aggression from lower-ranked individuals. During replacement of a senescent alpha, high-rankers need to detect a drop in fertility that makes it worthwhile for them to attempt a challenge. Policing workers can prevent a challenge while their mother is still sufficiently fertile, but will favor it after a drop in her fertility. The interests of sterile helpers converge with the gamergate's or queen's only as long as the latter is able to produce many offspring.

Testing Insect Perception of Fertility Signals

Variations in cuticular hydrocarbons give reliable information about the reproductive status of individuals in all the ants, wasps, and bees investigated thus far (Monnin 2006; Liebig and Peeters, in prep.). This information is useful for human investigators, but more evidence is needed to determine if insects use it to resolve their reproductive conflicts. Although individuals that are induced experimentally to start or stop laying eggs exhibit predictable modifications in their cuticular profiles, it is important for future research to demonstrate the pheromonal function with an artificial profile. More behavioral studies together with a better understanding of the significance of the physico-chemical properties of the different types of cuticular hydrocarbons are timely. Polar compounds on the cuticle should also be investigated (Dapporto, Dani, and Turillazzi 2008).

The cuticular hydrocarbons that co-vary with fertility are seldom the same across species. One may ask, why are all species systematically different in hydrocarbon chemistry? Is it just an epiphenomenon of hydrocarbon biosynthesis or is this diversity selected for (see Breed and Buchwald, this volume)? Although a cuticular layer of alkanes might be sufficient to prevent desiccation, it is not biosynthetically feasible to produce alkanes exclusively (R. Crewe, pers. comm.). There are many steps along the metabolic pathways involved in the synthesis of long-chained hydrocarbons (Howard and Blomquist 2005), and some of the enzymes that catalyze each step may result from differential gene expression (e.g., Ferveur and Jallon 1996).

Typically, about 50 different long-chained hydrocarbons occur in a typical ant species. Determining the specific cuticular compounds involved in the recognition of either nestmates or reproductives is a crucial step. Multivariate statistical tools can help to identify potential candidates, but do not tell us which of several molecules are behaviorally important. The most conspicuous difference may be misleading as individuals may respond to another compound or set of compounds. Systematic evaluation of the role of hydrocarbon classes would be useful. As an example, alkenes and methyl-groups are important carriers of information in nestmate discrimination in *Apis* (Dani et al. 2001; Dani et al. 2005). As a result of 3-D conformation, branched alkanes might be easier to distinguish than linear alkanes and thus more likely to be represented in a fertility signal.

Another approach to understanding the very basics of hydrocarbon recognition involves electro-antennograms (EAG). In *Pachycondyla inversa,* the major compound of the queen signal produced a significantly higher response in an EAG than other components of the profile (D'Ettorre, Heinze, 2004). Although this shows that the ants have a higher sensitivity to this compound, it does not explain how the information is processed and what the natural response to the compound is. EAGs cannot replace bioassays that show the result of the processing of olfactory information (the behavioral or physiological response). They give only limited information about the potential to perceive the respective substances and may suggest receptor specialization for certain compounds. In combination with bioassays EAGs can, however, help to isolate active compounds within a complex profile.

Cues of Colonial Identity and Individual Information about Fertility

Colonial recognition based on cuticular hydrocarbons has been demonstrated in a small number of ant species (e.g., Lahav, Soroker, and Hefetz 1999; Thomas et al. 1999; Wagner et al. 2000). In *Pachycondyla goeldii,* Denis, Blatrix, and Fresneau (2006) found that the cuticular hydrocarbons responsible for the best discrimination among ovarian development classes also yielded a clear discrimination among colonies. In *Camponotus floridanus,* the hydrocarbon profile typical for the colony was clearly present in addition to the reproductive profile in queens (Endler et al. 2004), and similar results were obtained in *Diacamma ceylonense* (Cuvillier-Hot et al. 2001). Hence, the cuticular profile contains dual information about colony membership and fertility status.

According to the Gestalt model, all nestmates share a colonial odor as a result of mixing their cuticular hydrocarbon profiles. However, as far as the various hydrocarbons that are correlated with fertility are concerned, the lack of gestalt is a reality. Furthermore, there are other intracolonial differences (behavioral subcaste, age, sex) encoded in the cuticular hydrocarbon profiles. How can these intracolonial differences be reconciled with the mixing of odors? Further studies are needed to verify that mutual exchange of hydrocarbons is always necessary to maintain a colony profile (a function claimed for the postpharyngeal gland, Lahav et al. 1998; Boulay et al. 2000), and the link with the genotype must be elucidated. Another possibility is that the hydrocarbons correlated with fertility are produced at a higher rate than other longer-chained hydrocarbons. This needs to be investigated empirically. Indirect support for this is the high frequency of self-grooming shown by new alphas in *S. peetersi* (Cuvillier-Hot et al. 2004). Self-grooming results in a transfer of hydrocarbons to the postpharyngeal gland via the basitarsal brush on the front legs, as shown in *Pachycondyla apicalis* (Hefetz et al. 2001). New alphas benefit from communicating their change in status as quickly as possible, and this can be achieved by removing cuticular hydrocarbons that no longer reflect their current physiological condition. It is a challenge for the future to understand how ants are able to extract multiple pieces of information from a single hydrocarbon profile. We need to identify the different active parts of the profile and separately demonstrate the ability of the ants to retrieve various kinds of information. Once this is achieved, the concept of "queen pheromone" can be usefully replaced by fertility pheromone.

Literature Cited

Ayasse, M., T. Marlovits, J. Tengö, T. Taghizadeh, and W. Francke. 1995. "Are there pheromonal dominance signals in the bumblebee *Bombus hypnorum* L. (Hymenoptera: Apidae)?" *Apidologie* 26: 163–180.

Bloch, G., and A. Hefetz. 1999. "Reevaluation of the role of mandibular glands in regulation of reproduction in bumblebee colonies." *Journal of Chemical Ecology* 25: 881–896.

Bonavita-Cougourdan, A., J.-L. Clément, and C. Lange. 1993. "Functional subcaste discrimination (foragers and brood-tenders) in the ant *Camponotus vagus* Scop.: Polymorphism of cuticular hydrocarbon patterns." *Journal of Chemical Ecology* 19: 1461–1477.

Bonavita-Cougourdan, A., G. Theraulaz, A. G. Bagnères, M. Roux, M. Pratte, E. Provost, and J.-L. Clément. 1991. "Cuticular hydrocarbons, social organization and ovarian development in a polistine wasp: *Polistes dominulus* Christ." *Comparative Biochemistry and Physiology B* 100: 667–680.

Boulay, R., A. Hefetz, V. Soroker, and A. Lenoir. 2000. "*Camponotus fellah* colony integration: Worker individuality necessitates frequent hydrocarbon exchanges." *Animal Behaviour* 59: 1127–1133.

Bourke, A. F. G. 1999. "Colony size, social complexity and reproductive conflict in social insects." *Journal of Evolutionary Biology* 12: 245–257.

Bourke, A. F. G., and N. R. Franks. 1995. *Social evolution in ants.* Princeton: Princeton University Press.

Brent, C., C. Peeters, V. Dietemann, R. Crewe, and E. Vargo. 2006. "Hormonal correlates of reproductive status in the queenless ponerine ant, *Streblognathus peetersi.*" *Journal of Comparative Physiology A* 192: 315–320.

Cuvillier-Hot, V., M. Cobb, C. Malosse, and C. Peeters. 2001. "Sex, age and ovarian activity affect cuticular hydrocarbons in *Diacamma ceylonense,* a queenless ant." *Journal of Insect Physiology* 47: 485–493.

Cuvillier-Hot, V., R. Gadagkar, C. Peeters, and M. Cobb. 2002. "Regulation of reproduction in a queenless ant: Aggression, pheromones and reduction in conflict." *Proceedings of the Royal Society of London B* 269: 1295–1300.

Cuvillier-Hot, V., A. Lenoir, R. Crewe, C. Malosse, and C. Peeters. 2004. "Fertility signaling and reproductive skew in queenless ants." *Animal Behaviour* 68: 1209–1219.

Cuvillier-Hot, V., A. Lenoir, and C. Peeters. 2004. "Reproductive monopoly enforced by sterile police workers in a queenless ant." *Behavioral Ecology* 15: 970–975.

Cuvillier-Hot, V., V. Renault, and C. Peeters. 2005. "Rapid modification in the olfactory signal of ants following a change in reproductive status." *Naturwissenschaften* 92: 73–77.

Dani, F. R., G. R. Jones, S. Corsi, R. Beard, D. Pradella, and S. Turillazzi. 2005. "Nestmate recognition cues in the honey bee: Differential importance of cuticular alkanes and alkenes." *Chemical Senses* 30: 477–489.

Dani, F. R., G. R. Jones, S. Destri, S. H. Spencer, and S. Turillazzi. 2001. "Deciphering the recognition signature within the cuticular chemical profile of paper wasps." *Animal Behaviour* 62: 165–171.

Dapporto, L., F. R. Dani, and S. Turillazzi. 2008 "Not only cuticular lipids: First evidence of differences between foundresses and their daughters in polar substances in the paper wasp *Polistes dominulus.*" Journal of Insect Physiology 54:89–95.

De Biseau J.-C., L. Passera, D. Daloze, and S. Aron. 2004. "Ovarian activity correlates with extreme changes in cuticular hydrocarbon profile in the highly polygynous ant, *Linepithema humile.*" *Journal of Insect Physiology* 50: 585–593.

Denis D., R. Blatrix, and D. Fresneau. 2006. "How an ant manages to display individual and colonial signals by using the same channel." *Journal of Chemical Ecology* 32: 1647–1661.

D'Ettorre, P., J. Heinze, and F. L. W. Ratnieks. 2004. "Worker policing by egg eating in the ponerine ant *Pachycondyla inversa.*" *Proceedings of the Royal Society of London B* 271: 1427–1434.

D'Ettorre P., J. Heinze, C. Schulz, W. Francke, and M. Ayasse. 2004. "Does she smell like a queen? Chemoreception of a cuticular hydrocarbon signal in the ant *Pachycondyla inversa.*" *Journal of Experimental Biology* 207: 1085–1091.

Dietemann, V., J. Liebig, B. Hölldobler, and C. Peeters. 2005. "Changes in the cuticular hydrocarbons of incipient reproductives correlate with triggering of worker policing in the bulldog ant *Myrmecia gulosa.*" *Behavioral Ecology and Sociobiology* 58: 486–496.

Dietemann, V., and C. Peeters. 2000. "Queen influence on the shift from trophic to reproductive eggs laid by workers of the ponerine ant *Pachycondyla apicalis.*" *Insectes Sociaux* 47: 223–228.

Dietemann, V., C. Peeters, and B. Hölldobler. 2005. "Role of the queen in regulating reproduction in the bulldog ant *Myrmecia gulosa:* Control or signalling?" *Animal Behaviour* 69: 777–784.

Dietemann, V., C. Peeters, J. Liebig, V. Thivet, and B. Hölldobler. 2003. "Cuticular hydrocarbons mediate recognition of queens and reproductive workers in the ant *Myrmecia gulosa.*" *Proceedings of the National Academy of Sciences USA* 100: 10341–10346.

Dillwith, J. W., T. S. Adams, and G. Blomquist. 1983. "Correlation of housefly sex-pheromone production with ovarian development." *Journal of Insect Physiology* 29: 377–386.

Dor, R., T. Katzav-Gozansky, and A. Hefetz. 2005. "Dufour's gland pheromone as a reliable fertility signal among honeybee *(Apis mellifera)* workers." *Behavioral Ecology and Sociobiology* 58: 270–276.

Downing, H. A., and R. L. Jeanne. 1985. "Communication of status in the social wasp *Polistes fuscatus* (Hymenoptera: Vespidae)." *Zeitschrift für Tierpsychologie* 67: 78–96.

Endler, A., B. Hölldobler, and J. Liebig. 2007. "Lack of physical policing and fertility cues in egg-laying workers of the ant *Camponotus floridanus.*" *Animal Behaviour* 74: 1171–1180.

Endler, A., J. Liebig, and B. Hölldobler. 2006. "Queen fertility, egg marking and colony size in the ant *Camponotus floridanus.*" *Behavioral Ecology and Sociobiology* 59: 490–499.

Endler, A., J. Liebig, T. Schmitt, J. Parker, G. Jones, P. Schreier, and B. Hölldobler. 2004. "Surface hydrocarbons of queen eggs regulate worker reproduction in a social insect." *Proceedings of the National Academy of Sciences USA* 101: 2945–2950.

Fan, Y. L., L. Zurek, M. J. Dykstra, and C. Schal. 2003. "Hydrocarbon synthesis by enzymatically dissociated oenocytes of the abdominal integument of the German cockroach, *Blatella germanica*." *Naturwissenschaften* 90: 121–126.

Ferveur J.-F., M. Cobb, and J.-M. Jallon. 1989. "Complex chemical messages in *Drosophila*. In R. Singh and N. Strausfeld, eds., *Neurobiology of sensory systems*, 397–409. London: Plenum Press.

Ferveur J.-F., and J.-M. Jallon. 1996. "Genetic control of male cuticular hydrocarbons in *Drosophila melanogaster*." *Genetical Research* 67: 211–218.

Fletcher, D. J., and M. S. Blum. 1983. "Regulation of queen number by workers in colonies of social insects." *Science* 219: 312–314.

Fletcher, D. J. C., and K. G. Ross. 1985. "Regulation of reproduction in eusocial Hymenoptera." *Annual Review of Entomology* 30: 319–343.

Forsyth, A. 1980. "Worker control of queen density in hymenopteran societies." *American Naturalist* 116: 895–898.

Foster, K. R. 2004. "Diminishing returns in social evolution: The not-so-tragic commons." *Journal of Evolutionary Biology* 17: 1026–1034.

Foster, K. R., and F. L. W. Ratnieks. 2001. "Convergent evolution of worker policing by egg eating in the honeybee and common wasp." *Proceedings of the Royal Society of London B* 268: 169–174.

Gobin, B., J. Billen, and C. Peeters. 1999. "Policing behaviour toward virgin egg layers in a polygynous ponerine ant." *Animal Behaviour* 58: 1117–1122.

Greene, M. J., and D. M. Gordon. 2003. "Social insects—Cuticular hydrocarbons inform task decisions." *Nature* 423: 32–32.

Hannonen, M., M. F. Sledge, S. Turillazzi, and L. Sundström. 2002. "Queen reproduction, chemical signaling and worker behaviour in polygyne colonies of the ant *Formica fusca*." *Animal Behaviour* 64: 477–485.

Hartmann, A., P. D'Ettorre, G. Jones, and J. Heinze. 2005. "Fertility signaling—The proximate mechanism of worker policing in a clonal ant." *Naturwissenschaften* 92: 282–286.

Hefetz, A., V. Soroker, A. Dahbi, M.-C. Malherbe, and D. Fresneau. 2001. "The front basitarsal brush in *Pachycondyla apicalis* and its role in hydrocarbon circulation." *Chemoecology* 11: 17–24.

Heinze, J. 2004. "Reproductive conflict in insect societies." *Advances in the Study of Behavior* 34: 1–57.

Heinze, J., B. Stengl, and M. F. Sledge. 2002. "Worker rank, reproductive status and cuticular hydrocarbon signature in the ant, *Pachycondyla inversa*." *Behavioral Ecology and Sociobiology* 52: 59–65.

Hölldobler, B., and S. H. Bartz. 1985. "Sociobiology of reproduction in ants." In B. Hölldobler and M. Lindauer, eds., *Experimental behavioral ecology and sociobiology,* 237–257. Stuttgart: Gustav Fischer Verlag.

Hölldobler, B., C. Peeters, and M. Obermayer. 1994. "Exocrine glands and the attractiveness of the ergatoid queen in the ponerine ant *Megaponera foetens.*" *Insectes Sociaux* 41: 63–72.

Howard, R. W., and G. J. Blomquist. 2005. "Ecological, behavioral, and biochemical aspects of insect hydrocarbons." *Annual Review of Entomology* 50: 371–393.

Kaib, M., B. Eisermann, E. Schoeters, J. Billen, S. Franke, and W. Franke. 2000. "Task-related variation of postpharyngeal and cuticular hydrocarbon compositions in the ant *Myrmicaria eumenoides.*" *Journal of Comparative Physiology A* 186: 939–948.

Keller, L., and P. Nonacs. 1993. "The role of queen pheromones in social insects: Queen control or queen signal?" *Animal Behaviour* 45: 787–794.

Kikuta, N., and K. Tsuji. 1999. "Queen and worker policing in the monogynous and monandrous ant, *Diacamma* sp." *Behavioral Ecology and Sociobiology* 46: 180–189.

Lahav, S., V. Soroker, and A. Hefetz. 1999. "Direct behavioral evidence for hydrocarbons as ant recognition discriminators." *Naturwissenschaften* 86: 246–249.

Lahav, S., V. Soroker, R. K. Vander Meer, and A. Hefetz. 1998. "Nestmate recognition in the ant *Cataglyphis niger:* Do queens matter?" *Behavioral Ecology and Sociobiology* 43: 203–212.

Lengyel, F., S. A. Westerlund, and M. Kaib. 2007. "Juvenile hormone III influences task-specific cuticular hydrocarbon profile changes in the ant *Myrmicaria eumenoides.*" *Journal of Chemical Ecology* 33: 167–181.

Liebig, J. 1998. "Eusociality, female caste dimorphism, and regulation of reproduction in the ponerine ant *Harpegnathos saltator* Jerdon." Ph.D. diss., Wissenschaft und Technik Verlag, Berlin.

Liebig, J., C. Peeters, and B. Hölldobler. 1999. "Worker policing limits the number of reproductives in a ponerine ant." *Proceedings of the Royal Society of London B* 266: 1865–1870.

Liebig, J., C. Peeters, N. J. Oldham, C. Markstädter, and B. Hölldobler. 2000. "Are variations in cuticular hydrocarbons of queens and workers a reliable signal of fertility in the ant *Harpegnathos saltator?*" *Proceedings of the National Academy of Sciences USA* 97: 4124–4131.

Lommelen E., C. Johnson, F. Drijfhout, J. Billen, T. Wenseleers, and B. Gobin. 2006. "Cuticular hydrocarbons provide reliable cues of fertility in the ant *Gnamptogenys striatula.*" *Journal of Chemical Ecology* 32: 2023–2034.

Monnin, T. 2006. "Chemical recognition of reproductive status in social insects." *Annales Zoologici Fennici* 43: 515–530.

Monnin, T., C. Malosse, and C. Peeters. 1998. "Solid-phase microextraction and cuticular hydrocarbon differences related to reproductive activity in the queenless ant *Dinoponera quadriceps.*" *Journal of Chemical Ecology* 24: 473–490.

Monnin, T., and C. Peeters. 1997. "Cannibalism of subordinates' eggs in the monogynous queenless ant *Dinoponera quadriceps.*" *Naturwissenschaften* 84: 499–502.

Monnin, T., and C. Peeters. 1999. "Dominance hierarchy and reproductive conflicts among subordinates in a monogynous queenless ant." *Behavioral Ecology* 10: 323–332.

Monnin, T., and F. L. W. Ratnieks. 2001. "Policing in queenless ponerine ants." *Behavioral Ecology and Sociobiology* 50: 97–108.

Oldroyd, B. P., L. A. Halling, G. Good, W. Wattanachaiyingcharoen, A. B. Barron, P. Nanork, and S. Wongsiri. 2001. "Worker policing and worker reproduction in *Apis cerana.*" *Behavioral Ecology and Sociobiology* 50: 371–377.

Ortius, D., and J. Heinze. 1999. "Fertility signaling in queens of a North American ant." *Behavioral Ecology and Sociobiology* 45: 151–159.

Peeters, C., and R. H. Crozier. 1988. "Caste and reproduction in ants: Not all mated egg-layers are 'queens.' " *Psyche* 95: 283–288.

Peeters, C., J. Liebig, and B. Hölldobler. 2000. "Sexual reproduction by both queens and workers in the ponerine ant *Harpegnathos saltator.*" *Insectes Sociaux* 47, 325–332.

Peeters, C., T. Monnin, and C. Malosse. 1999. "Cuticular hydrocarbons correlated with reproductive status in a queenless ant." *Proceedings of the Royal Society of London B* 266: 1323–1327.

Passera, L. 1984. *L'organisation Sociale des Fourmis.* Toulouse: Privat.

Ratnieks, F. L. W. 1988. "Reproductive harmony via mutual policing by workers in eusocial Hymenoptera." *American Naturalist* 132: 217–236.

Roux, E., L. Sreng, E. Provost, M. Roux, and J.-L. Clément. 2002. "Cuticular hydrocarbon profiles of dominant versus subordinate male *Nauphoeta cinerea* cockroaches." *Journal of Chemical Ecology* 28: 1221–1235.

Schal, C., V. Sevala, H. Young, and J. Bachmann. 1998. "Sites of synthesis and transport pathways of insect hydrocarbons: Cuticle and ovary as target tissues." *American Zoologist* 38: 382–394.

Seeley, T. D. 1985. *Honeybee ecology: A study of adaptation in social life.* Princeton: Princeton University Press.

Sledge, M. F., F. Boscaro, and S. Turillazzi. 2001. "Cuticular hydrocarbons and reproductive status in the social wasp *Polistes dominulus.*" *Behavioral Ecology and Sociobiology* 49: 401–409.

Sledge, M., I. Trinca, A. Massolo, F. Boscaro, and S. Turillazzi. 2004. "Variation in cuticular hydrocarbon signatures, hormonal correlates and establishment of reproductive dominance in a polistine wasp." *Journal of Insect Physiology* 50: 73–83.

Sommer, K., and B. Hölldobler. 1995. "Colony founding by queen association and determinants of reduction in queen number in the ant *Lasius niger.*" *Animal Behaviour* 50: 287–294.

Thomas, M. L., L. J. Parry, R. A. Allan, and M. A. Elgar. 1999. "Geographic affinity, cuticular hydrocarbons and colony recognition in the Australian meat ant *Iridomyrmex purpureus.*" *Naturwissenschaften* 86: 87–92.

Trabalon, M., M. Campan, P. Porcheron, J.-L. Clément, J.-C. Baehr, M. Morinière, and C. Joulie. 1990. "Relationships among hormonal changes, cuticular hydrocarbons, and attractiveness during the first gonadotropic cycle of the female *Calliphora vomitoria* (Diptera)." *General Comparative Endocrinology* 80: 216–222.

Tregenza, T., and N. Wedell. 1997. "Definitive evidence for cuticular pheromones in a cricket." *Animal Behaviour* 54: 979–984.

Tsuji, K., K. Egashira, and B. Hölldobler. 1999. "Regulation of worker reproduction by direct physical contact in the ant *Diacamma* sp. from Japan." *Animal Behaviour* 58: 337–343.

Vargo, E. L. 1998. "Primer pheromones in ants." In R. K. Vander Meer, M. D. Breed, K. E. Espelie, and M. L. Winston, eds., *Pheromone communication in social insects—Ants, wasps, bees, and termites*, 293–313. Boulder: Westview Press.

Vargo, E. L., and C. D. Hulsey. 2000. "Multiple glandular origins of queen pheromones in the fire ant *Solenopsis invicta.*" *Journal of Insect Physiology* 46: 1151–1159.

Visscher, P. K., and R. Dukas. 1995. "Honey bees recognize development of nestmates' ovaries." *Animal Behaviour* 49: 542–544.

Wagner, D., M. Tissot, W. Cuevas, and D. M. Gordon. 2000. "Harvester ants utilize cuticular hydrocarbons in nestmate recognition." *Journal of Chemical Ecology* 26: 2245–2257.

West-Eberhard, M. J. 1977. "The establishment of reproductive dominance in social wasp colonies." In *Proceedings of 8th International Congress of IUSSI, Wageningen*, 223–227.

Wicker, C., and J.-M. Jallon. 1995. "Hormonal control of sex pheromone biosynthesis in *Drosophila melanogaster.*" *Journal of Insect Physiology* 41: 65–70.

Wilson, E. O. 1971. *The insect societies.* Cambridge: Harvard University Press.

Winston, M. L., and K. N. Slessor. 1998. "Honey bee primer pheromones and colony organization: Gaps in our knowledge." *Apidologie* 29: 81–95.

CHAPTER ELEVEN

Vibrational Signals in Social Wasps: A Role in Caste Determination?

ROBERT L. JEANNE

SOCIAL INSECT COLONIES are characterized by specialization of their individual members whose diverse social activities are coordinated by means of cues and signals (Seeley 1995). It has long been recognized that the chemical mode dominates the signal channels used by social species (Wilson 1971). Mechanical signals, while probably less important than chemical signals, are also widely used and take a variety of forms, including stridulation, head-banging, piping, antennation, jerking, wing-buzzing, scraping, and drumming (Hill 2001; Hölldobler and Roces 2001).

Paper wasps (Vespidae) produce mechanical signals in a variety of contexts (Figure 11.1, Pratte and Jeanne 1984; Jeanne and Keeping 1995; Matsuura 1984). Mechanical signals are especially conspicuous in the independent-founding Polistinae belonging to the genera *Polistes, Mischocyttarus, Ropalidia,* and *Belonogaster.* Females produce these signals either by rapidly shaking the body while standing on the nest or by beating some part of the body against the nest or even a nestmate. Frequencies are typically in the range of 3–30 strikes/sec and may be of high enough amplitude to cause the nest itself to vibrate. In most cases they are issued in short bursts lasting a second or less, sometimes repeated regularly. Movements that strike the nest are often vigorous enough to produce sound audible to the human ear a meter or more from the nest (Keeping 1992; Pratte and Jeanne 1984). Because vespid wasps lack ears, these sounds are incidental and the energy must be perceived by colony members via vibration of the nest structure. The frequency and intensity at which they are produced, and the risk that the sounds produced may call

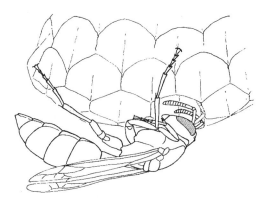

Figure 11.1. Antennal drumming by a *Polistes fuscatus* queen. The antennal flagella move in synchrony (see dotted outline of antennae) to strike the far rim of the larval cell over which she is poised. Drawn from a frame of 16-mm movie film. (Reproduced from Pratte and Jeanne 1984, with permission of the publisher.)

the attention of predators, suggest that these movements are costly to produce and that they are therefore likely to benefit their producers and can be assumed to be signals (Seeley 1995).

I have two aims in this chapter. The first is to review the occurrence of vibrational signals across the independent-founding wasps. The review is organized around the contexts in which the signals occur. Two contexts are recognized: brood care, particularly trophallactic contacts with larvae (feeding context), and non–brood care contexts (non-feeding context). Although some hornets and yellowjackets produce vibratory signals in similar contexts, because they are less well known the Vespinae are not considered here.

The second aim is to critically review hypotheses on the functions of vibratory signals in these wasps. Although several functions have been proposed, none is well supported, and the meaning of these signals remains mysterious. I propose a new hypothesis: that these vibrations are signals that bias caste development via biochemical changes and gene expression, either as stress inducers or as fertility indicators.

In this review I adopt the following terminology. If the vibration is produced by striking the nest with a part of the body, it is called *drumming*, modified by the body part used, if known (antennal drumming, gastral drumming). If it is not clear whether or not the nest is struck, and for forms where the nest is not struck but is shaken, I use the term *vibration* preceded by a modifier describing the form of the movement (lateral vibration, longitudinal body vibration).

Feeding Context: Vibrational Signals Associated with Brood Care

Vibratory signals produced in close association with trophallactic visits to larvae occur in most of the better-studied species of independent-founding polistine wasps. The behavior has been most thoroughly studied in *Polistes*, but analogous behavior occurs in at least three of the four other genera.

Polistes

Pratte and Jeanne (1984) showed that antennal drumming (AD) in *P. fuscatus* is associated with the feeding of prey liquid to larvae (Figure 11.1). When a founding-stage queen returns to the nest with prey, she malaxates it for several minutes, imbibing the juices, then distributes the solid residue to the larger larvae. She then moves from cell to cell in the nest, pausing to drum her antennae on the rim of each without contacting the brood, before moving on to the next cell and repeating. The dorsal surfaces of the antennal flagella synchronously strike the rim of the cell at a rate of 25–32 beats per second; each drum lasts about one second. When the drumming is performed especially vigorously, the wasp's entire body moves rapidly forward and backward in synchrony with each strike of the antennae (see Pratte and Jeanne 1984, Figure 11.1, for an illustration). After several minutes of this, the queen enter's each cell after drumming on it. If the cell contains a larva, she regurgitates a droplet of the prey liquid she has imbibed, touches it to the larva's mouthparts, and the larva ingests it. She repeats this sequence until each larva is visited and regurgitated to many times. Essentially the same behavior has been reported for *Polistes annularis, bellicosus, canadensis, carnifex, carolinus, erythrocephalus, exclamans, instabilis, jokahamae* (*f/jadwigae*), *metricus*, and *versicolor* (Pratte and Jeanne 1984).

Although *P. dominulus* lacks AD, females perform a more or less vigorous side-to-side shaking of the abdomen (Brillet, Tian-Chansky, and Le Conte 1999) that is virtually identical in form to the lateral vibration (LV) described for other *Polistes* (Savoyard et al. 1998; see below). The intensity of LV varies but may be strong enough to cause the entire body to shake, and sometimes the abdomen strikes the nest during vibration (Brennan 2007). Nearly 90% of the LVs were performed in the context of feeding larvae; the other 10% occurred in non-feeding contexts such as aggressive encounters with other adults (Brillet, Tian-Chansky, and Le Conte 1999; see below). Unlike *P. fuscatus*, a female *P. dominulus* performs LV as she begins to feed the solid bolus, continuing, but with reduced intensity, as she feeds the liquid. LVs are also sometimes performed during inspection of cells without feeding (Brillet et al. 1999).

Mischocyttarus

Females of the Neotropical wasp *Mischocyttarus drewseni* process and distribute solid food in the same way as *Polistes fuscatus* (Jeanne 1972). During the subsequent liquid-feeding phase, the female rapidly drums the gaster dorsoventrally (GD) against the nest for a fraction of a second prior to entering each cell. At the end of a round of feeding, the female occasionally performs a violent longitudinal body vibration (LBV) with its head inserted in an egg or larval cell, producing an audible rattle (Jeanne 1972). GD is also performed by *M. bimaculatus* and *M. mexicanus* (S. Suryanarayanan, pers. comm.).

Ropalidia

R. cincta females issue a vigorous wing buzz prior to entering a cell to feed solid food to the larva (Darchen 1976). After the solid is distributed, the females engage in trophallaxis with the larvae, again preceding each visit by a wing buzz. *R. marginata* females perform three kinds of body oscillations in the feeding context (S. Suryanarayanan, pers. comm.): a burst of wing-buzzing upon leaving a larval cell after feeding solid or liquid; a vigorous LBV, sometimes hitting the rim of the cell with the head or antennae; and a jerk of the body after exiting from a cell. While feeding liquid, *R. fasciata* females perform LBV with the head in the cell, the vibration lasting 10–30 seconds (Ito 1983). *R. revolutionalis* antennal drums like *Polistes fuscatus* (pers. obs.).

Belonogaster

In the first report of vibrational signaling in wasps, Roubaud (1916) described how, after feeding solid prey to the larvae, females of *B. juncea* rustle or vibrate the wings before entering each larval cell. For the same species, Tindo, Francescato, and Dejean (1997) describe a dorso-ventral movement of the gaster. *B. petiolata* performs an LBV with the body head-down and nearly perpendicular to the nest surface, occasionally causing an audible rattle (Keeping 1992; M. G. Keeping, pers. comm.). Several females, including those not feeding larvae, may perform LBV simultaneously and synchronously.

Several facts argue against a role for these signals in dominance interactions among adults. First, solitary foundresses perform the behavior in the absence of other adults (Brillet, Tian-Chansky, and Le Conte 1999; Harding and Gamboa 1998; Pratte and Jeanne 1984). Second, even when other adults are present, the behavior is not directed at them (Harding and Gamboa 1998). Finally, other adults on the nest do not respond in any way to the vibrations (Brillet, Tian-Chansky, and Le Conte 1999). On the other hand, there is ample evidence that the signals are directed at the brood, and particularly the larvae. First, they are closely temporally associated with adult-larva contacts. Second, the behavior does not appear in a developing colony until after the hatching of the first larvae (Brennan 2007; Brillet, Tian-Chansky, and Le Conte 1999; Jeanne 1972; Pratte and Jeanne 1984; S. Suryanarayanan, pers. comm.). On balance, the evidence supports the conclusion that these signals are directed at the larvae and not the adults.

The Non-feeding Context: Vibrations Not Associated with Brood Care

In *Polistes*, two forms of body oscillations occur outside the context of feeding larvae. The first is LV: the female shakes the gaster vigorously from side to side while standing or moving on the nest (Brillet, Tian-Chansky, and Le Conte 1999; Gamboa and Dew 1981; West-Eberhard 1969; see Savoyard et al. 1998, Figure 1, for an illustration). The posterior gastral sternites often brush the nest surface, producing a rustling or rattling sound audible a meter or more away (West-Eberhard 1969) and is sometimes vigorous enough to shake the entire comb (Savoyard et al. 1998). LV of essentially the same form has been reported for *Polistes annularis, bernardii, richardsi, chinensis, canadensis, dominulus, erythrocephalus, exclamans, flavus, metricus,* and *versicolor* (Brillet et al. 1999; Downing and Jeanne 1985; Esch 1971; Gamboa and Dew 1981; Hermann, Barron, and Dalton 1975; Ito 1995; Kasuya 1983; Strassmann 1981; West-Eberhard

1969). Mean frequencies are 15–21 strokes (i.e., half a cycle) per second and bursts last 0.3–0.7 seconds (Brillet, Tian-Chansky, and Le Conte 1999; Esch 1971; Savoyard et al. 1998). In *P. dominulus,* LV performed in this context is similar in form to LV performed in the feeding context (Brillet, Tian-Chansky, and Le Conte 1999). *Polistes*-like LVs have been reported for *Mischocyttarus angulatus* and *M. basimacula* (Ito 1993), but are absent in *M. drewseni* (Jeanne 1972). Dominant females of *Ropalidia cyathiformis* produce a rapid dorso-ventral or lateral movement of the gaster, or a single quick flap of the wings (Gadagkar and Joshi 1984). *Belonogaster juncea* shakes the gaster dorso-ventrally, sometimes causing the entire body to vibrate (Tindo, Francescato, and Dejean 1997).

The second form of body oscillation is similar in form to LV, but is a much slower abdominal movement. Described by Gamboa and Dew (1981) as "abdominal wagging" (AW), the female shakes her gaster horizontally at 3–7 oscillations per second as she walks over cells containing brood. It lasts 2–10 seconds and often produces a faint rustling sound (Gamboa and Dew 1981). It has been reported for *P. metricus, P. fuscatus,* and *P. stigma* (Gamboa and Dew 1981; Harding and Gamboa 1998; Suzuki 1996).

Whether LV and AW are qualitatively distinct in form is an open question. Some observations suggest they are points on a continuum (Brennan 2007). Vibrations intermediate in intensity between AW and LV have been observed in *P. metricus* (Gamboa and Dew 1981), and in *P. dominulus* LVs range continuously in intensity from extremely low (AW-like) to so vigorous that the entire body shakes (Brillet, Tian-Chansky, and Le Conte 1999).

Current Hypotheses on Function

Feeding-Context Vibrational Signals

One hypothesis is that vibrational signals in this context stimulate the larvae to release trophallactic salivary secretion (Darchen 1976; Keeping 1992; Roubaud 1916; Tindo, Francescato, and Dejean 1997). However, in species in which these trophallactic visits have been carefully observed, the direction of flow is to the larvae (Corn 1972; Ito 1983; Jeanne 1972; Röseler and Röseler 1989; Yamane 1971). In *P. fuscatus,* during a female's AD visits following distribution of solid food, she regurgitates a droplet then moves her head slowly into the cell until the droplet makes contact with the larva's mouthparts. The adult's mouthparts remain open and still, while those of the larva quiver slightly, suggesting that the larva is imbibing the fluid. The adult's gaster telescopes slowly inward, indicating that liquid is being forced out of the crop. This has been confirmed for two species using dye-colored prey items (Jeanne 1972; Pratte and Jeanne 1984). The pattern in which the feeding of solid prey to larvae is followed by an extended round of repeated contacts for liquid trophallaxis with the larvae is widespread if not universal in these wasps. Parsimony therefore suggests that larvae are being fed similarly in those genera that have not yet been carefully studied, and that reports that the trophallactic contacts are solicitations of larval saliva are incorrect.

Reports for several species that vibratory signaling occurs during the feeding of solid food and non-feeding inspections of brood cells (Darchen 1976; Brillet, Tian-Chiansky, and Le Conte 1999; Suryanarayanan, pers. comm.) also fail to support this hypothesis: females feeding solid material to larvae are unlikely to simultaneously solicit larval saliva to imbibe. Furthermore, males and newly eclosed adults of both sexes visit larvae and solicit trophallactic saliva (Jeanne 1972), but have never been reported to perform any kind of vibrational behavior. Thus, vibratory signals are not necessary to elicit larval saliva.

Taken together, these observations indicate that this hypothesis has little support and that vibrational signaling in this context does not function as a stimulus for the release of larval saliva.

An alternative hypothesis states that vibrational signaling in this context inhibits the release of salivary secretion by larvae. Pratte and Jeanne (1984) showed that larvae of *Polistes fuscatus* released significantly less saliva following AD than when not subjected to AD. They hypothesized that the inhibition prevents saliva from becoming mixed with and diluting the liquid food coming from the adult (Pratte and Jeanne 1984).

However, several observations raise doubts about this hypothesis as well. First, Pratte and Jeanne (1984) found AD in the feeding context to reduce trophallactic saliva release by less than 23% over controls, not a very strong inhibition. Second, vibrational signaling does not always accompany the feeding of liquid. Workers rarely perform it when feeding larvae, indicating that it is not a necessary signal in this context (Brillet, Tian-Chansky, and Le Conte 1999, Pratte and Jeanne 1984). Third, in *P. dominulus* LV is frequently performed during non-feeding inspection visits to brood cells (Brillet, Tian-Chansky, and Le Conte 1999). Finally, late in the colony cycle of *P. dominulus* feeding-context LVs are rare or absent, yet larvae are fed liquid normally (Brillet, Lian-Chansky, and Le Conte 1999).

Non–Feeding-Context Vibrational Signals

Compared to feeding-context signals, it has been more difficult to determine whether non–feeding-context signals are directed at the adults or the larvae. Again, there are two hypotheses on function.

First, there is some evidence that LV is a form of aggression toward other *Polistes* adult females. This evidence follows several lines. In a number of species, LV is performed most frequently and/or energetically by queens, less frequently by high-ranking subordinates, and rarely by workers (Brillet, Tian-Chiansky, and Le Conte 1999; Downing and Jeanne 1985; Gamboa and Dew 1981; Ito 1995; Kasuya 1983; Savoyard et al. 1998; Strassmann 1981; Theraulaz et al. 1992; West-Eberhard 1969, 1986). Similarly, abdominal wagging in *Polistes* is more often performed by queens than by subordinate foundresses or workers (Gamboa and Dew 1981). Furthermore, dominance hierarchies based on frequency of performance of LV match those based on aggressive interactions (Hughes, Beck, and Strassman 1987; but see Savoyard et al. 1998) and LV is often temporally associated with aggressive encounters (Brillet, Tian-Chiansky, and Le Conte 1999; Ito 1993; Theraulaz et al. 1992; West-Eberhard 1969, 1986; but see Downing and Jeanne 1985). In particular, in some species LV is most conspicuous during the founding period or after queen disappearance, when there is reproductive competition among females (West-Eberhard 1969). Subordinates and dominants react differently to the behavior (Gamboa and Dew 1981; Ito 1993; Theraulaz et al. 1992). Performance of an LV is sometimes directed at another adult on the nest, especially newly emerged females, subordinates, and workers (Brillet, Tian-Chansky, and Le Conte 1999) which may avoid individuals engaging in LV or leave the nest in response (West-Eberhard 1986). Finally, LV performance frequency is positively correlated with ovary development (Ito 1995).

A second hypothesis is that vibratory signals in the non-feeding context regulate the release of larval trophallactic saliva. Indeed, there is some evidence that these signals are directed at the larvae. LV and AW in *Polistes* may be performed by a solitary foundress or by a co-foundress when she is alone on a nest containing larvae (Downing and Jeanne 1985; Savoyard et al. 1998). In some cases LV and AW in *Polistes* spp. occur rarely or not at all until the larvae hatch (Downing and Jeanne 1985; Gamboa and Dew 1981; Savoyard et al. 1998). In *P. metricus*, the frequency of LV and the number of larvae in the nest are positively correlated (Gamboa and Dew 1981). The strongest evidence that LV is correlated with the presence of larvae comes from Savoyard et al.'s demonstration for *P. fuscatus* that removal of larvae from founding-stage, multiple-foundress nests significantly reduced the rate of lateral vibration. After the removed larvae were replaced, LV rate increased.

Based on patterns of occurrence, Harding and Gamboa (1998) hypothesized for *Polistes fuscatus* that AW stimulates release of larval saliva and LV inhibits it. Larvae released significantly less saliva immediately after an LV than 30 minutes after (Cummings, Gamboa, and Harding 1999). Noting that LVs cluster in the 2.5–minute interval prior to a foraging trip, Cummings et al. suggested that LV minimizes release of larval saliva during the foundress's absence when it might attract parasites and predators to the nest.

The case that non-feeding–context signals regulate release of larval saliva is less than compelling for several reasons. First, the experimental work with LV has shown only modest reductions (20–30%) in the amount of saliva yielded by larvae (Cummings et al. 1999), less than expected if the signal were an effective inhibitor. The reduced yield of saliva and the retraction of larvae in response to these signals (Cummings, Gamboa, and Harding 1999; Savoyard et al. 1998) could be generalized startle responses. Second, if these signals had to do simply with the mechanics of regulating flow of larval saliva, why should only the most dominant females use them? Third, there is no experimental evidence that AW stimulates release of larval saliva. Finally, why and how could AW and LV evolve to have opposite affects on the receiver, when they are so similar in form (Brillet, Tian-Chansky, and Le Conte 1999; Gamboa and Dew 1981)?

Although there is evidence that LV and AW are directed at larvae, the case against their being directed at adults as well is not compelling. There are reports of these signals being performed on multi-foundress nests containing only eggs (Hughes, Beck, and Strassman 1987; Strassmann 1981; Tindo, Francescato, and Dejean 1997), and LV is often closely associated with adult-adult aggression and/or is directed at other females. It is possible, therefore, that these signals have effects on both larvae and adults.

Vibrational Signals and Caste Determination

Do vibrational signals play a role in caste determination? Existing hypotheses on the function of these signals assume that the signals have releaser effects on behavior. I suggest that this is incorrect, and that they instead have a modulatory effect on growth, development, and reproductive physiology. This notion was first suggested by Brillet, Tian-Chansky, and Le Conte, whose observations of LV in *Polistes dominulus* led them to conclude that feeding-context signals can have little to do with the mechanics of adult-larva trophallaxis and to suggest instead that it somehow helps to prepare workers for their future status as dominated subordinates (Brillet, Tian-Chansky, and Le Conte 1999). They also suggested that vibrational signals performed in the non-feeding context maintain females in a subordinate, nonreproductive state.

I suggest that Brillet et al.'s hypothesis can be applied to all the independent-founding polistine wasps, that the various forms of vibrational signals performed in the feeding context are directed at the developing larvae, and that vibrations performed in the non-feeding context have effects on both the larvae and adults. Further, I hypothesize that all of these signals are part of the same functional continuum as physical attacks by dominant individuals on subordinates; that is, they act via the same pathways to have suppressive effects on reproductive physiology of the recipients. Finally, I argue that the signals initiate a cascade of biochemical events in the recipients, resulting in patterns of gene expression that reduce the reproductive potential of the recipient. The vibrations are transmitted through the nest carton to the adults standing on the nest as well as to the brood inside the cells, whereas dominance attacks transmit the signal directly onto the body of the recipient adult female.

The argument that vibrational signals directed at the larvae bias development toward worker-like adults rests on the assumption that some degree of pre-imaginal caste determination occurs in these wasps. The queen has an interest in producing early offspring that will behave as workers, and to this end she can exert influence on them both as larvae and as adults. Although most independent-founding polistines lack discrete morphological castes, there is evidence of behavioral, physiological, and morphological differences among adults that are traceable to the larval stage (Gadagkar et al. 1991; Hunt and Amdam 2005; reviewed by O'Donnell 1998). Eickwort (1969) found that females of *P. exclamans* on late-season nests fall into two groups based on quantity and quality of fat body. Workers had fat body that was thin, patchy, opaque, and yellowish, while gynes had a continuous thick blanket of whitish, translucent fat body. Keeping (2002) has shown for *B. petiolata* that gynes are significantly larger, differ in body shape, and have significantly more fat than workers. In the only direct test of the null hypothesis that all adult females are reproductively capable, Gadagkar et al. (1991) showed that when eclosing females of *Ropalidia marginata* were isolated and fed *ad-libitum,* only about half were capable of initiating nests and laying eggs. Finally, analysis of storage proteins in adult *Polistes* showed differences between females emerging early and late in the colony cycle that could be linked to caste differentiation in the pre-imaginal stage (Hunt, Buck, and Wheeler 2003). It should be cautioned that pre-imaginal influences appear to have only a biasing effect on caste in most of these wasps, leaving considerable flexibility in the adult stage for social and environmental factors to influence caste (Gadagkar et al. 1991; Mead and Gabouriaut 1993).

Current Hypotheses on Mechanisms of Pre-imaginal Caste Determination

The current consensus is that differential nutrition during the larval stage is the cause of pre-imaginal caste determination in social vespids (Hunt 1994; O'Donnell 1998; Wheeler 1986, 2003). Well-fed female larvae develop into gynes, whereas the less well-fed become workers (Hunt and Amdam 2005). This has some support from food supplementation studies. Compared with controls, colonies given ample nutrients produce higher frequencies of female offspring with gyne-like traits, including larger size, more fat body, and enhanced cold tolerance (Hunt and Dove 2002; Karsai and Hunt 2002; Mead et al. 1994; Miyano 1998; Rossi and Hunt 1988). Conversely, underfeeding produces smaller offspring (Karsai and Hunt 2002). It has been also suggested that workers produced early in the colony cycle are the result of poor larval nutrition due to a high larva/worker (L/W) ratio, while the later-eclosing gynes benefit from improved nutrition due to a declining L/W ratio (Gadagkar et al. 1991; Reeve 1991; West-Eberhard 1969).

Hunt and Amdam (2005) recently developed a model for the origin of eusociality in *Polistes* that is based on the developmental pathways already in place in the presumed bivoltine, solitary ancestor. They postulate that larvae diverge into one of two developmental pathways: worker-biased if they receive scanty food, gyne-biased if they receive abundant food. However, the quantity of nourishment received by larvae appears to be insufficient to account for observed patterns of development (Reeve 1991; West-Eberhard 1969) for atleast two reasons. First, rate of larval development, generally positively correlated with resource availability in insects (Nijhout 2003), does not correlate well with caste. In *Polistes* and *Ropalidia,* larval development times are shortest for the first few (worker) offspring, rise rapidly to a maximum for later-eclosing workers, then gradually decline to intermediate durations for the remainder of the colony cycle, when gynes are produced (Kojima 1989; Kudo 2003; Mead et al. 1994; Miyano 1983, 1990, 1998; Strassmann and Orgren 1983). Thus, larval growth rate of workers spans the range of fastest to slowest in the life of the colony, while gynes have intermediate development rates. This pattern is also found in tropical *Mischocyttarus* (Jeanne 1972), so it cannot be temperature related. Interestingly, virtually all the difference in development time between the earliest and later workers is concentrated in the fifth instar (*Polistes chinensis:* Miyano 1990), just as in *Apis mellifera* (Wang 1965). This may be a manifestation of the greater accumulation of hexameric storage proteins being laid down in the last instar (Hunt, Buck, and Wheeler 2003), as predicted by the Hunt and Amdam model.

Second, the first-produced workers not only develop more rapidly than any subsequently produced females, but paradoxically they are also the smallest (Karsai and Hunt 2002; Miyano 1983; Turillazzi 1980; Turillazzi and Conte 1981). Although Mead et al. (1994) and Kudo (2003) concluded that the rapid growth of the first larvae in *Polistes* is due to more intensive feeding, this is contradicted by supplemental feeding of larvae, which leads to both more rapid larval development and growth to larger, not smaller, adults (Kudo 2003; Miyano 1998; Rossi and Hunt 1988; West-Eberhard 1969). In other words, the sizes to which offspring grow do not correlate well with their rates of development as larvae. Furthermore, the larger females produced later in the colony cycle sometimes include workers that do not differ measurably in size from gynes (Miyano 1983).

In summary, quantity of food received by a larva appears to be insufficient to explain the patterns of offspring size, rate of development, and caste. The evidence suggests that the effect of nutrition must be modulated by at least one additional source of environmental input (Miyano

1983, 1990; Miyano and Hasegawa 1998; Reeve 1991). I suggest that the missing factor is the vibrational stimuli produced so prominently by these wasps.

The Mechanical-Switch Hypothesis: Possible Mode of Action

I postulate that vibrational signals performed in the context of larval feeding have developmental effects. How environmental signals induce gene expression leading to the development of castes in most eusocial Hymenoptera is not fully understood, but current evidence suggests that differences in food quality trigger changes in levels of neurohormones, juvenile hormone (JH), insulin, and ecdysone, which in turn may act as token stimuli that mediate the gene-expression differences leading to caste-specific developmental pathways (Nijhout 2003; Wheeler 2003). The vibrational signals transmitted through the nest to the larvae (Brennan 2007) early in the colony's development induce biochemical changes in the developing larvae that may ultimately trigger changes in gene expression, causing those larvae to develop worker traits. Vibration, when coupled with differences in quantity of nutrition received by the larvae, could give rise to the observed differences in development rates, sizes, and caste-specific traits. Thus, they may be modulator or inducer signals (Hölldobler 1999) in that they modify the developmental response of the larva to the amount of food it receives, or modulate the reproductive physiology of an adult female. This may explain why behavioral responses to these signals by larvae and adults are typically absent.

There is good evidence that various kinds of stressors, including mechanical, affect levels of biogenic amines in insects. *Tribolium castaneum* larvae subjected to tumbling in vials at 40 revolutions per minute for three days experienced a 191% increase in whole-body octopamine level over that of controls (Hirashima, Uenoi, and Eto 1992). Vibration stimulated pupation in *T. freemani* larvae housed in crowded conditions, suggesting that octopamine is involved in larval programming (Hirashima et al. 1995). Cockroaches, locusts, and crickets show elevated octopamine concentration in response to mechanical and other forms of stress (Davenport and Evans 1984; Orchard, Loughton, and Webb 1981; Woodring, Meier, and Rose 1988). Male *Drosophila virilis* show a significant elevation in dopamine concentration following shaking for 60 minutes (Rauschenbach et al. 1993). And worker honey bees subjected to dorso-ventral vibration by other workers experienced significant elevations in JH titer 15 to 30 minutes later (Schneider, Lewis, and Huang 2004).

Changes in biogenic amine levels in turn influence levels of hormones, including juvenile hormone, known to be involved in queen-worker caste differentiation during the larval stage (Rachinsky and Hartfelder 1990; Rachinsky et al. 1990). Octopamine and serotonin stimulate release of JH by the corpora allata in honey bee larvae, suggesting a role for biogenic amines in the regulation of caste determination in *Apis* (Rachinsky 1994). There is also direct evidence that stress can cause developmental change. Subjecting *Tribolium castaneum* larvae to mechanical stress caused a 54% reduction in larval growth to 54% of that of controls (Hirashima, Uenoi, and Eto 1992), and vibration at 100 Hz for two days reduced weight gain to 55% that of controls (Hirashima et al. 1993). Exposure of mice to low magnitude, 90-Hz vibration for 15 minutes per day over 15 weeks inhibited adipogenesis by 27% (Rubin et al. 2007).

These studies suggest that repeated vibration may be as effective an environmental signal as food quality in triggering a cascade of events linking biogenic amines, through hormones, to patterns of gene expression and development, and thus may play a role in caste determination in independent-founding social wasps.

In *Apis, Vespa,* and *Vespula,* the third larval instar (of five) is the critical stage at which queen- and worker-destined larvae begin to diverge developmentally (Ishay 1975; Evans and Wheeler 1999). In *A. mellifera* this is caused by a "nutritional switch" (Wheeler 1986) that leads to differences in gene expression in the two developing castes (Evans and Wheeler 1999; Hepperle and Hartfelder 2001). In *Polistes dominulus* the onset of feeding-context LV signals coincides with the eclosion of third-instar larvae (Brillet, Tian-Chansky, and Le Conte 1999). In *P. fuscatus,* although Pratte and Jeanne (1984) reported that AD in lab colonies appeared with the first-instar larvae, in field colonies it begins with the third (S. Suryanarayanan, pers. comm.). Given the reasonable assumption that in these wasps, as in *Apis* and the vespines, the third is the instar at which developmental paths diverge, this pattern supports the hypothesis that feeding-context signals bias development of larvae into workers. It also suggests that third-instar larvae issue some cue or signal that releases feeding-context signals, as found in *Apis mellifera* (Brillet, Tian-Chansky, and Le Conte 1999).

Brillet et al.'s longitudinal study of vibrational signals in *Polistes dominulus* showed that feeding-context LVs were most frequent when worker-destined larvae were being reared. LVs performed in the non-feeding context, on the other hand, increased sharply at about the time the first workers emerged (week five) and remained high through week 10, the end of Brillet et al.'s study. This pattern supports the hypothesis that these signals function in part to suppress reproductive maturation in adult worker offspring.

Although several studies have found little evidence that adults are the target of vibrational signals in this context (Downing and Jeanne 1985; Gamboa and Dew 1981; Savoyard et al. 1998) or that adults respond to the signals (Brillet, Tian-Chiansky, and Le Conte 1999; Esch 1971; Gamboa and Dew 1981; Savoyard et al. 1998), others link vibrational signals to physical dominance attacks. In *Mischocyttarus drewseni*, dominance attacks involving chewing on the body of a subordinate female were often accompanied by a rapid longitudinal body vibration (LBV), with the head of the dominant striking the subordinate (Jeanne 1972). On other occasions the dominant, while still facing a subordinate it had just dominated, drummed the gaster (GD) vigorously against the nest surface. Both LBV and GD are also used in the feeding context (see above), suggesting similar function whether directed at larvae or adults and supporting the notion that non-feeding–context signals are directed at both larvae and adults. Brillet et al. (1999) have linked non-feeding–context LVs to aggressiveness in *P. dominulus*, noting that they were often directed at newly eclosed workers. Although LV is apparently absent in *Ropalidia fasciata* (Ito 1983), the dominant female often mounts the back of a newly returned forager and solicits fluid from it, sometimes while vibrating the body violently enough to cause the forager's body to shake (Ito 1993).

If the mechanical-switch hypothesis is correct, vibrational signals in the non-feeding context may well affect development of both larvae and adults and need neither be directed at nor elicit a behavioral response by individuals of either to have their effect.

Conclusions

Vibrational signals in the independent-founding polistines, notwithstanding their conspicuousness and frequency of occurrence, have proven remarkably resistant to functional analysis. Part of the difficulty may be traced to our reluctance to think of functions other than releasers of behavioral responses in larvae and/or adults. The difficulty may also be due in part to the narrow focus to date on one or a few related species in attempts to understand function. The comparative review presented here suggests not only that these signals are widespread and variable across taxa, but highlights differences in details of temporal pattern, form, and association with other social behavior on the nest that do not fit current hypotheses based on releaser functions. The broader perspective taken here suggests instead that these movements may have become ritualized to have inducer, or modulatory, functions relating to development of caste differences. The hypothesis proposes in broad outline how vibrational signals may have developmental consequences for larvae and adults.

In socially complex Hymenoptera such as honey bees and many ants, the production of workers and gynes is ultimately regulated by queen pheromones. Proximate control is provided by environmental stimuli in the form of different chemical signals issued, respectively, to worker- and queen-destined larvae. Recent research, especially on bees and ants, has begun to piece together the biochemical pathways that bring about the differential gene expression leading to caste differentiation and to specialization within the worker caste (Evans and Wheeler 1999; Toth and Robinson 2007; Wheeler 2003). What is proposed here is that the biochemical machinery of the pathway is in place in even the most primitively eusocial species. The onset of vibrational signaling in the feeding context in *Polistes* coincides with the appearance of third-instar larvae in the nest (Brillet, Tian-Chansky, and Le Conte 1999), which suggests that it is the third larval instar that is sensitive to an environmental trigger, as in more complex bees and wasps. In the simple societies of the independent-founding wasps chemical triggers may be absent. Instead, vibrational signals may represent an alternative, albeit cruder, mechanism for biasing the development of the first offspring toward worker-like behavior and suppressing their reproductive function as adults.

In making the case for the effect of vibrational signals on the development of larvae, I have adopted the language of "parental manipulation" to describe how these signals manipulate the development of the larvae, possibly against their own interests. Inasmuch as the larvae are completely subject to the control of the adults on the nest in terms of how they are fed and otherwise treated, this seems a reasonable interpretation. Nevertheless, it is equally possible to interpret these vibrations as fertility signals that indicate the presence of a viable egglayer and that larvae and adults respond by adopting a developmental pathway that is likely to maximize their inclusive fitness under those conditions (Peeters and Liebig, this volume). Just as the weight of evidence supports such an assessment hypothesis for adult-adult interactions among social insects (Liebig, Monnin, and Turillazzi 2005; Peeters and Liebig, this volume; West-Eberhard 2003), it may well turn out to be the case for adult-brood interactions as well.

In adopting the comparative approach in this chapter I have made the simplifying assumption that the functions of vibrational signals are the same across the independent-founding polistine genera. There is the risk that this glosses over what may be real differences in function among species and genera. On the other hand, focusing on the differences can point the way to fruitful lines of research. A particularly interesting avenue for further study is whether these wasps all use vibrational signals in the same way or whether some will be shown to have pheromonal regulators of caste. The differences and anomalous cases among the species and genera reviewed here will provide the footholds and leverage necessary to tease apart the details of the function or functions of these conspicuous, yet puzzling, signals.

Acknowledgments

I thank George Gamboa, Teresa León, Sean O'Donnell, Sainath Suryanarayanan, and Ben Taylor for helpful discussion of the ideas developed herein, even though they do not necessarily agree with all of them. Research supported by the College of Agricultural and Life Sciences, University of Wisconsin, Madison, and by National Science Foundation grant BNS 77–04081.

Literature Cited

Brennan, B. J. 2007. "Abdominal wagging in the social paper wasp *Polistes dominulus:* Behavior and substrate vibrations." *Ethology* 113: 692–702.

Brillet, C., S. S. Tian-Chansky, and Y. Le Conte. 1999. "Abdominal waggings and variation of their rate of occurrence in the social wasp, *Polistes dominulus* Christ. I. Quantitative analysis." *Journal of Insect Behavior* 12: 665–686.

Corn, M. L. 1972. "Notes on the biology of *Polistes carnifex* (Hymenoptera: Vespidae) in Costa Rica and Colombia." *Psyche* 79: 150–157.

Cummings, D. L. D., G. J. Gamboa, and B. J. Harding. 1999. "Lateral vibrations by social wasps signal larvae to withhold salivary secretions (*Polistes fuscatus*, Hymenoptera: Vespidae)." *Journal of Insect Behavior* 12: 465–473.

Darchen, R. 1976. "*Ropalidia cincta,* guêpe sociale de la savane de Lamto (Côte-D'Ivoire) (Hym. Vespidae)." *Annales de la Societé Entomologique de France* 12: 579–601.

Davenport, A. K., and P. D. Evans. 1984. "Stress-induced changes in octopamine levels of insect haemolymph." *Insect Biochemistry* 14: 135–143.

Downing, H. A., and R. L. Jeanne. 1985. "Communication of status in the social wasp *Polistes fuscatus* (Hymenoptera: Vespidae)." *Zeitschrift für Tierpsychologie* 67: 78–96.

Eickwort, K. 1969. "Separation of the castes of *Polistes exclamans* and notes on its biology (Hym.: Vespidae)." *Insectes Sociaux* 16: 67–72.

Esch, H. 1971. "Wagging movements in the wasp *Polistes versicolor vulgaris* Bequaert." *Zeitschrift für vergleichende Physiologie* 72: 221–225.

Evans, J. D., and D. E. Wheeler. 1999. "Differential gene expression between developing queens and workers in the honey bee, *Apis mellifera.*" *Proceedings of the National Academy of Sciences USA* 96: 5575–5580.

Gadagkar, R., S. Bhagavan, K. Chandrashekara, and C. Vinutha. 1991. "The role of larval nutrition in pre-imaginal biasing of caste in the primitively eusocial wasp *Ropalidia marginata* (Hymenoptera: Vespidae)." *Ecological Entomology* 16: 435–440.

Gadagkar, R., and N. V. Joshi. 1984. "Social organization in the Indian wasp *Ropalidia cyathiformis* (Hymenoptera: Vespidae)." *Zeitschrift für Tierpsychologie* 64: 15–32.

Gamboa, G. J., and H. E. Dew. 1981. "Intracolonial communication by body oscillations in the paper wasp *Polistes metricus.*" *Insectes Sociaux* 28: 13–26.

Harding, B. J., and G. J. Gamboa. 1998. "The sequential relationship of body oscillations in the paper wasp, *Polistes fuscatus* (Hymenoptera: Vespidae)." *The Great Lakes Entomologist* 31: 191–194.

Hepperle, C., and K. Hartfelder. 2001. "Differentially expressed regulatory genes in honey bee caste development." *Naturwissenschaften* 88: 113–116.

Hermann, H., R. Barron, and L. Dalton. 1975. "Spring behavior of *Polistes exclamans* (Hymenoptera: Vespidae: Polistinae)." *Entomological News* 86: 173–178.

Hill, P. S. M. 2001. "Vibration and animal communication: A review." *American Zoologist* 41: 1135–1142.

Hirashima, A., T. Nagano, and M. Eto. 1993. "Stress-induced changes in the biogenic amine levels and larval growth of *Tribolium castaneum* Herbst." *Bioscience Biotechnology and Biochemistry* 57: 2085–2089.

Hirashima, A., R. Takeya, E. Taniguchi, and M. Eto. 1995. "Metamorphosis, activity of juvenile-hormone esterase and alteration of ecdysteroid titres: Effects of larval density and various stress on the red flour beetle, *Tribolium freemani* Hinton (Coleoptera: Tenebrionidae)." *Journal of Insect Physiology* 41: 383–388.

Hirashima, A., R. Uenoi, and M. Eto. 1992. "Effects of various stressors of larval growth and whole-body octopamine levels of *Tribolium castaneum*." *Pesticide Biochemistry and Physiology* 44: 217–225.

Hölldobler, B. 1999. "Multimodal signals in ant communication." *Journal of Comparative Physiology* A 184: 129–141.

Hölldobler, B., and F. Roces. 2001. "The behavioral ecology of stridulatory communication in leaf-cutting ants." In L. A. Dugatkin, ed., *Model systems in behavioral ecology: Integrating conceptual, theoretical, and empirical approaches*, 92–109. Princeton: Princeton University Press.

Hughes, C. R., M. O. Beck, and J. E. Strassmann. 1987. "Queen succession in the social wasp, *Polistes annularis*." *Ethology* 76: 124–132.

Hunt, J. H. 1994. "Nourishment and social evolution in wasps *sensu lato*." In J. H. Hunt and C. A. Nalepa, eds., *Nourishment and evolution in insect societies*, 211–244. Boulder: Westview Press.

Hunt, J. H., and G. V. Amdam. 2005. "Bivoltinism as an antecedent to eusociality in the paper wasp genus *Polistes*." *Science* 308: 252, 264–267.

Hunt, J. H., N. A. Buck, and D. E. Wheeler. 2003. "Storage proteins in vespid wasps: Characterization, developmental pattern, and occurrence in adults." *Journal of Insect Physiology* 49: 785–794.

Hunt, J. H., and M. A. Dove. 2002. "Nourishment affects colony demographics in the paper wasp *Polistes metricus*." *Ecological Entomology* 27: 467–474.

Ishay, J. 1975. "Caste determination by social wasps: Cell size and building behaviour." *Animal Behaviour* 23: 425–431.

Ito, Y. 1983. "Social behaviour of a subtropical paper wasp, *Ropalidia fasciata* (F.): Field observations during founding stage." *Journal of Ethology* 1: 1–14.

———. 1993. *Behaviour and social evolution of wasps: The communal aggregation hypothesis*. Oxford: Oxford University Press.

———. 1995. "Notes on social behavior and ovarian condition in *Polistes canadensis* (Hymenoptera: Vespidae) in Panama." *Sociobiology* 26: 247–257.

Jeanne, R. L. 1972. "Social biology of the Neotropical wasp *Mischocyttarus drewseni*." *Bulletin of the Museum of Comparative Zoology* 144: 63–150.

Jeanne, R. L., and M. G. Keeping. 1995. "Venom spraying in *Parachartergus colobopterus*: A novel defensive behavior in a social wasp (Hymenoptera: Vespidae)." *Journal of Insect Behavior* 8: 433–442.

Karsai, I., and J. H. Hunt. 2002. "Food quantity affects traits of offspring in the paper wasp *Polistes metricus* (Hymenoptera: Vespidae)." *Environmental Entomology* 31: 99–106.

Kasuya, E. 1983. "Behavioral ecology of Japanese paper wasps, *Polistes* spp. IV. Comparison of ethograms between queens and workers of *P. chinensis antennalis* in the ergonomic stage." *Journal of Ethology* 1: 34–45.

Keeping, M. G. 1992. "Social organization and division of labour in colonies of the polistine wasp, *Belonogaster petiolata.*" *Behavioral Ecology and Sociobiology* 31: 211–224.

———. 2002. "Reproductive and worker castes in the primitively eusocial wasp *Belonogaster petiolata* (DeGeer) (Hymenoptera: Vespidae): Evidence for pre-imaginal differentiation." *Journal of Insect Physiology* 48: 867–879.

Kojima, J. 1989. "Growth and survivorship of preemergence colonies of *Ropalidia fasciata* in relation to foundress group size in the subtropics (Hymenoptera: Vespidae)." *Insectes Sociaux* 36: 197–218.

Kudo, K. 2003. "Growth rate and body weight of foundress-reared offspring in a paper wasp, *Polistes chinensis* (Hymenoptera: Vespidae): No influence of food quantity on the first offspring." *Insectes Sociaux* 50: 77–81.

Liebig, J., T. Monnin, and S. Turillazzi. 2005. "Direct assessment of queen quality and lack of worker suppression in a paper wasp." *Proceedings of the Royal Society B* 272: 1339–1344.

Matsuura, M. 1984. "Comparative biology of the five Japanese species of the genus *Vespa* (Hymenoptera: Vespidae)." *The Bulletin of the Faculty of Agriculture, Mie University* 69: 1–131.

Mead, F., and D. Gabouriaut. 1993. "Post-eclosion sensitivity to social context in *Polistes dominulus* Christ females (Hymenoptera: Vespidae)." *Insectes Sociaux* 40: 11–20.

Mead, F., C. Habersetzer, D. Gabouriaut, and J. Gervet. 1994. "Dynamics of colony development in the paper wasp *Polistes dominulus* Christ (Hymenoptera: Vespidae): The influence of prey availability." *Journal of Ethology* 12: 43–51.

Miyano, S. 1983. "Number of offspring and seasonal changes of their body weight in a paperwasp, *Polistes chinensis antennalis* Perez (Hymenoptera: Vespidae), with reference to male production by workers." *Researches on Population Ecology* 25: 198–209.

———. 1990. "Number, larval durations and body weights of queen-reared workers of a Japanese paper wasp, *Polistes chinensis antennalis* (Hymenoptera: Vespidae)." *Natural History Research* 1: 93–97.

———. 1998. "Amount of flesh food influences the number, larval duration, and body size of first brood workers, in a Japanese paper wasp, *Polistes chinensis antennalis* (Hymenoptera: Vespidae)." *Entomological Science* 1: 545–549.

Miyano, S., and E. Hasegawa. 1998. "Genetic structure of the first brood of workers and mating frequency of queens in a Japanese paper wasp, *Polistes chinensis antennalis.*" *Ethology Ecology and Evolution* 10: 79–85.

Nijhout, H. F. 2003. "The control of body size in insects." *Developmental Biology* 261: 1–9.

O'Donnell, S. 1998. "Reproductive caste determination in eusocial wasps (Hymenoptera: Vespidae)." *Annual Review of Entomology* 43: 323–346.

Orchard, I., B. G. Loughton, and R. A. Webb. 1981. "Octopamine and short term hyperlipaemia in the locust." *General and Comparative Endocrinology* 45: 175–180.

Pratte, M., and R. L. Jeanne. 1984. "Antennal drumming behavior in *Polistes* wasps (Hymenoptera: Vespidae)." *Zeitschrift für Tierpsychologie* 66: 177–188.

Rachinsky, A. 1994. "Octopamine and serotonin influence on corpora allata activity in honey bee *(Apis mellifera)* larvae." *Journal of Insect Physiology* 40: 549–554.

Rachinsky, A., and K. Hartfelder. 1990. "Corpora allata activity, a prime regulating element for caste-specific juvenile hormone titre in honey bee larvae *(Apis mellifera carnica)*." *Journal of Insect Physiology* 36: 189–194.

Rachinsky, A., C. Strambi, A. Strambi, and K. Hartfelder. 1990. "Caste and metamorphosis: Hemolymph titers of juvenile hormone and ecdysteroids in last instar honeybee larvae." *General and Comparative Endocrinology* 79: 31–38.

Rauschenbach, I. Y., L. I. Serova, I. S. Timochina, N. A. Chentsova, and L. V. Schumnaja. 1993. "Analysis of differences in dopamine content between two lines of *Drosophila virilis* in response to heat stress." *Journal of Insect Physiology* 39: 761–767.

Reeve, H. K. 1991. *Polistes.* In K. G. Ross and R. W. Matthews, eds., *The social biology of wasps,* 99–148. Ithaca: Cornell University Press.

Röseler, P. F., and I. Röseler. 1989. "Dominance of ovariectomized foundresses of the paper wasp *Polistes gallicus.*" *Insectes Sociaux* 36: 219–234.

Rossi, A. M., and J. H. Hunt. 1988. "Honey supplementation and its developmental consequences evidence for food limitation in a paper wasp *Polistes metricus.*" *Ecological Entomology* 13: 437–442.

Roubaud, E. 1916. "Recherches biologiques sur les guêpes solitaires et sociales d'Afrique. La genèse de la vie sociale et l'évolution de l'instinct maternel chez les vespides." *Annales des Sciences Naturelles* 10: 1–160.

Rubin, C. T., E. Capilla, Y. K. Luu, B. Busa, H. Crawford, D. J. Nolan, V. Mittal, C. J. Rosen, J. E. Pessin, and S. Judex. 2007. "Adipogenesis is inhibited by brief, daily exposure to high-frequency, extremely low-magnitude mechanical signals." *Proceedings of the National Academy of Sciences USA* 104: 17879–17884.

Savoyard, J. L., G. J. Gamboa, D. L. D. Cummings, and R. L. Foster. 1998. "The communicative meaning of body oscillations in the social wasp, *Polistes fuscatus* (Hymenoptera: Vespidae)." *Insectes Sociaux* 45: 215–230.

Schneider, S. S., L. A. Lewis, and Z. Y. Huang. 2004. "The vibration signal and juvenile hormone titers in worker honeybees, *Apis mellifera.*" *Ethology* 110: 977–985.

Seeley, T. D. 1995. *The wisdom of the hive.* Cambridge: Harvard University Press.

Strassmann, J. E. 1981. "Wasp reproduction and kin selection: Reproductive competition and dominance hierarchies among *Polistes annularis* foundresses." *Florida Entomologist* 64: 74–88.

Strassmann, J. E., and M. C. F. Orgren. 1983. "Nest architecture and brood development times in the paper wasp, *Polistes exclamans* (Hymenoptera: Vespidae)." *Psyche* 90: 237–248.

Suzuki, T. 1996. "Natural history and social behaviour of the cofoundresses in a primitively eusocial wasp, *Polistes stigma* (Fabricius) (Hymenoptera: Vespidae), in India: A case study." *Japanese Journal of Entomology* 64: 35–55.

Theraulaz, G., J. Gervet, G. Thon, M. Pratte, and S. S. Tian-Chanski. 1992. "The dynamics of colony organization in the primitive eusocial wasp *Polistes dominulus* Christ." *Ethology* 91: 177–202.

Tindo, M., E. Francescato, and A. Dejean. 1997. "Abdominal vibrations in a primitively eusocial wasp *Belonogaster juncea juncea* (Vespidae: Polistinae)." *Sociobiology* 29: 255–261.

Toth, A. L., and G. E. Robinson. 2007. "Evo-Devo and the evolution of social behavior." *Trends in Genetics* 23: 334–341.

Turillazzi, S. 1980. "Seasonal variations in the size and anatomy of *Polistes gallicus* (L.) (Hymenoptera: Vespidae)." *Monitore Zoologico Italiano* 14: 63–75.

Turillazzi, S., and A. Conte. 1981. "Temperature and caste differentiation in laboratory colonies of *Polistes foederatus* (Kohl) (Hymenoptera: Vespidae)." *Monitore Zoologico Italiano* 15: 275–297.

Wang, D. I. 1965. "Growth rates of young queen and worker honeybee larvae." *Journal of Apicultural Research* 4: 3–5.

West-Eberhard, M. J. 1969. "The social biology of polistine wasps." *Miscellaneous Publications, Museum of Zoology, University of Michigan* 140: 1–101.

———. 1986. "Dominance relations in *Polistes canadensis* (L.), a tropical social wasp." *Monitore Zoologico Italiano* 20: 263–281.

———. 2003. *Developmental plasticity and evolution.* New York: Oxford University Press.

Wheeler, D. E. 1986. "Developmental and physiological determinants of caste in social Hymenoptera: Evolutionary implications." *The American Naturalist* 128: 13–33.

———. 2003. "One hundred years of caste determination in Hymenoptera." In T. Kikuchi, N. Azuma, and S. Higashi, eds., *Genes, behaviors and evolution of social insects,* 35–53. Sapporo: Hokkaido University Press.

Wilson, E. O. 1971. *The insect societies.* Cambridge: Harvard University Press.

Woodring, J. P., O. W. Meier, and R. Rose. 1988. "Effect of development, photoperiod, and stress on octopamine levels in the house cricket, *Acheta domesticus." Journal of Insect Physiology* 34: 759–766.

Yamane, S. 1971. "Daily activities of the founding queens of two *Polistes* wasps, *P. snelleni* and *P. biglumis* in the solitary stage (Hymenoptera: Vespidae)." *Kontyu* 39: 203–218.

CHAPTER TWELVE

Convergent Evolution of Food Recruitment
Mechanisms in Bees and Wasps

JAMES C. NIEH

THE STUDY OF FORAGING activation has played a crucial role in the theoretical development of sociobiology. In particular, explorations of the foraging and recruitment behavior of social bees and wasps have provided the
groundwork for much of the theory on central place foraging. Advances in
our understanding of foraging activation in bees (Apiformes) and social
wasps (Vespidae) show fascinating convergent similarities within and between these groups, suggesting that they have found similar solutions to
the problems of group foraging. Despite detailed studies of recruitment
in individual taxa (Dornhaus and Chittka 2004; Dyer 2002; Nieh 2004;
Raveret Richter 2000), the parallels between these different groups have
not been fully explored. In this chapter I compare and briefly, though not
exhaustively, summarize what is known about food recruitment in bees
and social wasps. The focus is on recent reviews and papers which are divided into two categories: (1) foraging activation (nest-based recruitment)
and (2) local enhancement (information provided in the field, outside the
nest). I also refer to nest-based mechanisms as "information center" mechanisms involving information transfer at a central location on or near the
nest site. In general, I hope to stir up lively debate and future studies, particularly over two hypotheses that the shared behavior of foraging through
flight is a basis for the parallel evolution of similar forms of (1) odor trail
communication in bees and wasps and (2) acoustic communication in the
corbiculate bees.

Phylogeny

The phylogenies of bees and stinging wasps (Hymenoptera, Aculeata) provide important information for understanding the evolution of foraging activation. However, as the phylogenies are not fully resolved, it is premature to map traits onto uncertain topologies. For example, there is controversy about the number of times that eusociality has evolved in the Apinae and whether stingless bee or orchid bees are the closest sister group to the honey bees (Cameron 1993). Similarly, the correct phylogenetic topology of the wasps is debated (Carpenter 2003). All social wasps are in the Vespidae with the exception of the primitively eusocial wasp, *Microstigmus comes,* which is in the Sphecidae. Within the Vespidae, sociality has evolved multiply (Schmitz and Moritz 1998). Furthermore, it is clear that sociality and thus cooperative group foraging have evolved independently between bees and wasps (Wcislo and Tierney 2007), and multiply within bees (Danforth et al. 2006) and wasps (Hines et al. 2007). Thus, even within the Apidae, some aspects of recruitment communication are likely to have evolved independently and not all similarities in communication mechanisms point to a similarly behaving ancestor.

Recruitment

Foraging Activation

Foraging activation is an increase in the probability of an individual leaving the nest to search for resources as a result of information received (at the nest) from successful foragers. This information can come from nestmates or non-nestmates (as in an information center), consist of cues (evolutionarily basal) or signals (evolutionarily derived), and indicate the general availability of resources or their specific location. Intranidal (within-nest) or at-nest communication of food location is rare in bees and undocumented in wasps.

Benefits of recruitment communication can depend on resource density. Specifically, honey bee location communication is advantageous if patches are variable, poor, and few, but not when resources are densely distributed (model results, Dornhaus et al. 2006). Thus communicating specific food location may not always be beneficial, even in species with this ability because additional nearby resources may be missed. Only honey bees and some species of stingless bees are known to communicate specific resource location at the nest (Nieh 2004; von Frisch 1967). Few studies have tested the ability of wasps to recruit to a specific location and to date there is no evidence that they can do so (Raveret Richter 2000).

Aggregations

An untested, though intriguing, possibility is that foraging activation exists among clumped nests of solitary bees. Mutual stimulation between solitary nesting bees could lead to foraging activation. Nesting aggregations occur among all taxonomic groups of bees, particularly soil nesters. Potts and Willmer (1998) report close spacing of up to 304 nests/m² with a nearest neighbor distance of 25.2±1.1 mm in the solitary ground nester, *Halictus rubicundus* (Halictinae). Given this population density, near neighbors could potentially monitor each other's departures. However, departure synchrony, or an increased rate of nest departures after the return of a successful forager, is not sufficient to demonstrate foraging activation because these effects may also arise from circadian rhythms and times of food availability on previous days. Some of these confounding factors could be eliminated with feeder studies (conducted during seasonal food dearth) using individually-marked bees whose daily foraging patterns are documented before, during, and after food is offered at unpredictable times.

At solitary bee nesting aggregations, cues such as floral odor adhering to returning foragers and, to a limited extent, visual and acoustic information could elicit foraging activation. Halictine visual acuity has not been measured, but may be similar to that of bumblebees, which can resolve a 2 cm object from 82 cm away (angular acuity of 0.36 cycles/degree, Macuda et al. 2001), and worker honey bees (0.26 cycles/degree, Srinivasan and Lehrer 1988). Such resolution should be sufficient to allow visual detection of near-neighbors exiting nests. It is not known if sounds produced by exiting Halictine foragers can activate neighbor foraging, but this hypothesis is testable.

Some wasp species exhibit foraging activation. Hrncir, Mateus, and Nascimento (2007) demonstrated foraging activation in the social swarm-founding wasp, *Polybia occidentalis,* by showing that newcomers only arrived at feeders after researchers trained foragers to the feeders. As with bees, there is no data on foraging activation in solitary wasp nest aggregations. However, solitary digger wasps *(Cerceris arenaria)* form dense nesting aggregations of up to 136.4 nests/m² over a 3.6 m² area. Moreover, wasps preferred to stay in their natal nests and in the natal nesting area, thus creating the potential for increased relatedness among neighbors (Polidori et al. 2006).

Mechanisms of Information Transfer

Several multimodal mechanisms of information transfer can activate foraging in social bees and social wasps. It is useful to consider these in detail, because information sources such as excitatory motions of returning foragers, colony resource levels, trophallaxis (food exchange), and olfactory cues (food scent) are likely basal, whereas olfactory signals (recruitment pheromones) and functionally referential communication are thought to be more derived (von Frisch 1967). For example, successful foragers of all eusocial corbiculate bees exhibit *increased movement rates* upon returning inside the nest. Increased food quality results in increased velocity and acceleration of movements by recruiting honey bees (Dyer 2002) and stingless bees (Schmidt, Zucchi, and Barth 2006). Bumblebees *(Bombus terrestris)* also perform excitatory runs inside the nest when returning from good food sources, and foragers increase their average speed when colony honey stores are experimentally depleted (Dornhaus and Chittka 2005). Given that such excitatory responses to food are widely observed in many insects, these behaviors are possibly basal (von Frisch 1967)

In wasps, Naumann (1970) reported a "departure dance" in which a rapidly running wasp forager was licked and antennated by nestmates. It would be useful to determine if these forager motions follow a similar pattern to that observed in recruiting stingless bees, which run in zigzag patterns interspersed with sudden turns (Nieh 2004). Stingless bee (Nieh 2004) and honey bee recruits (Rohrseitz and Tautz 1999) also frequently contact recruiters.

Excitatory buzzing runs during wasp swarming offers another parallel between bees and wasps. Although swarming is distinct from food gathering, it transfers information about a resource location, the new nest site. In stingless bees and honey bees, swarming and foraging use many of the same guidance and communication mechanisms (Roubik 1989). Excitatory buzzing runs are seen throughout the swarm in many wasp species (Naumann 1975), similar to the behavior of buzz runners in swarming honey bees (Seeley et al. 2003).

Successful foragers returning to the nest can produce *acoustic signals* in bumblebees *(B. terrestris;* Oeynhausen and Kirchner 2001), stingless bees (Nieh 2004), and honey bees (Dyer 2002). In these groups, thoracic muscle contractions can generate sound and vibrations that could increase forager conspicuousness to nestmates; however, thoracic muscle contractions can occur silently (Heinrich 1984) and thus a non-exclusive route for the ritualization of an acoustic recruitment signal is the buzz of foragers as they fly away from the nest. Wasp foragers are not known to produce foraging-related acoustic signals, although wasps produce a wide variety of vibrational signals including alarm tapping (Jeanne and Keeping 1995).

Trophallaxis between nestmates can result in foraging activation, particularly if successful foragers offer their food to nestmates. For example, trophallaxis increases after the return of a successful forager in *Megalopta* bees (Halictidae; Wcislo and Gonzalez 2006), honey bees (von Frisch 1967), and stingless bees (Hrncir et al. 2000). Returning foragers of the facultatively social *Megalopta genalis* and *M. ecuadoria* (Halictidae) regularly give nectar to nestmates. All females can participate in foraging, with the second oldest female usually making the most foraging trips (Wcislo and Gonzalez 2006). Thus, foraging activation following food exchange is possible. In other cases, such as *Xylocopa sulcatipes* (Xylocopinae), typically only one female bee forages and other females receive nectar trophallactically from her. This would not lead to foraging activation, but Velthius (1987) observed two-female nests in which both females foraged and could thus potentially activate each other.

In honey bees, trophallaxis from dancing foragers provides information about the odor and sweetness of the nectar, thus contributing toward a forager's decision to visit the advertised food source (Farina and Wainselboim 2005). No studies have shown that trophallaxis alone leads to foraging activation in naïve honey bees, although one suspects it may lead to foraging reactivation in experienced foragers. Similarly, no stingless bee studies have systematically examined the possibility of food alertment due solely to trophallaxis.

Trophallaxis and grooming activity can increase after a *Mischocyttarus* or *Polistes* forager returns to the nest, and one or more foragers may then leave (Jeanne 1972). However, we do not know if these forager departures occur because of food received from a successful forager. Aggression can activate wasp foraging, and provides an interesting parallel to aggressive behavior in bees of dominant *Megalopta* females. In the wasp *Polybia occidentalis*, O'Donnell (2001) found a positive correlation between foraging and the rate of being bitten, with some workers leaving to begin foraging activities immediately after being bitten. Here, the aggressed wasp does not offer food to the aggressor, as in the *Megalopta*, but the aggressed wasp does leave the nest, presumably to obtain food. Biting may be an example of foraging activation, but it is unclear if the biting wasps had recently discovered food.

Stored food levels, as assayed through gustation, touch, and olfaction, provide an information reservoir that can modulate foraging activation. Communal food provisioning can allow individual foragers to assess colony food levels and learn about resource availability. Food levels alone can inform nestmates of colony need and alter foraging, thus modulating foraging activation. For example, honey bee colonies increase pollen foraging when pollen levels decrease (Fewell and Winston 1992). The rate of bumble bee *(B. terrestris)* exits increases and successful returning foragers spend more time running excitedly when honey pots are depleted than when they are full (Dornhaus and Chittka 2005). We know little about the effects of communal food provisioning on foraging in other bee groups. This is somewhat surprising given that communal provisioning is widespread among bees such as Anthophoridae (genus *Exomalopsis*; Michener 1974); Halictidae, (genus *Lasioglossum*; Richards, French, and Paxton 2005); and Euglossini (limited cross-provisioning in *Eulaema nigrita*; Zucchi, Sakagami, and Camargo 1969).

Non-food related stores might also affect foraging decisions, although the influence of non-food supplies on colony foraging has received less attention than food stores. In the orchid bee, *Euglossa townsendi* (Euglossini), resin is reused from old cells and taken from resin dumps created by foragers near the nest entrance (Augusto and Garófalo, 2004). Whether resin foraging activation occurs remains to be determined. In *Eulaema nigrita* (Euglossini), foragers created separate piles of building materials (resin, mud, and feces) that were used communally to seal cracks in the nest and to complete brood cell construction (Santos and Garófalo 1994). It is unknown if non-food stores can influence Euglossine colony foraging. In stingless bees, foragers are known to collect salt, water, urine, feces, resins, bark, leaves, and mud (Lorenzon and Matrangolo 2005). In honey bees, Nakamura and Seeley (2006) found that *Apis mellifera* foragers perform waggle dances for resin sources deep inside the nest where the resin is typically used, thus facilitating resin use and direct sensing of its need by collectors. Similar studies in stingless bees would provide useful comparative information. Like bees, social wasps need to forage for food, water, and nest materials (Raveret Richter 2000).

Olfactory cues and signals can activate foraging. A cue such as food odor can reactivate experienced honey bee foragers to visit their former feeding site (Reinhard et al. 2004). Whether food odor alone can lead to foraging reactivation in other social bees deserves investigation. The bumblebee *B. terrestris* releases tergal gland pheromone during the excitatory movements of a successful returning forager (Dornhaus and Chittka 2004). One primary function of these movements may be to disperse the recruitment pheromone. Recruiting honey bees also produce volatile compounds, and a synthetic blend of these compounds increased the number of bees exiting the hive (Thom et al. 2006). Stingless bees may also use intranidal recruitment pheromones, although this awaits experimental evidence.

Olfactory cues can also lead to wasp foraging activation. Overmeyer and Jeanne (1998) showed that inexperienced *Vespula germanica* foragers prefer to visit feeders with the same scent as that carried back to the nest by successfully foraging nestmates. Foragers based their preference on scent alone because the authors eliminated visual local enhancement, unlike previous studies. There is no evidence that wasps produce an intranidal recruitment pheromone, but no published studies have examined this possibility.

Forager temperature inside the nest may contribute toward foraging activation. Returning foragers have intranidal thorax temperatures that are elevated over ambient air temperature (ΔTth) and correlate with collected sugar concentrations in honey bees (Stabentheiner 2001) and stingless bees (Nieh and Sánchez 2005). The function of elevated ΔT_{th} has not been determined, but it may keep thoracic flight musculature at higher temperatures and thus facilitate a more rapid return to higher quality food (flight facilitation hypothesis). Other testable hypotheses are that it attracts potential recruits to foragers advertising high quality food (attraction hypothesis), or enhances the release of food odors or foraging activation pheromones (odor signal modulation hypothesis). Research using artificially heated bees could help to distinguish between these different hypotheses. Successful *B. terrestris* foragers returning from a rich sucrose solution produce a foraging activation pheromone; the resulting higher body temperature could enhance recruitment pheromone release. We do not know if this species has elevated intranidal ΔT_{th} corresponding to food quality; but in *B. wilmattae*, ΔT_{th} is correlated with sucrose concentration in bees feeding within a foraging arena (Nieh et al. 2006).

Food carbohydrate levels can also affect wasp thoracic temperatures; for example, *Paravespula vulgaris* foragers increased ΔT_{th} when feeding on more concentrated sucrose solution (Kovac and Stabentheiner 1999). Currently, we do not know if ΔT_{th} elevation at the feeder persists at the nest and is thus a potential foraging cue for either species. This would be worth investigation. Moreover, the wasp data suggest that we should also examine the effect of pollen protein quality on social bee ΔT_{th}. Parallel effects may exist in wasps and bees, particularly if thermal hypothesis 1 is correct and the primary function of elevated temperatures is not signaling but facilitating flight.

Functionally referential signals occupy a behavioral continuum and are defined by two key features: they are stimulus-class specific (specific to the environmental information, event, or item being signaled) and context independent (the sender signals and the receiver behaves appropriately without the direct presence of what is signaled). Thus the honey bee waggle dance is functionally referential because it is (1) specific to the communication of resource location and (2) communicates spatial location and elicits appropriate receiver responses without the signaler and receiver being at the communicated location. In social insects, the information-rich honey bee waggle dance dominates the concept of functionally referential communication. Yet functionally referential communication, as understood in other animal systems, occurs in forms such as predator-specific alarm calls in ground squirrels (Blumstein 1999) and chicken food calls (Evans and Evans 1999). Thus the potential encoding of food quality in stingless bee sound pulses (Hrncir, Barth, and Tautz 2006) may also be an example of functionally referential communication. Researchers could demonstrate this if receivers that do not receive a food sample (context independence requirement) respond appropriately to this sound-encoded food quality information.

In stingless bees, the evidence for functionally referential communication is primarily based on correlations between sound pulse duration and food distance (Nieh 2004). Direct evidence is required to demonstrate functionally referential communication because it is also possible that these sounds activate foraging without communicating distance. Functionally referential foraging communication has not been found or sought in wasps.

Local Enhancement

Local enhancement is the facilitation of learning resulting from an individual's attention being drawn to a locale and then reinforced with a reward (Roberts 1941). In social insects, this can occur when the presence of an individual high quality food attracts another individual, who thereby obtains a food reward. Local enhancement is a subset of *social facilitation*, which Wilson (1971) defined as "behavior initiated or increased by the action of another individual." Olfactory, visual, and acoustic information could lead to social insect local enhancement, although studies to date have not tested acoustic local enhancement. Local evolutionary enhancement toward visual and olfactory cues and piloting are likely basal evolutionary conditions. Odor trail and target only olfactory signaling may have evolved subsequently.

We do not know if local enhancement occurs in solitary bees; I therefore focus on social bees. Social insects are champion associative learners when it comes to food and local enhancement is thus likely to be more widespread than is currently documented. In general, visual local enhancement has been neglected in bee research. One exception is provided by Slaa, Wassenberg, and Biesmeijer (2003), who reported that newly recruited foragers of the stingless bee *Trigona amalthea* exhibited visual local enhancement, approaching nestmates on feeders. No studies have yet demonstrated visual local enhancement in bumblebees, honey bees, or non-corbiculate social bees.

Unlike bee research, wasp studies have focused more on visual local enhancement. Investigators used odor-extracted posed wasps and controlled for olfactory local enhancement by counting choices made in the absence of other live foragers. In this study, *Vespula germanica, V. consobrina*, and *Polistes fuscatus* foragers were attracted to extracted (odor-free) posed wasps on feeders or flowers over feeders or flowers without posed wasps. In *V. maculifrons*, attraction was density dependent because foragers were attracted to other foragers on a closely spaced feeder, but not to foragers on a widely spaced array (Raveret Richter 2000). Similarly, *Polybia occidentalis* newcomers were visually attracted to extracted wasp dummies placed on a feeder (Hrncir, Mateus, and Nascimento 2007).

Wasps can be attracted to the visual or olfactory presence of other wasps on food. Investigators trapped *V. germanica* and *V. maculifrons* foragers in pierced clear plastic containers that allowed meat bait and potential forager odors to escape and found that foragers were attracted to these baits over control baits with meat alone (Raveret Richter 2000). *Polybia occidentalis* and *P. diguetana* foragers were attracted to caterpillar baits occupied by a live forager over an unoccupied bait (Raveret Richter 2000). However, *V. germanica* were not attracted to potential odor marks deposited after 50 or 100 feeder visits (Raveret Richter 2000), therefore we do not know if wasps can deposit odors to mark good resources.

Piloting is a form of local enhancement that can combine vision and olfaction. In social insects, piloting is perhaps the most basal recruitment strategy for an individual to lead one or more nestmates to a resource, yet it has been difficult to directly demonstrate piloting in flying insects. Honey bee scouts use piloting to help guide the swarm to their new home. The swarm relies on the conspicuous visual behavior of a relatively few fast fly-ing "streaker" bees. Sealing the Nasanov glands of these guide bees does not impair swarm guidance, and thus honey bees evidently use vision, not olfaction, in swarm piloting (Beekman, Fathke, and Seeley 2006). Piloting has not been found in honey bee food recruitment (Riley et al. 2005). In stingless bees, Aguilar, Fonseca, and Biesmeijer (2005) found a close tem-poral synchrony in the arrival times of foragers and newcomers (recruits) in *Trigona corvina* and *Plebeia tica,* evidence that suggests piloting. Odor release may facilitate piloting. Kerr (1994) hypothesized that some meliponines deposit aerial odor trails, creating an "odor tunnel" as they fly to the food source during windless conditions under a dense forest canopy. To date, no studies have tested this hypothesis.

Partial piloting, in which foragers lead nestmates part of the distance to the food source rather than the entire distance, may occur accidentally if recruits lose track of the recruiter (a potential basal state), but may also be a consistent strategy, as is hypothesized for several species of *Melipona* (Kajobe and Echazarreta 2005). To date, no studies have directly demon-strated the existence of partial piloting (Nieh 2004), relying instead on in-direct evidence that recruiters need some contact with recruits as they leave the nest (to communicate direction) and that recruiter and recruits often do not arrive in synchrony, as would be expected in complete pi-loting. Some form of piloting may exist in wasps, but this remains to be tested. As with some stingless bees, *Apoica pallens* wasps are hypothesized to use aerially released pheromones to help guide swarms, but this re-mains untested (Howard et al. 2002).

Target-only odor marking (i.e., odor-marking of the food source alone) is widespread among the social bees (Stout and Goulson 2001). I adopt this term to distinguish target-only odor marking from odor trails. The evolutionary precursor to food odor-marking may lie in nest entrance odors that help to orient returning bees. Nest entrance orientation marks have been found in solitary bees (Guédot et al. 2006), bumblebees (*B. oc-cidentalis;* Cameron et al. 1999), honey bees (*Apis mellifera;* von Frisch 1967), stingless bees (Schmidt, Zucchi, and Barth 2005), and wasps (*Vespula vulgaris;* Butler, Fletcher, and Watler et al. 1969).

Some stingless bee species (Schmidt, Zucchi, and Barth 2005) and bum-blebees (in foraging arenas; Schmitt, Lübke, and Franke 1991) can deposit attractive target-only odor marks by walking. In certain species, these "foot-print" odor marks may be odor cues (cuticular hydrocarbons) or odor signals that are not specific to food. For example, the stingless bee *Nannotrigona testaceicornis* deposits attractive olfactory compounds by walking on food sources. These compounds are equally attractive as those deposited by bees walking at the nest entrance (Schmidt, Zucchi, and Barth 2005). On natural flowers, Eltz (2006) demonstrated that *Bombus pascuorum* foragers deposited lipid footprints with the same composition as cuticular hydrocarbons found on other parts of the foragers' bodies. Similarly, wasp foragers (*Vespa crabro* and *Vespula vulgaris*) can deposit cuticular hydrocarbons to create an intranidal orientation trail (Steinmetz, Sieben, and Schmoltz 2002).

Target-only odor marks deposited on food can also be aversive, marking food sources that have been depleted and should thus be avoided (Stout and Goulson 2001). *Xylocopa virginica texana* (carpenter bee) foragers visiting passion flowers deposited repellant chemical marks on depleted flowers (Frankie and Vinson 1977). Recently, Saleh and Chittka (2006) have shown that foragers can learn to positively or negatively associate these bee-deposited odors with rewarding or unrewarding flowers, respectively. Thus, the same odor can act as an attractant or a repellant. Depositing cuticular hydrocarbons may be an inevitable consequence of multiple visitations to a rewarding food source. It would not be surprising if several species of stingless bees and bumblebees were able to associate food quality (handling time, sugar quality, and travel cost) with tarsal deposits of cuticular hydrocarbons. The ability of bees to learn associatively can enhance the flexibility of their behavioral repertoire without necessitating the evolution of specific olfactory foraging signals. Cues can sometimes be sufficient.

Odor trails are a form of local enhancement because they draw a recruit's attention to a rewarding location. In many cases, odor trails are observationally associated with some degree of piloting (stingless bees, Nieh 2004; wasps, Jeanne 1981). Some stingless bee species can produce an odor trail consisting of odor droplets deposited a few meters apart on vegetation between a feeder and the nest, even up a vertical substrate such as a tower. Some meliponine species also deposit partial odor trails consisting of odor marks deposited in decreasing spatial density extending from the feeder in the direction of the nest, but not the entire distance to the nest (Nieh 2004). In at least two meliponine species, these partial odor trails are polarized, allowing foragers who enter the odor trail to determine the correct endpoint (the food source) without first traveling to one of the endpoints. Stingless bees deposit these odor trails by briefly landing, often on the edge of leaf or twig, and rubbing their mandibles against the substrate. Schorkopf et al. (2007) showed that this odor trail is composed of labial gland secretions in *Trigona recursa*. To date, odor trail studies have investigated only recruitment to nectar sources, although meliponine odor trails may also indicate resources such as pollen, resin, and building materials.

Bumblebee (*B. terrestris*) workers deposit short odor trails between the nest and the nest entrance (tested in a 100 cm diameter foraging arena, Cederberg 1977). The Amazonian bumblebee, *B. transversalis*, clears short trails (2–3 m) that extend from the nest and facilitate nest material gathering. Foragers followed the trail by keeping their antennae just above the substrate surface and crawling forward, moving in a sinusoidal fashion. This is reminiscent of how ants follow odor trails. Such a trail may be useful in nature when foragers negotiate obstacles to find the entrances of their subterranean nests. It is not known if these trails are odor-marked (Cameron et al. 1999). *B. impatiens* foragers also deposited odor trails within a foraging arena and were able to use these trails, in darkness, to walk to food sources (Chittka et al. 1999). Similarly, Steinmetz et al. reported that wasp foragers (*Vespa crabro* and *Vespula vulgaris*) can use odor trails to navigate through dark entrance tunnels within the nest. As in some stingless bees (Schmidt, Zucchi, and Barth 2005), these intranidal wasp odor trails may consist of cuticular hydrocarbons deposited by walking foragers (Steinmetz, Sieben, and Schmoltz 2002).

Some wasp species use odor trails to guide swarms or to assist orientation within nest cavities. Although odor trails are not known to be involved in wasp foraging (Raveret Richter 2000), foraging in many tropical species with large colonies (and thus a potential need for mass recruitment) remains to be studied. Moreover, the similarities between how wasps and stingless bees deposit odor trails are fascinating, because flying insects must confront the same problems of how to deposit and orient toward an odor trail that consists of small odor deposits widely spaced.

Substrate-deposited exocrine gland secretions are involved in the formation of some wasp swarm clusters and in the subsequent guidance of the swarm to a new location. Swarming *Polybia sericea* adults dragged their gasters over leaves and other substrates around the swarm cluster, depositing a substance similar in odor to an exocrine gland at the base of the fifth gastric sternite. Only wasps in a swarming behavioral state were attracted to extracts of this gland when it was smeared onto filter papers (Jeanne 1981). We do not know if wasp trails are polarized. Brazilian *Synoeca septentrionalis* wasp workers chewed and licked leaves on swarm routes (Jeanne, Downing, and Post 1983), a behavior with clear similarities to meliponine odor-trail marking (Schorkopf et al. 2007). Given that

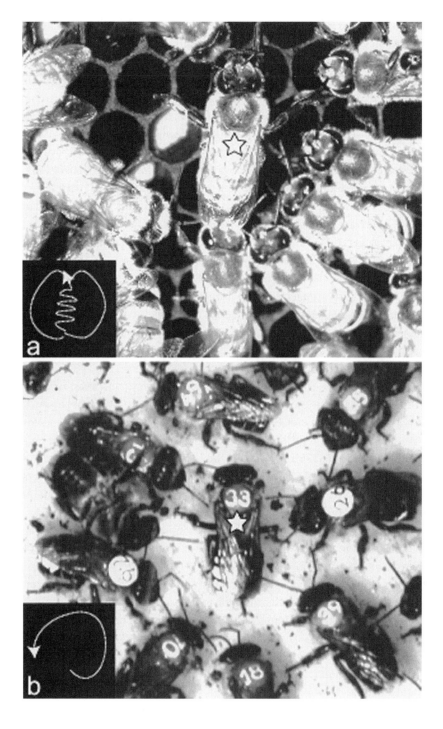

swarm founding evolved four separate times in the Vespidae, abdominal rubbing to mark the route to a new nest is widespread and has evolved independently multiple times (Smith, O'Donnell, and Jeanne 2002).

Ritualization of Excitatory Movements and Sounds

Esch (1967) proposed that the honey bee waggle dance may be a ritualized representation of the outbound and inbound flights of a forager to a food source outside the nest. I would add the possibility that these circular motions, found in honey bees and stingless bees, evolved from the ritualization of the cycle of entry, unloading, and exit *within the nest*. In the honey bee round dance and waggle dance and in the semicircular spins executed by foragers of some stingless bee species, one may see the generalized looping motion of the forager entering, unloading, and then exiting the nest (Figure 12.1). In addition, exiting acoustic cues from the flight departure sound or from pre-flight warm-up sounds (von Frisch 1967) may have evolved into signals. For example, recruiting *M. panamica* foragers return through the narrow nest entrance facing away from the entrance, and then must turn toward it to exit, all the while producing loud buzzing sounds (Nieh 2004). The repetition of these motions and sounds may have been ritualized into the spinning recruitment motions and buzzes found in some social bees.

   More generally, this hypothesis could be extended to the evolution of non-referential excitatory motions in bees and wasps. At the most basic level, a forager of a social bee or wasp colony returns to the nest, unloads her food, and then returns to gather more food, thus describing some form of loop at the nest. Agitation and excitement upon the return of a successful forager is thought, in many cases, to be expressed through various running, spinning, and zigzag motions inside the nest (von Frisch 1967). Contact with nestmates, perhaps initially through trophallaxis, or even accidentally

Figure 12.1. Recruitment communication movements of highly social bees inside the nest. Signal sender is indicated with a star and is surrounded by signal receiverts. A: Honey bee waggle dance (movement pattern shown in inset). B:Stingless bee *(Melipona panamica)* dance phase (movement shown in inset). In both cases, the signal sender is shown just before she is about to turn. In both species, the signaler is producing sounds by vibrating her wings and thorax and is attended by followers (signal receivers, photos by J. Nieh).

during this process, may have become ritualized because such contact provided increased nestmate exposure to information that good food sources can be gathered as well as information about that resource (odor, quality, and resource type), thus increasing colony foraging efficiency.

Conclusions and Future Directions

In both social bees and social wasps we know that visual local enhancement and odor trail communication exist, although odor trails have thus far been documented only for swarming, not foraging, in wasps. The convergent evolution of food recruitment mechanisms in both groups may be due, in part, to the shared characteristics of foraging through flight, colonial life, and excellent associative learning.

The ability to follow a flying forager and the difficulty of leaving a persistent odor trail in the air give rise to common solutions. Descriptions of piloting in bees and swarm-following in wasps suggest that slowing flight or increasing the conspicuousness of flight behaviors (meliponine zig zag flights, Esch 1967) may facilitate following. Moreover, depositing substrate-borne odor trails (odor droplets widely spaced each few meters) may enhance piloting. In ants, odor trails may have evolved from tandem-running, a form of piloting (Hölldobler and Wilson 1990). Bee and wasp odor trails may thus be more derived than piloting behavior, although supporting evidence awaits studies on a wider variety of species.

Colonial life has led to the evolution of sophisticated chemical signaling in bees and wasps. Nonetheless, we have yet to fully explore the role of odor cues. For example, the activating effect of food odor in bees and wasps may have evolved convergently because the association between food odor and foraging success is an important field cue, and thus extends easily into the nest, where it can cue nestmates to the success of compatriots. We now know that cuticular hydrocarbons, which facilitate nest and nestmate recognition, are also deposited on food and can therefore be learned by foragers as cues—guideposts to the rewarding and unrewarding.

A virtually unexplored area is the role of associative learning in local enhancement. Wasp and bee foragers can easily learn the appearance of food. Do they also learn that nestmate presence signals food? Work on relatively well-studied bee groups such as honey bees and bumblebees has not yet examined the role of visual local enhancement, something better documented in wasps. On the other hand, very little is known about the possibility of location-specific recruitment in wasps. Our understanding of the

Table 12.1. Foraging-related information transfer mechanisms in bees and social wasps

Category	Bees	Social wasps
Foraging activation		
Solitary aggregations	?	?
Trophallaxis	Halictidae,[1] Meliponini,[2] Apini	Mischocyttarini,[1*] Polistini[1*]
Excitatory nest behaviors	Bombini, Meliponini, Apini	Epiponini* (Polybia,[3] Protopolybia[4])
Referential communication	Meliponini,[5] Apini	?
Local enhancement		
On resource	Meliponini	Vespidae, Polistini,* Epiponini*
Piloting	Meliponini,[6] Apini[7]	Vespidae[8]
Partial piloting	Meliponini	?
Target-only odor marking	Xylocopini, Meliponini, Bombini, Apini	?
Partial odor trails	Meliponini	?
Complete odor trails	Bombini	Epiponini,[9*] Vespidae[10]

* Belongs to the subfamily Polistinae (Arévalo et al. 2004)

1. Not known if this behavior leads to foraging activation.

2. Not known if this behavior leads to foraging activation, but likely.

3. Inferred from foraging activation in *Polybia occidentalis*.

4. Excitatory behavior observed for successful foragers, but increase in foraging activation not documented.

5. Correlations between the distance to food and recruitment sound pulse duration reported in several species (no direct tests).

6. Indirect evidence.

7. Only known in the context of swarm guidance.

8. Likely exists in the context of swarm guidance.

8. Suggested.

9. Guidance within the nest and short trails (a few meters) extending from the nest.

10. Guidance within the nest.

evolution of recruitment communication would benefit greatly from pursuing these questions. Thus, in many respects, our understanding of social wasp and bee foraging activation is at a similar and tantalizingly incomplete stage.

There are many gaps in our knowledge of social bee and wasp foraging (Table 12.1), but I have selected five particular questions that will enhance our understanding of basal and derived recruitment mechanisms. In some cases, I have listed suggested study species based on preliminary published observations or biological characteristics of these species.

1. Is there neighbor foraging activation in nest aggregations of solitary bees or wasps?
2. Do wasps perform excitatory foraging activation behaviors at the nest?
3. Foraging activation based on the scent of collected food has been demonstrated in wasps and bumblebees, and foraging reactivation in honey bees. Do foraging activation and reactivation via food scent occur in the Halictidae, Meliponini (*Frieseomelitta silvestrii, F. flavicornis,* and *F. freiremaiai*), or Euglossini (*Euglossa townsendi* and *Eulaema nigrita*)?
4. Can some wasp species (*Polybia occidentalis, Brachygastra mellifica, B. lecheguana,* or *V. germanica*) recruit nestmates to a specific location? If so, do they use mechanisms analogous to those known for social bees?
5. The captivating honey bee waggle dance provides an astonishing example of functionally referential communication. However, do other, perhaps less information-rich, examples of functionally referential communication such as food quality or alarm signaling exist in bees (*Melipona panamica, M. quadrifasciata,* and *M. seminigra*) and wasps (*Polybia occidentalis*)?

Literature Cited

Aguilar, I., A. Fonseca, and J. C. Biesmeijer. 2005. "Recruitment and communication of food source location in three species of stingless bees (Hymenoptera, Apidae, Meliponini)." *Apidologie* 36: 313–324.

Arévalo, E., Y. Zhu, J. M. Carpenter, and J. E. Strassmann. 2004. "The phylogeny of the social wasp subfamily Polistinae: Evidence from microsatellite flanking sequences, mitochondrial COI sequence, and morphological characters." *BMC Evolutionary Biology* 4: 8–24.

Augusto, S. C., and C. A. Garófalo. 2004. "Nesting biology and social structure of *Euglossa (Euglossa) townsendi* Cockerell (Hymenoptera, Apidae, Euglossini)." *Insectes Sociaux* 51: 400–409.

Beekman, M., R. L. Fathke, and T. D. Seeley. 2006. "How does an informed minority of scouts guide a honeybee swarm as it flies to its new home?" *Animal Behaviour* 71: 161–171.

Blumstein, D. M. 1999. "The evolution of functionally referential alarm com-munication: Multiple adaptations; multiple constraints." *Evolution of Communication* 3: 135–147.

Butler, C. G., D. J. C. Fletcher, and D. Watler. 1969. "Nest entrance marking with pheromones by the honey bee, *Apis mellifera*, and by a wasp, *Vespula vulgaris*." *Animal Behaviour* 17: 142–147.

Cameron, S. A. 1993. "Multiple origins of advanced eusociality in bees inferred from mitochondrial DNA sequences." *Proceedings of the National Academy of Sciences of the USA* 90: 8687–8691.

Cameron, S. A., J. B. Whitfield, M. Cohen, and N. Thorp. 1999. "Novel use of walking trails by the Amazonian bumble bee, *Bombus transversalis* (Hymenoptera: Apidae)." In G. W. Byers, R. H. Hagen, and R. W. Brooks, eds., *Entomological contributions in memory of Byron A. Alexander,* 187–193. University of Kansas Natural History Museum Special Publication 24 Lawrence, Kansas.

Carpenter, J. M. 2003. "On molecular phylogeny of Vespidae (Hymenoptera) and the evolution of sociality in wasps." *American Museum Novitates* 3389: 1–20.

Cederberg, B. 1977. "Evidence for trail marking in *Bombus terrestris* Wor (Hymenoptera, Apidae)." *Zoon* 5: 143–146.

Chittka, L., N. M. Williams, H. Rasmussen, and J. D. Thomson. 1999. "Navigation without vision: Bumblebee orientation in complete darkness." *Proceedings of the Royal Society of London B* Biological Sciences: 266: 45–50.

Danforth, B. N., J. Fang, and S. Sipes. 2006. "Analysis of family-level relation-ships in bees (Hymenoptera: Apiformes) using 28S and two previously unexplored nuclear genes: CAD and RNA polymerase II." *Molecular Phylogenetics and Evolution* 39: 358–372.

Dornhaus, A., and L. Chittka. 2004. "Information flow and foraging decisions in bumble bees *(Bombus* spp.)." *Apidologie* 35: 183–192.

———. 2005. "Bumble bees *(Bombus terrestris)* store both food and informa-tion in honeypots." *Behavioral Ecology* 16: 661–666.

Dornhaus, A., F. Klügl, C. Oechslein, F. Puppe, and L. Chittka. 2006. "Benefits of recruitment in honey bees: Effects of ecology and colony size in an individual-based model." *Behavioral Ecology* 17: 336–344.

Dyer, F. C. 2002. "The biology of the dance language." *Annual Review of Entomology* 47: 917–949.

Eltz, T. 2006. "Tracing pollinator footprints on natural flowers." *Journal of Chemical Ecology* 32: 907–915.

Esch, H. 1967. "Die Bedeutung der Lauterzeugung für die Verständigung der stachellosen Bienen." *Zeitschrift für Vergleichende Physiologie* 56: 408–411.

Evans, C. S., and L. Evans. 1999. "Chicken food calls are functionally referential." *Animal Behaviour* 58: 307–319.

Farina, W. M., and A. J. Wainselboim. 2005. "Trophallaxis within the dancing context: A behavioral and thermographic analysis in honeybees *(Apis mellifera)." Apidologie* 36: 43–47.

Fewell, J. H., and M. L. Winston. 1992. "Colony state and regulation of pollen foraging in the honey bee, *Apis mellifera* L." *Behavioral Ecology and Sociobiology* 30: 387–393.

Frankie, G. W., and S. B. Vinson. 1977. "Scent marking of passion flowers in Texas by females of *Xylocopa virginica texana." Journal of the Kansas Entomological Society* 50: 613–625.

Guédot, C., T. L. Pitts-Singer, J. S. Buckner, J. Bosch, and W. P. Kemp. 2006. "Olfactory cues and nest recognition in the solitary bee *Osmia lignaria." Physiological Entomology* 31: 110–119.

Heinrich, B. 1984. "Learning in invertebrates." In P. Marler and H. S. Terrace, eds., *The biology of learning*, 135–147. New York: Springer-Verlag.

Hines, H. M., J. H. Hunt, T. K. O'Connor, J. J. Gillespie, and S. A. Cameron. 2007. "Multigene phylogeny reveals eusociality evolved twice in vespid wasps." *Proceedings of the National Academy of Sciences USA* 104: 3295–3299.

Hölldobler, B., and E. O. Wilson. 1990. *The ants.* Cambridge: Belknap Press of Harvard University Press.

Howard, K. J., A. R. Smith, S. O'Donnell, and R. L. Jeanne. 2002. "Novel method of swarm emigration by the epiponine wasp, *Apoica pallens* (Hymenoptera Vespidae)." *Ethology, Ecology, and Evolution* 14: 365–371.

Hrncir, M., F. G. Barth, and J. Tautz. 2006. "Vibratory and airborne-sound signals in bee communication (Hymenoptera)." In S. Drosopoulous and M. F. Claridge, eds., *Insect sounds and communication.* New York: Taylor & Francis Group.

Hrncir, M., S. Jarau, R. Zucchi, and F. G. Barth. 2000. "Recruitment behavior in stingless bees, *Melipona scutellaris* and *M. quadrifasciata.* II. Possible mechanisms of communication." *Apidologie* 31: 93–113.

Hrncir, M., S. Mateus, and F. S. Nascimento. 2007. "Exploitation of carbohydrate food sources in *Polybia occidentalis:* Social cues influence foraging decisions in swarm-founding wasps." *Behavioral Ecology and Sociobiology* 61: 975–983.

Jeanne, R. L. 1972. "Social biology of the Neotropical wasp *Mischocyttarus drewseni." Bulletin of the Museum of Comparative Zoology of Harvard University* 144: 63–150.

———. 1981. "Chemical communication during swarm emigration in the social wasp *Polybia sericea* (Olivier)." *Animal Behaviour* 29: 102–113.

Jeanne, R. L., H. A. Downing, and D. C. Post. 1983. "Morphology and function of sternal glands in polistine wasps (Hymenoptera: Vespidae)." *Zoomorphology* 103: 149–164.

Jeanne, R. L., and M. G. Keeping. 1995. "Venom spraying in *Parachartergus colobopterus:* A novel defensive behavior in a social wasp (Hymenoptera: Vespidae)." *Journal of Insect Behavior* 8: 433–442.

Kajobe, R., and C. M. Echazarreta. 2005. "Temporal resource partitioning and climatological influences on colony flight and foraging of stingless bees (Apidae; Meliponini) in Ugandan tropical forests." *African Journal of Ecology* 43: 267–275.

Kerr, W. E. 1994. "Communication among *Melipona* workers (Hymenoptera: Apidae)." *Journal of Insect Behavior* 7: 123–128.

Kovac, H., and A. Stabentheiner. 1999. "Effect of food quality on the body temperature of wasps *(Paravespula vulgaris)*." *Journal of Insect Physiology* 45: 183–190.

Lorenzon, M. C., and C. A. Matrangolo. 2005. "Foraging on some nonfloral resources by stingless bees (Hymenoptera, Meliponini) in a Caatinga region." *Brazilian Journal of Biology* 65: 291–298.

Macuda, T., R. J. Gegear, T. M. Laverty, and B. Timney. 2001. "Behavioural assessment of visual acuity in bumblebees *(Bombus impatiens)*." *Journal of Experimental Biology* 204: 559–564.

Michener, C. D. 1974. *The social behavior of the bees.* Cambridge: Harvard University Press.

Nakamura, J., and T. D. Seeley. 2006. "The functional organization of resin work in honeybee colonies." *Behavioral Ecology and Sociobiology* 60: 339–349.

Naumann, M. G. 1970. *The nesting behavior of Protopolybia pumila in Panama.* Lawrence: University of Kansas.

———. 1975. "Swarming behavior: Evidence for communication in social wasps." *Science* 189: 642–644.

Nieh, J. C. 2004. "Recruitment communication in stingless bees (Hymenoptera, Apidae, Meliponini)." *Apidologie* 35: 159–182.

Nieh, J. C., A. Leon, S. A. Cameron, and R. Vandame. 2006. "Hot bumble bees at good food: Thoracic temperature of feeding *Bombus wilmattae* foragers is tuned to sugar concentration." *Journal of Experimental Biology* 209: 4185–4192.

Nieh, J. C., and D. Sánchez. 2005. "Effect of food quality, distance, and height on thoracic temperatures in a stingless bee, *Melipona panamica*." *Journal of Experimental Biology* 208: 3933–3943.

O'Donnell, S. 2001. "Worker biting interactions and task performance in a swarm-founding eusocial wasp *(Polybia occidentalis,* Hymenoptera: Vespidae)." *Behavioral Ecology* 12: 353–359.

Oeynhausen, A., and W. H. Kirchner. 2001. "Vibrational signals of foraging
    bumblebees *(Bombus terrestris)* in the nest." In *Proceedings of the Meeting
    of the European Sections of IUSSI*, 25–29. Berlin:

Overmeyer, S. L., and R. L. Jeanne. 1998. "Recruitment to food by the German yel-
    lowjacket, *Vespula germanica*." *Behavioral Ecology and Sociobiology* 42: 17–21.

Polidori, C., M. Casiraghi, M. Di Lorenzo, B. Valarani, and F. Andrietti. 2006.
    "Philopatry, nest choice, and aggregation temporal-spatial change in the
    digger wasp *Cerceris arenaria* (Hymenoptera: Crabronidae)." *Journal of
    Ethology* 24: 155–163.

Potts, S. G., and P. Willmer. 1998. "Compact housing in built-up areas: Spatial
    patterning of nests in aggregations of a ground-nesting bee." *Ecological
    Entomology* 23: 427–432.

Raveret Richter, M. 2000. "Social wasp (Hymenoptera: Vespidae) foraging
    behavior." *Annual Review of Entomology* 45: 121–150.

Reinhard, J., M. V. Srinivasan, D. Guez, and S. W. Zhang. 2004. "Floral scents
    induce recall of navigational and visual memories in honeybees." *Journal of
    Experimental Biology* 207: 4371–4381.

Richards, M. H., D. French, and R. J. Paxton. 2005. "It's good to be queen:
    Classically eusocial colony structure and low worker fitness in an obligately
    social sweat bee." *Molecular Ecology* 14: 4123–4133.

Riley, J. R., U. Greggers, A. D. Smith, D. R. Reynolds, and R. Menzel. 2005.
    "The flight paths of honeybees recruited by the waggle dance." *Nature* 435:
    205–207.

Roberts, D. 1941. "Imitation and suggestion in animals." *Bulletin of Animal
    Behaviour* 1: 11–19.

Rohrseitz, K., and J. Tautz. 1999. "Honey bee dance communication: Waggle
    run direction coded in antennal contacts?" *Journal of Comparative
    Physiology A:* 463–470.

Roubik, D. W. 1989. *Ecology and natural history of tropical bees.* New York:
    Cambridge University Press.

Saleh, N., and L. Chittka. 2006. "The importance of experience in the inter-
    pretation of conspecific chemical signals." *Behavioral Ecology and
    Sociobiology* 61: 215–220.

Santos, M. L., and C. A. Garófalo. 1994. "Nesting biology and nest re-use of
    Eulaema nigrita (Hymenoptera: Apidae, Euglossini)." *Insectes Sociaux* 41:
    99–110.

Schmidt, V. M., R. Zucchi, and F. G. Barth. 2005. "Scent marks left by
    Nannotrigona testaceicornis at the feeding site: Cues rather than signals."
    Apidologie 36: 285–291.

———. 2006. "Recruitment in a scent trail laying stingless bee (*Scaptotrigona*
    aff. *depilis*): Changes with reduction but not with increase of the energy
    gain." *Apidologie* 37: 487–500.

Schmitt, U., G. Lübke, and W. Franke. 1991. "Tarsal secretions mark food sources in bumblebees (Hymenoptera: Apidae)." *Chemoecology* 2: 35–40.

Schmitz, J., and R. F. A. Moritz. 1998. "Molecular phylogeny of Vespidae (Hymenoptera) and the evolution of sociality in wasps." *Molecular Phylogenetics and Evolution* 9: 183–191.

Schorkopf, D. L., S. Jarau, W. Francke, R. Twele, R. Zucchi, M. Hrncir, V. M. Schmidt, M. Ayasse, and F. G. Barth. 2007. "Spitting out information: *Trigona* bees deposit saliva to signal resource locations." *Proceedings of the Royal Society of London. Series B: Biological Sciences* 274: 895–898.

Seeley, T. D., M. Kleinhenz, B. Bujok, and J. Tautz. 2003. "Thorough warm-up before take-off in honey bee swarms." *Naturwissenschaften* 90: 256–260.

Slaa, E. J., J. Wassenberg, and J. C. Biesmeijer. 2003. "The use of field-based social information in eusocial foragers: Local enhancement among nest-mates and heterospecifics in stingless bees." *Ecological Entomology* 28: 369–379.

Smith, A. R., S. O'Donnell, and R. L. Jeanne. 2002. "Evolution of swarm communication in eusocial wasps (Hymenoptera: Vespidae)." *Journal of Insect Behavior* 15: 751–764.

Srinivasan, M. V., and M. Lehrer. 1988. "Spatial acuity of honeybee vision and its spectral properties." *Journal of Comparative Physiology* A 162: 159–172.

Stabentheiner, A. 2001. "Thermoregulation of dancing bees: Thoracic temperature of pollen and nectar foragers in relation to profitability of foraging and colony need." *Journal of Insect Physiology* 47: 385–392.

Steinmetz, I., S. Sieben, and E. Schmoltz. 2002. "Chemical trails used for orientation in nest cavities by two vespine wasps, *Vespa crabo* and *Vespula vulgaris.*" *Insectes Sociaux* 49: 354–356.

Stout, J. C., and D. Goulson. 2001. "The use of conspecific and interspecific scent marks by foraging bumblebees and honeybees." *Animal Behaviour* 62: 183–189.

Thom, C., D. C. Gilley, J. Hooper, and H. E. Esch. 2006. "The scent of the dance: Honey bee waggle dancers produce volatile compounds that affect foraging behavior." In *IUSSI 2006 Congress.* Washington, DC:

Velthuis, H. H. W. 1987. "The evolution of sociality: Ultimate and proximate factors leading to primitive social behavior in carpenter bees." *Experientia Supplementum* 54: 405–430.

von Frisch, K. 1967. *The dance language and orientation of bees.* Cambridge: Belknap Press of Cambridge University Press.

Wcislo, W. T., and V. H. Gonzalez. 2006. "Social and ecological contexts of trophallaxis in facultatively social sweat bees, *Megalopta genalis* and *M. ecuadoria* (Hymenoptera, Halictidae)." *Insectes Sociaux* 53: 220–225.

Wcislo, W. T., and S. M. Tierney. 2007. "Evolution of communal behavior in bees and wasps: An alternative to eusociality." In J. H. Fewell, ed., *Organization of insect societies: From genome to socio-complexity*, INSERT WHEN KNOWN. Cambridge: Harvard University Press.

Wilson, E. O. 1971. *The insect societies.* Cambridge: Belknap Press of Harvard University Press.

Zucchi, R., S. F. Sakagami, and J. M. F. Camargo. 1969. "Biological observations on a neotropical parasocial bee, *Eulaema nigrita*, with a review of the biology of the Euglossinae. A comparative study." *Journal of the Faculty of Science, Hokkaido Univ. (VI, Zoology)* 17: 271–380.

CHAPTER THIRTEEN

The Organization of Social Foraging in Ants: Energetics and Communication

FLAVIO ROCES

FORAGING IN SOCIAL INSECTS is a complex process that results from individual decisions and their integration at the colony level. Even though it is tempting to consider a social insect colony as a unit that collectively "decides" about, for instance, the selection of a given food source, a colony's foraging response ultimately arises from the decisions made by each individual worker. Individual foragers are expected to behave so as to maximize food delivery at the colony level because we can assume the foraging performance of a colony, measured as the delivery rate of food, correlates with colony fitness.

The analysis of individual foraging performance and its extrapolation to account for overall colony foraging is complicated by the fact that most of the resources collected by each individual worker are not for its own consumption, but fed to the brood or nestmates, or stored. Put into economic terms, the costs and benefits of the foraging decisions of individual workers are paid for and received by different workers. Moreover, returning foragers deliver to the colony two commodities: food and information. Workers could conceivably favor the delivery of one of these commodities at the expense of the other; that is, workers may spend less time collecting at a food source, thus reducing their food delivery rate, but allocate more effort to information transfer, promoting recruitment of nestmates and increasing the performance of the entire colony.

The trade-off between these two competing decisions—either to spend more time at a discovered food source collecting larger loads or to spend less time in order to return earlier to the colony and recruit nestmates—is illustrated in Figure 13.1. In one hypothetical extreme, a forager that discovers a high-reward food source should quickly return to the colony in order to deliver the appropriate information without investing time loading at the food source. At the other extreme, each worker could exploit the food source so as to individually maximize its food delivery rate. It is clear that when this dual aspect is kept in mind, the optimal policies for delivering each of the commodities might be different, and that the maximization of information transfer is incompatible with a concomitant maximization of food delivery at the level of the individual.

There is ample evidence that foraging social insects (honey bees, wasps, ants) often return to the nest with partial loads (Josens, Farina, and Roces 1998; Núñez 1966; Pflumm 1975; Roces 1990a; Roces and Núñez 1993). For "central place foragers," optimal foraging theory predicts the collection of partial loads when animals feed in a patch where intake rate decreases with time (for instance, due to resource depression), or when the net yield of energy increases at a diminishing rate due to increased foraging costs

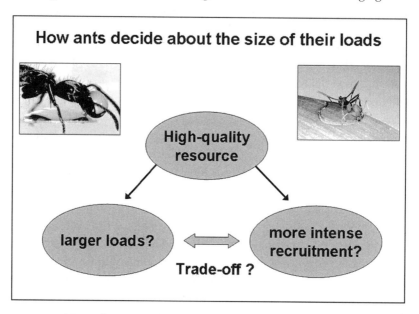

Figure 13.1. Upon discovery of an attractive food source, worker decisions are expected to be the outcome of a trade-off between two competing decisions: either to spend longer times at a discovered food source, thus collecting larger loads, or to spend shorter times in order to return earlier to the colony for recruitment of nestmates via information transfer.

(food search, transport, etc.). When collecting at *ad libitum* sources, animals should load themselves maximally (Orians and Pearson 1979).

Contrary to this expectation, social insect foragers often return with partial loads even when collecting food at *ad libitum* sources, as observed in nectar-feeding ants and leaf-cutting ants. This pattern does not match the expectations of optimality theory. It seems that social insect workers forage in a suboptimal way, at least based on considerations of performance at the level of the individual. However, social insect foragers may compromise their individual performance to increase colony performance. The extreme strategy of time-saving at the source for information transfer is perhaps what we observe in leaf-cutting ants when scout workers, upon discovery of a new food source, do not start cutting fragments but instead return quickly to the nest, laying chemical trails to recruit other nestmates. Workers only start cutting leaves when a foraging column has been established (Jaffé and Howse 1979).

In order to understand the extent to which individual decisions are the outcome of a trade-off between these two competing tendencies (i.e., maximizing loading and maximizing information transfer), we need to know how workers assess resource characteristics and translate this assessment into the decision to continue loading or to return to the nest for information transfer. Worker decisions are also expected to depend on colony status; that is, they do not only influence, but are influenced by the decisions of other colony members.

In this chapter, I explore the rules used by ant workers during food collection, concentrating on the trade-off between food collection and information transfer. This trade-off is a major factor influencing worker ant foraging decisions. I discuss evidence for this hypothesis using ant species with disparate foraging ecology: nectar-feeding ants and leaf-cutting ants. These ants differ in their criteria for food selection, loading capacity, and costs of food acquisition (liquid intake vs. leaf-cutting). Finally, I briefly discuss how the empirical work fits into the current framework of foraging theory and comment on some shortcomings of the theoretical arguments used thus far to understand social insect foraging.

Nectar-Feeding Ants: When to Leave an Attractive Source and Return to the Nest?

Nectar-feeding ants repeatedly visit renewable resources such as extrafloral nectaries or aphid colonies, yet the criteria used by workers to evaluate resource quality and the rules used to decide when to leave the patch have been only partially identified. Such renewable resources usually offer nectar at flow rates much lower than the maximal intake rate of individual ants. This means that ants have to wait for the nectar to be produced. Under such conditions it may be relevant for ants not to spend a long time to fill their crops but to stop feeding earlier and return to the nest with partial nectar loads. Even under *ad libitum* conditions, workers often return to the nest with partially filled crops (Josens and Roces 2000; Mailleux, Detrain, and Deneubourg 2000), with the extent of filling being a function of food quality (Josens, Farina, and Roces 1998).

What are the characteristics of the food source that determine the workers' decision to stop feeding? In a laboratory study, Schilman and Roces (2003) tested whether the nectar flow rate is the decisive parameter controlling foraging decisions. *Camponotus rufipes* foragers were trained to visit an artificial food patch providing 20% sucrose solution either *ad libitum* or at controlled flow rates, varying from 0.118 to 2.36 µl/min. These flow rates simulate the conditions faced by workers when visiting plant extrafloral nectaries. Ants adjusted their visit times to the different flow rates, so that the time spent at the feeder decreased with increasing nectar flow rates. The volume of nectar collected increased with increasing nectar flow rates, and workers were observed to return to the nest with partially filled crops.

To investigate the rules used by ants to decide when to depart from the patch, experienced workers (i.e., those that collected nectar at a given flow rate over four visits) were confronted in the fifth visit with a depleted patch, and the time spent there before leaving was recorded. Ants accustomed to low nectar flow rates waited at the patch longer than did ants accustomed to higher flow rates. Their waiting times in tests in which the patch was depleted was similar to that during training. This suggests that ants, like vertebrates, can measure time intervals and, as a departure rule, they use an estimate of time that depends on the flow rate previously experienced. In addition, the rate of feeding attempts at the depleted patch increased with increasing flow rates; that is, it varied as a function of the previously experienced nectar flow rate. These results indicate that ant workers, much like honeybees, quantitatively assess the nectar flow rate (Núñez 1970; Wainselboim, Roces, and Farina 2002), and that they arrive at the patch with an expectation about the nectar flow rate.

Assessment of nectar flow rate implies that individuals have to be able to compute the collected volume over the time spent while drinking. Two phenomena underlie behavioral timing in animals: a *phase sense*, which refers to the ability of animals to anticipate events that recur at a fixed time of the day, and an *interval sense*, which refers to the ability to respond at a defined time after an event occurred (Gallistel 1990). Temporal learning in honey bees (Wahl 1932) and nectar-feeding ants (Harrison and Breed 1987) demonstrate that insects have a phase sense and can be conditioned to search for food at a certain time of the day. There is also experimental evidence suggesting an interval sense in honey bees (Farina 2000; Núñez 1966, 1982; Seeley 1989; Wainselboim, Roces, and Farina 2002, 2003), but the mechanisms involved remain unknown.

Why don't ants completely fill their crops when collecting nectar provided at lower rates? Ants could simply stay longer at the source, fill their crops, and maximize their gross energy gain per trip. At the lowest nectar flow rate, a medium-sized worker should spend on average one hour to fill its crop; however, such a response was never recorded. In some cases, *Camponotus* workers were observed to return to the nest with less than 20% of their maximal crop capacity. Considering the low costs of load carrying (Schilman and Roces 2005, 2006), it seems unlikely that returning to the nest with a partially-filled crop results from minimizing transport costs and thereby maximizing energy returns during foraging, as suggested for foraging honey bees (Moffatt 2000; Schmid-Hempel, Kacelnik, and Houston 1985; Varjú and Núñez 1991). It is important to note that partially-loaded workers do not give up the source, because they return to it after unloading the nectar at the nest, and lay chemical trails along their way. The question of whether the observed partial loading in *Camponotus* ants contributes to an increase in the rate of information exchange at the nest (i.e., giving information about the exploited source or receiving information about alternative ones) remains to be investigated.

How is the assessment of nectar availability translated into communication signals? Recent studies on *Lasius niger* investigated the criteria used by ants to assess the amount of nectar available, as well as the rules governing their decision to lay a recruitment trail (Mailleux, Deneubourg, and Detrain 2000, 2003; Mailleux, Detrain, and Deneubourg 2005). When scouts discovered nectar droplets with volumes exceeding the capacity of their crop, most of them collected maximal crop loads and immediately returned to the nest laying a recruitment trail. In contrast, when smaller nectar droplets were encountered, several scouts ingested the amount found and continued exploring the area for additional nectar. If unsuccessful, they returned to the nest without laying a trail. It appears that the key criterion that regulates recruitment behavior of scouts is their ability to ingest their own "desired" volume, which acts as a threshold triggering trail-laying, and is an all-or-non response.

In these investigations, workers found either nectar droplets of different volumes or nectar that needed to be sucked through a cotton-wool cork inserted in a capillary. This latter manipulation made the nectar harder to collect and resulted in longer drinking times (5 minutes on average). However, workers collected the same nectar loads as they did at *ad libitum* nectar droplets, suggesting that such sources were treated by ants, despite their increased feeding times, as unlimited patches. If this argument is correct, the results of Mailleux and colleagues (2003) are not in conflict with our results on partial crop-filling in *Camponotus* ants (Josens, Farina, and Roces 1998; Josens and Roces 2000; Schilman and Roces 2003, 2006), but may correspond to conditions that, while occurring in nature, do not include low rates of nectar production as they occur in plant extrafloral nectaries (Dreisig 2000). It is for this low range of nectar production that a compromise between spending time for further loading or returning to the nest for information should be expected. Regrettably, no such studies have been done to date on nectar-feeding ants combining controlled rates of nectar availability and analysis of recruitment behavior. The pioneer studies on honey bees which used controlled rates of nectar flow (Núñez 1970, 1971; Núñez and Giurfa 1996) and a relatively straightforward way to quantify crop loads and information exchange in honey bees (via dance-behavior, food transfer, and begging behavior) might inspire comparative research on nectar-feeding behavior in ants and provide new insights about the organization of collective foraging in social insects in general.

Leaf-Cutting Ants: Individual Cutting Rules and Information Transfer

Leaf-cutting ants are the dominant herbivores in the neotropics. They cut vegetation into small fragments, which they transport to the nest, on which the ants rear a symbiotic fungus. The fungus garden represents the sole food source of the developing brood. Adult workers obtain a large proportion (more than 90%) of their energy requirements from the plant sap of the harvested material. Since the harvested leaf fragments are incorporated into the fungus garden and not directly consumed by the workers, it is conceivable that ants' plant preferences are driven in part by the requirements of the fungus. On the one hand, ant workers may prefer plants that support maximal rates of fungus growth, more or less irrespective of the attractiveness of the plant sap being imbibed during the harvesting process. On the other hand, workers may decide about the quality of a given resource based on the immediate availability of energy to support their foraging activity.

Leaf-cutting ants appear to be well suited to explore the rules underlying the organization of collective foraging. At the individual level, research has focused on the rules workers use to decide about the size of the leaf fragment to be cut (reviewed by Roces 2002). At the colony level, a number of studies have provided detailed information about plant selection and foraging activity of colonies in the field (e.g., Howard 1988), and hypotheses have been developed for the evolution of foraging decisions (Burd and Howard 2005a, 2005b).

While foraging, scout workers from a leaf-cutting ant colony search for suitable resources and, upon discovery, decide whether a given resource is worth communicating to other nestmates or not. The acceptance of a plant is mainly based on chemical and physical features of its leaves. If the source is attractive, workers may decide to lay a chemical trail while returning to the nest, or to cut a fragment and carry it back to the nest. It has been observed repeatedly that polymorphic leaf-cutting ant foragers frequently harvest leaf pieces that correlate in mass to that of the foraging individual (Lutz 1929). This correlation is probably a result of the geometric method of leaf cutting. Workers anchor on the leaf edge by their hind legs and pivot around them while cutting arcs out of the leaves. Therefore, the load-size is determined by a fixed reach of an individual and dependent on worker body size. However, not all workers cut fragments of maximal size, suggesting a more flexible mechanism of load-size determination in leaf-cutting ants.

That foraging workers indeed use flexible rules during cutting is nicely illustrated with the following observations. It is known that there is a negative correlation between leaf area density (leaf mass/leaf area) and the size of the harvested fragment (Roces 2002); that is, the harder the leaf, the smaller the fragment cut. This pattern arises from the worker's decision about the cutting angles to be described after assessment of the leaf resistance. Figure 13.2 shows a worker of the leaf-cutting ant *Atta sexdens* cutting a pseudoleaf of Parafilm arranged as follows. The worker has first to cut along a 3 mm-wide margin of one-layered Parafilm, then along a two-layered one, and finally, when approaching again the boundary between the different layers, along a one-layered Parafilm again (Figure 13.2A). The resistance of a single Parafilm layer corresponds to that of a tender leaf, while the two-layered Parafilm stands for a harder (thicker) leaf. The ant started with the usual cutting trajectory, describing an ample arc as expected for tender leaves. However, it severed the two-layered pseudoleaf with a sharper angle, as expected for harder leaves (Figure 13.2B), thus not following the initial cutting trajectory (dotted line in Figure 13.2). Particularly interesting is the point at which the ant again reached the one-layered pseudoleaf. Instead of cutting straight ahead to finish the cut (Figure 13.2C), it continued cutting (following an unexpected convex trajectory), and after a final turn severed the fragment (Figure 13.2D).

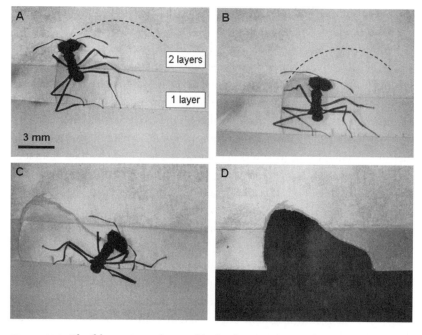

Figure 13.2. Flexible cutting rules used by leaf-cutting ant workers (*Atta sexdens*). The ant cuts a pseudoleaf of Parafilm, composed of a 3mm-wide margin of one layer, and a central two-layered Parafilm. The dotted line indicates the cutting trajectory expected for one-layered Parafilm (Photos by H. Heilmann).

Do leaf-cutting ant workers use versions of flexible cutting rules to trade-off loading for information transfer, as previously discussed for nectar-feeding ants? The extent to which the decision to transfer information about a food discovery influences individual cutting behavior was analyzed under controlled experimental conditions (Roces 1993; Roces and Núñez 1993). The rationale of the experimental approach was as follows: in independent assays, scout workers of the leaf-cutting ant *Acromyrmex lundi* were first exposed to droplets of scented sugar solution of either of two concentrations: 1% or 10% sucrose. Scouts detected these droplets and returned to the colony, leaving a chemical recruiting trail. When the recruited workers arrived, they encountered not sugar solution, but sheets of Parafilm impregnated with the same scent yet containing no sugar. Since all recruited workers found the same standardized material (Parafilm as a pseudoleaf), differences in cutting behavior among workers initially recruited to either 1% or 10% sucrose must be the result of the information transmitted by the scout worker that initially found the sugar solution.

Several behavioral parameters of the recruited workers were observed to depend on the information about food quality they received from a single recruiting scout. Workers recruited by scouts that found 10% sucrose are referred to as "10%-workers," and those recruited by scouts that found 1% sucrose as "1%-workers," but it should remain clear that both groups encountered a sugar-free pseudoleaf at the source. It was observed that 10%-workers cut smaller fragments of Parafilm, returned to the nest at higher velocities, and displayed more active recruiting behavior (by laying chemical trails) than 1%-workers. In another experiment, recruited workers could only collect small pieces of paper of the same size. They still traveled at a faster pace if they had been recruited by 10%-scouts, indicating that the differences in speed were not caused by differences in loading. Greater velocity did not compensate for the reduction in fragment size: 10%-workers, despite their higher velocity, showed a lower delivery rate to the nest than 1%-workers (Roces and Núñez, 1993).

These results contradicted the most obvious functional hypotheses and led to a vivid discussion (Clark 1994; Kacelnik 1993; Roces 1994a; Ydenberg and Schmid-Hempel 1994). It was first surprising that workers cut fragments of different size, since all found a similar source (Parafilm). Moreover, a single Parafilm-leaf represents a non-depleting patch for the individual ant, and animals should load themselves maximally when collecting at *ad libitum* sources (Orians and Pearson 1979). But particularly puzzling was the observation that workers recruited to 10%-sucrose, the more attractive source, cut smaller fragments than workers recruited to the less-attractive source.

The Information-Transfer Hypothesis

Roces and Núñez (1993) advanced a hypothesis to account for these *a priori* unexpected results. It was called the "information-transfer hypothesis" and it was based on the original hypothesis that Núñez (1982) developed to explain partial crop-filling and seemingly suboptimal foraging behavior in honey bees. The information-transfer hypothesis states that a newly discovered food source motivates a worker to shorten its loading time and return not completely filled/loaded to the nest in order to recruit additional nestmates. A worker seems, therefore, to sacrifice its individual delivery rate in order to return earlier to the colony for further recruitment. Its performance as a food collector is therefore reduced, but the colony as a whole, which represents the level at which natural selection in eusocial insects predominantly takes place, increases its harvesting rate due to the recruited workers that participate in the resource-gathering activity. This hypothesis was consistent with the higher travel speed and the reduced leaf fragment size in 10%-workers. By cutting smaller fragments, 10%-workers saved cutting time and by increasing speed they saved travel time. This hypothesis could also explain the more intense trail-marking behavior observed in 10%-workers.

Further support for the information-transfer hypothesis was obtained by analyzing the behavior of recruited workers in more detail, specifically addressing the changes in worker responses after the perception of recruitment signals (Roces 1993). The behavioral responses of recruited workers have rarely been investigated in ants (Beckers, Deneubourg, and Goss 1992), despite the fact that they are responsible for the amplification and establishment of a recruitment process. In laboratory experiments, attention has been centered on the behavior of recruited workers that found a standardized resource after being recruited to sugar solutions of different quality, as follows. *Acromyrmex lundi* scout workers were first exposed to droplets of scented sucrose solution of different concentrations. Once at the source, recruited workers encountered, not sugar solution, but standardized paper discs impregnated with the same scent. Paper discs were either untreated or had sugar added in order to increase their quality. Comparison between assays using untreated and sugar-coated discs of the same weight allowed a distinction between workers' responses based on the information they obtained during recruitment (which was similar for workers recruited to a given sucrose solution), and those that resulted from their actual evaluation of the disc quality (untreated or sugared disc).

Several effects of the information about resource quality transferred by a single scout ant were observed in the behavioral responses of recruited workers. First, the running speed of a recruited worker to the feeding site and the speed carrying a paper disc back to the colony were positively correlated with the concentration of the sucrose solution that the initial scout found. In other words, travel speed of recruited workers to and from the nest depended on the information about resource quality they received during recruitment. As intuitively expected, disc-laden workers ran slower than outbound workers. This reduction in speed, however, could not be attributed to the effects of the load itself, since workers collecting discs of the same weight, but with sugar added, ran at the same speed as outbound, unladen workers.

Second, workers collecting sugared discs were observed to reinforce the chemical trail on their way to the nest. The percentage of trail-layers was higher when workers were recruited to the 10% sugar solution than to the 1% solution, although the collected discs were identical. These facts indicate that the workers' own evaluation of resource quality, exhibited as trail-reinforcement, was not absolute and depended on the perception of recruitment signals. This information-dependent assessment of resource quality was evident when the workers' running speed was considered. The observed reduction in speed after the collection of a non-sugared disc suggests that workers, through the information received during recruitment, may have expected to find a resource of satisfactory quality, and that their expectation was not fulfilled. They were willing to collect the non-sugared discs, but were reluctant to reinforce the chemical trail. Third, the load itself had no effect on velocity because workers collecting sugared discs of the same weight showed no reduction in velocity, and trail reinforcement occurred (Roces 1993).

Why do recruited workers run faster when they have been informed about a richer source? The adaptive value of this response might be related to a rapid transmission of information about the newly discovered food source since it allows a faster build-up of workers. At the initial phases of trail development, saving time would be of great importance in order to monopolize a food source as soon as possible. The evaluation of resource quality by recruited workers and, therefore, the probability of reinforcing the chemical trail are, in part, dependent on the information they receive; that is, recruits seem to partially rely on the decisions of the first successful ants to amplify a recruitment process. As mentioned above, *Acromyrmex lundi* workers recruited to rich sources cut smaller fragments than those recruited to poorer ones, although both groups found the same standardized pseudoleaf at the source. Recruits' decisions appear to be "channeled" by those of scouts, and it is an open question as to whether this non-democratic decision-making system (Jaffé and Villegas 1985) may be related to both the complexity of food assessment by leaf-cutting ants and the need for rapid responses to monopolize newly discovered sources.

Such complexity in individual decision making indeed affects the development of colony-level foraging patterns, as demonstrated under controlled laboratory conditions by comparing the foraging behavior of satiated versus harvesting-deprived, "hungry" workers (Roces and Hölldobler 1994). Harvesting-deprived workers cut smaller leaf fragments than satiated workers, a fact that appears counterintuitive at the individual level. They also showed higher recruitment rates than satiated workers. The harvesting of small fragments was observed only during the initial phases of the recruitment process, when information about the discovery needs to be transferred, indicating that workers traded off carrying performance for information transfer.

Leaf-Cutting Ants: Cooperative Load Transport and Information Flow

In several leaf-cutting ant species, the leaf fragments are not carried to the nest by the workers that cut them. The harvested fragments may be dropped to the ground or initially transported a short distance and then dropped or passed to other nestmates for further carriage. This phenomenon has been named the *transport chain*. Considerations about optimal foraging strategies can, therefore, go beyond the focus on flexible cutting rules, as discussed above, to encompass the adaptive value of such sequential cooperation for food carriage. The existence of transport chains in leaf-cutting ants was first described for *Atta sexdens rubropilosa* and *Atta cephalotes* (Fowler and Robinson 1979; Hubbell et al. 1980). More recently, transport chains were reported in *Atta colombica*, in which approximately 20% of the harvested fragments are transported in that way (Anderson and Jadin 2001; Hart and Ratnieks 2001).

The adaptive value of the formation of transport chains was investigated in grass-cutting ants, *Atta vollenweideri*. The complete foraging process can be easily analyzed in this ground-foraging species (Röschard and Roces 2003a, 2003b). It was observed that on long foraging trails, more than half of the fragments were transported by chains; besides the cutter, generally two or three carriers transported the load sequentially.

The kind of decisions workers face when harvesting grasses is summarized in Figure 13.3. First, cutters decide about the size (length) of the grass blade to be cut. This aspect has not been investigated so far. Second, carriers decide whether the carried fragment should be dropped or carried further to the nest. It remains unclear what variables motivate workers to drop their fragments, leading to the formation of transport chains. Ants might decide to drop fragments that are not sufficiently attractive, thus rejecting them. But since all dropped fragments are retrieved again, this appears to be very unlikely. Dropping might occur because of a mismatch between body and fragment size (i.e., either the carrier is too small

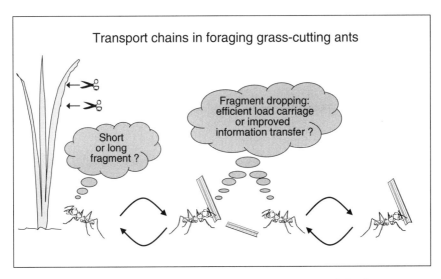

Figure 13.3. Instances of decision making during harvesting and the formation of transport chains in grass-cutting ants *(Atta vollenweideri)*.

for the fragment or the fragment too large to be carried). This seems plausible when the detrimental effects of large loads on transport rates are taken into account (Röschard and Roces 2002). In fact, sequential transport via transport chains in *Atta vollenweideri* leads to a slightly better size-matching between worker and load, which suggests that the formation of transport chains might improve the delivery rate of the involved fragments. Finally, transport chains may be formed to improve information transfer along the trail because more workers will be informed about the kind of resource being harvested either by direct transfers or upon finding a dropped fragment on the trail. Due to new recruitment, improved information transfer may lead to an increased overall rate of resource transportation. Based on this information-transfer hypothesis, the behavioral response of transferring fragments, either directly or indirectly, may have been selected for because it has positive effects on the information flow.

Transport Chains: Maximization of Leaf Delivery Rate?

Fragment dropping may allow cutters to quickly return to the selected plant for further harvesting. While this argument may account for the dropping behavior of cutters, why do carriers drop the fragments once more and walk back toward the source? Is the sequential transport via a transport chain faster than the transport by single carriers, thus enhancing colony-wide material intake rates? For the sake of simplicity, I would like to term these arguments as the "economic-transport hypothesis." It should be noted that *economic* in this context refers to the maximization of the transportation speed of a leaf fragment (Lutz 1929), which at the colony level may result in an increased overall rate of resource transportation. Maximization of leaf transportation has been proposed as a foraging criterion for workers of three leaf-cutting ant species that transfer loads or cache fragments on the ground (Anderson and Jadin 2001; Fowler and Robinson 1979; Hubbell et al. 1980). Direct leaf transfer between *Atta colombica* workers, which occurs for only 9% of the transported fragments, resulted in a higher transportation speed after transfer, although the transferred fragments did not travel faster than those not transferred because of the transfer delays (Anderson and Jadin 2001). In another study on the same species, however, fragments recovered from a cache were transported back to the nest more slowly than normally foraged leaf fragments

(Hart and Ratnieks 2001), so that the adaptive value of such response remains obscure. In *Atta vollenweideri,* transport time of fragments carried by a chain was 25% longer (on average 8 minutes longer) than that of fragments carried by a single worker all the way to the nest. This was due primarily to both the waiting time of the dropped fragments and the handling time by the subsequent foragers. Thus, in terms of foraging time and material transport rates, sequential transport by chains was less efficient than transport by single carriers (Röschard and Roces 2003a).

Transport Chains: Improved Information Transfer?

Do transport chains at least accelerate the transfer of information about the plant species actually harvested, even if they do not increase transport rate? Several processes may contribute to the acceleration of information transfer and to the rapid build-up of worker numbers at the discovered source. First, fragment dropping after a given distance may allow cutting workers to quickly go back to the harvesting plant, making it easier for them to find the source again by following the freshly deposited pheromone trail (Fowler and Robinson 1979; Hubbell et al. 1980). Second, moving along a short trail sector during foraging may enable workers to locally reinforce the pheromonal marking better than when walking all the way to the nest. Third, fragments dropped on the trail may act as "signal posts." It has been shown that leaf-cutting ant foragers learn the odors of the harvested resources and that worker responses at the patch depend on what nestmates are currently transporting on the trail (Howard et al. 1996; Roces 1990b, 1994b). The odor of the fragment on the trail might function as a stimulus for olfactory learning. Most foragers were observed to antennate the dropped fragments they found, even those workers that continued on their way to the cutting site without loads. Thus, outgoing foragers may obtain information about the harvested resources both by contacting laden nestmates along the trail or by finding a dropped fragment on the trail. This information may lead outgoing workers to search for a particular plant species, thus leading workers from the trail to the new plant.

   The predictions of the two competing hypotheses discussed above were tested in a field study (Röschard 2002). The economic-transport hypothesis predicts that workers transfer fragments because of a mismatch between load size and body size, making transport therefore inefficient. As a consequence, the probability of occurrence of transport chains would be expected to depend on fragment size and not on fragment quality. The probability should be higher for larger fragments that are difficult to carry. Alternatively, the information-transfer hypothesis states that the behavioral response of transferring fragments has been selected for because of its positive effect on the information flow. By dropping the load, a worker may be able to return earlier to the foraging site, which would reinforce the chemical trail, and thus enhance recruitment. In addition, the transferred fragments could themselves act as cues, giving information about the plant that is worth harvesting. This hypothesis predicts that the formation of transport chains should strongly depend on fragment quality and not on fragment size.

To distinguish between these alternatives, workers from a field colony were presented with standardized paper fragments that differed either in size or in quality (sugared vs. tannin-rich fragments). The occurrence of transport chains was quantified by following individually-marked grass fragments all the way to the nest. Results demonstrated that neither an increase in fragment mass nor in fragment length modifies the probability of occurrence of transport chains. In addition, transport time by transport chains took longer than by a single carrier. The frequency of occurrence of transport chains increased with increasing fragment quality, however, independent of the fragment's size. In addition, high-quality fragments were transferred after shorter distances than less attractive ones (i.e., attractive loads were dropped more frequently and after a shorter distance). Taken together, these results strongly suggest that transport chains increase the information flow at the colony level rather than the economic load carriage at the individual level.

Conclusions and Future Directions

The first quantitative studies investigating partial loading in social insect foragers were done in honeybees (Núñez 1966, 1970, 1971, 1982). These investigations led to the formulation of the information-transfer hypothesis, which argues that returning earlier to the nest (with partial loads) would increase the probability of information exchange. The relevant behavioral evidence was easily observable and quantifiable in honey bees because besides the conspicuous dancing (information transfer), getting information was evident in the forager's begging behavior (Núñez 1970).

An alternative hypothesis to account for partial loading in honey bees was developed in the framework of optimal foraging theory, using Núñez's published data (Kacelnik, Houston, and Schmid-Hempel 1986; Schmid-Hempel, Kacelnik, and Houston 1985). Surprisingly, the authors explicitly ignored any reference to information and recruitment, and modeled honey bee foraging responses based exclusively on individual optimization criteria, arguing that honeybees maximize individual efficiency by not filling their crops. A key argument in these models was the assumption that, because of the large energy costs of flying among flowers while foraging, honey bees experience diminishing returns as they load a fuller crop. However, energetic measurements of locomotion and load carriage in honey bees did not clearly support this assumption (Balderrama, Almeida de Balderrama, and Núñez 1992; Moffatt 2000; Moffatt and Núñez 1997). For nectar-feeding ants, which also often return with partial loads, the costs of load carriage were observed to be relatively low (Fewell 1988; Fewell et al. 1996; Schilman and Roces 2005, 2006), suggesting that time, and not energy costs, may underlie the selection of partial crop loads.

To date, central-place foraging theory has been unable to provide a compelling explanation for the occurrence of partial loading in foraging social insects. The first models discussed foragers' decisions in terms of energetic considerations at the individual level, without any reference to information exchange. There is quantitative evidence for honeybees as well as for leaf-cutting ants regarding the relevance of information transfer for workers' foraging behavior, although other processes, such as the interactions between above-ground and below-ground foragers, may have contributed to the evolution of load-size determination in ants (Burd and Howard 2005a, 2005b). Experimental data on flexible load-size determination and information transfer in leaf-cutting ants (Roces and Núñez, 1993) called renewed attention to the issue of partial loading in foraging social insects, and although not formalized in a model, the role of information transfer in modulating foragers' responses was then explicitly acknowledged in theoretical accounts (Kacelnik 1993; Ydenberg and Schmid-Hempel 1994). It is, therefore, surprising that more than 20 years after the development of the first model discussing partial crop-filling in honey bees, which was based on central-place foraging theory (Schmid-Hempel, Kacelnik, and Houston 1985), more recent models asking the same question restricted their focus to the benefits of partial crop loads, and the resulting time-savings it might confer in terms of collecting information (Dornhaus et al. 2006), and omitted analysis of the benefits that partial crop loads might provide in terms of earlier information transfer, for example via dance behavior.

These considerations show that we are far from understanding the regulation of foraging behavior. As a consequence, empiricists interested in social insect foraging are encouraged to design and perform experiments focusing on the precise mechanisms underlying individual decision making and social communication. Theorists should develop models that, based on empirical results, have heuristic value and suggest ways to test their predictions. Both empiricists and theorists are invited to rise to the challenge of integrating theoretical arguments with empirical research. Considering the bulk of information on foraging social insects that already exists, it seems a promising and rewarding endeavor.

Literature Cited

Anderson, C., and J. L. V. Jadin. 2001. "The adaptive benefit of leaf transfer in *Atta colombica.*" *Insectes Sociaux* 48: 404–405.

Balderrama, N., L. O. Almeida de Balderrama, and J. A. Núñez. 1992. "Metabolic rate during foraging in the honeybee." *Journal of Comparative Physiology B* 162: 440–447.

Beckers, R., J. L. Deneubourg, and S. Goss. 1992. "Trail laying behaviour during food recruitment in the ant *Lasius niger* (L.)." *Insectes Sociaux* 39: 59–72.

Burd, M., and J. J. Howard. 2005a. "Central-place foraging continues beyond the nest entrance: The underground performance of leaf-cutting ants." *Animal Behaviour* 70: 737–744.

———. 2005b. "Global optimization from suboptimal parts: Foraging sensu lato by leaf-cutting ants." *Behavioral Ecology and Sociobiology* 59: 234–242.

Clark, C. W. 1994. "Leaf-cutting ants may be optimal foragers." *Trends in Ecology and Evolution* 9: 63.

Dornhaus, A., E. J. Collins, F. X. Dechaume-Moncharmaont, A. I. Houston, N. R. Franks, and J. M. McNamara. 2006. "Paying for information: Partial loads in central place foragers." *Behavioral Ecology and Sociobiology* 61: 151–161.

Dreisig, H. 2000. "Defense by exploitation in the Florida carpenter ant, *Camponotus floridanus*, at an extrafloral nectar resource." *Behavioral Ecology and Sociobiology* 47: 274–279.

Farina, W. M. 2000. "The interplay between dancing and trophallactic behavior in the honey bee *Apis mellifera.*" *Journal of Comparative Physiology A* 186: 239–245.

Fewell, J. H. 1988. "Energetic and time costs of foraging in harvester ants, *Pogonomyrmex occidentalis.*" *Behavioral Ecology and Sociobiology* 22: 401–408.

Fewell, J. H., J. F. Harrison, J. R. B. Lighton, and M. D. Breed. 1996. "Foraging energetics of the ant, *Paraponera clavata.*" *Oecologia* 105: 419–427.

Fowler, H. G., and S. W. Robinson. 1979. "Foraging by *Atta sexdens* (Formicidae: Attini): Seasonal patterns, caste and efficiency." *Ecological Entomology* 4: 239–247.

Gallistel, C. R. 1990. *The organization of learning.* Cambridge: The MIT Press.

Harrison, J. M., and M. D. Breed. 1987. "Temporal learning in the giant tropical ant, *Paraponera clavata.*" *Physiological Entomology* 12: 317–320.

Hart, A. G. and F. L. W. Ratnieks. 2001. "Leaf caching in the leafcutting ant *Atta colombica*. Organizational shifts, task partitioning and making the best of a bad job." *Animal Behaviour* 62: 227–234.

Howard, J. J. 1988. "Leafcutting ant diet selection: Relative influence of leaf chemistry and physical features." *Ecology* 69: 250–260.

Howard, J. J., M. L. Henneman, G. Cronin, J. A. Fox, and G. Hormiga. 1996. "Conditioning of scouts and recruits during foraging by a leaf-cutting ant, *Atta colombica.*" *Animal Behaviour* 52: 299–306.

Hubbell, S. P., L. K. Johnson, E. Stanislav, B. Wilson, and H. Fowler. 1980. "Foraging by bucket-brigade in leaf-cutter ants." *Biotropica* 12: 210–213.

Jaffé, K., and P. E. Howse. 1979. "The mass recruitment system of the leaf cutting ant, *Atta cephalotes* (L.)." *Animal Behaviour* 27: 930–939.

Jaffé, K., and G. Villegas. 1985. "On the communication systems of the fungus-growing ant *Trachymyrmex urichi.*" *Insectes Sociaux* 32: 257–274.

Josens, R. B., W. M. Farina, and F. Roces. 1998. "Nectar feeding by the ant *Camponotus mus:* Intake rate and crop filling as a function of sucrose concentration." *Journal of Insect Physiology* 44: 579–585.

Josens, R. B., and F. Roces. 2000. "Foraging in the ant *Camponotus mus:* Nectar intake and crop filling depend on colony starvation." *Journal of Insect Physiology* 46: 1103–1110.

Kacelnik, A. 1993. "Leaf-cutting ants tease optimal foraging theorists." *Trends in Ecology and Evolution* 8: 346–348.

Kacelnik, A., A. I. Houston, and P. Schmid-Hempel. 1986. "Central-place foraging in honey bees: The effect of travel time and nectar flow on crop filling." *Behavioral Ecology and Sociobiology* 19: 19–24.

Lutz, F. E. 1929. "Observations on leaf-cutting ants." *American Museum Novitates* 388: 1–21.

Mailleux, A.-C., J.-L. Deneubourg, and C. Detrain. 2000. "How do ants assess food volume?" *Animal Behaviour* 59: 1061–1069.

————. 2003. "Regulation of ants' foraging to resource productivity."
    Proceedings of the Royal Society of London B 270: 1609–1616.

Mailleux, A.-C., C. Detrain, and J.-L. Deneubourg. 2005. "Triggering and per-
    sistence of trail-laying in foragers of the ant *Lasius niger.*" *Journal of Insect
    Physiology* 51: 297–304.

Moffatt, L. 2000. "Changes in the metabolic rate of the foraging honeybee:
    Effect of the carried weight or of the reward rate?" *Journal of Comparative
    Physiology A* 186: 299–306.

Moffatt, L., and J. A. Núñez. 1997. "Oxygen consumption in the foraging hon-
    eybee depends on the reward rate at the food source." *Journal of
    Comparative Physiology B* 167: 36–42.

Núñez, J. A. 1966. "Quantitative Beziehungen zwischen den Eigenschaften
    von Futterquellen und dem Verhalten von Sammelbienen." *Zeitschrift für
    vergleichende Physiologie* 53: 142–164.

————. 1970. "The relationship between sugar flow and foraging and recruiting
    behaviour of honey bees (*Apis mellifera* L.)." *Animal Behaviour* 18: 527–538.

————. 1971. "Beobachtungen an sozialbezogenen Verhaltensweisen von
    Sammelbienen." *Zeitschrift für Tierpsychologie* 28: 1–18.

————. 1982. "Honeybee foraging strategies at a food source in relation to its
    distance from the hive and the rate of sugar flow." *Journal of Apicultural
    Research* 21: 139–150.

Núñez, J. A., and M. Giurfa. 1996. "Motivation and regulation of honey bee
    foraging." *Bee World* 77: 182–196.

Orians, G. H., and N. E. Pearson. 1979. "On the theory of central place forag-
    ing." In D. J. Horn, G. R. Stairs, and R. D. Mitchell, eds., *Analysis of eco-
    logical systems,* 155–177. Columbus: Ohio State University Press.

Pflumm, W. 1975. "Einflu;sz verschiedener Zuckerwasser-Konzentrationen und
    der Anwesenheit von Artgenossen auf das Sammelverhalten von Wespen
    (Hymenoptera: Vespidae)." *Entomologia Germanica* 2: 7–21.

Roces, F. 1990a. "Leaf-cutting ants cut fragment sizes in relation to the dis-
    tance from the nest." *Animal Behaviour* 40: 1181–1183.

————. 1990b. "Olfactory conditioning during the recruitment process in a
    leaf-cutting ant." *Oecologia* 83: 261–262.

————. 1993. "Both evaluation of resource quality and speed of recruited leaf-
    cutting ants (*Acromyrmex lundi*) depend on their motivational state."
    Behavioral Ecology and Sociobiology 33: 183–189.

————. 1994a. "Cooperation or individualism: How leaf-cutting ants decide on
    the size of their loads." *Trends in Ecology and Evolution* 9: 230.

————. 1994b. "Odour learning and decision-making during food collec-
    tion in the leaf-cutting ant *Acromyrmex lundi.*" *Insects Sociaux* 41:
    235–239.

————. 2002. "Individual complexity and self-organization in foraging by leaf-cutting ants." *Biological Bulletin* 202: 306–313.

Roces, F., and B. Hölldobler. 1994. "Leaf density and a trade-off between load-size selection and recruitment behavior in the ant *Atta cephalotes.*" *Oecologia* 97: 1–8.

Roces, F., and J. A. Núñez. 1993. "Information about food quality influences load-size selection in recruited leaf-cutting ants." *Animal Behaviour* 45: 135–143.

Röschard, J. 2002. "Cutters, carriers and bucket brigades—Foraging decisions in the grass-cutting ant Atta vollenweideri." Ph.D. diss., Fakultät für Biologie, Julius-Maximilians-Universität, Würzburg, Germany.

Röschard, J., and F. Roces. 2002. "The effect of load length, width and mass on transport rate in the grass-cutting ant *Atta vollenweideri.*" *Oecologia* 131: 319–324.

————. 2003a. "Cutters, carriers and transport chains: Distance-dependent foraging strategies in the grass-cutting ant *Atta vollenweideri.*" *Insectes Sociaux* 50: 237–244.

————. 2003b. "Fragment-size determination and size-matching in the grass-cutting ant *Atta vollenweideri* depend on the distance from the nest." *Journal of Tropical Ecology* 19: 647–653.

Schilman, P. E., and F. Roces. 2003. "Assessment of nectar flow rate and memory for patch quality in the ant *Camponotus rufipes.*" *Animal Behaviour* 66: 687–693.

————. 2005. "Energetics of locomotion and load carriage in the nectar feeding ant, *Camponotus rufipes.*" *Physiological Entomology* 30: 332–337.

————. 2006. "Foraging energetics of a nectar-feeding ant: Metabolic expenditure as a function of food-source profitability." *Journal of Experimental Biology* 209: 4091–4101.

Schmid-Hempel, P., A. Kacelnik, and A. I. Houston. 1985. "Honeybees maximize efficiency by not filling their crop." *Behavioral Ecology and Sociobiology* 17: 61–66.

Seeley, T. D. 1989. "Social foraging in honey bees: How nectar foragers assess their colony's nutritional status." *Behavioral Ecology and Sociobiology* 24: 181–199.

Varjú, D., and J. Núñez. 1991. "What do foraging honeybees optimize?" *Journal of Comparative Physiology A* 169: 729–736.

Wahl, O. 1932. "Neue Untersuchungen über das Zeitgedächtnis der Bienen." *Zeitschrift für vergleichende Physiologie* 16: 529–589.

Wainselboim, A. J., F. Roces, and W. M. Farina. 2002. "Honeybees assess changes in nectar flow within a single foraging bout." *Animal Behaviour* 63: 1–6.

————. 2003. "Assessment of food source profitability in honeybees *(Apis mel-lifera):* How does disturbance of foraging activity affect trophallactic behaviour?" *Journal of Comparative Physiology A* 189: 39–45.

Ydenberg, R., and P. Schmid-Hempel. 1994. "Modelling social insect foraging." *Trends in Ecology and Evolution* 9: 491–493.

PART THREE

Neurogenetic Basis of Social Behavior

ROBERT E. PAGE JR.

HOW DO INSECT SOCIETIES evolve? The evolutionary process requires that alternative alleles at gene loci located within the cells of sometimes-sterile individuals (workers) change in frequency as a consequence of colony-level selection for organization of a colony with respect to division of labor. But the effects of alternative alleles of individual genes must act across many levels of biological organization including gene and protein interactions (epistasis), gene and protein signaling within and between cells, development, neural physiology and anatomy, and stimulus response systems that integrate the behavior of individuals cohabiting a nest. The evolution of changes at these levels must occur against a functional background with origins in solitary living and must effect developmental changes against that background. Bert Hölldobler has spent his life studying the complex social structure of ants. His elegant behavioral and anatomical studies have fascinated us with the diverse social patterns of ants and the incredible anatomical and physiological adaptations that accompany them. The powerful effects of colony-level selection on sterile castes are replete throughout his major work, *The Ants*, and raise questions of what are social traits, can we find signatures of their evolution in developmental patterns of the individuals that comprise the social units, and can we reconstruct the evolution of complex social traits from the gene to the society?

The authors of the five chapters in this section examine the mechanisms of social organization across levels of organization from genes through neuroanatomy and physiology. In Chapter 14, Jürgen Gadau and Greg Hunt discuss the genetic architectures underlying social traits and argue for the need to study the genetic basis and regulation of behavior that results in reproductive and nonreproductive life histories. They also make an appeal for comparative studies with less advanced social species and between social and solitary species that are closely related, as well as comparative studies across greater phylogenetic distances. In Chapter 16, however, Robert Page, Timothy Linksvayer, and Gro Amdam, argue that the signatures of the history of social evolution can be found by studying development in highly social species. They propose an explanation for the evolution of division of labor that is entrenched in the evolution of development. The gonotropic cycle that is widespread in solitary insects has been co-opted by colony-level selection to generate division of labor while the temporal orchestration of hormones involved in reproductive maturation have undergone a heterochronic shift. Early activation of the reproductive network results in workers expressing maternal behavior soon after emergence as adults, hence providing an explanation for the most fundamental component of eusociality: stay home and work in the maternal nest.

In Chapters 15, 17, and 18 behavioral and neural plasticity, and sensory response systems and behavior, are discussed. In Chapter 15, Ricarda Scheiner and Joachim Erber present the relationships between response thresholds to stimuli and behavioral plasticity in honey bees. They address the specific questions: (1) Do bees that differ in their activities within the colony have different thresholds for specific sensory stimuli? (2) What are the neurophysiological mechanisms that control sensory thresholds for specific stimuli? (3) How do changes in the environment or within the colony modulate sensory thresholds? (4) Is there a genetic bias that affects response thresholds for specific stimuli? Wulfila Gronenberg and Andre Riveros inquire into whether social insects have evolved a "social brain" in Chapter 17. The social brain hypothesis of primates proposes that certain areas of the brains of the more social species are enlarged, a consequence of adaptation to complex social life. They explore the anatomy of the brains of social insects and look for evidence of selection in the social environment, however, they suggest that the signature of such selection might actually be a reduction in brain size, a consequence of a reduced behavioral repertoire because of division of labor.

The final chapter of this section is by Guy Bloch, who looks at the timing of behavior and its synchronization with a biological clock. Bloch proposes that circadian behavioral plasticity in social groups is a mechanism to temporally coordinate activities among nestmates and that the integration of the circadian clock into a colony-level pattern is crucial for efficient colony-level behavior. He proposes that we should look to circadian models of Drosophila and the mouse to help us understand the chronobiology of social insects.

CHAPTER FOURTEEN

Behavioral Genetics in Social Insects

JÜRGEN GADAU

GREG J. HUNT

BEHAVIORAL CHANGES or pre-adaptations of individual organisms must have been a prerequisite for the evolution of sociality in insects. For example, individuals had to first tolerate the presence of others in the nest and then had to give up their own reproduction and help others to rear their offspring, leading to a reproductive division of labor. As societies grew and became more complex, more intricate systems of division of labor between nonreproductive individuals or workers evolved. We argue in this chapter that in order to completely understand how sociality and complex social organization evolved and to construct realistic models of social evolution (i.e., evolutionary analysis), it is necessary to study the genetic basis and regulation of social behaviors that result in a reproductive and nonreproductive division of labor (i.e., mechanistic analysis). Once we have a better understanding of the genetic architecture and regulation of these behavioral traits, we will be able to infer what genetic changes have taken place in the transition from a solitary to a social lifestyle by following the genetic and regulatory changes that accompanied these transitions.

Subsequently, a behavioral genetic analysis of clades that show a solitary/social polymorphism for lifestyles, as in some communal or primitively eusocial bees, will help to differentiate between genetic changes essential for the evolution of eusociality and other changes that were adopted into already existing eusocial life histories to fine tune colony efficiency of more complex societies with thousands of workers. This differentiation is impossible if we only study highly derived eusocial insects. Our final goal will be to compare the genetic architecture of social behavior between phylogenetically distant insect lineages like ants, bees, and termites. This should enable us to determine whether convergent genetic and regulatory changes have taken place during the evolution of sociality in these independent lineages or whether each lineage found an alternative solution to the same problems faced by highly eusocial colonies.

Division of Labor—The Essence of Social Insect Lifestyle

Behavioral genetics in social insects has focused primarily on the genetic basis of division of labor among nonreproductive individuals. Most studies have used the honey bee *Apis mellifera,* but scattered studies in other social insects showed similar results (Snyder 1992; Stuart and Page 1991). In honey bees, as in other social insects, it has been documented that genotypic variation among workers leads to differences in the probability of individual workers of the same colony to perform a specific task. Allelic variation at the DNA level that leads to predictable behavioral variability of workers can therefore be used to identify the genes responsible for task preferences of workers. In contrast, very little is known about the genetic basis for the reproductive division of labor, the situation in which individuals make the ultimate altruistic sacrifice by foregoing reproduction and becoming workers. We have genes, or rather Quantitative Trait Loci (QTL), for division of labor among workers, and also genes that are differentially expressed between task groups (for a summary of these genes see Table 1 in Robinson Grozinger, and Whitfield 2005; Hunt et al. 2007; Page, Linksvayer, and Amdam this volume), but no legitimate claims have been made so far for an "altruism gene."

The reason we have been unable to find an altruism gene may be in part because caste determination in most social insects is based on environmental cues and each individual larva is reproductively totipotent. Therefore, we would not expect to see genetic variation at the DNA level for a reproductive division of labor in social insect populations; that is, allelic variation for the probability of becoming a queen or worker (see Volny and Gordon 2002; Julian et al. 2002; and Helms-Cahan and Keller 2003, for examples of genetic caste determination). Theoretically, any gene that leads to the preferential development into a queen would sweep quickly to fixation, or strong selection on unlinked genes would lead to the evolution of a suppressor gene that keeps such a gene under control. To understand the genetic architecture of reproductive caste determination in social insects we need a different scheme than the one used in studies of division of labor among workers. This approach must reveal how different environmental factors trigger different developmental programs leading to reproductive or nonreproductive phenotypes, which are associated with completely different behavioral programs, such as egg laying versus brood care, foraging, or aggressive versus submissive behavior in interactions (Jeanne, this volume). Hence, to understand the genetic architecture of reproductive division of labor or to understand which genes are involved in the different behavioral repertoires of queens and workers, we need to move up one organizational level in our analysis. Instead of associating genotypic variation with the phenotype, we have to analyze the differences in gene regulation and gene expression between the two castes.

This approach has been championed in honey bees and it has been shown that many genes are differentially expressed between queen larvae and worker larvae (Evans and Wheeler 2000; Christino et al. 2006). But there are limitations for identifying regulatory genes that initiate caste differentiation (i.e., the genes that interact with the environment to determine the fate of a still totipotent larva) because, as is the case with sex determination, such signals are probably very ephemeral. Maybe the only way to find the regulatory genes is to look at species with large genetic effects on caste determination analogous to the identification of the sex determination locus using diploid males. From an evolutionary point of view, the unknown regulatory genes could be called altruism genes because they control reproductive division of labor. These genes, however, respond in a predetermined way to the environment they find themselves in, so it is really the environment or the behavior of the workers feeding larvae that determines the reproductive status of an individual (Linksvayer 2006, 2007; Linksvayer and Wade 2005).

As mentioned above, another angle for identifying genes involved in the regulation of castes is to use social insect species with genetic caste determination or major disturbances of the environmental caste determination system. This approach is analogous to the use of mutants in *Drosophila* to reveal the underlying genetics of major genes influencing early embryonic development because genes that underlie genetic caste determination must influence early reproductive development. Potential systems that could be used to test this approach can be found in ants (e.g., *Pogonomyrmex* spp., Julian et al. 2002; Volny and Gordon 2002; Helms-Cahan and Keller 2003; *Acanthomyops*, Umphreys 2006) or stingless bees (*Melipona quadrifasciatus,* Kerr 1950). In honey bees the genetic basis of the "anarchistic bees" could be studied (Montague and Oldroyd 1998), but the mechanism allowing workers to reproduce during the presence of a fertile queen is most likely independent from the regulatory genes influencing the worker-queen dichotomy. A detailed genetic analysis of the basis of worker reproduction might nevertheless provide a fruitful avenue toward an understanding of the genetic architecture of reproductive caste determination. The results of such an endeavor into the genetic basis of reproductive division of labor may prove to be as exciting and revealing as the studies on division of labor in a colony's workforce. Interestingly, research on the reproductive groundplan hypothesis for the evolution of division of labor among workers (West-Eberhardt 1987; Amdam et al. 2004; Guidugli et al. 2005; Hunt et al. 2007) suggest that these seemingly disparate lines of research might coalesce because both honey bee foraging behavior and female reproductive development may be influenced by the same underlying regulatory networks (Amdam et al. 2006).

We now briefly review the considerable progress that has been made in the field of behavioral genetics of social insects (see also Bloch and Jeanson and Deneubourg, this volume). The focus on the genetics of division of labor has shifted significantly over time from finding genes or markers associated with task preferences of workers, toward an understanding of how the modification of ancestral and preexistent gene regulatory networks in solitary ancestors can explain colony-level phenotypes like division of labor. Interestingly, this shift was championed by a nongeneticist 20 years ago (West-Erberhardt 1987) and had a precursor as much as 50 years ago (Bier 1954), but evaluations of these claims had to wait until we had the appropriate genetic tools (Amdam et al. 2006; Page et al., this volume).

The Beginning of Behavioral Genetics in Social Insects

The first studies on the genetic basis of individual behavior of social insects were conducted in bees using artificial selection for hygienic behavior (Rothenbuhler 1964a, b) and pollen hoarding (Hellmich 1985), or through the association of morphological or genetic (allozyme) markers representing different patri- or matrilines, with specific tasks performed preferentially by each line even when raised in the same colony; Robinson and Page 1988, 1989). To our knowledge, selection studies in social insects have been restricted to honey bees, but we could learn a lot by applying selective breeding to species like bumblebees, termites, or other social insects that can be bred in the laboratory. In addition to the information about genetic variation maintained in populations for specific behaviors and the contribution of additive genetic variance in behavioral traits, selected lines in honey bees have become important tools in mapping QTL to search for the genes underlying behavioral variation.

During the last decade it has become possible to construct linkage maps for social insects (e.g., Hunt and Page 1995; Gadau et al. 2001; Sirvio et al. 2006) and to map genomic regions that have a significant effect on the observed behavior (e.g., Hunt et al. 1995; Hunt et al. 1998; Chandra et al. 2001; Lapidge, Odroyd, and Spivak 2002; Page et al. 2000; Rueppell et al. 2006; Arechavaleta-Velasco and Hunt 2004). For the honey bee, we now have an annotated genome sequence (The Honeybee Genome Sequencing Consortium 2006) and tools to screen and manipulate this genome *in vivo* are becoming increasingly available. Significant advances in the understanding of gene effects on behavioral and neurobiological traits will be made possible by combining innovative methods from neuroscience with marker-association studies and gene-expression analyses. Especially important for the future will be to understand the interplay between genetic variation and social interactions (genotype by social-environment interactions), because all workers of a colony are capable of performing each task during their adult life yet each individual performs only a specific set of tasks at any particular time, and in such a way as to respond appropriately to stimuli that depend on colony state. Therefore, one major open question is: How do the genotypic differences among individuals in a colony result in physiological or gene regulatory differences that in turn produce the observed adaptive system of division of labor?

Studies in the honey bee, where we have more molecular tools than we do for other social insects, will provide the opportunity to understand the genetic architecture of division of labor and reproduction in a highly eusocial species. However, in order to understand exactly how the underlying genetic architecture has changed during the transition from solitary to social organization, we need to compare these results with taxa that harbor both solitary and social species. Another option would be to contrast honey bees with taxa like termites or ants that have evolved eusociality independently, or we could compare honey bees to related solitary or primitively social species such as euglossine bees or bumblebees.

The quantitative genetic studies of social behavior are reviewed below and we consider what linkage mapping and genomics has added to our understanding of the genetics and evolution of social behavior. We also discuss how these and other genomic tools might help us to unravel the evolution of naturally occurring behavioral variability and to identify genes influencing division of labor. The defensive behavior and dichotomy between pollen and nectar foragers in honey bees provide examples of how linkage mapping can lead to the identification of specific genes for the division of labor in social insects. Interactions between genes and social environments and epigenetic effects complicate the picture of bee behavior, but candidate genes, once identified, will help us to better understand these complexities.

Characterizing the genetic architecture of social behavior and social life histories raises important questions, and the answers may help us to understand how complex social systems evolved. For example, what drives the evolution of an unusually high recombination rate in the honey bee and the leaf-cutter ant *Acromyrmex echinatior*? Is an elevated recombination rate related to the social life history? How is the expression of genes that influence social behaviors coordinated? What are the influences of epigenetic and epistatic effects on social behaviors? How does the social and abiotic environment modify the expression of genes? Finally, what do comparative studies from other model organisms tell us about the genes that nature has selected to regulate division of labor?

Case Studies—Genetic Architecture of Social Behavior

Hygienic behavior was probably the first social behavior that was analyzed using classic quantitative genetics. Rothenbuhler (1964a, b) conducted an elegant series of experiments which lead him to conclude that this behavior was controlled by two genes that influenced uncapping behavior and two genes that influenced removal behavior. Although often cited in textbook examples of genetic effects on behavior, Rothenbuhler's conclusions about gene number, based on the distribution of colony phenotypes in a small population, were unwarranted (Moritz 1988). More recently, a study using a QTL approach found seven putative QTL influencing this behavior (Lapidge, Odroyd, and Spivak 2002). Similar studies in natural populations of bumblebees have revealed not only multiple loci for resistance against a specific parasite but also more complex interactions (e.g., epistasis) of the loci underlying this partial resistance (Wilfert, Gadaeu, and Schmid-Hempel 2007a).

Division of Labor in Honey Bees

Social Hymenoptera typically exhibit age polyethism in which younger individuals perform tasks inside the nest and older individuals forage or act as defenders outside the nest. Analyses of distribution of allozyme genotypes or color markers of worker bees performing different tasks showed that virtually all the tasks that are influenced by age-based behavioral development are also influenced by the genotype of the worker bee (Frumhoff and Baker 1988; Breed, Robinson, and Page 1990; Page and Robinson 1991; Calderone and Page 1992). But how does natural selection act upon groups and individuals to maintain this genetic variation, and what is its significance for colony fitness? Determining the actual genes involved in this process may teach us something about the evolutionary history of division of labor and the significance of gene effects on specific behaviors. There are two principally different approaches toward finding the genes involved in a phenotype. *Forward genetics* seeks to go from the observed phenotype to the gene, whereas a *candidate gene approach* (reverse genetics) tries to capitalize on previous knowledge of gene functions in other species to show that a gene with a particular phenotypic effect in species A has a similar effect in species B, with species A always being a genetically well-characterized model organism (e.g., *Drosophila melanogaster* or *Caenorhabditis elegans*). Both approaches have advantages and disadvantages.

The Candidate Gene Approach: From Gene to Behavior

A tremendous amount of information is already available for the function of genes in model organisms such as *Drosophila* or *Caenorhabditis*. The reverse genetics approach for identifying genes that influence social behaviors has been to use this information to choose orthologous genes in social insects to study their effects on a specific behavior (Fitzpatrick et al. 2004, Robinson and Barron this volume). The strategy is to choose a gene that has a large effect on behavior or signaling pathways in a model organism, and study the ortholog in a social insect either by manipulating its expression levels (e.g., silencing using RNA interference; RNAi) or measuring expression (mRNA or protein) levels and correlating expression with behavioral phenotypes. If it is possible to detect allelic variation for that particular gene in the species being studied, it would also be possible to analyze the effect of a particular allele on the behavior of interest, but it seems that this has yet to be done in social insects.

The candidate gene approach has been applied most extensively to honey bees by Robinson and his collaborators and much of this work has been reviewed here in previous chapters. One intriguing example comes from studies on the foraging gene (*for*) from *Drosophila*, which encodes a cGMP-dependent protein kinase, or PKG. Kinases are important for activating other proteins and PKG has been shown to influence feeding-related behaviors and other processes in a variety of model systems (reviewed by Sokolowski 2001; Ben-Shahar 2005). The honey bee ortholog, *AmFOR*, has been mapped and cloned. Allelic variation of *for* in *Drosophila* is responsible for the rover/sitter behavioral polymorphisms in wild populations. The polymorphic behavior is characterized by different search strategies in larval foraging behavior, which is tied to differential expression of *for* (Sokolowski 2001). *AmFOR* in honey bees might influence the age of onset for foraging (Ben-Shahar et al. 2002; Ben-Shahar et al. 2003). Foragers and undertaker bees have higher levels of *AmFOR* mRNA in their brains than nurse bees, and application of 8-Br-cGMP, which increases cGMP levels and activates PKG, results in earlier foraging. However, this does not constitute ironclad proof because the effect could also be due to toxicity, and application of 8-Br-cGMP did not increase expression of foraging-related genes, which were activated by the juvenile hormone analog methoprene (Whitfield et al. 2006). It is unclear whether the expression of this gene is causally involved in the naturally observed phenotypic variation of honey bee foraging behavior, as it is in *Drosophila*, but what does seem clear is that *AmFOR* influences positive phototaxis, which may be a critical stimulus for initiating foraging (Ben-Shahar 2005).

QTL Mapping: Going from Behavior to the Gene

The forward genetics approach is to map chromosomal regions that affect the trait and search through the genes in the region, which are then considered candidate genes. For Hymenopteran insects, the QTL method was first used to map genes that influence foraging resource choice (pollen versus nectar) in honey bees (Hunt et al. 1995; Page et al. 2000; Page et al., this volume, Scheiner and Erber, this volume). The QTL method also has been applied to map genes influencing behaviors that result in reproductive isolation in *Nasonia* (Gadau, Page, and Werren 1999;) as well as other traits related to foraging division of labor in honey bees (Rueppell et al. 2004a, 2004b, 2006), immune defense in bumblebees (Wilfert, Gadau, and Schmid-Hempel 2007a), and defensive behavior in honey bees (Hunt et al. 1998; Arechavaleta-Velasco and Hunt 2004).

To illustrate the QTL method, let us consider defensive behavior (see Page et al., this volume, for other QTL studies in honey bee social behavior). Honey bee defensive behavior has been used as an example of altruism because individuals die after stinging. Natural selection has resulted in large phenotypic variation for tendency to sting when relevant stimuli are presented to a bee colony. This variation presumably is the result of allelic variation of specific genes interacting with the environment, which includes social interactions. Highly defensive African-derived honey bees respond much more quickly to potential threats and respond with about five times as many stings in a leather target compared to low-defensive European-derived honey bees (reviewed by Breed et al. 2004). The organization of defensive behavior within a colony is complex because genotype-by-environment interactions occur through the interactions of individuals of divergent genotypes. Individual African-derived bees (AHB) are much more persistent at guarding the nest entrance than European honey bees (EHB). AHB are more persistent at guarding if co-fostered with high proportions of other African-derived bees than they are in colonies with high proportions of EHB, but "gentle" European nestmates do not respond in the same way (Hunt et al. 2003). In addition, bees with high-defensive genotypes efficiently recruit low-defensive genotypes to sting (Guzmán-Novoa et al. 2004). The situation is further complicated by a paternal effect on colony stinging response. Colonies composed of hybrids with African-derived fathers are as defensive as the African type, but the reciprocal cross, involving hybrids with European-derived fathers, are intermediate in defensive behavior (Guzmán-Novoa et al. 2005). This strongly suggests that there is a large epigenetic component to colony defensive behavior.

In colonial insects, genes influencing behavioral traits can be mapped at either the colony or individual level by associating the inheritance of DNA marker alleles with phenotypes. Colony-level behavioral responses were used to map QTL that influence stinging-behavior of honey bees using crosses between African- and European-derived stocks (Hunt et al. 1998). QTL influencing colony-level traits are most efficiently mapped using backcross colonies with queens that share the same haploid father, each mated to a drone from an F1 queen. A linkage map can be constructed based on the inheritance of marker-alleles in the haploid fathers of the colonies, and specific marker-alleles of the fathers can be associated with colony phenotype (such as numbers of stings) to map colony-level QTL.

Individual behavior can be used to map QTL or to confirm behavioral effects of previously mapped QTL. For example, we would expect that a random sample of backcross workers would segregate 1:1 for each random DNA marker, unless a viability effect is linked to the marker. However, samples of bees performing a particular behavior would show a deviation from 1:1 if a linked gene is influencing the likelihood of performing that behavior. In the case of defensive behavior, collections of individuals that were among the first five to ten bees to sting a leather patch presented at the hive entrance showed an overrepresentation of the defensive-parent allele at a marker linked to *sting1*, as compared to control samples in two independent studies (Guzmán-Novoa et al. 2002), thus confirming the influence of *sting1* on individual stinging behavior; and a similar study confirmed the effects of *sting1*, *sting2*, and *sting3* on guarding behavior (Arechavaleta-Velasco, Hunt, and Emore 2003).

Another interesting case that used the association of genetic markers with a distinct colony organization (monogyny versus polygyny) to understand the genetic basis of this difference is the red imported fire ant *Solenopsis invicta*. In the last few decades, researchers observed within the introduced North American *S. invicta* population the evolution and rapid spread of a polygynous form. Polygynous colonies outcompete the monogynous form whenever these two forms compete. Although polygyny is known from the native range of *S. invicta* in South America, the polygynous colonies of North American differ in many respects from the South American polygynous colonies (e.g., number of queens). Ross and colleagues have dissected the behavioral, ecological, evolutionary, and genetic details behind this difference in queen number over the last 20 years (Ross and Fletcher 1985; Ross and Keller 1998; Krieger and Ross 2002). This example highlights how a strong correlation between a certain allozyme genotype and colony phenotype can ultimately lead to the isolation of the underlying gene. In *S. invicta* a novel behavior during colony development, the re-adoption and tolerance of additional reproductives into an existing queen-right colony, can ultimately lead to population subdivision and might potentially mark the beginning of a new species. This novel behavior has been linked to an allelic substitution at the gene Gp-9, a putative pheromone binding protein (Krieger and Ross 2005). Although it is not yet understood in detail how this allelic substitution led to the observed behavioral differences, the current results suggest that it might have to do with a change in cues that a mated queen of the polygynous, versus monogynous, form emits and how workers of the different forms react to these cues (Ross and Keller 2002). Further studies revealed that the same allele linked to polygyny in *S. invicta* is also present and correlated with polygyny in other *Solenopsis* species that are social polymorphic (i.e., have both monogynous and polygynous colonies; Krieger and Ross 2005).

QTL or association mapping has one advantage over the candidate gene approach in that it directly identifies the loci that influence natural variation in behavior and provides a quantitative estimate of their effects. The major drawback so far is that QTL mapping does not identify a specific gene. How can we go from mapped location to the key genes that natural selection is operating on? In species with a genome sequence we can try to narrow the search by using physical mapping information. For example, confidence intervals for the location of defensive behavior QTL were approximated by taking the region of the linkage group within 1.5 LOD-value of the peak value. This provided a roughly 97% confidence interval (CI) in which to search for candidate genes. These CIs were quite large in terms of linkage map distance, consisting of about 40 centimorgans (cM); however, the amount of physical distance within the CIs was not too large (about one megabase of DNA) because of the high recombination rate in the honey bee (Hunt and Page 1995; Solignac et al. 2004) and the CIs each contained about 40 genes. The number of genes was even less than what was expected in a random 40 cM interval because the local recombination rates were even higher than average within these defensive-behavior CIs (Hunt et al. 2007). In contrast to the honey bee results, lower recombination rates in the mouse and the fruit fly have the effect that we would expect—roughly 500 and 2,000 genes, respectively, in 40 cM CIs. Another intriguing finding is that the two most highly social insects for which we have linkage maps, the honey bee and the leaf-cutting ant *Acromyrmex echiniator,* appear to have the highest recombination rates known for any metazoan. Aside from the benefit of localizing genes for social traits, these results raise the question of whether there is an adaptive significance for high recombination rates in social insects, or whether it could be a byproduct of colony life related to lower effective population sizes (see Gadau et al. 2000; Sirvio et al. 2006; Beye et al. 2006, Wilfert et al. 2007b).

In thinking about the sorts of genes that would be involved in the evolution of cooperative nest defense, we might expect that genes affecting central nervous system (CNS) activity or development, or genes involved in sensory tuning, would be likely candidates (reviewed by Hunt 2007). Predicted genes annotated by the Honeybee Genome Consortium (2006) within the sting QTL CIs were analyzed by performing BLASTp searches against the non-redundant protein database (NCBI website http://www .ncbi.nlm.nih.gov). This provided information on homologs and identified protein domains. Putative functions of homologs and orthologs (genes in *Drosophila* and humans) revealed that many of the 50 genes in the *sting1* CI had functions in the development and functioning of the CNS. The *sting2* and *sting3* CIs had 60 and 16 genes, respectively, and contained genes that could influence sensory tuning—an arrestin (*AmARR4*), a GABA$_B$-R1 receptor, and the homer protein). Arrestin modulates perception of both visual and olfactory stimuli. The GABA$_B$-R1 receptor is a subunit of the receptor that is a key inhibitory neurotransmitter, and homer expression is known to respond to synaptic signaling. Expression studies showed that some of the candidate genes were up-regulated in highly defensive bees (Hunt et al. 2007), but some unlinked genes were also more likely to be up-regulated in defensive bees, suggesting that gene expression may be generally higher in defensive bees. Techniques such as fine-scale mapping, expression studies, pharmacology, and manipulation of expression in vivo are now required to prove connections of genes with aggressive behavior.

Aside from the fine-scale mapping problem, another major disadvantage of QTL studies is that they will only identify genes that have genetic variation that produces observable phenotypic variation. In order to identify genes that are differentially regulated between task groups but are not necessarily different in sequence, we need to use expression studies.

Expression Studies: Microarrays and Quantitative Real-time PCR (Polymerase Chain Reaction)

Partial sequences of genes from EST projects can be used to design microarrays for measuring the expression levels of many genes on a single chip. About 25% of roughly 4,000 to 4,500 genes on microarrays that were probed with RNA from honey bee brains differed in expression between nurses and foragers (Whitfield, Cziko, and Robinson 2003). Usually, honey bee nurses are young and foragers are the oldest bees in a colony, but in this study colony manipulations controlled for age effects by producing young, precocious foragers and unusually old nurse bees. Surprisingly, Whitfield et al. found that most of the variance in gene expression between these two task groups was associated with behavior rather than age. Another microarray experiment made use of treatments with queen-mandibular gland pheromone (QMP). QMP has an impact on the behavioral transition from nurse to forager and this microarray study corroborated the previous results by showing that most of the same genes differed in expression in the predicted direction following QMP treatment (Grozinger et al. 2003). However, about a third of genes analyzed on the chip were differentially regulated between the two treatment groups in this latter study, making interpretation of the roles of individual genes difficult. This underscores the fact that both microarray studies and QTL studies result in many false positives—genes identified that do not influence the behavior of interest and must be eliminated in follow-up studies.

Gene expression may also be correlated with short-duration behaviors such as comb-building, guarding, and removal of corpses (undertaking or hygienic behavior). A recent study found that age-matched comb builders differed from guards and undertakers for 248 and 32 cDNAs at $p > 0.01$ and 0.001, respectively, which is roughly five times more genes than would be expected by chance. But no differences between guards and undertakers were observed over the expected number of false positives (Cash et al. 2005). Small expression differences throughout the brain or larger differences localized to small regions of the brain could be missed in microarray studies, so perhaps it is not surprising that no differences were found between undertakers and guards. Another explanation could be that in general the same genes influence both hygienic behavior and guarding. One study involving just two colonies did show that undertakers guarded more than workers previously identified as comb builders or food storers, and that both undertakers and guards frequented the lower portions of the hive (Trumbo, Huang, and Robinson 1997).

Microarrays were recently used to look at natural variation in the rate of behavioral development that exists between two honey bee subspecies, A. m. ligustica and A. m. mellifera (Whitfield et al. 2006). Principal component analyses identified key genes that were primarily regulated by behavioral differences between subspecies and these can be considered as candidate genes that either influence, or are influenced by, behavioral development. One surprising outcome of this research was that hive-restricted bees had gene expression that was virtually indistinguishable from foragers. This shows that the brain gene expression that has been observed so far as a signature of foraging behavior is independent of flight, foraging experience, and exposure to light! This study helps to tease out the factors of environment, hormones, and signaling molecules on gene expression in foragers. Methoprene treatment was found to influence forager-like expression patterns more than treatments related to two candidate genes. In general, the corroboration between honey bee microarray studies appears to be higher than studies involving human psychiatric disorders, presumably because bees are simpler systems (see Miklos and Maleszka 2004). Mapping QTL that vary in populations and influence behavior, combined with expression analyses with microarrays, may suggest regulatory connections between QTL and specific genes. Such a study could be extended to potentially mapping genes (as "eQTL") that influence the expression of any gene on the array and help to reveal the underlying gene network.

Conclusions and Future Directions

Sequencing genomes continues to become cheaper. The genome sequences of three *Nasonia* species (Hymenoptera, Chalcidoidea) are now available, which will allow direct comparisons between the genetic architecture of behavioral traits in social and solitary Hymenopteran species. This should provide valuable insights into how certain genes take on new roles in modulating division of labor (Nelson et al. 2007). Beyond that, we hope/predict that in the near future other social insect genomes (e.g., Apinae, Isoptera, Formicidae) will become available both for a genomic comparison of convergently evolved social systems and for comparison of related taxa that differ in the degree of sociality; for example, honey bees (*Apis* spp.), stingless bees (Meliponini), bumblebees (Bombini), and orchid bees (Euglossini). These genomes and the comparison between their expression patterns are essential to the reconstruction of genetic changes associated with the evolution of eusociality. Genotyping arrays will greatly accelerate the integration of QTL mapping and genome sequence and provide powerful tools for population genetics, which in turn will help to identify areas in the genome that were recently (e.g., during the evolution of social behavior) under selective pressure. Arrays that can query thousands of single nucleotide polymorphisms (SNPs) in many individuals will be most suitable for these types of studies (Fan, Chee, and Gunderson 2006). Another approach that will become increasingly important is massively parallel sequencing to identify SNPs or for transcriptional profiling. This technique can produce an incredible amount of sequences in short fragments at low cost (Leamon, Braverman, and Rothberg 2007).

Analyses of the expression levels of candidate genes with microarrays or qRT-PCR will continue to be important for learning more about gene effects in social insects and to integrate genetic variation with gene regulation. Experimental manipulation of gene expression will also be essential to test for effects of candidate genes on phenotypes. Some results from QTL mapping in honey bee and other hymenoptera indicate that epistasis is probably a common feature of behavioral traits, but it is difficult to have a test that has enough power to identify these effects because all pairwise combinations of markers need to be tested for effects on the phenotype. A more powerful way to study these regulatory networks may be to observe what happens to expression within the network of genes when one gene is experimentally down-regulated.

More complex or nontraditional genetics such as epistasis, indirect genetic effects (Linksvayer 2006, 2007; Linksvayer and Wade 2005), or epigenetic effects (Haig 2000; Queller 2001) may prove to be very important for the evolution of sociality because they can alter the selective environment of genes and hence may be crucial factors in the early evolution of sociality or the expression of facultative social behavior. Thompson et al. (2006) interpret the extension of the major royal jelly protein-family in honey bees as a possible example for gene regulation via a sib-social effect because it allows workers to determine caste development through differential feeding of larvae. Hence the fate of a female larva in honey bees— whether she will become a worker or queen—crucially depends on genes expressed in her sibs.

To really understand the evolution of social behavior we need to know more about the ancestral status of the novel social phenotypes, such as the precursor of a nectar foraging or pollen foraging bee, and what genes influenced these behaviors. However, to do so, we first need to have a better understanding of the genetic architecture of behaviors that are unique or novel in their regulation in a highly social insect. Once we understand the genetic architecture and identify the genes underlying these social behaviors in highly eusocial species, we can discover how these genes fit into regulatory networks in the solitary relatives of social insects, and how they have changed during the evolution to sociality.

Literature Cited

Amdam, G. V., A. Csondes, M. K. Fondrk, and R. E. Page Jr. 2006. "Complex social behavior derived from maternal reproductive traits." *Nature* 439: 76–78.

Arechavaleta-Velasco, M. E., and G. J. Hunt. 2004. "Binary trait loci that influence honey bee guarding behavior." *Annals of the Entomological Society of America* 97: 177–183.

Arechavaleta-Velasco, M. E., G. J. Hunt, and C. Emore. 2003. "Quantitative trait loci that influence the expression of guarding and stinging behaviors of individual honey bees." *Behavior Genetics* 33: 357–364.

Ben-Shahar, Y. 2005. "The foraging gene, behavioral plasticity, and honeybee division of labor." *Journal of Comparative Physiology A* 191: 987–994.

Ben-Shahar, Y., H. Leung, W. Pak, M. Sokolowski, and G. Robinson. 2003. "cGMP-dependent changes in phototaxis: A possible role for the foraging gene in honey bee division of labor." *Journal of Experimental Biology* 59: 269–278.

Ben-Shahar, Y., A. Robichon, M. B. Sokolowski, and G. E. Robinson. 2002. "Influence of gene action across different time scales on behavior." *Science* 296: 741–744.

Bier, K. 1954. "Über den Saisondimorphismus der Oogenes von Formica rufa rufo-pratensis minor Gssw. Und dessen Bedeutung für die Kastendetermination." *Biologisches Zentralblatt* 73: 170–190.

Breed, M. D., G. E. Robinson, and R. E. Page Jr. 1990. "Division of labor during honey bee colony defense." *Behavioral Ecology and Sociobiology* 27: 395–401.

Beye, M., I. Gattermeier, M. Hasselmann, T. Gempe, M. Scheioett, J. F. Baines, D. Schlipalius, F. Mougel, C. Emore, O. Rueppell, A. Sirviö, E. Guzmán-Novoa, G. Hunt, M. Solignac, and R. E. Page Jr. 2006. "Exceptionally high levels of recombination across the honey bee genome." *Genome Research* 16: 1339–1344.

Calderone, N. W., and R. E. Page Jr. 1992. "Effects of interactions among genotypically diverse nestmates on task specialization by foraging honey bees (Apis mellifera)." *Behavioral Ecology and Sociobiology* 30: 219–226.

Cash, A. C., C. W. Whitfield, N. Ismail, and G. E. Robinson. 2005. "Behavior and the limits of genomic plasticity: Power and replicability in microarray analysis of honeybee brains." *Genes, Brain, and Behavior* 4: 267–271.

Christino, A. S., F. M. F. Nunes, C. H. Lobo, M. M. G. Bitondi, Z. L. P. Simões, L. da Fontoura Costa, H. M. G. Lattorff, R. F. A. Moritz, J. D. Evans, and K. Hartfelder. 2006. "Caste development and reproduction: A genome-wide analysis of hallmarks of insect eusociality." *Insect Molecular Biology* 15: 703–714.

Evans, J. D., and D. E. Wheeler. 2000. "Expression profiles during honeybee caste determination." *Genome Biology* 2: 6.

Fan, J.-B., Chee, M. S., and K. L. Gunderson. 2006. "Highly parallel genomic assays." *Nature Reviews Genetics* 7: 632–644.

Frumhoff, P. C., and J. Baker. 1988. "A genetic component to division of labour within honey bee colonies." *Nature* 333: 358–361.

Gadau, J., C. U. Gerloff, N. Krüger, H. Chan, P. Schmid-Hempel, A. Wille, and R. E. Page Jr. 2001. "A linkage analysis of sex determination in *Bombus terrestris* (L.) (Hymenoptera: Apidea)." *Journal of Heredity* 87: 234–242.

Gadau, J., R. E. Page Jr., and J. H. Werren. 1999. "Mapping of hybrid incompatibility loci in Nasonia." *Genetics* 153: 1731–1741.

Gadau, J., R. E. Page Jr., J. H. Werren, and P. Schmid-Hempel. 2000. "Genome organization and social evolution in Hymenoptera." *Naturwissenschaften* 87: 87–89.

Guidugli, K. R., A. Nascimento, G. V. Amdam, A. R. Barchuk, S. Omholt, Z. L. P. Simoes, and K. Hartfelder. 2005. "Vitellogenin regulates hormonal dynamics in the worker caste of a social insect." *FEBS Letters* 579: 4961–4965.

Guzmán-Novoa, E., G. J. Hunt, R. E. Page, J. L. Uribe-Rubio, D. Prieto-Merlos, and F. Becerra-Guzmán. 2005. "Paternal effects on the defensive behavior of honey bees. *Journal of Heredity* 63: 1–5.

Guzmán-Novoa, E., G. J. Hunt, J. L. Uribe, C. Smith, and M. E. Arechavaleta-Velasco. 2002. "Confirmation of QTL effects and evidence of genetic dominance of honey bee defensive behavior: Results of colony and individual behavioral assays." *Behavior Genetics* 32: 95–102.

Guzmán-Novoa, E., G. J. Hunt, J. L. Uribe-Rubio, and D. Prieto-Merlos. 2004. "Genotypic effects of honey bee *(Apis mellifera)* defensive behavior at the individual and colony levels: The relationship of guarding, pursuing and stinging." *Apidologie* 35: 14–24.

Haig, D. 2000. "The kinship theory of genomic imprinting." *Annual Review of Ecological Systems* 31: 9–32.

Hunt, G. J. 2007. "Flight and fight: A comparative view of the neurophysiology and genetics of honey bee defensive behavior." *Journal of Insect Physiology* 53: 399–410.

Hunt, G. J., G. V. Amdam, D. Schlipalius, C. Emore, N. Sardesai, C. E. Williams, O. Rueppell, E. Guzmán-Novoa, M. Arechavaleta-Velasco, S. Chandra, M. K. Fondrk, M. Beye, and R. E. Page Jr. 2007. "Behavioral genomics of honeybee foraging and nest defense." *Naturwissenschaften* 94: 247–267.

Hunt G. J., A. M. Collins, R. Riviera, R. E. Page Jr., and E. Guzmán-Novoa. 1999. "Quantitative trait loci for honeybee alarm pheromone production." *Journal of Heredity* 90: 585–589.

Hunt, G. J., E. Guzmán-Novoa, M. K. Fondrk, and R. E. Page Jr. 1998. "Quantitative trait loci for honeybee stinging behavior and body size." *Genetics* 148: 1203–1213.

Hunt, G. J., E. Guzmán-Novoa, J. L. Uribe-Rubio, and D. Prieto-Merlos. 2003. "Genotype by environment interactions in honey bee guarding behavior." *Animal Behaviour* 66: 469–477.

Hunt, G. J., and R. E. Page Jr. 1995. "A linkage map of the honeybee, *Apis mellifera*, based on RAPD markers." *Genetics* 139: 1371–1382.

Hunt G. J., R. E. Page Jr., M. K. Fondrk, and C. J. Dullum. 1995. "Major quantitative trait loci affecting honeybee foraging behavior." *Genetics* 141: 1537–1545.

Julian, G. E., J. H. Fewell, J. Gadau, R. A. Johnston, and D. Larrabee. 2002. "Genetic determination of queen caste in an ant hybrid zone. *Proceedings of the National Academy of Sciences USA* 99: 8157–8160.

Kerr, W. E. 1950. "Genetic determination of castes in the genus *Melipona*." *Genetics* 35: 143–152.

Krieger, M. J. B., and K. G. Ross. 2002. "Identification of a major gene regulating complex social behavior." *Science* 295: 328–332.

———. 2005. "Molecular evolutionary analyses of the odorant-binding protein gene Gp-9 in fire ants and other *Solenopsis* species." *Molecular Biology and Evolution* 22: 2090–2103.

Lapidge, K. L., B. P. Odroyd, and M. Spivak, M. 2002. "Seven suggestive quantitative trait loci influence hygienic behavior of honey bees." *Naturwissenschaften* 89: 565–568.

Leamon, J. H., M. S. Braverman, and J. M. Rothberg. 2007. "High-throughput, massively parallel DNA sequencing technology for the era of personalized medicine. *Gene Therapy and Regulation* 3: 15–31.

Linksvayer, T. A. 2006. "Direct, maternal, and sibsocial genetic effects on individual and colony traits in an ant." *Evolution* 60: 2552–2561.

———. 2007. "Ant species differences determined by epistasis between brood and worker genomes." *PLoS One* 2: e994.

Linksvayer, T. A., and M. J. Wade. 2005. "The evolutionary origin and elaboration of sociality in the aculeate Hymenoptera: Maternal effects, sib-social effects, and heterochrony." *Quarterly Review of Biology* 80: 317–336.

Miklos, G. L. G., and R. Maleszka. 2004. "Microarray reality checks in the context of a complex disease." *Nature Biotechnology* 22: 615–621.

Moritz, R. F. A. 1988. "A reevaluation of the two-locus model for hygienic behaviour in honeybees (*Apis mellifera* L)." *Journal of Heredity* 79: 257–262.

Page Jr., R. E., M. K. Fondrk, G. J. Hunt, E. Guzmán-Novoa, M. A. Humphries, K. Nguyen, and A. Greene. 2000. "Genetic dissection of honeybee (*Apis mellifera* L.) foraging behavior. *Journal of Heredity* 91: 474–479.

Page Jr., R. E., and G. E. Robinson. 1991. "The genetics of division of labor in honey bee colonies." In *Advances in insect physiology,* 117–169. New York: Academic Press.

Queller, D. C. 2001. "Theory of genomic imprinting conflict in social insects." *BMC Evolutionary Biology* 3: 15.

Ross, K. G., and D. J. C. Fletcher. 1985. "Comparative study of the genetic and social structure in two forms of the fire ant *Solenopsis invitca* (Hymenoptera: Formicidae)." *Behavioral Ecology and Sociobiology* 17: 349–356.

Ross, K. G., and L. Keller. 1998. "Genetic control of social organization in an ant." *Proceedings of the National Academy of Sciences USA* 95: 14232–14237.

———. 2002. "Experimental conversion of colony social organization by manipulation of worker genotype composition in fire ants *(Solenopsis invicta)*." *Behavioral Ecology and Sociobiology* 51: 287–295.

Rothenbuhler, W. C. 1964a. "Behaviour genetics of nest cleaning in honey bees. I. Responses of four inbred lines to disease-killed brood." *Animal Behaviour* 12: 578–583.

———. 1964b. "Behavior genetics of nest gleaning in honey bees. IV. Responses of F_1 and backcross generations to disease-killed brood." *American Zoologist* 4: 111–123.

Sokolowski, M. B. 2001. "*Drosophila:* Genetics meets behavior." *Nature Reviews Genetics* 2: 879–892.

Thompson, G. J., R. Kucharski, R. Maleszka, and B. P. Oldryod. 2006. "Towards a molecular definition of worker sterility: Differential gene expression and reproductive plasticity in honey bees." *Insect Molecular Biology* 15: 637–644.

Trumbo, S. T., Z.-Y. Huang, and G. E. Robinson. 1997. "Division of labor between undertaker specialists and other middle-aged workers in honey bee colonies." *Behavioral Ecology and Sociobiology* 41: 151–163.

Volny, V. P., and D. M. Gordon. 2002. "Genetic basis for queen-worker dimorphism in a social insect." *Proceedings of the National Academy of Sciences USA* 99: 6108–6111.

Whitfield, C. W., M. R. Band, M. F. Bonaldo, C. G. Kumar, L. Liu, J. R. Pardinas, H. M. Robertson, M. B. Soares, and G. E. Robinson. 2002.

"Annotated expressed sequence tags and cDNA microarrays for studies of brain and behavior in the honey bee." *Genome Research* 12: 555–566.

Whitfield, C. W., Y. Ben-Shahar, C. Brillet, I. Leoncinii, D. Crauser, Y. LeConte, S. Rodriguez-Zas, and G. E. Robinson. 2006. "Genomic dissection of behavioral maturation in the honey bee." *Proceedings of the National Academy of Sciences USA* 103: 1668–1675.

Whitfield, C. W., A.-M. Cziko, and G. E. Robinson. 2003. "Gene expression profiles in the brain predict behavior in individual honey bees." *Science* 302: 296–299.

Wilfert, L., J. Gadau, and P. Schmid-Hempel. 2007a. "The genetic architecture of immune defense and reproduction in male *Bombus terrestris* bumble bees." Evolution 61: 804–815.

———. 2007b. "Sociality and genomic recombination." *Heredity* 98: 189–197.

———. 2007c. "Variation in genomic recombination rates among animal taxa and the case of social insects." *Heredity* 98: 189–197.

CHAPTER FIFTEEN

Sensory Thresholds, Learning, and the Division of Foraging Labor in the Honey Bee

RICARDA SCHEINER

JOACHIM ERBER

HONEY BEE COLONIES DISPLAY an amazingly complex social life. Individuals can selectively respond to a large diversity of sensory stimuli with behaviors that contribute to the survival of their hivemates. In addition to division of reproductive labor, honey bees show age-dependent division of labor. Whereas young bees work in the center of the nest, tending the brood or the queen, older bees are involved in the production of honey and the maintenance of the hive. Older bees work in the periphery of the nest as guards and later leave the hive for foraging trips. Within the group of foragers, which are of similar age, another form of division of labor occurs. Some bees forage for pollen, others collect nectar, a number collect water, and a few collect propolis. Division of labor in honey bees is flexible and resilient; the behavior of individuals can change considerably according to the actual needs of a colony on time scales that range from minutes to weeks.

All these myriad behavioral interactions are controlled by the relatively small nervous system of each individual. This complex control of social behavior requires sensory organs that receive and filter signals of different modalities. Higher-order interneurons must evaluate relevant signals and select appropriate behavioral responses that are executed by the complex motor system of the individual bee. At first sight, any attempt to analyze the neuronal mechanisms underlying the complex organization of social behavior in bees seems to be a hopeless venture. Nevertheless, there are a number of hypotheses on the mechanisms underlying social behavior that can be directly tested with methods developed by behavioral and neural sciences. One example of this is the *response threshold model*. This theory attempts to explain division of labor in an insect colony by differences in task-related response thresholds of individuals. The response threshold model provides neurobiologists with a number of testable hypotheses on the behavioral, cellular, and molecular levels. In the last few years, significant advances have been made in the analyses of sensory response thresholds and their relationships to different behaviors in bees. Most of these analyses focus on division of foraging labor. These experiments suggest that foraging labor is controlled by only a few simple rules at the level of the nervous system, which are translated into complex social interactions within a honey bee colony. The hypotheses on sensory thresholds derived from experiments with foraging labor can now also be tested with other social behaviors.

Response Thresholds and Division of Labor

Division of labor in a colony is a complex process which involves different levels of information processing. The individual has to follow rules which determine the behavior it has to perform at a specific time in a defined sequence under the given boundary conditions. The individual has to asses constantly complex environmental and social stimuli which control its behavior. Under natural conditions it is impossible for the experimenter to control the sensory stimuli that are perceived by an individual. Therefore, it is necessary to study these processes under reduced and controlled laboratory conditions. Honey bees are very suitable for such experiments because they show complex behaviors such as learning and orientation even under very restrained laboratory conditions.

Division of labor within a honey bee colony is assumed to be based on response thresholds for specific stimuli (Page and Robinson 1991; Robinson 1992; Beshers and Fewell 2001). According to this hypothesis, response thresholds differ among individuals; an individual with a low response threshold for a specific sensory stimulus responds to low stimulus intensities by initiating the associated task, while other bees with higher thresholds are unaffected by this stimulus. Division of labor works because some individuals have a low response threshold for a stimulus that is associated with one task, but have high response thresholds for stimuli that are associated with other tasks. It is possible in the laboratory to measure response thresholds of bees that perform different tasks in the colony. The measurement of response thresholds in the laboratory establishes correlations between behavior in the field and the properties of sensory systems, but not causal relations. Causal effects of sensory thresholds can be analyzed by observing the effects on division of labor in the colony after manipulating the sensory thresholds of individuals (e.g., by pharmacological treatments). The threshold hypothesis leads to a number of interesting questions that can be analyzed under controlled laboratory conditions.

1. Do bees that differ in their activities within the colony have different thresholds for specific sensory stimuli?
2. What are the neurophysiological mechanisms that control sensory thresholds for specific stimuli?
3. How do changes in the environment or within the colony modulate sensory thresholds?
4. Is there a genetic bias that affects response thresholds for specific stimuli?

Some of these questions have been addressed experimentally in the last few years. Response thresholds for stimuli of different modalities can be determined by measuring how individuals respond to increasing intensities of sensory stimuli. Averaged responses to different stimulus intensities can be calculated for a number of individuals. These intensity-response functions describe the responsiveness of a group of individuals to different intensities of a specific stimulus.

These intensity-response curves can be used to estimate the response threshold or the sensitivity of a group of individuals for a specific stimulus. For some stimulus modalities it is easy to define the lowest stimulus intensity that evokes a response in an individual (e.g., responses to pollen or to an odor, see below). In these cases, the response threshold is the lowest stimulus intensity that produces a significant behavioral response in an individual. For other stimulus modalities, like gustatory or visual stimuli, even very low stimulus intensities can evoke behavioral responses, which depend on the stimulus intensity. For these stimuli, a response threshold often cannot be estimated and therefore needs to be defined differently.

An example of the measurement of responsiveness and the estimation of response thresholds for gustatory stimuli in two groups of bees is shown in Figure 15.1. In this experiment, the bees were harnessed in small tubes, and it was tested whether proboscis extension occurred during stimulation of the antennae with water or different concentrations of sucrose (Page,

Erber, and Fondrk 1998). In both groups the percentage of bees responding with proboscis extension increased with increasing concentrations of
sucrose. These concentration-response curves can be used to estimate response thresholds by determining the sucrose concentration that evokes a
response in, for example, 75% of the bees (Figure 15.1). The sucrose concentration that elicits a response in 75% of the individuals with low thresholds is approximately 20 times lower than the equivalent concentration in
the other group (response thresholds: 0.16% vs. 3.5% sucrose). The sucrose sensitivity of these bees can be estimated by calculating the inverse

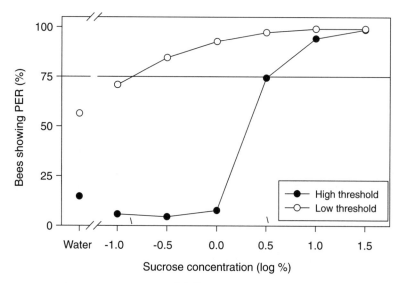

Figure 15.1. Responses to water and different sucrose concentrations of bees
with high and low gustatory response thresholds. The abscissa shows the water
stimulation and the log of the sucrose concentrations tested. The ordinate
displays the percentage of bees showing proboscis extension response (PER).
Bees of both groups become more responsive with increasing sucrose
concentrations. The concentration-response curves can be used to estimate
response thresholds by determining the sucrose concentration that, for
example, evokes a response in 75% of the bees. The two groups presented here
differ in their 75% response thresholds approximately by the factor 20. In bees
with low gustatory threshold, the 75% response appears at about 1.6% sucrose
(log −0.8). In bees with high sucrose threshold, the 75% response lies at about
3.5% sucrose (log 0.55). Data taken from Scheiner et al. 2005.

of the sucrose concentration (1/concentration) at the defined threshold. In the example shown in Figure 15.1, the sensitivity for the group with the lower threshold is about 20 times higher than that for the other group.

Gustatory Response Thresholds and Division of Foraging Labor

Bees display a clear division of foraging labor by collecting nectar, pollen, water, or propolis. Pollen and non-pollen foragers represent different groups of bees that perform different tasks in the social context of the hive. As these two groups are present during most of the foraging season, they can be collected in large numbers at the hive entrance and tested for differences in response thresholds in the laboratory.

The proboscis extension response (PER) is a very robust behavior for determining sensory thresholds for water or sucrose solutions (for review see Scheiner, Page, and Erber 2004). Response thresholds to sucrose have been shown to correlate with division of foraging labor (Figure 15.2A). On average, pollen foragers show high responsiveness to water and low concentrations of sucrose. They have lower response thresholds for sucrose than nectar foragers, which are less responsive to these gustatory stimuli (Page, Erber, and Fondrk 1998; Scheiner, Erber, and Page 1999; Scheiner, Page, and Erber 2001b; Scheiner, Barnert, and Erber 2003). Water collectors are similarly sensitive for sucrose to pollen foragers. Bees collecting both pollen and nectar are rather insensitive for sucrose stimuli (Pankiw and Page 2000).

A good indicator of individual responsiveness is the gustatory response score (GRS). The GRS of a single bee comprises the number of proboscis extensions the bee shows during the consecutive stimulation with water and increasing concentrations of sucrose. Pollen and nectar foragers differ in their relative distributions of bees with high and low gustatory thresholds, which are indicated by the GRSs (Figure 15.2B). Although the distributions of GRSs overlap between nectar and pollen foragers, a higher proportion of nectar foragers have a low GRS (i.e., high thresholds), while pollen foragers comprise a higher frequency of bees with a high GRS (i.e., low thresholds).

Responsiveness to sucrose depends on the age of the bee. Gustatory response thresholds are high in newly emerged bees and decrease strongly with age (Figure 15.2C; Pankiw and Page 1999). Individual sucrose response thresholds measured in 1-week-old bees are a good predictor of the

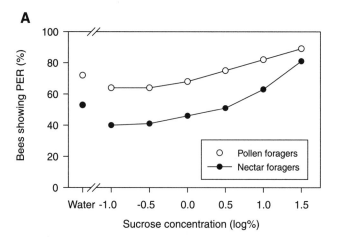

A

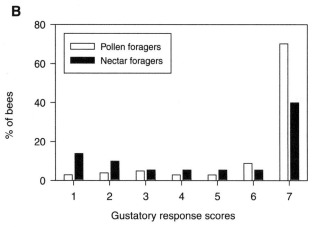

B

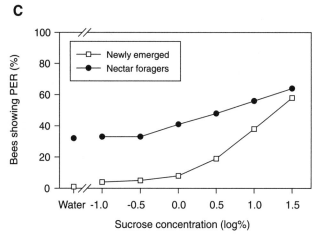

C

foraging behavior these bees will adopt later in life (Pankiw and Page 2000). Bees with the lowest threshold to sucrose at 1 week will later collect water. Those with slightly higher thresholds will later collect pollen. Individuals with high thresholds to sucrose will collect nectar or both pollen and nectar. Bees with the highest thresholds for sucrose in their first week of adult life are later found to return empty from foraging trips. These experiments demonstrate that sucrose thresholds of preforaging bees correlate with the materials that they collect as foragers (Pankiw and Page 2000).

Apparently the sucrose receptors (Haupt 2004) of pollen and nectar foragers differ in their thresholds to sensory stimuli. To understand the neural mechanisms underlying different sensory thresholds it is also necessary to analyze other types of sensory receptors in these two groups of bees. There is good experimental evidence that the thresholds of sucrose receptors correlate with the thresholds of other sensory receptors. Sensitivity for water often correlates positively with sensitivity to sucrose (Page, Erber, and Fondyk 1998), although water and sucrose stimuli are perceived by different antennal sensilla (Haupt 2004). Response thresholds for pollen can be measured under laboratory conditions by stimulating the antennae with pollen mixtures and registering the PER (Figure 15.3A). Bees with a low

---

Figure 15.2. A. Gustatory response curves of pollen and nectar foragers. The abscissa shows the water stimulation and the log concentrations of the different sucrose stimuli tested. The ordinate displays the percentage of bees showing the proboscis extension response (PER). The percentage of pollen foragers that respond with proboscis extension to stimulation with different sucrose concentrations is generally greater than that of nectar foragers. Data taken from Scheiner et al. 2003. B. Distribution of gustatory response scores (GRSs) in pollen and nectar foragers. Each bee was stimulated with water and six sucrose concentrations. The GRS of an individual represents the number of proboscis responses to these seven stimuli and can therefore range between 0 (no response to any stimulus) and 7 (responses to all stimuli tested) for each individual. The distributions of the GRS lead to different median GRS values for pollen and nectar foragers (pollen foragers $x_{med} = 7$, nectar foragers $x_{med} = 5$). Data taken from Scheiner et al. 2003. C. Gustatory response curves of newly emerged bees and of nectar foragers. The axes of the diagram are identical to those in 2A. Nectar foragers are more responsive to water and the different sucrose concentrations tested than newly emerged bees.

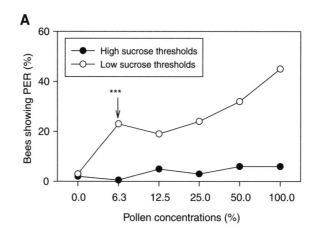

A

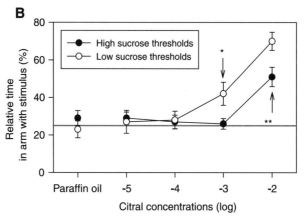

B

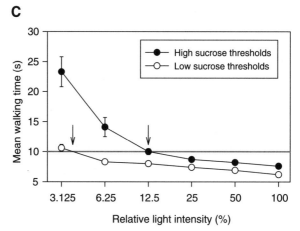

C

response threshold for sucrose have a low response threshold for pollen and respond with the PER to a stimulus that contains only about 6% pollen. Bees with higher thresholds for sucrose do not show significant responses even to pure pollen. Sucrose, pollen, and water stimuli are perceived by different sensory receptors, but they all belong to the same modality—gustation. The correlations between sensitivities for these stimuli might therefore be a consequence of the modality-specific neuroarchitecture of the insect brain. For a better understanding of the principles of

Figure 15.3. Correlations between gustatory response thresholds and response thresholds for other stimulus modalities. A. Responsiveness to different pollen concentrations of bees with high or low sucrose response thresholds. The abscissa shows the different relative pollen concentrations. The ordinate shows the percentage of bees showing the proboscis extension response (PER) when pollen mixtures were applied to the antennae of the bees. Individuals with high sucrose response thresholds were more responsive to pollen than bees with low sucrose thresholds. The response threshold for pollen of bees with low sucrose thresholds is indicated by asterisks (°°°$p \leq 0.001$; two-tailed Fisher Exact Probability Test). Data taken from Scheiner et al. 2004. B. Responsiveness to odors of bees with high or low sucrose response thresholds measured in a four-armed olfactometer. The abscissa displays the different odor concentrations that were produced by diluting citral with paraffin oil. The ordinate shows the relative time individuals spent in the odor-containing arm compared to the other arms of the olfactometer. Bees with high sucrose response thresholds were more sensitive to citral than bees with low sucrose thresholds. Bees with low sucrose thresholds had an olfactory response threshold of 10^{-3} (°$p \leq 0.05$, two-tailed Fisher Exact Probability Test). Those with high sucrose thresholds had an olfactory response threshold that was 10-fold higher (°°$p \leq 0.01$, two-tailed Fisher Exact Probability Test). Data taken from Scheiner et al. 2004. C. Visual responsiveness of bees with high or low sucrose response thresholds measured in a phototaxis arena. The abscissa shows the different relative light intensities which were produced with neutral density filters in front of green (520 nm) light emitting diodes. The ordinate displays the mean times that bees need to walk across the arena to a light source with a defined intensity. The walking times decrease with increasing light intensities. For comparison of the different thresholds for light in the two groups of bees, a threshold criterion of 10 s walking time is indicated. Bees with low sucrose thresholds need lower light intensities to reach the walking time threshold of 10 s, while bees with high gustatory thresholds need higher light intensities. Data taken from Erber et al. 2006.

neuronal mechanisms controlling sensory thresholds, it is necessary to analyze the correlations between gustatory sensitivities and sensitivity for olfactory or visual signals.

Olfactory Thresholds and the Division of Labor

Olfactory signals govern the world of social insects (Hölldobler and Wilson 1990). They control and trigger social interactions within the colony, they are the basis for nestmate recognition, and they help to identify food sources. On the basis of a response threshold model for division of labor we would expect that response thresholds for odors in the laboratory differ in insects that perform different tasks that are controlled by olfactory signals. Many experiments conducted in the last few years support this hypothesis.

Response thresholds for odors can be determined effectively in honey bees using an olfactometer, olfactory learning assays, or extracellular electrophysiological recordings from the antenna (electroantennograms). Response thresholds for queen pheromone, for example, have been determined in an olfactometer in which the bees walk toward an odor source containing the pheromone. The thresholds increase with age, with young bees (≤5 days) exhibiting the lowest response threshold for queen pheromone (Pham-Delègue et al. 1991).

Masterman et al. (Masterman, Smith, and Spivak 2000; Masterman et al. 2001) found a genetic component that influences response thresholds for the specific odor of chalkbrood. Hygienic bees had lower response thresholds for the odor of chalkbrood than nonhygienic bees, which indicates a genetic basis for this behavior.

Alarm pheromone, together with visual alert signals, rapidly induces stinging responses in bees. Individuals in a colony differ in their response thresholds to alarm pheromone, with bees working as guards or having recently worked as guards being most likely to sting at lower levels of defensive stimuli. Sensitivity for alarm pheromone can be assayed by observing the amount of wing flickering after pheromone application (Collins and Rothenbuhler 1978). This assay demonstrated that response thresholds to alarm pheromone decrease with age. Response thresholds for alarming signals can be also measured by waiving a suede leather patch above the top of the frames. This induces stinging and defensive behaviors in a colony. Differences in individual response thresholds for this signal have been shown for European and Africanized bees (Hunt et al. 2003). Taken together, these experimental findings demonstrate that olfactory response thresholds correlate with the tasks that a bee performs.

Sensory Thresholds across Modalities

Few studies have analyzed sensory thresholds of the same individuals for different sensory modalities. Sensory modalities that are probably involved in the division of labor include taste, olfaction, vision, and touch. Sensory sensitivities for different modalities in an insect could hypothetically be controlled by neuromodulatory mechanisms that are common to all modalities (*common sensitivity control*). In this case, we would expect that an insect that is sensitive for one stimulus modality (e.g., taste) is also sensitive for another modality (e.g., vision). Alternatively, sensory sensitivities could be controlled separately for each modality and perhaps even for specific submodalities (*specific sensitivity control*). In that case, we would expect individual sensitivity for a specific stimulus quality (e.g., a specific pheromone) but not for other sensory signals.

In honey bees, response thresholds for sucrose were compared with thresholds for odors and for visual stimuli. Significant correlations between sensitivities across modalities were found. Olfactory thresholds for citral were tested in bees in an olfactometer after measuring their gustatory responsiveness. Bees with low sucrose response thresholds had an olfactory threshold that was one order of magnitude smaller than the threshold for bees that were less sensitive to sucrose (Figure 15.3B; Scheiner, Page, and Erber 2004). The thresholds for visual stimuli were tested in a phototaxis assay (Erber, Hoormann, and Scheiner 2006). Bees with low sucrose response thresholds had phototactic response thresholds that were approximately three times smaller than the phototactic thresholds of bees with high sucrose thresholds (Figure 15.3C). Taken together, these experiments demonstrate that sensory thresholds correlate across modalities. Bees that are sensitive to sucrose are also sensitive to water, pollen, odor, and light. To date, there has been no unequivocal experimental evidence for a stimulus-specific tuning of thresholds.

Modulation of Sensory Thresholds

Behavioral response thresholds for different stimulus modalities display a remarkable degree of plasticity. Response thresholds for sucrose, for example, change greatly in nectar foragers during the season, whereas the thresholds of pollen foragers remain almost constant (Scheiner, Barnert, and Erber 2003). These changes in response thresholds of nectar foragers are probably related to the changing variety of nectar-offering flowers throughout the foraging season. In addition to environmental stimuli, pheromones released by hivemates can change response thresholds. Both brood pheromone and queen pheromone affect response thresholds to sucrose in a complex way (Pankiw and Page 2001, 2003; Pankiw 2004).

These experiments demonstrate that sensory thresholds can be modulated by different stimuli in the environment and within the hive. The mechanisms behind this modulation of behavior still have to be identified, but there is some evidence that hormones and biogenic amines are part of these mechanisms. The hormone that has received most attention in studies on honey bee behavior in recent years is juvenile hormone, produced by the *corpora allata* (Robinson 1985; Huang, Robinson, and Borst 1994; Schulz, Sullivan, and Robinson 2002). Levels of juvenile hormone increase before initiation of foraging (Jassim, Huang, and Robinson 2000; Elekonich et al. 2001) and treatment with the juvenile hormone analog methoprene leads to precocious foraging (for review see Bloch, Wheeler, and Robinson 2002). There is also experimental evidence that juvenile hormone induces behavioral changes by changing the sensitivity of individuals. Response thresholds to alarm pheromone can be reduced and the number of guard bees increased by the application of methoprene (Robinson 1987; Sasagawa, Sasaki, and Okada 1989). Response thresholds to sucrose also decrease after topical application of methoprene (Pankiw and Page 2003).

Biogenic amines are important messenger substances that mediate a large number of cellular and physiological functions in vertebrates and invertebrates. In the honey bee, the biogenic amines octopamine, tyramine, dopamine, and serotonin have important behavioral functions (for review, see Scheiner, Baumann, and Blenau 2006). Octopamine can decrease response thresholds for gustatory stimuli, water vapor, visual stimuli, and odors (Mercer and Menzel 1982; Braun and Bicker 1992; Barron, Schulz, and Robinson 2002; Scheiner et al. 2002; Pankiw and Page 2003; Spivak et al. 2003). Some of the behavioral changes induced by octopamine and juvenile hormone are probably related, because octopamine can stimulate juvenile hormone release under *in vitro* conditions (Rachinsky 1994), and juvenile hormone treatment can increase levels of octopamine in the antennal lobes (Schulz, Sullivan, and Robinson 2002). Similar to octopamine, tyramine can also decrease gustatory response thresholds (Scheiner, Baumann, and Blenau 2002). Whereas octopamine and tyramine are only present in very small quantities in the bee brain, dopamine and serotonin have been found in considerably higher concentrations (for review, see Scheiner, Baumann, and Blenau 2006). The behavioral effects of dopamine are often very different and sometimes functionally antagonistic to those of octopamine and tyramine in the honey bee. Injections of dopamine, for example, can lead to an increase in response thresholds for sucrose and water vapor (Macmillan and Mercer 1987; Blenau and Erber 1998). Serotonin similarly appears to act functionally antagonistically to octopamine in the gustatory and visual systems (Erber, Kloppenburg, and Scheidler 1993; Erber and Kloppenburg 1995; Kloppenburg and Erber 1995; Blenau and Erber 1998).

The signaling pathways by which juvenile hormone and biogenic amines modulate response thresholds are not yet known; however, there is evidence that the modulation of sensory response thresholds involves both cyclic AMP-dependent and cyclic GMP-dependent signaling cascades. Gustatory response thresholds, for example, can be decreased by application of 8-Br-cAMP, an activator of cAMP-dependent protein kinase (Scheiner et al. 2003). Visual response thresholds are related to *Amfor*, a gene that encodes a cGMP protein kinase (=PKG) in the honey bee, and can be decreased by feeding of 8-Br-cGMP (Ben-Shahar et al. 2003).

Sensory Thresholds and Learning

Sensory thresholds for different modalities determine how stimuli are perceived and evaluated by a honey bee. The perception of gustatory stimuli, for example, plays a key role when bees forage for nectar, pollen, or water. A bee has to identify the foraging resource, it has to evaluate its relative value, and it has to associate the location of the resource with signals like odor, color, shape, and characteristic landmarks. During this complex process, gustatory stimuli will be the major information source determining whether a bee accepts a resource and stores its characteristics in memory for further visits. A bee with a low threshold for sucrose might learn the characteristics of a food source with a low nectar concentration and store the information in memory, while a bee with a high sucrose threshold might not respond to the food source at all. It can be hypothesized that response thresholds for sensory stimuli could affect learning and the exploitation of available food resources differing in nectar concentration.

It is known from a number of experiments that bees display many different forms of non-associative and associative learning (for review, see Menzel and Müller 1996; Page and Erber 2002; Scheiner, Page, and Erber 2004). The relationship between thresholds and learning can be analyzed under controlled conditions in the laboratory. Nonassociative habituation occurs when a bee is repeatedly stimulated at the antenna with a low sucrose concentration. After a number of stimulations the bee will eventually stop responding (Braun and Bicker 1992; Scheiner 2004). Studies in other phyla demonstrated that the perceived stimulus strength determines the course of habituation (Thompson and Spencer 1966). In bees we should therefore expect differences in habituation of the PER in individuals that differ in sucrose thresholds. Recent experiments have indeed shown that individuals with high response thresholds to sucrose showed fast habituation of the PER (Scheiner 2004), while bees with a low sucrose response threshold habituated very slowly. Based on the results from laboratory experiments, it can be predicted that free-flying animals with high sucrose response thresholds, such as most nectar foragers, will rapidly stop probing a food source of low nectar concentration with the proboscis.

During a foraging trip, a free-flying bee can associate the characteristics of a profitable food source with a reward provided by that source. The odor, color, and shape of a food source are rapidly learned by a bee, in addition to landmarks that help to locate the resource. Under controlled laboratory conditions, bees can quickly learn to associate an odor with a reward (Bitterman et al. 1983, Menzel and Müller 1996), they can be conditioned to antennal tactile cues (Erber, Pribbenow, et al. 1997; Erber, Kierzek, et al. 1998), and their antennal motor activity can be conditioned operantly (Kisch and Erber 1999). In all these learning paradigms, bees are rewarded by small amounts of sucrose, which they can take up with the proboscis. Thresholds for sensory stimuli are involved in several stages of these learning processes. The sucrose stimulus is first perceived by antennal taste hairs, which can elicit proboscis extension if the stimulus exceeds the threshold of those taste hairs (Haupt 2004). The stimulus is then perceived by taste hairs on the proboscis, which control the ingestion of food and have higher thresholds than antennal taste hairs (Scheiner et al. 2005). During learning, the sensory thresholds for olfactory and tactile cues will determine the perception of those sensory signals that will be associated with the reward. Sensory thresholds should, therefore, influence associative learning processes in bees.

There is now good experimental evidence for this hypothesis. Bees with low response thresholds to sucrose learn faster and reach a higher asymptote of the acquisition function than bees with higher response thresholds (Scheiner, Erber, and Page 1999; Scheiner, Page, and Erber 2001a, b, 2004; Scheiner et al. 2001; Scheiner, Barnert, and Erber 2003). In addition, retrieval in tests several hours or days after conditioning is higher in bees with low response thresholds to sucrose than in bees with higher thresholds (Scheiner, Erber, and Page 1999; Scheiner, Page and Erbert 2001a, b, 2004). Pollen foragers, which are on average more responsive to sucrose than nectar foragers, perform better than nectar bees in associative tactile and olfactory learning (Scheiner, Erber, and Page 1999; Scheiner, Page, and Erber 2001; Scheiner, Barnert, and Erber 2003).

Older bees generally learn better than younger bees (Pham-Delègue, DeJong, and Masson 1990; Ray and Ferneyhough 1997). This correlates with respective changes in response thresholds during life (Pankiw and Page 1999). The associative learning performance during the foraging season also correlates with response thresholds for sucrose stimuli. On average, learning performance is high when bees are sensitive to sucrose, which is usually the case at the end of the foraging season. At this time, sucrose response thresholds of pollen and nectar foragers are similarly low, and differences in the associative learning performance of these groups disappear (Scheiner, Barnert, and Erber 2003).

Based on controlled laboratory experiments, one can conclude that sensory response thresholds should have a major influence on the foraging behavior of bees. Bees that are sensitive to sucrose will test a potential food source frequently with the proboscis even if it has low sugar concentrations. In the laboratory, such a bee can be conditioned successfully even with a water reward. In the field, these bees could function as water collectors. Bees with a high sucrose threshold only accept high sucrose concentrations as a reward. They will soon stop probing a food source with a low nectar yield. These bees could function as nectar collectors, bringing back to the hive nectar with high sugar concentrations. These hypotheses derived from laboratory experiments are still to be tested in the field.

Sensory Thresholds, Learning, and the Division
of Foraging Labor

The continuous provision of a colony with different types of food is an extremely complex task. Foraging has to be adapted to short-, medium-, and long-term changes within the colony and in the environment. There is good experimental evidence that the modification of response thresholds is an excellent way of controlling foraging labor on different time scales. Sucrose response thresholds measured in the laboratory depend on the genetic background of an individual, on the presence of brood in the colony, on pheromone signals within the colony, on weather conditions, on season, on the quality of available food sources, and on the actual state of an individual. These factors influence the perception of sucrose stimuli and, we assume, the perception of other stimulus modalities because the thresholds for different modalities are correlated (Figure 15.4).

In our view, the evaluation of a food source is dominated by gustatory stimuli that are perceived by the antennae. Both the response threshold of the individual bee and the sugar concentration of the food source will affect the learning of sensory cues and the formation of memory. There is experimental evidence that the difference between the concentration of a sucrose reward and the sucrose response threshold of an individual determines the acquisition function during learning (Scheiner, Erber, and Page 1999; see also Figure 15.4). A bee can potentially become a water collector when it is highly sensitive to gustatory stimuli and if it accepts a water stimulus as a reward. The bee learns the characteristics of this resource by associating them with a water reward. As a consequence of this learning process, the bee will return successfully to the water source for foraging. A bee can potentially become a pollen forager when it accepts low amounts of sugar solution as a reward and associates the features of a pollen source with this reward. A bee with a high threshold for sucrose can potentially become a nectar forager, because it only accepts high concentrations of sucrose as a reward. Such a bee will learn the sensory cues of a nectar source and will bring back to the hive nectar of high concentration. We hypothesize that the variance in sensory thresholds is a necessary prerequisite for bees to learn different types of food or water sources.

Because bees differ in their sensory response thresholds, their learning and their memory formation depend on the sugar concentration of the reward (Scheiner et al. 2005). Bees with high response thresholds need a

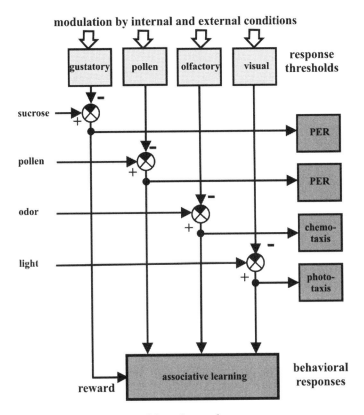

Figure 15.4. Schematic diagram of the relations between sensory response thresholds across modalities and behavioral responses. In our hypothesis, the behavioral responses to gustatory, pollen, olfactory, and visual stimuli depend both on the sensory response thresholds for a specific stimulus and the intensity of the perceived stimulus. If the difference between stimulus intensity and the response threshold of an individual for this modality is greater than 0, the behavior is performed. The occurrence of the proboscis extension response (PER), for example, depends on the difference between the response threshold for sucrose or for pollen and the intensity of the respective stimuli. The difference between the concentration of an odor stimulus and the olfactory response threshold of an individual determines the response of the bee toward the odor in an olfactometer. The difference between the intensity of a light source and the individual visual threshold determines the phototactic response of a bee toward a light source. Gustatory stimuli function as rewards in associative learning. If the difference between gustatory response threshold and sucrose concentration of the reward is very high, the probability that an individual learns a pollen source, an odor, or possibly a visual stimulus is very high. In contrast, a small difference between sucrose concentration of the reward and the gustatory response threshold of a bee will result in poor learning.

reward with a high sucrose concentration to show the same retrieval as bees with low sucrose thresholds that were rewarded with only a low concentration of sucrose. At the moment, it is not clear to what degree associative learning is influenced by the sensory thresholds for other stimulus modalities. In this respect, the diagram shown in Figure 15.4 summarizes our present working hypothesis.

The consequences of these results found under laboratory conditions have to be tested in experiments with free-flying animals that differ in their foraging activity and in their gustatory sensitivity. The laboratory findings suggest that there should be differences in learning performance and memory formation among bees that collect water, pollen, or nectar. These differences in the different forms of learning have, in our view, an important affect on the division of labor within a colony.

Literature Cited

Barron, A. B., D. J. Schulz, and G. E. Robinson. 2002. "Octopamine modulates responsiveness to foraging-related stimuli in honey bees *(Apis mellifera)*." *Journal of Comparative Physiology* A 188: 603–610.

Ben Shahar, Y., H.-T. Leung, W. L. Pak, M. B. Sokolowski, and G. E. Robinson. 2003. "cGMP-dependent changes in phototaxis: A possible role for the *foraging* gene in honey bee division of labor." *Journal of Experimental Biology* 206: 2507–2515.

Beshers, S. N., and J. H. Fewell. 2001. "Models of division of labor in social insects." *Annual Review of Entomology* 46: 413–440.

Bitterman, M. E., R. Menzel, A. Fietz, and S. Schäfer. 1983. "Classical conditioning of proboscis extension in honeybees *(Apis mellifera)*." *Journal of Comparative Physiology* 97: 107–119.

Blenau, W., and J. Erber. 1998. "Behavioural pharmacology of dopamine, serotonin and putative aminergic ligands in the mushroom bodies of the honeybee *(Apis mellifera)*." *Behavioural Brain Research* 96: 115–124.

Bloch, G., D. E. Wheeler, and G. E. Robinson. 2002. "Endocrine influences on the organization of insect societies." *Hormones, Brain and Behavior* 3: 195–235.

Braun, G., and G. Bicker. 1992. "Habituation of an appetitive reflex in the honeybee." *Journal of Neurophysiology* 67: 588–598.

Collins, M., and W. C. Rothenbuhler. 1978. "Laboratory test of the response to an alarm chemical, isopentyl acetate, by *Apis mellifera*." *Annals of the Entomological Society of America* 71: 906–909.

Elekonich, M. M., D. J. Schulz, G. Bloch, and G. E. Robinson. 2001. "Juvenile hormone levels in honey bee (*Apis mellifera* L.) foragers: Foraging experience and diurnal variation." *Journal of Insect Physiology* 47: 1119–1125.

Erber, J., J. Hoormann, and R. Scheiner. 2006. "Phototactic behaviour correlates with gustatory responsiveness in honey bees (*Apis mellifera* L.)." *Behavioural Brain Research* 174: 174–180.

Erber, J., S. Kierzek, E. Sander, and K. Grandy. 1998. "Tactile learning in the honeybee." *Journal of Comparative Physiology* A 183: 737–744.

Erber, J., and P. Kloppenburg. 1995. "The modulatory effects of serotonin and octopamine in the visual system of the honey bee (*Apis mellifera* L.): I. Behavioral analysis of the motion-sensitive antennal reflex." *Journal of Comparative Physiology* A 176: 111–118.

Erber, J., P. Kloppenburg, and A. Scheidler. 1993. "Neuromodulation by serotonin and octopamine in the honeybee: Behavior, neuroanatomy and electrophysiology." *Experientia* 49: 1073–1083.

Erber, J., B. Pribbenow, K. Grandy, and S. Kierzek. 1997. "Tactile motor learning in the antennal system of the honeybee (*Apis mellifera*)." *Journal of Comparative Physiology* A 181: 355–365.

Haupt, S. S. 2004. "Antennal sucrose perception in the honey bee (*Apis mellifera*): Behaviour and electrophysiology." *Journal of Comparative Physiology* A 190: 735–745.

Hölldobler, B., and E. O. Wilson. 1990. *The ants.* Berlin, Heidelberg: Springer-Verlag.

Huang, Z. Y., G. E. Robinson, and D. W. Borst. 1994. "Physiological correlates of division of labor among similarly aged honey bees." *Journal of Comparative Physiology* A 174: 731–739.

Hunt, G. J., E. Guzmán-Novoa, J. L. Uribe-Rubio, and D. Prieto-Merlos. 2003. "Genotype-environment interactions in honeybee guarding behaviour." *Animal Behaviour* 66: 459–467.

Jassim, O., Z. Y. Huang, and G. E. Robinson. 2000. "Juvenile hormone profiles of worker honey bees, *Apis mellifera,* during normal and accelerated behavioural development." *Journal of Insect Physiology* 46: 243–249.

Kisch, J., and J. Erber. 1999. "Operant conditioning of antennal movements in the honey bee." *Behavioural Brain Research* 99: 93–102.

Kloppenburg, P., and J. Erber. 1995. "The modulatory effects of serotonin and octopamine in the visual system of the honey bee (*Apis mellifera* L.): II. Electrophysiological analysis of motion-sensitive neurons in the lobula." *Journal of Comparative Physiology* A 176: 119–129.

Macmillan, C. S., and A. R. Mercer. 1987. "An investigation of the role of dopamine in the antennal lobes of the honeybee, *Apis mellifera.*" *Journal of Comparative Physiology* A 160: 359–366.

Masterman, R., R. Ross, K. Mesce, and M. Spivak. 2001. "Olfactory and behavioral response thresholds to odors of diseased brood differ between hygienic and non-hygienic honey bees (*Apis mellifera* L.)." *Journal of Comparative Physiology* A 187: 441–452.

Masterman, R, B. H. Smith, and M. Spivak. 2000. "Brood odor discrimination abilities in hygienic honey bees (*Apis mellifera* L.) using proboscis extension reflex conditioning." *Journal of Insect Behavior* 13: 87–101.

Menzel, R., and U. Müller. 1996. "Learning and memory in honeybees: From behavior to neural substrates." *Annual Review of Neuroscience* 19: 379–404.

Mercer, A. R., and R. Menzel. 1982. "The effects of biogenic amines on conditioned and unconditioned responses to olfactory stimuli in the honeybee *Apis mellifera*." *Journal of Comparative Physiology* 145: 363–368.

Page, R. E., and J. Erber. 2002. "Levels of behavioral organization and the evolution of division of labor." *Naturwissenschaften* 89: 91–106.

Page, R. E., J. Erber, and M. K. Fondrk. 1998. "The effect of genotype on response thresholds to sucrose and foraging behavior of honey bees (*Apis mellifera* L.)." *Journal of Comparative Physiology* A 182: 489–500.

Page, R. E., and G. E. Robinson. 1991. "The genetics of division of labor in honey bee colonies." *Advances in Insect Physiology* 23: 117–169.

Pankiw, T. 2004. "Worker honey bee pheromone regulation of foraging ontogeny." *Naturwissenschaften* 91: 178–181.

Pankiw, T., and R. E. Page. 1999. "The effect of genotype, age, sex, and caste on response thresholds to sucrose and foraging behavior of honey bees (*Apis mellifera* L.)." *Journal of Comparative Physiology* A 185: 207–213.

———. 2000. "Response thresholds to sucrose predict foraging behavior in the honey bee (*Apis mellifera* L.)." *Behavioral Ecology and Sociobiology* 47: 265–267.

———. 2001. "Genotype and colony environment affect honeybee (*Apis mellifera* L.) development and foraging behavior." *Behavioral Ecology and Sociobiology* 51: 87–94.

———. 2003. "Effect of pheromones, hormones, and handling on sucrose response thresholds of honey bees (*Apis mellifera* L.)." *Journal of Comparative Physiology* A 189: 675–684.

Pham-Delègue, M. H., R. De Jong, and C. Masson. 1990. "Effect de l'age sur la réponse conditionnée d'extension du proboscis chez l'abeille domestique." *Comptes rendus de l'Académie des sciences Paris/Life sciences* 310: 527–532.

Pham-Delègue, M.-H., J. Trouiller, E. Bakchine, B. Roger, and C. Masson. 1991. "Age dependency of worker bee response to queen pheromone in a four-armed olfactometer." *Insectes Sociaux* 38: 283–292.

Rachinsky, A. 1994. "Octopamine and serotonin influence on *corpora allata* activity in honey bee (*Apis mellifera*) larvae." *Journal of Insect Physiology* 40: 549–554.

Ray, S., and B. Ferneyhough. 1997. "The effects of age on olfactory learning and memory in the honeybee *Apis mellifera*." *NeuroReport* 8: 789–793.

Robinson, G. E. 1985. "Effects of a juvenile hormone analogue on honey bee foraging behaviour and alarm pheromone production." *Journal of Insect Physiology* 31: 277–282.

———. 1987. "Modulation of alarm pheromone perception in the honey bee: Evidence for division of labor based on hormonally regulated response thresholds." *Journal of Comparative Physiology A* 160: 613–619.

———. 1992. "Regulation of division of labor in insect societies." *Annual Review of Entomology* 37: 637–665.

Sasagawa, H., M. Sasaki, and I. Okada. 1989. "Hormonal control of the division of labour in adult honeybees (*Apis mellifera* L.). I. Effect of methoprene on *corpora allata* and hypopharyngeal gland, and its a-glucosidase activity." *Applied Entomology and Zoology* 24: 66–77.

Scheiner, R. 2004. "Responsiveness to sucrose and habituation of the proboscis extension response in honey bees." *Journal of Comparative Physiology A* 190: 727–733.

Scheiner, R., M. Barnert, and J. Erber. 2003. "Variation in water and sucrose responsiveness during the foraging season affects proboscis extension learning in honey bees." *Apidologie* 34: 67–72.

Scheiner, R., A. Baumann, and W. Blenau. 2006. "Aminergic control and modulation of honeybee behaviour." *Current Neuropharmacology* 4: 259–276.

Scheiner, R., J. Erber, and R. E. Page. 1999. "Tactile learning and the individual evaluation of the reward in honey bees (*Apis mellifera* L.)." *Journal of Comparative Physiology A* 185: 1–10.

Scheiner, R., A. Kuritz-Kaiser, R. Menzel, and J. Erber. 2005. "Sensory responsiveness and the effects of equal subjective rewards on tactile learning and memory of honeybees." *Learning and Memory* 12: 626–635.

Scheiner, R., U. Müller, S. Heimburger, and J. Erber. 2003. "Activity of protein kinase A and gustatory responsiveness in the honey bee (*Apis mellifera* L.)." *Journal of Comparative Physiology A* 189: 427–434.

Scheiner, R., R. E. Page, and J. Erber. 2001a. "Responsiveness to sucrose affects tactile and olfactory learning in preforaging honey bees of two genetic strains." *Behavioural Brain Research* 120: 67–73.

———. 2001b. "The effects of genotype, foraging role, and sucrose responsiveness on the tactile learning performance of honey bees (*Apis mellifera* L.)." *Neurobiology of Learning and Memory* 76: 138–150.

———. 2004. "Sucrose responsiveness and behavioral plasticity in honey bees (*Apis mellifera*)." *Apidologie* 35: 133–142.

Scheiner, R., S. Plückhahn, B. Öney, W. Blenau, and J. Erber. 2002. "Behavioural pharmacology of octopamine, tyramine and dopamine in honey bees." *Behavioural Brain Research* 136: 545–553.

Scheiner, R., A. Weiß D. Malun, and J. Erber. 2001. "Learning in honey bees with brain lesions: How partial mushroom-body ablations affect sucrose responsiveness and tactile learning." *Animal Cognition* 4: 227–235.

Schulz, D. J., J. P. Sullivan, and G. E. Robinson. 2002. "Juvenile hormone and octopamine in the regulation of division of labor in honey bee colonies." *Hormones and Behavior* 42: 222–231.

Spivak, M., R. Masterman, R. Ross, and K. A. Mesce. 2003. "Hygienic behavior in the honey bee (*Apis mellifera* L.) and the modulatory role of octopamine." *Journal of Neurobiology* 55: 341–354.

Thompson, R. F., and W. A. Spencer. 1966. "Habituation: A model phenomenon for the study of neuronal substrates of behavior." *Psychological Review* 73: 16–43.

CHAPTER SIXTEEN

Social Life from Solitary Regulatory Networks: A Paradigm for Insect Sociality

ROBERT E. PAGE JR.

TIMOTHY A. LINKSVAYER

GRO V. AMDAM

HOW DO COMPLEX social systems evolve? What are the evolutionary and developmental building blocks of division of labor and specialization, the hallmarks of insect societies? In this chapter we describe research into the evolution and development of division of labor in the honey bee (*Apis mellifera*). In solitary insects, shifts during life history between reproductively active and inactive states are associated with widespread changes in physiological state. In honey bees, variation in the physiological state of workers is also associated with variation in behavior. We suggest that worker behavioral specialization and division of labor are based on the modification of regulatory networks underlying shifts in reproductive state.

We begin by describing how studies of the phenotypic and genetic architecture underlying pollen hoarding in honey bees led us to propose a link between worker behavioral specialization and reproductive state. Next we describe how hormonal systems underlie associations between reproductive state and behavior in solitary insects, and how evolutionary adoption of these regulators is a plausible foundation for honey bee worker behavior. This view is summarized in the reproductive ground plan hypothesis of social evolution, which explains the link between worker behavior and reproductive state. Finally, we broadly consider the evolution of eusociality to elucidate how adoption of ancestral genetic, developmental, and reproductive physiological machineries can be of general importance for emergence of advanced social behavior.

The Evolution of Division of Labor in Honey Bees

Division of labor among worker honey bees is based predominantly on age, with individuals progressing through a series of tasks from in-hive tasks to foraging (Robinson 1992). Further specialization occurs among foragers for pollen or nectar collection. The cumulative efforts of pollen and nectar foragers determine colony pollen and nectar stores. Page and Fondrk (1995) conducted two-way (bidirectional) selection for the amount of surplus pollen stored in the comb (pollen hoarding; see also Hellmich, Kulinceric, and Rothenbuhler 1985). After just three generations, colonies of the high pollen hoarding strain contained about six times more pollen, demonstrating a strong response to selection. With subsequent generations of selection, Page and coworkers studied individual behavioral and physiological traits that changed as a result of selection on the colony-level phenotype. This enabled them to look for mechanisms at different levels of biological organization that causally underlie the differences in the colony-level phenotype (Page and Erber 2002).

One dramatic change that arose was in the age at which bees initiated foraging behavior. High-strain bees (workers from the high pollen hoarding strain) initiate foraging about 10 days earlier in life than low-strain bees (Pankiw and Page 2001). High-strain bees are more likely to specialize on collecting pollen while low-strain bees are more likely to specialize on nectar (Page and Fondrk 1995; Fewell and Page 2000; Pankiw and Page 2001). High-strain bees are also more likely to collect water, and when they collect nectar, they accept nectar with lower sugar content than do bees of the low strain. Low-strain bees are also much more likely to return empty from foraging trips (Page, Erber, and Fondrk 1998).

Differences in forager pollen load sizes between strains arise through their dissimilar responses to pollen foraging stimuli. Fewell and Winston (1992) showed that colonies respond to changes in quantities of stored pollen by altering the allocation of foraging effort to pollen collection. When presented with additional stored pollen beyond what had already been stored, colonies responded with a reduction in the number of pollen foragers and the sizes of the pollen loads. The opposite effect on foraging behavior was observed when stored pollen was removed. Colonies, therefore, maintain the amount of stored pollen around a regulated set point. Studies by Dreller, Page, and Fondrk (1999) and Dreller and Tarpy (2000) demonstrated that foragers directly assess the amount of pollen stored in the combs and adjust their foraging behavior accordingly (also see Vaughan and Calderone 2002). The mechanism appears to involve the assessment of empty cells near the areas of the nest where larvae and pupae are located. Therefore, the regulatory mechanism underlying pollen storage involves individual assessment of stored pollen and individual "decisions" with respect to what to collect on a foraging trip (Fewell and Page 2000). High-strain colonies regulate set point centers around much larger quantities of stored pollen than do low-strain colonies. Therefore, high-strain bees have a threshold for stored pollen (or empty cells near the brood) that is different from low-strain bees. When co-fostered in an unselected wild-type colony, high-strain bees perceive the amount of stored pollen as being below their optimal set point while the low-strain bees perceive it as above theirs. As a result, high-strain bees are more likely to forage for pollen and low-strain bees are more likely to forage for nectar.

High- and low-strain bees also respond differently to changes in pollen and brood stimuli in colonies. Young larvae and hexane rinses of young larvae, which extract pheromones, stimulate pollen-specific foraging behavior, while stored pollen acts as an inhibitor (Pankiw, Page, and Fondrk 1998). Pankiw and Page (2001) co-fostered high- and low-strain bees in colonies with high- and low-pollen hoarding stimuli. High stimulus colonies were experimentally manipulated to contain less stored pollen and more larvae than the low stimulus colonies. Foragers in the high stimulus colonies were more likely to collect pollen, collected larger loads of pollen, and, consequently, collected smaller loads of nectar independent of whether they were of the high or low strain. High-strain bees, however, had a larger difference in foraging behavior between treatments, demonstrating a genotype-by-environment interaction, where high-strain bees were more sensitive to the foraging stimulus environment than the low-strain bees.

Such changes in foraging behavior are expected consequences of bidirectional selection on pollen hoarding. However, high-strain bees are also more likely to forage for water than are low-strain bees (Page, Erber, and Fondrk 1998), and when they collect nectar they accept nectar with lower sugar concentrations. There was no obvious physiological or behavioral mechanism to explain these relationships until Page and colleagues looked at the responses of pollen and nectar foragers to sucrose solutions. Bees respond reflexively to antennal stimulation with sucrose by extending the proboscis. Page et al. used a series of solutions with increasing sucrose concentrations to determine the sucrose responses of wild-type pollen and nectar foragers. The results were surprising: pollen foragers were more likely than nectar foragers to respond to water and lower concentrations of sucrose.

The sucrose sensitivity of pollen and nectar foragers might be related to the physiological status of the bees. Pollen foragers could be comparatively depleted of blood sugars as a result of their foraging activity and, therefore, more responsive. However, sucrose sensitivity might also be a property of the neural states of the animals, which then in turn result in differences in foraging behavior. Thereby, sucrose sensitivity would be an indicator of potential foraging behavior. To distinguish between these two alternative hypotheses, high- and low-strain bees were tested for sucrose sensitivity when they were no more than a week old, before they initiated foraging (Page, Erber, and Fondrk 1998). High-strain bees were more responsive to sucrose solutions and water at this early age, suggesting that selection for pollen hoarding had changed a fundamental property of the sensory-response system with consequences at the level of foraging behavior. Subsequent studies have shown that differences in water and sucrose responses exist between the selected strains at adult emergence, 2 to 3 weeks before the bees initiate foraging (Pankiw and Page 1999).

If water and sucrose responses are indicators of differences in neural states related to nectar and pollen foraging, then it should be possible to also assay wild-type bees when they emerge as adults and predict their foraging behavior 2 to 3 weeks later. Pankiw and Page (2000) tested wild-type bees for their responses to water and sucrose when they were less than a week old. Bees were marked for individual identification, placed back into their colony, and returning foragers were collected and their foraging loads analyzed. Bees that were the most responsive to water and sucrose solutions when they were 5 days old were the most likely to collect water on a foraging trip. The next most responsive group collected pollen, followed by both pollen and nectar, nectar exclusively, and the least responsive group was most likely to return to the nest empty (Figure 16.1). Thus, responses to sucrose and water can be said to be reliable indicators of the neural states of bees and used to predict foraging behavior. This result has been confirmed in additional, independent studies by assaying newly emerged bees (Pankiw 2003; Pankiw et al. 2004).

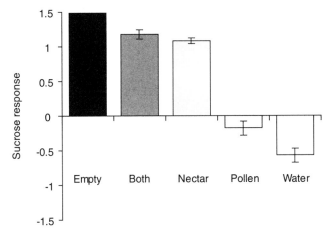

Figure 16.1. Sucrose responses of 1-week-old bees predict their foraging behavior later in life. The x-axis shows the foraging material collected by the bees when they have reached foraging age. The y-axis shows the lowest sucrose concentrations (Log10) at which 1-week-old bees responded with proboscis extension. Bees with the highest sucrose responsiveness (i.e., the lowest threshold) at young age are more likely to later forage for water or pollen. Individuals with low sucrose responsiveness (i.e., a high threshold) are more likely to later collect nectar, nectar and pollen, or to return empty (data from Pankiw and Page 1999).

In general, bees with high responsiveness to sucrose, like the high-strain bees and pollen foragers, learn faster and reach a higher asymptote of learning than bees with that are less responsive (Scheiner, Erber, and Page 1999; Scheiner, Page, and Erber 2001a, b, 2004; Scheiner et. al. 2001; Scheiner, Barnert, and Erber 2003). This is because learning performance is related to the evaluation of the sucrose stimuli used during conditioning, which can be measured as a response threshold to sucrose solution (Page, Erber, and Fondrk 1998; Pankiw and Page 2000; Scheiner, Erber, and Page 1999; Scheiner, Page, and Erber, 2001b; Scheiner, Barnert, and Erber 2003; Scheiner and Erber, this volume). In accordance with these findings, high-strain bees and pollen foraging wild-type bees perform better on tactile and olfactory associative learning tests than do low-strain bees and nectar foragers (Scheiner, Erber, and Page 1999; Scheiner, Page, and Erber 2001a, b).

Responsiveness to sucrose also correlates with locomotor activity when bees first emerge as adults. Humphries, Fondrk, and Page (2005) tested locomotion in newly emerged wild-type bees by measuring their walking activity in an enclosed arena, and then determined their response to sucrose using the proboscis extension response protocols. The more active bees were also more responsive to sucrose. High-strain bees, furthermore, were more active than low-strain bees, consistent with the results from wild-type bees.

Sensory sensitivity and activity levels in response to stimuli associated with food, mating, and oviposition sites change through the reproductive cycle of solitary insects (reviewed by Amdam et al. 2004). Such associations motivated studies on correlations between the reproductive physiology and behavior of worker bees. Worker bees from the high-strain group have larger ovaries (more ovarioles per ovary) than do workers from the low-strain group (Amdam et al. 2006). Wild-type workers that forage for pollen likewise have more ovarioles per ovary than do those that collect nectar. Wild-type bees that return empty from foraging trips ("unsuccessful" foragers) have the fewest ovarioles (Figure 16.2), in accordance with abovementioned trait associations of the low-strain workers. Wild-type workers with larger ovaries forage earlier in life than those with smaller ovaries (Amdam et al. 2006). Wild-type workers with more ovarioles are more responsive to low concentration sucrose solutions than those with fewer ovarioles (Tsuruda, Amdam, and page forthcoming), thus linking the whole suite of traits discussed above with ovary size and, thereby, the full phenotypic syndrome of high-strain bees. Ovariole number is determined during larval development, about 3 to 5 days after hatching. Therefore, events that take place during this period that result in variation in ovariole numbers in workers shape the subsequent behavior of worker honey bees. This is manifested in specialization and division of labor and, in the absence of the queen, oogenesis and oviposition.

Genetic Architecture of Pollen Hoarding

Genetic mapping studies have been used to elucidate the genetic basis of the phenotypic differences between the high and low pollen hoarding strains. These studies revealed four major quantitative trait loci (QTL) that explain a significant amount of the phenotypic variance for pollen hoarding

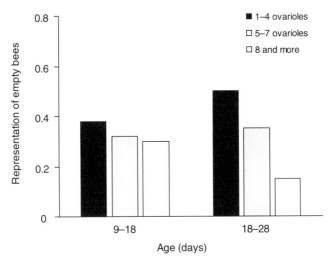

Figure 16.2. The relationship between ovariole number and the probability of returning empty to the nest from presumably the first foraging trip. Bees with the fewest ovarioles are more likely to return empty both in the group of young bees (9–18 days old) and in the group of older workers of more typical foraging age (18–28 days old). The difference is less apparent in the first group, probably because of a higher probability of randomly obtaining empty nonforagers when sampling from a population of workers younger than the typical foraging age of bees.

and foraging behavior (Hunt et al. 1995; Page et al. 2006; Rueppell et al. 2004; Rueppell, Pankiw, and Page 2004). The genetic architecture of pollen hoarding and foraging behavior is complex (Figure 16.3). All QTL have pleiotropic effects on multiple traits associated with pollen and nectar foraging, thus providing an explanation for the correlative association of this set of traits. They are also richly epistatic, interacting with one another in complex ways. All individual QTL and most of their interactions affect pollen and nectar load sizes. All individual QTL also affect concentration of nectar collected. The *pln1* region is especially interesting because it has a demonstrated direct effect on all behavioral traits. The combination of these QTL studies and the completed honey bee genome sequence and annotation provide informed candidates for future studies of the genetic basis for variation in pollen hoarding and foraging behavior. A recent analysis proposed that positional candidate genes involved in endocrine signaling provide the most coherent explanation for the syndromes (Hunt et al. 2007).

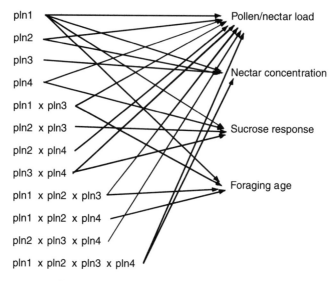

Figure 16.3. Complex genetic architecture of traits associated with foraging behavioral differences between the high and low pollen-hoarding strains of Page and Fondrk (1995). Arrows indicate significant effects involving the four major Quantitative Trait Loci *(pln1–pln4)*. Interactions between QTL indicate epistasis and effects on multiple traits indicate pleiotropy.

Hormonal Signaling Cascades

The suite of traits that vary with foraging behavior include ovary size, temporal behavioral development, sensory modulation, and motor response systems. The genetic architecture of this set of traits suggests an interactive regulatory network that operates on biological systems at multiple levels of organization in a time scale of days and weeks, thus making hormonal signaling cascades prime causal candidates for modulating the differences between pollen and nectar foragers.

Classical Endocrine Factors

Ecdysone and juvenile hormone (JH) are key hormonal modulators of insect behavior (Hartfelder 2000). Ecdysone is produced by the prothoracic gland during larval and pupal development, and by the ovary during the adult stage. JH is a growth hormone produced by the *corpora allata* of insects (Hagenguth and Rembold 1978). JH has been hypothesized to play an important role in honey bee division of labor by pacing age-related changes in behavior, especially the transition to foraging (Robinson 1992; Robinson and Vargo 1997). Many studies have demonstrated elevated blood titers of JH in foragers relative to bees that perform tasks in the nest (e.g., Robinson 1987; Huang and Robinson 1992; Huang, Robinson, and Borst 1994; Sullivan, Jassim, et al. 2000; Sullivan, Fahrbach, et al. 2003). Treatment with the JH analog methoprene results in bees initiating foraging behavior earlier in life (for review see Bloch, Sullivan, and Robinson 2002) and increases sucrose responsiveness in young bees (Pankiw and Page 2003), suggesting that JH plays a role in sensory modulation.

Overall, JH correlates with age-based changes in honey bee behavior and sensory sensitivity, but does it pace behavioral development? Sullivan et al. (2000, 2003) removed the *corpora allata* from newly emerged bees. The allatectomized workers initiated foraging at about the same time as the control bees, suggesting no effect on the transition to foraging. In another study, worker honey bees from the high and low pollen hoarding strains initiated foraging at different ages and also differed in JH titer at adult emergence; however, their JH titer was not different 12 days later, just prior to the initiation of foraging (Schulz et al. 2004). Thus, it is clear that JH is not necessary for behavioral development, but that treatments with JH and JH analog nonetheless have behavioral effects.

Endocrine Effects of Vitellogenin

Vitellogenin provides a possible alternative endocrine pathway for the development of pollen foraging. Vitellogenin is a major yolk precursor in many insects (Babin et al. 1999) and is also the most abundant hemolymph protein in worker bees that perform tasks in the nest prior to foraging (Engels and Fahrenhorst 1974; Fluri, Sabatini, et al. 1981; Fluri, Lüscher, and Gerig 1982). Recent studies have shown that vitellogenin gene activity suppresses the JH titer of worker bees (Guidugli et al. 2005). Conversely, JH is known to suppress the synthesis of honey bee vitellogenin at onset of foraging (Pinto, Bitondi, and Simões 2000). These data suggest that the two proteins are linked in a positive feedback loop via a mutual ability to suppress each other. Amdam and Omholt (2003) hypothesized that foraging behavior is initiated when vitellogenin titer drops below a certain threshold level. The feedback action of JH on vitellogenin could be a reinforcing mechanism that causes workers to become behaviorally and physiologically locked into the forager stage.

In support of Amdam and Omholt's hypothesis, Nelson et al (2007) found that reduction of vitellogenin gene activity by RNA interference (RNAi) caused bees to forage earlier in life. Amdam et al. (2006) demonstrated that vitellogenin RNAi increases the sucrose responsiveness of worker bees, and suggested that honey bee vitellogenin modulated behavior and sensory sensitivity via a signaling pathway that includes JH as a downstream feedback element.

Honey bee vitellogenin is produced by the abdominal fat body, but evidence suggests that this protein triggers responses in other cell types (Guidugli et al. 2005), implying that vitellogenin itself can be classified as a hormone. The documented effects of JH and JH analog treatments, therefore, can be understood as results of suppressed vitellogenin action (Amdam et al. 2006).

Reproductive Ground Plan—A Synthesis

Associations between foraging behavior and traits such as vitellogenin level, ovary size, and rates of behavioral development suggest that division of labor and particularly foraging specialization in honey bees are derived from the reproductive regulatory networks of solitary ancestors. Amdam et al. (2004) proposed that the suite of traits associated with foraging behavior and their underlying genetic architecture were part of a reproductive regulatory network (see also West-Eberhard 1987b, 1996). In solitary insects, different stages of the female reproductive cycle (pre-vitellogenesis, vitellogenesis, oviposition, and brood care) are linked and involve coupled physiological and behavioral changes (Finch and Rose 1995). JH and ecdysone are key hormones controlling vitellogenesis in many insect species (e.g., Socha et al. 1991; Hiremath and Jones 1992; Brownes 1994); in addition, they regulate behavioral transitions associated with changes in reproductive state, such as the shift from foraging for nectar in previtellogenic females to protein foraging in vitellogenic individuals, as occurs in the mosquito *Culex nigripalpus* (Hancock and Foster 2000). JH and ecdysone also modulate changes in sensory perception, locomotor activity, and reproductive physiology (Zera and Bottsford 2001)—traits that have been shown to be different in workers from the high and low pollen hoarding strains and in wild-type pollen and nectar foragers.

In solitary insects, hormonal effects on reproductive traits typically act in mature adults following a pre-reproductive phase where the animals may enter diapause or aestivate and disperse (Hartfelder 2000). In honey bees, however, these hormonal signals seem to have shifted in time (Amdam et al. 2004), occurring in the late pupal stages where they activate the production of vitellogenin (Barchuk, Bitondi, and Simões 2002). Differential amplitude of JH titers are observed in newly emerged high and low pollen hoarding bees where high-strain workers have higher titers of JH (Schulz et al. 2004). This elevated titer correlates with a higher level of vitellogenin mRNA and a higher vitellogenin hormone titer in the blood (Amdam et al. 2004). Compared to the low-strain bees, workers of the high pollen hoarding strain have more ovarioles, which already show an active previtellogenic ovarian phenotype at adult emergence (Amdam et al. 2006). It has been proposed that if such documented markers of JH and ecdysone action are present early in honey bee adult life (Figure 16.4),

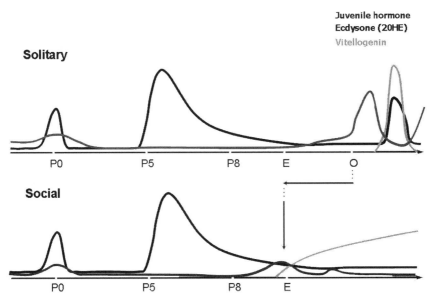

Figure 16.4. A time course of blood hormones and vitellogenin titers from early to late pupal stages (P0–P8) through emergence (E) and into mature adults with activated ovaries (O) in solitary insects (upper panel), compared to honey bee development (lower panel) (from Pinto et al. 2000; Barchuk et al. 2002). Amdam et al. (2004) hypothesized that the spikes of hormone titers linked to O in solitary insects has shifted in time in social insects and is homologous with the increases in titer observed at E in honey bee workers.

then pleiotropic effects on behavior may have shifted from later life-stages as well (Amdam et al. 2004), as demonstrated by the differences in sensory responses and locomotor activity of high- and low-strain bees and the correlation of locomotor and sensory responses in wild-type workers.

The recent finding that ovariole number correlates with sensory responsiveness in wild-type bees (Tsuruda, Amdam, and page forthcoming), and the known association between such sensory responses and foraging behavior 2 to 3 weeks later, suggest that gonotropic events in young bees have persistent effects on adult behavior. High-strain bees and pollen foragers seem to be similar to ancestral gono-active females, while low-strain bees and nectar foragers are like gono-inactive females. These insights have been summarized in the "reproductive ground plan" hypothesis of social evolution (West-Eberhard 1987b, 1996; Amdam et al. 2004). The hypothesis proposes that the genetic and hormonal networks that govern reproductive development, physiology, and behavior in solitary species represent a fundamental regulatory system with the capacity to serve as the basis for the evolution of social phenotypes. We discuss next how evolutionary modification of pre-existing developmental, endocrine, and behavioral building blocks can lead to the evolutionary origin and elaborate on the two traits fundamental to eusociality: sib-care and queen-worker caste dimorphism.

Evolution of Eusociality

Origin of Eusociality

It is commonly accepted that sib-care behavior expressed by helper females toward the sibling brood is homologous with and evolutionarily derived from maternal care behavior expressed toward offspring (West-Eberhard 1987a; Alexander, Noonan and Crespi 1991). In the heterochrony model for the origin of eusociality, sib-care behavior expressed in helpers is the result of the modified, early expression of genes for maternal care (Linksvayer and Wade 2005). In the ancestral condition, maternal care behavior is expressed as one of the final steps in a coordinated series of physiological and behavioral changes that occur through reproductive development (West-Eberhard 1996). In the derived condition, the timing of expression of maternal care behavior is altered so that this behavior is expressed pre-reproductively toward siblings instead of post-reproductively toward offspring. Under this model, the evolution of the capacity for females to provide care pre-reproductively toward their siblings is a first step in the evolutionary origin of eusociality from subsociality (Linksvayer and Wade 2005). The next step involves the regulation of the timing of expression of genes for maternal care behavior so that eusocial colonies produce both helper females and fully-reproductive females.

As discussed above, in both queen and worker honey bees there is a shift in the timing of hormonal signals involved in activating ovaries relative to solitary insects, from post- to pre-emergence (Amdam et al. 2004). Because behavior and gonotropic cycle are linked, this shift may be related to the heterochronous shift in behavior hypothesized for the origin of sib-care. That is, the evolution of the capacity of females to provide care pre-reproductively may be associated with a shift in the timing of hormonal signals in all females (i.e., both reproductive "queen" phenotypes and helper "worker" phenotypes) so that the hormonal shift observed in highly social honey bees may be the result of ancient evolutionary events. Early ovary activation and vitellogenesis may also increase the reproductive potential of young queens and provide young worker bees with a source of protein that can be converted into larval food (Amdam et al. 2003, 2004). Thus, timing of hormonal signals and the physiological and behavioral responses observed in highly social honey bees may be the result of evolutionary modification associated both with the origin of eusociality as well as more recent evolutionary elaboration associated with increased colony size and social complexity. Comparative studies using other eusocial, as well as subsocial, aculeate Hymenoptera will elucidate these hypotheses.

Elaboration of Eusociality

After the origin of eusociality, among-colony selection would likely favor the evolutionary divergence of helper and reproductive phenotypes, if the result was a more efficient division of labor. However, this divergence is initially constrained because of their common genetic basis; maternal care and sib-care are, theoretically, influenced by the same set of genes, so the two traits cannot evolve independently (Linksvayer and Wade 2005). Therefore, evolutionary modification of the genetic basis of maternal and sib-care through gene duplication (Gadagkar 1997), or more simply caste-specific gene expression, can enable these phenotypes to diverge. Yet even in highly social taxa such as honey bees, many genes have pleiotropic effects on queen and worker phenotypes, which is expected given that these phenotypes are derived from common genetic, physiological, and behavioral building blocks (West-Eberhard 1996; Amdam et al. 2004, 2006; Linksvayer and Wade 2005).

Polyphenisms such as reproductive caste in social insects are thought to be derived from phenotypically plastic traits, using preexisting physiological and endocrine developmental mechanisms (Nijhout 2003). The evolution of discrete castes involves the elaboration and conversion of preexisting phenotypic plasticity to phenotypic differences between castes. This occurs in part through the evolutionary modification of endocrine and developmental mechanisms that are sensitive to environmental conditions (Wheeler 1986; West-Eberhard 1987a, 1996). Just as the timing of expression of maternal care behavior is affected by both intrinsic and extrinsic factors, so are traits associated with reproductive caste such as ovary size and body size. Larval genes affect developmental responses to environmental conditions (such as nutritional quality and quantity). For example, the sensitivity of the developmental switch of caste determination is influenced by genes affecting the endocrine response to nutritional signals (Wheeler 1986). Additionally, the environmental conditions of developing larvae are determined by the social milieu of the colony provided by nestmates, and this social environment is influenced by genes expressed in sibling brood, sibling helpers, and the queen (Linksvayer and Wade 2005; Linksvayer 2006). The evolution of distinct developmentally canalized queen and worker phenotypes then involves evolutionary fine-tuning of both the social environment and the developmental response.

In subsocial animals with extended maternal care, the evolution of maternal and brood phenotypes has been considered as a co-evolutionary process (Wolf and Brodie 1998; Agrawal, Brodie, and Brown 2001; Kölliker, Brodie, and Moore 2005). After the origin of eusociality, a third class of social partners, adult helper females, is added, so that social insect phenotypes are influenced by the genomes of three types of interacting social partners: brood, workers, and queens. The evolution of phenotypes in eusocial colonies (e.g., those associated with reproductive caste) can be considered as the co-evolution of queen, worker, and brood phenotypes (Linksvayer and Wade 2005; Linksvayer 2006).

Conclusion

We suggest here that complex social behavior as found in eusocial insects is derived from reproductive regulatory networks common to all insects. Small changes in the timing of expression of maternal care behavior may be all that is needed to form reproductive and nonreproductive phenotypes, the basis of eusociality. Additional evolutionary modification of hormonal networks regulating development, reproduction, and maternal care, and tuning of the developmental environment through modification of the behavior of social partners, may produce the amazingly diverse and complex insect societies that we continue to admire.

Acknowledgments

Parts of this chapter are modified from Page et al. (2006). Research presented here was funded primarily by the National Science Foundation, National Institutes of Health, and the United States Department of Agriculture to REP and the Norwegian Research Council to GVA.

Literature Cited

Agrawal, A. F., E. D. Brodie III, and J. Brown. 2001. "Parent-offspring coadaptation and the dual genetic control of maternal care." *Science* 292: 1710–1712.

Alexander, R. D., K. M. Noonan, and B. J. Crespi. 1991. "The evolution of eusociality." In P. W. Sherman, J. U. M. Jarvis, and R. D. Alexander, eds., *The biology of the naked mole rat*, 3–44. Princeton: Princeton University Press.

Amdam, G. V., A. Csondes, M. K. Fondrk, and R. E. Page. 2006. "Complex social behavior derived from maternal reproductive traits." *Nature* 439: 76–78.

Amdam, G. V., K. Norberg, A. Hagen, and S. W. Omholt. 2003. "Social exploitation of vitellogenin." *Proceedings of the National Academy of Sciences USA* 100: 1799–1802.

Amdam, G. V., K. Norberg, M. K. Fondrk, and R. E. Page. 2004. "Reproductive ground plan may mediate colony-level selection effects on individual foraging behavior in honey bees." *Proceedings of the National Academy of Sciences USA* 101: 11350–11355.

Amdam, G. V., and S. W. Omholt. 2003. "The hive bee to forager transition in honeybee colonies: The double repressor hypothesis." *Journal of Theoretical Biology* 223: 451–464.

Babin, P. J., J. Bogerd, F. P. Kooiman, W. J. A. Van Marrewijk, and D. J. Van der Horst. 1999. "Apolipophorin II/I, Apolipoprotein B, Vitellogenin, and Microsomal Triglyceride Transfer Protein Genes Are Derived from a Common Ancestor." *Journal of Molecular Evolution* 49: 150–160.

Barchuk, A. R., M. .M. G. Bitondi, and Z. L. P. Simões. 2002. "Effects of juvenile hormone and ecdysone on the timing of vitellogenin appearance in hemolymph of queen and worker pupae of *Apis mellifera*." *Journal of Insect Science* 2: 1.

Bloch, G., J. P. Sullivan, and G. E. Robinson. 2002. "Juvenile hormone and circadian locomotor activity in the honey bee *Apis mellifera*." *Journal of Insect Physiology* 48: 1123–1131.

Brownes, M. 1994. "The regulation of the yolk protein genes, a family of sex differentiation genes in *Drosophila melanogaster*." *Bio Essays* 16: 745–752.

Dreller, C., R. E. Page, and M. K. Fondrk. 1999. "Regulation of pollen foraging in honeybee colonies: Effects of young brood, stored pollen, and empty space." *Behavioral Ecology and Sociobiology* 45: 227–233.

Dreller, C., and D. R. Tarpy. 2000. "Perception of the pollen need by foragers in a honeybee colony." *Animal Behaviour* 59: 91–96.

Engels, W., and H. Fahrenhorst. 1974. "Alters- und kastenspezifische Veräbderungen der Haemolymph-Protein-Spektren bei *Apis mellificia*." *Roux's* Arch. 174: 285–296.

Fewell, J. H., and R. E. Page. 2000. "Colony-level selection effects on individual and colony foraging task performance in honeybees, *Apis mellifera* L." *Behavioral Ecology and Sociobiology* 48: 173–181.

Fewell, J. H., and M. L. Winston. 1992. "Colony state and regulation of pollen foraging in the honey bee, *Apis mellifera* L." *Behavioral Ecology and Sociobiology* 30: 387–393.

Finch, C. E., and M. R. Rose. 1995. "Hormones and the physiological architecture of life history evolution." *Quarterly Review of Biology* 70: 1–52.

Fluri, P., M. Lüscher, H. Wille, and L. Gerig, L. 1982. "Changes in weight of the pharyngeal gland and haemolymph titres of juvenile hormone, protein and vitellogenin in worker honey bees." *Journal of Insect Physiology* 28: 61–68.

Fluri, P., A. Sabatini, M. A. Vecchi, and H. Wille. 1981. "Blood juvenile hormone, protein and vitellogenin titres in laying and non-laying queen honeybees." *Journal of Apicultural Research* 20: 221–225.

Gadagkar, R. 1997. "The evolution of caste polymorphism in social insects: Genetic release followed by diversifying evolution." *Journal of Genetics* 76: 167–179.

Guidugli, K. R., A. M. Nascimento, G. V. Amdam, A. R. Barchuk, R. Angel, S. W. Omholt, Z. L. P. Simões, and K. Hartfelder. 2005. "Vitellogenin regulates hormonal dynamics in the worker caste of a eusocial insect." *FEBS Letters* 579: 4961–4965.

Hagenguth, H., and H. Rembold. 1978. "Identification of juvenile hormone-3 as the only JH homolog in all developmental stages of the honey bee." *Zeitschrift für Naturforschung C* 33: 847–850.

Hancock, R. G., and W. A. Foster. 2000. "Exogenous juvenile hormone and methoprene, but not male accessory gland substances or ovariectomy, affect the blood/nectar choice of female *Culex nigiripalpus* mosquitoes." *Medical and Veterinary Entomology* 14: 373–382.

Hartfelder, K. 2000. "Insect juvenile hormone: From 'status quo' to high society." *Brazilian Journal of Medical and Biological Research* 33: 157–177.

Hellmich, R. L., J. M. Kulincevic, and W. C. Rothenbuhler. 1985. "Selection for high and low pollen hoarding in honey bees." *Journal of Heredity* 76: 155–158.

Hiremath, S., and D. Jones. 1992. "Juvenile hormone regulation of vitellogenin in the gypsy moth, *Lymantria Dispar:* Suppression of vitellogenin mRNA in the fat body." *Journal of Insect Physiology* 38: 461–474.

Huang, Z.-Y., and G. E. Robinson. 1992. "Honeybee colony integration: Worker-worker interactions mediate hormonally regulated plasticity in division of labor." *Proceedings of the National Academy of Sciences USA* 89: 11726–11729.

Huang, Z.-Y., G. E. Robinson, and D. W. Borst. 1994. "Physiological correlates of division of labor among similarly aged honey bees." *Journal of Comparative Physiology A* 174: 731–739.

Humphries, M. A., M. K. Fondrk, and R. E. Page. 2005. "Locomotion and the pollen hoarding behavioral syndrome of the honey bee (*Apis mellifera* L.)." *Journal of Comparative Physiology A* 191: 669–674.

Hunt, G. J., R. E. Page, M. K. Fondrk, and C. J. Dullum. 1995. "Major quantitative trait loci affecting honey bee foraging behavior." *Genetics* 141: 1537–1545.

Hunt, G. J., G. V. Amdam, D. Schlipalius, C. Emore, N. Sardesai, C. E. Williams, O. Rueppell, E. Guzmán-Novoa, M. Arechavaleta-Velasco, S. Chandra, M. K. Fondrk, M. Bege, and R. E. Page. 2007. "Behavioral genomics of honeybee foraging and nest defense." *Naturwissenschaften* 94: 247–267.

Kölliker, M., E. D. Brodie III, and A. J. Moore. 2005. "The coadaptation of parental supply and offspring demand." *American Naturalist* 166: 506–516.

Linksvayer, T. A. 2006. "Direct, maternal, and sibsocial genetic effects on individual and colony traits in an ant." *Evolution* 60: 2552–2561.

Linksvayer, T. A., and M. J. Wade. 2005. "The evolutionary origin and elaboration of sociality in the aculeate Hymenoptera: Maternal effects, sib-social effects, and heterochrony." *Quarterly Review of Biology* 80: 317–336.

Nelson, C. M., K. E. Ihle, M. K. Fondrk, R. E. Page, and G. V. Amdam. 2007. "The gene vitellogenin has multiple coordinating effects on social organization." *PLoS Biology* 5: 673–677.

Nijhout, H. F. 2003. "Development and evolution of adaptive polyphenisms." *Evolution and Development* 5: 9–18.

Page, R. E., J. Erber, and M. K. Fondrk. 1998. "The effect of genotype on response thresholds to sucrose and foraging behavior of honey bees (*Apis mellifera* L.)." *Journal of Comparative Physiology* A 182: 489–500.

Page, R. E., and J. Erber. 2002. "Levels of behavioral organization and the evolution of division of labor." *Naturwissenschaften* 89: 91–106.

Page, R. E., and M. K. Fondrk. 1995. "The effects of colony-level selection on the social organization of honey bee (Apis mellifera L.) colonies: Colony-level components of pollen hoarding." *Behavioral Ecology and Sociobiology* 36: 135–144.

Page, R. E., R. Scheiner, J. Erber, and G. V. Amdam. 2006. "The development and evolution of division of labor and foraging specialization in a social insect (*Apis mellifera* L.)." *Current Topics in Developmental Biology* 74: 253–286.

Pankiw, T. 2003. "Directional change in a suite of foraging behaviors in tropical and temperate evolved honey bees (*Apis mellifera* L.)." *Behavioral Ecology and Sociobiology* 54: 458–464.

Pankiw, T., and R. E. Page. 1999. "The effect of genotype, age, sex, and caste on response thresholds to sucrose and foraging behavior of honey bees (*Apis mellifera* L.)." *Journal of Comparative Physiology* A 185: 207–213.

Pankiw, T., and R. E. Page. 2000. "Response thresholds to sucrose predict foraging division of labor in honeybees." *Behavioral Ecology and Sociobiology* 47: 265–267.

Pankiw, T., and R. E. Page. 2001. "Genotype and colony environment affect honeybee (*Apis mellifera* L.) development and foraging behavior." *Behavioral Ecology and Sociobiology* 51: 87–94.

Pankiw, T., and R. E. Page. 2003. "Effects of pheromones, hormones, and handling on sucrose response thresholds of honey bees (*Apis mellifera* L.)." *Journal of Comparative Physiology* A 189: 675–684.

Pankiw, T., R. E. Page, and M. K. Fondrk. 1998. "Brood pheromone stimulates pollen foraging in honey bees (*Apis mellifera*)." *Behavioral Ecology and Sociobiology* 44: 193–198.

Pankiw, T., M. Nelson, R. E. Page, and M. K. Fondrk. 2004. "The communal crop: Modulation of sucrose response thresholds of preforaging honey bees with incoming nectar quality." *Behavioral Ecology and Sociobiology* 55: 286–292.

Pinto, L. Z., M. M. G. Bitondi, and Z. L. P. Simões. 2000. "Inhibition of vitellogenin synthesis in *Apis mellifera* workers by a juvenile hormone analogue, pyriproxyfen." *Journal of Insect Physiology* 46: 153–160.

Robinson, G. E. 1987. "Regulation of honey bee age polyethism by juvenile hormone." *Behavioral Ecology and Sociobiology* 20: 329–338.

Robinson, G. E. 1992. "Regulation of division of labor in insect societies." *Annual Review of Entomology* 37: 637–665.

Robinson, G. E., and E. L. Vargo. 1997. "Juvenile hormone in adult eusocial Hymenoptera: Gonadotropin and behavioral pacemaker." *Archives of Insect Biochemistry and Physiology* 35: 559–583.

Rueppell, O., T. Pankiw, D. Nielson, M. K. Fondrk, M. Beye, and R. E. Page. 2004. "The genetic architecture of the behavioral ontogeny of foraging in honeybee workers." *Genetics* 167: 1767–1779.

Rueppell, O., T. Pankiw, and R. E. Page. 2004. "Pleiotropy, epistasis and new QTL: The genetic architecture of honey bee foraging behavior." *Journal of Heredity* 95: 481–491.

Scheiner, R., M. Barnert, and J. Erber. 2003. "Variation in water and sucrose responsiveness during the foraging season affects proboscis extension learning in honey bees." *Apidologie* 34: 67–72.

Scheiner, R., J. Erber, and R. E. Page. 1999. "Tactile learning and the individual evaluation of the reward in honey bees (*Apis mellifera* L.)." *Journal of Comparative Physiology A* 185: 1–10.

Scheiner, R., R. E. Page, and J. Erber. 2001a. "Responsiveness to sucrose affects tactile and olfactory learning in preforaging honey bees of two genetic strains." *Behavioural Brain Research* 120: 67–73.

———. 2001b. "The effects of genotype, foraging role, and sucrose responsiveness on the tactile learning performance of honey bees (*Apis mellifera* L.)." *Neurobiology of Learning and Memory* 76: 138–150.

———. 2004. "Sucrose responsiveness and behavioral plasticity in honey bees (*Apis mellifera*)." *Apidologie* 35: 133–142.

Scheiner, R., A. Weiß, D. Malun, and J. Erber. 2001. "Learning in honey bees with brain lesions: How partial mushroom-body ablations affect sucrose responsiveness and tactile learning." *Animal Cognition* 4: 227–235.

Schulz, D. J., T. Pankiw, M. K. Fondrk, G. E. Robinson, and R. E. Page. 2004. "Comparison of juvenile hormone hemolymph and octopamine brain titers in honey bees (Hymenoptera: Apidae) selected for high and low pollen hoarding." *Annals of the Entomological Society of America* 97: 1313–1319.

Socha, R., J. Sula, D. Kodrík, and I. Gelbic. 1991. "Hormonal control of vitellogenin synthesis in *Pyrrhocoris apterus* (L.) (Heteroptera)." *Journal of Insect Physiology* 37: 805–816.

Sullivan, J. P., O. Jassim, S. E. Fahrbach, and G. E. Robinson. 2000. "Juvenile hormone paces behavioral development in the adult worker honey bee." *Hormones and Behavior* 37: 1–14.

Sullivan, J. P., S. E. Fahrbach, J. F. Harrison, E. A. Capaldi, J. H. Fewell, and G. E. Robinson. 2003. "Juvenile hormone and division of labor in honey bee colonies: Effects of allatectomy on flight behavior and metabolism." *Journal of Experimental Biology* 206: 2287–2296.

Tsurada, J., G. V. Amdam, and R. E. Page. In Press. "Sensory response system of social behavior tied to female reproductive tracts." *PLOS One.*

Vaughan, D. M., and N. W. Calderone. 2002. "Assessment of pollen stores by foragers in colonies of the honey bee, *Apis mellifera* L." *Insectes Sociaux* 49: 23–27.

West-Eberhard, M. J. 1987a. "The epigenetical origins of insect sociality." In J. Eder and H. Rembold, eds., *Chemistry and biology of social insects,* 369–372. Munchen: Verlag J. Peperny.

West-Eberhard, M. J. 1987b. "Flexible strategy and social evolution." In Y. Itô, J. L. Brown, and J. Kikkawa, eds., *Animal societies: Theories and fact,* 35–51. Tokyo: Japan Scientific Societies Press.

West-Eberhard, M. J. 1996. "Wasp societies as microcosms for the study of development and evolution." In S. Turillazzi and M. J. West-Eberhard, eds., *Natural history and evolution of paper-wasps,* 290–317. New York: Oxford University Press.

Wheeler, D. E. 1986. "Developmental and physiological determinants of caste in social Hymenoptera: Evolutionary implications." *American Naturalist* 128: 13–34.

Wolf, J. B., and E. D. Brodie III. 1998. "The coadaptation of parental and offspring characters." *Evolution* 52: 299–308.

Zera, A. J., and J. Bottsford. 2001. "The endocrine genetic basis of life history variation: The relationship between the ecdysteroid titer and morph specific reproduction in the wing polymorphic cricket *Gryllus firmus.*" *Evolution* 55: 538–549.

CHAPTER SEVENTEEN

Social Brains and Behavior—Past and Present

WULFILA GRONENBERG

ANDRE J. RIVEROS

SOCIAL BEHAVIOR IS AMONG the most remarkable of evolutionary advances (Maynard-Smith and Szathmary 1995), and group life is a phylogenetically widespread behavioral phenomenon that has led to numerous changes at the individual level. Because the nervous system generates and controls behavior, the evolution of social life presumably incorporates both previous and current adaptations of the neural substrate, including structural and physiological changes of the brain. The consequence easiest to measure is a change in brain size.

In the vertebrate literature since the 1950s, the "social brain hypothesis" has emerged as an alternative to the ecological hypothesis' explanation of the allometrically unexpected large brains of primates ("encephalization," Jerison 1973; Barton and Harvey 2000). The social brain hypothesis suggests that the intelligence of primates, especially their increased neocortex size, is primarily an adaptation to the special complexities of primate social life (Dunbar 2003). This theory is supported by the correlation between social complexity (often measured as group size) and neocortex size in many taxa such as primates (Dunbar and Bever, 1998), bats (Barton and Dunbar 1997), dolphins (Marino 1996), and birds (Burish, Kueh, and Wang 2004). The basic argument of the social brain hypothesis is that the need to deal with complex information required for social organization is the main selective pressure favoring an increase in neocortex size (Dunbar and Bever 1998; Adolphs 2003).

Relevance of the Social Brain Hypothesis
for Social Insects

Although some of the most advanced animal societies are found among insects, a theory comparable to the social brain hypothesis in vertebrates does not exist for social insects. This probably reflects the major difference in the patterns of social evolution: in mammals, evolution results in individualized societies where social interactions require the recognition of an individual's behavior and a memory of previous encounters (de Waal and Tyak 2003). Hence, social vertebrates require large brain capacities for processing additional and more complex information related to cognitive aspects of their social life. In contrast, in large eusocial colonies social insect evolution has led to a substantial reduction in individual behavioral repertoires associated with individual specialization. Nevertheless, it traditionally has been accepted that social interactions and communication require advanced neuronal processing power in social insects as well. For example, generating or "reading" a honey bee waggle dance certainly requires additional sensory, learning, and motor skills in comparison to solitary insects.

Task specialization is the hallmark of insect societies. One would expect a concomitant specialization of social insect brains and, as specialized members of advanced social insect species are less pluripotent than solitary insects, the size of their brains may actually be reduced compared to solitary insects of the same size. We propose therefore that two opposing tendencies affect brain size evolution in social insects (Figure 17.1). As insects advance from solitary to communal life we expect an increase in brain size similar to that found in social vertebrates. However, the evolution of more complex societies with an increase in individual task specialization should lead to a decrease in brain size.

Despite this general tendency for brain-computing power requirement, evolution does not act on overall brain size *per se* in either vertebrates or insects. Instead, evolution increases brain components that underlie specific sensory, motor, or cognitive skills particularly important in a social environment and in turn reduce brain components that underlie tasks that the respective individuals do not have to perform. For example, on the one hand, workers in general may reduce neural tissue involved in courtship and reproduction, and a leaf-cutting ant worker dedicated to tending fungus may not need sensory and motor skills and the brain components

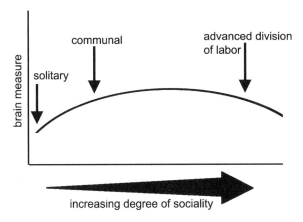

Figure 17.1. Sketch indicating the hypothetical correlation between brain measures (e.g., overall brain size, size of particular brain compartments, or number of neurons) and degree of a species' social organization. As basal social systems evolve from solitary insects, the brain is thought to increase in size to allow processing of social signals and more complex interactions. In species with advanced division of labor, individuals can be highly specialized on certain tasks and thus do not need the full behavioral repertoire of the species. The reduction in repertoire is hypothesized to go along with reduction in brain measures.

required for outside foraging. On the other hand, the sophisticated cognitive abilities of some social insects require that brain components underlying advanced social behavior such as communication should be enlarged or otherwise enhanced. We would, therefore, expect differential expansion or reduction of particular brain components in different social insect castes and subcastes and in species with more or less worker task specialization.

How does brain size vary with sociality in insects? In the mid-18th century, a general trend was discovered for basal taxa across the animal kingdom (most of which are solitary) to have smaller brains compared to more advanced crown groups. The evolutionary increase of brain size and complexity in general allows animals to better cope with and exploit complex environments (Bernays and Wcislo 1994; Roth and Dicke 2005). Accordingly, advanced aculeate Hymenoptera, which include all social Hymenoptera, have considerably larger brains or brain components than the more basal Hymenoptera (symphyta; Figure 17.2c; Flögel 1878; von Alten 1910; Hanström 1928).

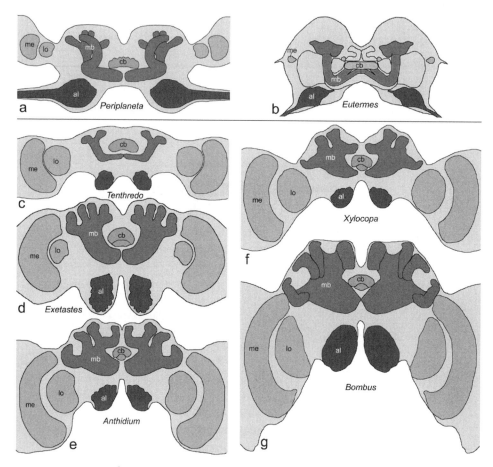

Figure 17.2. Schematic drawings of brains of a cockroach (a), a termite (b), a sawfly (c), an ichneumonid wasp (d), a leaf-cutting bee (e), a carpenter bee (f) and a bumblebee (g). Major brain neuropils indicated: medulla (me), lobula (lo), antennal lobe (al), central body (cb), mushroom body (mb). Not drawn to scale. Based on (a) Strausfeld et al. 1998; (b) Kühnle 1913; (c-f) von Alten 1910.

Besides phylogenetic correlations, relative brain size also correlates with body size, such that brains of larger animals generally make up a smaller portion of overall body weight. This would suggest that brains are relatively more costly for smaller species or individuals, as maintaining brain tissue is energetically costly (Aiello and Wheeler 1995; Laughlin, de Ruyter van Steveninck, and Anderson 1998; Laughlin 2001). This relationship has been repeatedly examined in vertebrates (e.g., Jerison 1969) and also holds for

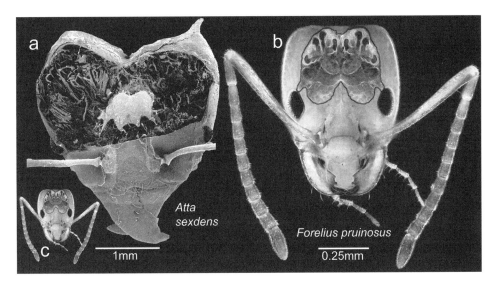

Figure 17.3. Head and brain size in a relatively large ant (*Atta sexdens*) major worker and a relatively small ant (*Forelius pruinosus*). (a): Scanning electron micrograph of opened head capsule; (b): montage of head photograph and brain micrograph; image b is reduced in c to match magnification of a. Note different scale bars in a and c versus b.

social insects (e.g., ants; Cole 1985; Jaffe and Perez 1989; Wehner, Tsukasa, and Isler 2007). The brain of the small ant *Forelius pruinosus* (Figures 17.3b, 17.3c) makes up a substantial part (37%) of the overall head volume, whereas in a larger worker of the ant *Atta sexdens* (Figure 17.3a) the brain is just a small fraction (less than 5%) of the head volume. Obviously, differences in brain size are even more pronounced when comparing large social bees (e.g., honey bees or bumblebees, Figure 17.4a, 17.4c) with ants (Figure 17.4d, 17.4f). Generally, brains of social insects range between less than 0.4% (large bumblebees) to more than 5% of body weight (small ants; Figure 17.3b), with the highest published number of 31% for an ant (Jaffe and Perez 1989). For comparison, in mice, the brain weighs about 3.2% of the body; in cows, roughly. 0.1%; and in fin whales, less than 0.01%.

Structure and Function of Social Insect Brains

Studies on insect brains mostly deal with general neurobiological aspects and only a small portion address social questions. Of the social insects, honey bees (Figure 17.3a) have been the primary focus of

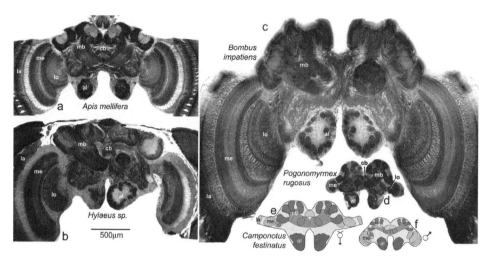

Figure 17.4. Hymenopteran brains, frontal views. (a) Honey bee worker (*Apis mellifera*); (b) yellow-faced bee (solitary bee, *Hylaeus sp.*); (c) bumblebee (*Bombus impatiens,* large worker) (montage of two sections); (d) seed harvesting ant (*Pogonomyrmex rugosus,* worker); (e) male (f) carpenter ant (*Camponotus festinatus*). In (b) to (d) left and right half of the images represent different depths. Optic lobes lamina (la), medulla (me), lobula (lo); antennal lobe (al); central body (cb); mushroom body (mb). osmium-stained material (a–d); sections shown in (b) courtesy of Birgit Ehmer. Note that optic lobes are small in ants.

neuro-ethological studies because of their elaborate behavioral repertoires, including communication and learning, their year-round availability, and their brain size, which is amenable to physiological studies. Fewer studies have been concerned with the brains of other social insects, such as wasps (Figure 17.2c, 17.2d, 17.4b), ants (Figure 17.4d–f), and termites (Figure 17.2b), and what we do know about their brains is mostly from early descriptive anatomical studies (e.g., Flögel 1878; von Alten 1910). Except for a comparative study on the physiological bases for color vision in different bee genera (Peitsch 1992) and electrophysiological work on neurons involved in a specialized mandible reflex in ants (Gronenberg 1995; Just and Gronenberg 1999), neurophysiological experiments have been performed only on honey bee brains (e.g., Homberg 1985; Hammer 1993; Galizia and Menzel 2001).

The brain (in the wider sense) comprises a matrix of neuronal processes into which several distinct compartments are embedded: the optic lobes process visual information, the antennal lobes process olfactory information, and the mushroom bodies and the central complex are higher integrative centers. While there are some quantitative differences among the brains of bees, ants, and social wasps, the principal design of their brains is similar (Figure 17.2, 17.4). One distinction in most ant workers is that the eyes and visual parts of the brain are reduced (Figure 17.4d–f). The anatomy of termite brains differs significantly from that of Hymenoptera; workers lack eyes and visual neuropiles (Figure 17.2b). Termite brains more closely resemble those of cockroaches (Figure 17.2a; Hanström 1928).

As mentioned above, overall brain size may give some clues as to learning capabilities and cognitive abilities in related species when corrected for the body-size trend (Rensch 1956). Particular brain components, such as the cerebral cortex in mammals, show a better correlation with behavioral or cognitive abilities than overall brain size (Rensch 1956; Jerison 1973). A comparison of Hymenopteran brains (Figure 17.2c–g) shows that the visual and olfactory brain centers can be quite substantial in the basal symphyta (Figure 17.1c), while the mushroom body is much more prominent in aculeate crown groups and in particular in social Hymenoptera (Figure 17.1g).

Sensory Processing

Social insects are best known for their extensive pheromone communication, which, besides multiple pheromone glands, requires an advanced olfactory system to decode not only ordinary odors but also pheromone messages to bring about appropriate behavior (Peeters and Liebig, this volume). Briefly, general odors and pheromones are perceived by sensory neurons in the antennae, which send their axons into the antennal lobes (Figure 17.2, 17.4). Antennal lobes comprise spherical subunits called glomeruli that represent particular aspects of an odor or pheromone components (Kleineidam and Rössler, this volume). While pheromone communication is paramount across social insects, pheromone processing pathways are not well understood in social insects (Peeters and Liebig, this volume) and have been studied mainly in moths (Christensen and Hildebrand 2002; but see Kleineidam and Rössler, this volume).

The use of pheromones is a hallmark of social insect communication, but there are also sophisticated behaviors in some social insects that are based on visual cues in part or entirely. Visual orientation and landmark recognition are particularly important for social insects that rely on navigation to exploit often remote food sources and return to central nests; bees and wasps find their nests visually (Tinbergen 1958). The use of landmarks and the polarization pattern of the sky has been most extensively studied in honey bees (Wehner and Menzel 1990; Collett and Collett 2000) and some ants (e.g., Hölldobler 1980; Fukushi 2001; Banks and Srygley 2003; Wehner 2003). Honey bees "report" to nestmates the estimated distances and directions to food sources during the waggle dance (von Frisch 1967) and use visual flow field pattern to estimate flight velocity, distances flown, and angles with respect to the sun position (Esch and Burns 1996; Si, Srinivasan, and Zhang 2003; Chittka 2004). Novice bees have to learn and to remember their environment, starting with orientation flights at successively increasing distances from the nest until they actually start foraging (Capaldi and Dyer 1999), and ant foragers remember their nest environment in the same way (Jander 1957; Rosengren 1971). Honey ants (*Myrmecocystus*) likewise rely on visual cues when judging the strength of opponent colonies during tournaments (Hölldobler and Wilson 1990). Recently, it has been discovered that paper wasps (*Polistes dominulus, P. fuscatus*) are able to discriminate individual facial patterns (Tibbetts 2002, 2004). This ability is involved in nestmate recognition or assessment of nestmate dominance and is reminiscent of primate face recognition.

As suggested by the many different visual cues used by social insects, processing of visual information is complex and requires major neuronal resources. Bees, wasps, and some ants have prominent optic lobes (the lamina, the medulla, and the lobula, respectively), whose architecture has been known since the late nineteenth century (Kenyon 1896; reviewed by Strausfeld 1989). Visual interneurons show complex responses such as color-opponent (Hertel 1980; Yang, Lin, and Hung 2004) or movement-specific responses (Ibbotson 1991), and response properties of individual neurons may change depending on where in the visual field a stimulus occurs or how often it has been presented (Paulk and Gronenberg 2005). Nevertheless, we are still far from being able to explain even simple visual behaviors in terms of their underlying neuronal substrate at any level of the brain.

Advanced Multisensory Integration

Two brain components are involved in higher order processing, the central body and the mushroom bodies. The central body (Figure 17.2, 17.4) is thought to coordinate movement of left and right legs (Strauss and

Heisenberg 1993) and is particularly elaborate in insects that perform complex leg movements (Strausfeld 1999) such as comb- or nest-building. Navigational control has been suggested as another function, based on the responses of some central complex neurons to polarized light (Vitzthum, Muller, and Homberg 2002) and the correlation of some central body neurons' discharge frequency with flight activity (Homberg 1994). Recent evidence suggests that it may also be involved in visual pattern recognition and short-term memory (Liu et al. 2006). The central complex appears to have connections to all major parts of the brain and in honey bees its neurons respond to stimuli of different modalities, including visual, mechanosensory, gustatory, and olfactory (Homberg 1985). The central complex is large in Hymenoptera, but there is no evidence that its size or structure correlate with social organization *per se.*

Mushroom bodies are central structures in the arthropod brain that process multimodal information. They were first described in solitary and social Hymenoptera by Dujardin (1850), who suggested that these structures endowed the insects with free will (as opposed to instinctive actions). In general, mushroom bodies were thought to be associated with the sophistication of behavioral repertoires and with intelligence, and later with learning and memory, by many authors in the late nineteenth and early twentieth centuries (reviewed by Strausfeld et al. 1998). These ideas were based on anatomical comparison of mushroom bodies, particularly between solitary and social Hymenoptera (von Alten 1910; Pietschker 1911; Ehmer and Hoy 2000; reviewed by Strausfeld et al. 1998).

Mushroom bodies (Figure 17.5) are composed of many intrinsic neurons (about 170,000 in honey bees; Witthöft 1967) called Kenyon cells. Their cell bodies reside around neuropiles referred to as calyces. The calyx is composed of the Kenyon cells' dendrites and terminal branches of neurons, which supply input to and modulate the Kenyon cells. In bees, ants, and social wasps, and also in cockroaches, the calyces are enlarged and form cup-shaped structures (Figures 17.2, 17.4, 17.5a); in sawflies the calyces are shallow (Figure 17.2c). In a comparative study, Jaffe and Perez (1989) reported that ant species with more advanced social systems had mushroom bodies that were "better developed." However, their study estimated brain volumes with little accuracy and showed several conflicting trends. The current hypothesis (Farris and Roberts 2005) is that mushroom body elaboration coincides with rich behavioral repertoires, but not necessarily with a species' social advancement.

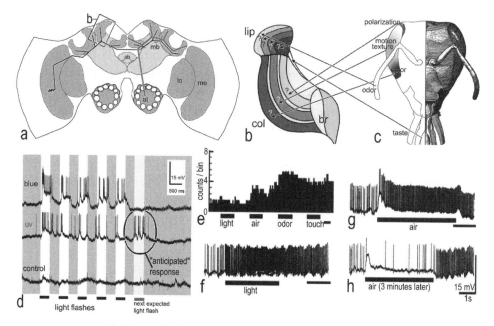

Figure 17.5. Mushroom bodies. a: Visual input from the medulla (me) and the lobula (lo) and olfactory input from the antennal lobe (al) converge onto the mushroom body's (mb) calyx. B: The calyx—enlargement from (a)—comprises the lip, collar (col) and basal ring (br) regions, which receive segregated inputs from the sensory structures shown in C. C: different parts of the eye support polarization vision, motion/texture discrimination, and color vision. It is not known if these different sensory qualities are processed by different layers in the collar *(question marks)*. Taste information is processed by a narrow band on the inner face of the calyx, and olfactory information projects to the lip region. The function of the different layers in the lip is unknown *(question marks)*. D: Complex response of a visual input neuron to the mushroom body calyx recorded in a bumblebee (courtesy of Angelique Paulk). After a series of five flashes of ultraviolet (UV) light, this neuron anticipates another stimulus: it shows a response (circle) even though no stimulus is presented. This response is specific for UV light, it does not occur with blue light stimulation. E–G: Responses of an output neuron from the mushroom body. Initially (E) the neuron does not respond to light, but after several stimulations (not shown), light triggers a strong, long-lasting excitatory response (F). G: Air currents originally evoke a similar kind of excitation, but after three minutes the same stimulus evokes an inhibition followed by an excitation after the stimulus is switched of (H).

In most insects, the calyx receives olfactory input from the antennal lobes (Figure 17.5a–c; Strausfeld et al. 1998). In advanced Hymenoptera, but not in other insects, the mushroom bodies also receive massive visual input from the medulla and the lobula (Figure 17.5a–c; Mobbs 1982; Gronenberg 1999; Ehmer and Gronenberg 2002). This ability is assumed to contribute to their advanced visual orientation and learning abilities. Visual and olfactory input is segregated within the calyx (Mobbs 1982; Gronenberg 1999) and visual input is again represented by different classes of input neurons that terminate in distinct layers (Figure 17.5b). Behavioral studies (Lehrer 1998) suggest that different visual submodalities such as color, movement, and pattern are served by different regions of the eye (Figure 17.5c). It is therefore possible that layers in the calyx' collar represent these different visual submodalities (Figure 17.5b, 17.5c; Ehmer and Gronenberg 2002), but no physiological evidence is available to support or reject this idea.

In ants, visual input to the mushroom bodies is less prominent and less differentiated as compared to social bees and wasps (Ehmer and Gronenberg 2004). Instead, in ants, olfactory mushroom body input is more pronounced and is layered in the calyx' lip region (Figure 17.5b, 17.5c; Gronenberg and Lopez-Riquelme 2004). This corresponds to the greater biological significance of olfaction for most ant species in comparison to bees or wasps.

While the calyx is an exclusive input region, the mushroom body peduncle and lobes give rise to output to and receive additional input from the surrounding brain areas. The current theory is that different kinds of sensory input to the calyx can modulate the information processing performed by the mushroom bodies in a context-specific way (Strausfeld 2002). Mushroom body neurons probably take into account the immediate history and stimulus context and may reflect, or be involved in, learning processes (Mauelshagen 1993; Grünewald 1999). Given the complex response characteristics of mushroom body neurons (Figure 17.5d–h), we are far from understanding what exactly the mushroom bodies do even in "model insects" (e.g., *Drosophila melanogaster* or *Apis mellifera*).

In basal termites (e.g., *Zootermopsis*) the mushroom bodies are similar to those of cockroaches (Figure 17.1a, 17.1b). Cockroaches and termites are phylogenetically related and have large mushroom body lobes (Kühnle 1913; Burling-Thompson 1916; Hanström 1928; Howse and Williams 1969), indicating that they comprise a large number of Kenyon cells. This would allow termites to process more information in parallel, perhaps resulting in more context-specific responses. However, one has to take into account that termite brains are small in absolute terms. A termite workers' brain occupies less space compared to an ant of similar head size (Figure 17.2b). Interestingly, the most derived termite taxa (Cubotermitidae and Apicotermitidae) have particularly small Kenyon cell numbers (Howse and Williams 1969; Figure 17.6). This supports the idea that division of labor in advanced social insects may go along with a decrease of mushroom body size as the behavioral complexity of the individual worker may be reduced analogous to ants and bees.

The mushroom bodies play important roles in learning and memory. Cooling or chemically ablating the mushroom bodies of honey bees can interrupt the formation of associative memory (Erber, Masuhr, and Menzel 1980; Komischke et al. 2005), and the involvement of mushroom bodies in place memory has been explored by lesion experiments in cockroaches (Mizunami, Weibrecht, and Strausfield 1998). Many neurons probing the mushroom bodies show context-specific responses or activity that reflects the

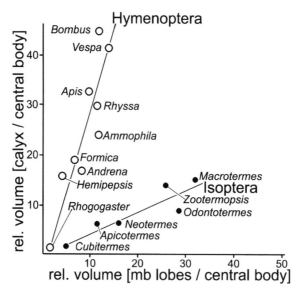

Figure 17.6. Relative size of the mushroom body calyx (ordinate) and lobes (abscissa) in different Hymenopteran and Isopteran (termite) species. Mushroom body sizes shown relative to the size of the central body, not of the overall brain. Based on data by Howse and Williams (1969).

history of stimulus sequences (Figure 17.5d–h) which would be important for learning (Li and Strausfeld 1997), or show response changes that coincide with ongoing learning acts (Mauelshagen 1993; Grünewald 1999).

Taken together, the anatomical, physiological, and behavioral facts all suggest the involvement of mushroom bodies in the processing of context-specific information that is important for navigation, foraging, and learning behavior. Among the Hymenoptera, mushroom bodies are much more prominent in aculeate crown groups, in particular in social species, than they are in the stem group symphyta, where mushroom bodies feature small calyces and thin lobes (Figure 17.2c). Thus mushroom body size and sophistication go along with the complexity of a species' behavioral repertoire, similar to the increase in neocortex size found in advanced primates (reviewed in Dunbar 2003). However, the finding that the most socially advanced termites have fewer Kenyon cells than more basal species suggests that the trend of brain size reduction in highly specialized individuals (Figure 17.1) also applies to mushroom bodies.

Division of Labor and Brain Specialization

In principle, division of labor should translate to a reduction in brain capacity (overall brain size, number or complexity of neurons) as specialized individuals need to perform fewer behaviors and may not require a brain as pluripotent as generalists or solitary insects. In advanced social insects this would lead to a reversal in the general correlation between body and brain size. Further, it has been shown for some insect species that specialists can make faster and more correct decisions than generalists, presumably because fewer choices require less brain capacity (Bernays and Funk 1999) or less elaborate mushroom bodies (Farris and Roberts 2005). Specialists' reduced need for neuronal computations and the high metabolic costs of brain tissue maintenance probably explain why honey bee queens have smaller brains than workers. Queens are highly specialized for reproduction and, since they found their colonies with the help of thousands of workers, they do not need to perform many of the tasks that honey bee workers do. The reduced mushroom body size in higher termites or in highly social ant species with polymorphic worker castes (Jaffe and Perez 1989) may be explained by the same principle; namely, that in species with a more advanced division of labor individual workers need only be able to perform reduced behavioral repertoires (Figure 17.1).

Besides pheromones and tactile communication, little is known about control of the social aspects of behavior. A particularly remarkable visual capacity has been recently discovered in paper wasps *(Polistes dominulus)*, where face markings signal an individual's position in the dominance hierarchy (Tibbetts 2004). Even more strikingly, in *Polistes fuscatus,* colony members can recognize each other individually by their different face markings (Tibbetts 2002). The underlying neuronal mechanisms are not known and it appears that no additional brain volume is required to perform the visual face recognition task (Gronenberg, Ash and Tibbetts 2008). In contrast, multiple foundresses in paper wasps do show a slight enlargement of particular brain substructures compared to solitary foundresses (Ehmer, Reeve, and Roy 2001).

Differences in brain size that determine the behavioral repertoires of reproductives, workers, soldiers, and so forth are established during larval development and are fixed during the adult stage, although some brain plasticity also occurs during the adult stage. In particular: in bumblebees (Figure 17.7), honey bees (Figure 17.8a; Withers, Fahrbach, and Robinson 1993; Durst, Eichmuller, and Mencel 1994), and carpenter ants (Figure 17.7, 17.8c;

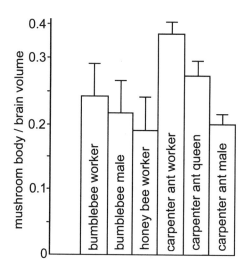

Figure 17.7. Relative mushroom body size (mean and standard deviation) in bumblebee *(Bombus impatiens)* workers and males; honey bee *(Apis mellifera)* workers; and carpenter ant *(Camponotus festinatus)* workers, queens, and males. All differences within species statistically significant.

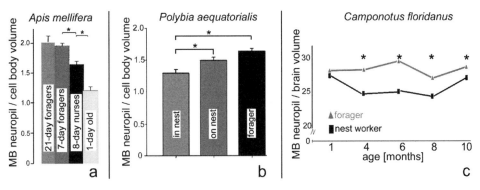

Figure 17.8. Mushroom body plasticity in (a) honey bees, (b) paper wasps, and (c) carpenter ants. In honey bees, the relative volume of mushroom body (MB) neuropil increases with age and is also larger in foragers compared to nurses of the same age (a). In paper wasps, the relative mushroom body neuropil volume of inside-nest workers is significantly smaller than it is in on-nest workers and foragers (b). In carpenter ants, the relative mushroom body neuropile volume is significantly larger in foragers aged 4 months and older compared to same-age inside-nest workers. Based on data by Withers, Fahrbach, and Robinson 1993 (a); O'Donnell et al. 2004 (b); and Gronenberg, Heeren, and Hölldobler 1996 (c).

Gronenberg, Heeren, and Hölldobler 1996) in all three species inside-nest workers have smaller mushroom bodies than foragers. The same has been described for brains of social insects that have a less highly advanced division of labor (*Polybia* paper wasps), where the mushroom bodies are relatively larger in foragers compared to nest workers (Figure 17.8b; O'Donnell, Donlan, and Jones 2004). The hypothesis in all these cases is that less sensory information needs to be integrated and less demanding tasks have to be solved by workers inside the restricted nest, as compared to their foraging sisters who have to navigate the environment and remember the location of nest and food sources. Specialists also might have enlarged brain components that serve the specific task that the individual is specialized for, which is common across animal species in general (reviewed for vertebrates, Dukas 2004) but has not been studied in social insects.

Behavioral Transitions and Brain Plasticity

Learning and the modification of behavior by experience is advantageous for any animal, but it may be particularly important for social central-place foragers. Unlike many other insects, these foragers need to remember their nest, which they usually revisit many times per day, and in many cases they have to find and revisit food sources using different visual, olfactory, proprioceptive, and/or magnetic cues, relying on some kind of representation of space. They also have to recognize their nestmates (i.e., learn a chemical profile). In addition, honey bees communicate distances and directions to food sources via their dance language; therefore both dancer and follower need to store this information. It is not surprising that insect learning and memory were demonstrated initially in honey bees (von Frisch 1914). Moreover, as colonies can be long-lived, workers have to adapt to changing food requirements and supply conditions. This behavioral flexibility is controlled by the brain, which suggests that social insect brains may be characterized by pronounced neuronal plasticity. Indeed, most studies on insect brain plasticity have been performed on social insects (other than fruit flies).

We have already mentioned studies showing that mushroom bodies are larger in foragers than in nest workers. In such cases the difference is the result of brain plasticity: young workers first work inside the nest and later become foragers (temporal polyethism). This behavioral transition goes along with an increase in mushroom body neuropil while the rest of the brain does not change significantly. The same transition has been shown recently for minor workers of several species of *Pheidole* ants. As the ants get older and mature, workers add more tasks to their behavioral repertoires (Seid and Traniello 2006); these changes are paired with an increase in the overall size and an enlargement of particular synapses of their mushroom bodies (Seid and Traniello 2005). In honey bees this increase in mushroom body volume has two components: one is developmentally (hormonally) controlled and independent of sensory experience and allows the nest worker to mature and get ready to perform foraging tasks ("experience-expectant" plasticity, Fahrbach et al. 1998); the other results from foraging experience and requires that bees actually fly out and learn to navigate the environment (Withers, Fahrbach, and Robinson 1993, 1995; Fahrbach et al. 1998).

A reverse kind of brain plasticity has been found in harvester ants (*Pogonomyrmex rugosus* and *Messor pergandei*). As in most ants, virgin females of these species have larger eyes and optic lobes than workers because they rely on vision during their nuptial flight. Once they are mated, have shed their wings, and are prepared for their underground life as queens, their optic lobes shrink considerably (Julian and Gronenberg 2002). This process likely helps save energy that would otherwise be required for the maintenance of metabolically expensive brain tissue (Laughlin, de Ruyter van Steveninck, and Anderson 1998). The same phenomenon has been found in another ant species *(Harpegnathos saltator)* in which workers can become functional substitute queens (gamergates), a transformation that goes along with a reduction of the optic lobes without an increase in other parts of the brain (Gronenberg and Liebig 1999).

Conclusions and Future Directions

In social insects, the overall behavioral complexity of a colony results from a combination of the individual workers' brain capabilities and some as yet unknown behavioral capacities that emerge from the fact that these individuals (hence their brains) can communicate with each other. Therefore, important information can be distributed across several brains and does not need to be stored and processed all within one brain. Counterintuitively, an individuals' brain might decrease in overall size with a species' increasing degree of sociality because an increase in the division of labor results in greater individual specialization. Such an expected general decrease in brain capacity in social insects (sketched in Figure 17.1) has not yet been shown. Good systems to examine this process should include closely related solitary and social species, as well as facultative social species. Alternatively, one might compare brains of ant or termite species with distinct soldier castes where smaller workers have much larger behavioral repertoires than soldiers (Brown and Traniello 1998).

In primates, the size of the cerebral cortex increases with social complexity because innovation, tool use, social learning, and other social interactions require enhanced brain storage and processing capacities (Reader and Laland 2002). Do analogous requirements exist in social insects? How are social insect brains equipped to support social communication? We do not know, though we can expect that olfactory processing will be involved since social insects employ more kinds of pheromones than do solitary insects. Nestmate recognition and the signaling of reproductive or dominance status are important for the cohesion of a colony, and in most cases studied these are based on chemical cues, often involving cuticular hydrocarbons (Beekman 2004; Lorenzi et al. 2004). Therefore, the antennal lobes of social insects, which process olfactory and pheromone information, are usually large. Most social insects also communicate by tactile information and one might expect the mechanosensory centers to be large. Thus far no studies have been done comparing antennal lobes or mechanosensory centers in solitary and social insects. Such research would perhaps reveal specializations of the olfactory systems that allow social insects to discriminate small variations in very complex hydrocarbon mixtures.

Most of the known or expected differences between the brains of solitary and social insects relate to certain sensory abilities or requirements. Nothing is known about possible differences with respect to the motor output. Do bees and wasps require particular motor skills when building comb or complex nests? Do ants that collectively retrieve large prey or food items require particular motor control? When weaver ants form long-living chains and layers comprising hundreds of individuals, or when army ants build bivouacs composed of millions of ants, do they rely on particular motor coordination? We also know little about the assignment of different tasks to different individuals, other than the sensory threshold hypotheses covered by multiple chapters in this volume.

Most of what we have said so far suggests that social insects do not have a specific "social brain," in contrast to mammals such as primates. While the mushroom bodies are particularly well developed in the social insects analyzed so far (termites, ants, social wasps, and bees), the significance of this correlation has yet to be revealed and studies need to be extended to other social insects (e.g., social aphids or thrips). Volumetric and anatomical measurements will only give some hints at differences between solitary and social brains. Just as the physical size of a computer or integrated circuit does not reveal much about its capacity, one will have to look at functional differences when comparing brains. New techniques, such as imaging with voltage-sensitive or calcium-sensitive dyes (Galizia et al. 2000) or the application of multi-channel electrophysiological recording techniques (Lei, Christensan, and Hildebrand 2004), may reveal aspects of social insect brains that are as yet hidden. Such properties might not be structural, but may affect the plasticity, adaptability, or synchronization of the brains of many members in a group. The difficulty with physiological measurements is that generally the animals cannot be examined in their natural context. Hence, if any particular aspects of brain activity are enhanced by or require social contact, even the most sophisticated physiological techniques may not reveal them if they rely on invasive procedures.

We may be able to gain more insights into how insect brains control social behavior using advanced molecular and genetic techniques like immunostaining, RNAi, or caged second messengers. We are only beginning to understand how hormones and neuromodulators control brain functions. In honey bees, biogenic amines, which act as neurotransmitters or neuromodulators, are involved in the control of division of labor (Schulz and Robinson 1999, 2001). Likewise, the brains of ant workers (*Pheidole dentata*) contain more biogenic amines as they mature (Seid and Traniello 2005). What does neuromodulation mean at the level of individual neurons? Are the neurons' effects enhanced? Do they exert control over more postsynaptic neurons? Or do postsynaptic neurons reduce their sensitivity to the neuromodulators such that connections become more specific (pruned)? The latter is suggested (for neurons in the mushroom bodies) by studies on honey bees (Farris, Robinson, and Fahrbach 2001) and ants (Seid et al. 2005), however, more studies using different approaches are needed to understand what happens at the neuronal level when social insects mature, learn, and make new experiences.

Literature Cited

Adolphs, R. 2003. "Investigating the cognitive neuroscience of social behavior." *Neuropsychologia* 41: 119–126.

Aiello, L. C., and P. Wheeler. 1995. "The expensive tissue hypothesis: The brain and the digestive system in human and primate evolution." *Current Anthropology* 36: 199–221.

Banks, A. N., and R. B. Srygley. 2003. "Orientation by magnetic field in leaf-cutter ants, *Atta colombica* (Hymenoptera: Formicidae)." *Ethology* 109: 835–846.

Barton, R. A., and R. I. M. Dunbar. 1997. "Evolution of the social brain." In A. Whiten and R. W. Byrne, eds., *Machiavellian intelligence II*, 240–263. Cambridge: Cambridge University Press.

Barton, R. A., and P. H. Harvey. 2000. "Mosaic evolution of brain structure in mammals." *Nature* 405: 105–108.

Beekman, M. 2004. "Is her majesty at home?" *Trends in Ecology and Evolution* 19: 505–506.

Bernays, E. A., and D. Funk. 1999. "Specialists make faster decisions than generalists: Experiments with aphids." *Proceedings of the Royal Society London B* 266: 1–6.

Bernays, E. A., and W. T. Wcislo. 1994. "Sensory capabilities, information processing, and resource specialization." *Quarterly Review of Biology* 69: 187–204.

Brown, J., and J. F. A. Traniello. 1998. "Regulation of brood-care behavior in the dimorphic castes of the ant *Pheidole morrisi* (Hymenoptera: Formicidae): Effects of caste ratio, colony size, and colony needs." *Journal of Insect Behavior* 11: 209–219.

Burish, M. J., H. Y. Kueh, and S. S. Wang. 2004. "Brain architecture and social complexity in modern and ancient birds." *Brain, Behavior and Evolution* 63: 107–124.

———. 1916. "The brain and the frontal gland of the castes of the 'white ant,' *Leucotermes flavipes, kollar.*" *Journal of Comparative Neurology* 26: 553–603.

Capaldi, E. A., and F. C. Dyer. 1999. "The role of orientation flights on homing performance in honeybees." *Journal of Experimental Biology* 202: 1655–1666.

Chittka, L. 2004. "Dances as windows into insect perception." *PLoS Biology* 2: 898–900.

Cole, B. J. 1985. "Size and behavior in ants: Constraints and complexity." *Proceedings of the National Academy of Science USA* 82: 8548–8551.

Collett, M., and T. S. Collett. 2000. "How do insects use path integration for their navigation?" *Biological Cybernetics* 83: 245–259.

de Waal, F. B. M., and P. L. Tyack. 2003. *Animal social complexity: Intelligence, culture, and individualized societies.* Cambridge: Harvard University Press.

Dujardin, F. 1850. "Mémoire sur le système nerveux des insects." *Annales Des Sciences Naturelles Zoologie* et Biologie 14:195–206.

Dukas, R. 2004. "Evolutionary biology of animal cognition." *Annual Review of Ecology, Evolution and Systematics* 35: 347–374.

Dunbar, R. I. M. 2003. "The social brain: Mind, language, and society in evolutionary perspective." *Annual Review of Anthropology* 32: 163–181.

Dunbar, R. I. M., and J. Bever. 1998. "Neocortex size predicts group size in carnivores and some insectivores." *Ethology* 104: 695–708.

Durst, C., S. Eichmüller, and R. Menzel. 1994. "Development and experience lead to increased volume of subcompartments of the honeybee mushroom body." *Behavioral and Neural Biology* 62: 259–263.

Ehmer, B., and W. Gronenberg. 2002. "Segregation of visual input to the mushroom bodies in the honey bee *(Apis mellifera).*" *Journal of Comparative Neurology* 45: 362–373.

Ehmer, B., and W. Gronenberg. 2004. "Mushroom body volumes and visual interneurons in ants: Comparison between sexes and castes." *Journal of Comparative Neurology* 469: 198–213.

Ehmer, B., and R. Hoy. 2000. "Mushroom bodies of vespid wasps." *Journal of Comparative Neurology* 416: 93–100.

Ehmer, B., H. K. Reeve, and R. R. Roy. 2001. "Comparison of brain volumes between single and multiple foundresses in the paper wasp *Polistes dominulus.*" *Brain, Behavior and Evolution* 57: 161–168.

Erber, J., T. Masuhr, and R. Menzel. 1980. "Localization of short-term memory in the brain of the bee, *Apis mellifera.*" *Physiological Entomology* 5: 343–358.

Esch, H., and J. E. Burns. 1996. "Distance estimation by foraging honeybees." *Journal of Experimental Biology* 199: 155–162.

Fahrbach, S. E., D. Moore, E. A. Capaldi, S. M. Farris, and G. E. Robinson. 1998. "Experience-expectant plasticity in the mushroom bodies of the honeybee." *Learning and Memory* 5: 115–123.

Farris, S. M., and N. S. Roberts. 2005. "Coevolution of generalist feeding ecologies and gyrencephalic mushroom bodies in insects." *Proceedings of the National Academy of Science USA* 102: 17394–17399.

Farris, S. M., G. E. Robinson, and S. E. Fahrbach. 2001. "Experience- and age-related outgrowth of intrinsic neurons in the mushroom bodies of the adult worker honey bee." *Journal of Neuroscience* 21: 6395–6404.

Flögel, J. H. L. 1878. "Über den einheitlichen Bau des Gehirns in den verschiedenen Insektenordnungen." *Zeitschrift für wissenschaftliche Zoologie (Suppl)* 30: 556–592.

Fukushi, T. 2001. "Homing in wood ants, *Formica japonica.*" *Journal of Experimental Biology* 204: 2063–2072.

Galizia, C. G., A. Küttner, J. Joerges, and R. Menzel. 2000. "Odour representation in honeybee olfactory glomeruli shows slow temporal dynamics: An optical recording study using a voltage-sensitive dye." *Journal of Insect Physiology* 46: 877–886.

Galizia, C. G, and R. Menzel. 2001. "The role of glomeruli in the neural representation of odours: Results from optical recording studies." *Journal of Insect Physiology* 47: 115–130.

Gronenberg, W. 1995. "The fast mandible strike in the trap-jaw ant *Odontomachus:* Temporal properties and morphological characteristics." *Journal of Comparative Physiology A* 176: 391–398.

———. 1999. "Modality-specific segregation of input to ant mushroom bodies." *Brain, Behavior and Evolution* 54: 85–95.

Gronenberg, W., S. Heeren, and B. Hölldobler. 1996. "Age-dependent and task-related morphological changes in the brain and the mushroom bodies of the ant, *Camponotus floridanus.*" *Journal of Experimental Biology* 119: 2011–2019.

Gronenberg, W., and J. Liebig. 1999. "Smaller brains and optic lobes in reproductive workers of the ant *Harpegnathos.*" *Naturwissenschaften* 86: 343–345.

Gronenberg, W., and G. O. López-Riquelme. 2004. "Multisensory convergence in the mushroom bodies of ants and bees." *Acta Biologica Hungarica* 55: 31–37.

Grünewald, B. 1999. "Physiological properties and response modulations of mushroom body feedback neurons during olfactory learning in the honeybee, *Apis mellifera*." *Journal of Comparative Physiology* 185: 565–576.

Hammer, M. 1993. "An identified neuron mediates the unconditioned stimulus in associative olfactory learning in honeybees." *Nature* 366: 59–63.

Hanström, B. 1928. "Über das Gehirn von *Termopsis nevadensis* und *Phyllium pulchrifolium* nebst Beiträgen zur Phylogenie der Corpora Pedunculata der Arthropoden." *Zeitschrift für Morphologie und Ökologie der Tiere* 19: 732–773.

Hertel, H. 1980. "Chromatic properties of identified inter neurons in the optic lobes of the bee *Apis mellifera*." *Journal of Comparative Physiology* A 137: 215–232.

Hölldobler, B. 1980. "Canopy orientation: A new kind of orientation in ants." *Science* 210: 86–88.

Hölldobler, B., and E. O. Wilson. 1990. *The ants.* Cambridge: Belknap Press.

Howse, P. E., and J. L. D. Williams. 1969. "The brains of social insects in relation to behaviour." *Proceedings of the Sixth Congress of the International Union for the Study of Social Insects*, Bern: 59–64.

Ibbotson, M. R. 1991. "Wide-field motion-sensitive neurons tuned to horizontal movement in the honeybee *(Apis mellifera)*." *Journal of Comparative Physiology* A 168: 91–102.

Jaffe, K., and E. Perez. 1989. "Comparative study of brain morphology in ants." *Brain, Behavior and Evolution* 33: 25–33.

Jander, R. 1957. "Die optische Richtungsorientierung der roten Waldameise *(Formica rufa* L.)." *Zeitschrift für vergleichende Physiologie* 40: 162–238.

Jerison, H. J. 1969. "Brain evolution and dinosaur brains." *American Naturalist* 103: 575–588.

———. 1973. *Evolution of the brain and intelligence.* New York: Academic Press.

Julian, G. E, and W. Gronenberg. 2002. "Smaller brains in queen ants." *Brain, Behavior and Evolution* 60: 152–164.

Just, S., and W. Gronenberg. 1999. "The control of mandible movements in the ant *Odontomachus*." *Journal of Insect Physiology* 45: 231–240.

Kenyon, F. C. 1896. "The brain of the bee: A preliminary contribution to the morphology of the nervous system of the Arthropoda." *Journal of Comparative Neurology* 6: 133–210.

Komischke, B., J.-C. Sandoz, D. Malun, and M. Giurfa. 2005. "Partial unilateral lesions of the mushroom bodies affect olfactory learning in honeybees *Apis mellifera* L." *European Journal of Neuroscience* 21: 477–485.

Kühnle, K. F. 1913. "Das Gehirn, die Kopfnerven und die Kopfdrüsen des gemeinen Ohrwurms (*Forficula auricularia* L.) mit Bemerkungen über die Gehirne und Kopfdrüsen eines Springschwanzes (*Tomocerus flavescens* Tullb.), einer Termitenarbeiterin *(Eutermes peruanus)* und der indischen Stabheuschrecke *(Dixippus morosus)."* *Jena Zeitschr* 50: 147–276.

Kuwabara, M. 1957. "Bildung des bedingten Reflexes von Pavlovs Typus bei der Honigbiene, *Apis mellifera.*" *Journal of the Faculty of Science (Zoology), Hokkaido University.* 13: 458–464.

Laughlin, S. B. 2001. "Energy as a constraint on the coding and processing of sensory information." *Current Opinion in Neurobiology* 11: 475–480.

Laughlin, S. B., R. R. de Ruyter van Steveninck, and J. C. Anderson. 1998. "The metabolic cost of neural information." *Nature Neuroscience* 1: 36–41.

Lehrer, M. 1998. "Looking all around: Honeybees use different cues in different eye regions." *Journal of Experimental Biology* 201: 3275–3292.

Lei, H., T. A. Christensen, and J. G. Hildebrand. 2004. "Spatial and temporal organization of ensemble representations for different odor classes in the moth antennal lobe." *Journal of Neuroscience* 24: 11108–11119.

Li, Y., and N. J. Strausfeld. 1997. "Morphology and sensory modality of mushroom body extrinsic neurons in the brain of the cockroach, *Periplaneta americana.*" *Journal of Comparative Neurology* 387: 631–650.

Lorenzi, M. C., M. F. Sledge, P. Laiolo, E. Sturlini, and S. Turillazzi. 2004. "Cuticular hydrocarbon dynamics in young adult *Polistes dominulus* (Hymenoptera: Vespidae) and the role of linear hydrocarbons in nestmate recognition systems." *Journal of Insect Physiology* 50: 935–941.

Marino, L. 1996. "What dolphins can tell us about primate evolution." *Evolutionary Anthropology* 5: 81–86

Mauelshagen, J. 1993. "Neural correlates of olfactory learning paradigms in an identified neuron in the honeybee brain." *Journal of Neurophysiology* 69: 609–625.

Maynard-Smith, J., and E. Szathmáry. 1995. *The major Transitions in evolution.* New York: Oxford University Press.

Mizunami, M., J. M. Weibrecht, and N. J. Strausfeld. 1998. "Mushroom bodies of the cockroach: Their participation in place memory." *Journal of Comparative Neurology* 402: 520–537.

Mobbs, P. G. 1982. "The brain of the honeybee *Apis mellifera.* I. The connections and spatial organisation of the mushroom bodies." *Philosophical Transactions of the Royal Society of London B* 298: 309–354.

O'Donnell, S., N. A. Donlan, and T. A. Jones. 2004. "Mushroom body structural change is associated with division of labor in eusocial wasp workers (*Polybia aequatorialis,* Hymenoptera: Vespidae)." *Neuroscience Letters* 356: 159–162.

Paulk, A., and W. Gronenberg. 2005. "Color and motion sensitive cells in the bee brain." *Society of Neuroscience Abstracts* 31.

Pietschker, H. 1911. "Das Gehirn der Ameise." *Jenaische Zeitschrift fur Naturwissenschaft* 47: 43–117.

Reader, S. M., and K. N. Laland. 2002. "Social intelligence, innovation, and enhanced brain size in primates." *Proceedings of the National Academy of Science USA* 99: 4436–4441.

Rensch, B. 1956. "Increase of learning capabilities with increase of brain size." *American Naturalist* 90: 81–95.

Rosengren, R. 1971. "Route fidelity, visual memory and recruitment behaviour in foraging wood ants of the genus *Formica* (Hymenoptera, Formicidae)." *Acta Zoologica Fennica* 133: 1–106.

Roth, G., and U. Dicke. 2005. "Evolution of the brain and intelligence." *Trends in Cognitive Sciences* 9: 250–257.

Schulz, D. J., and G. E. Robinson. 1999. "Biogenic amines and division of labor in honey bee colonies: Behaviorally related changes in the antennal lobes and age-related changes in the mushroom bodies." *Journal of Comparative Physiology A* 184: 481–488.

———. 2001. "Octopamine influences division of labor in honey bee colonies." *Journal of Comparative Physiology A* 187: 53–61.

Seid, M. A., M. Kristen, K. M. Harris, and J. F. A. Traniello. 2005. "Age-related changes in the number and structure of synapses in the lip region of the mushroom bodies in the ant *Pheidole dentata*." *Journal of Comparative Neurology* 488: 269–277.

Seid, M. A., and J. F. A. Traniello. 2005. "Age-related changes in biogenic amines in individual brains of the ant *Pheidole dentata*." *Naturwissenschaften* 92: 198–201.

Seid, M. A., and J. F. A. Traniello. 2006. "Age-related repertoire expansion and division of labor in *Pheidole dentata* (Hymenoptera: Formicidae): A new perspective on temporal polyethism and behavioral plasticity in ants." *Behavioral Ecology and Sociobiology* 60: 631–644.

Si, A., M. V. Srinivasan, and S. Zhang. 2003. "Honeybee navigation: Properties of the visually driven 'odometer.'" *Journal of Experimental Biology* 206: 1265–1273.

Strausfeld N. J. 1989. "Beneath the compound eye: Neuroanatomical analysis and physiological correlates in the study of insect vision." In D. G. Stavenga and R. C. Hardie, eds., *Facets of vision*, 318–359. Heidelberg, NY: Springer.

Strausfeld N. J. 1999. "A brain region in insects that supervises walking." In M. D. Blinker, ed., *Progress in brain research*, 123: 273–284.

Strausfeld N. J. 2002. "Organization of the honey bee mushroom body: Representation of the calyx within the vertical and gamma lobes." *Journal of Comparative Neurology* 450: 4–33.

Strausfeld, N. J., L. Hansen, Y. Li, R. S. Gomez, and K. Ito. 1998. "Evolution, discovery, and interpretation of arthropod mushroom bodies." *Learning and Memory* 5: 11–37.

Strauss, R., and M. Heisenberg. 1993. "Higher control center of locomotor behavior in the *Drosophila* brain." *Journal of Neuroscience* 13: 1852–1861.

Tibbetts, E. A. 2002. "Visual signals of individual identity in the paper wasp *Polistes fuscatus.*" *Proceedings of the Royal Society London B* 269: 1423–1428.

———. 2004. "Complex social behavior can select for variable visual features: A case study in *Polistes* wasps." *Proceedings of the Royal Society London B* 271: 1955–1960.

Tinbergen, N. 1958. *Curious naturalists.* Garden City: Doubleday.

Vitzthum, H., M. Müller, and U. Homberg. 2002. "Neurons of the central complex of the locust Schistocerca gregaria are sensitive to polarized light." *Journal of Neuroscience* 22: 1114–1125.

von Alten, H. 1910. "Zur Phylogenie des Hymenopterengehirns." *Jenaische Zeitschrift für Naturwissenschaft* 46: 511–590.

von Frisch, K. 1914. "Der Farbensinn und Formensinn der Biene." *Zoologische Jahrbücher (Physiologie)* 37: 1–238.

Von Frisch. 1967. *Dance language and orientation of honeybee.* Cambridge: Harvard University Press.

Wehner, R. 2003. "Desert ant navigation: How miniature brains solve complex tasks." *Journal of Comparative Physiology A* 189: 579–588.

Wehner, R., and R. Menzel. 1990. "Do insects have cognitive maps?" *Annual Review of Neuroscience* 13: 403–414.

Withers, G. S., S. E. Fahrbach, and G. E. Robinson. 1993. "Selective neuroanatomical plasticity and division of labour in the honeybee." *Nature* 364: 238–240.

Witthöft, W. 1967. "Absolute anzahl und verteilung der zellen im hirn der honigbiene." *Zeitschrift für Morphologie der Tiere* 61: 160–184.

Yang, E.-C., H.-C. Lin, and Y. S. Hung. 2004. "Patterns of chromatic information processing in the lobula of the honeybee, *Apis mellifera L.*" *Journal of Insect Physiology* 50: 913–925.

CHAPTER EIGHTEEN

Plasticity in the Circadian Clock and the
Temporal Organization of Insect Societies

GUY BLOCH

SOCIALLY MEDIATED PLASTICITY in circadian rhythms is a mechanism
likely to contribute to temporal coordination and social integration in insect societies. My aim in this chapter is to briefly summarize current
knowledge on the circadian system of honey bees and to review three lines
of evidence in favor of socially mediated plasticity in circadian rhythms: (1)
social factors synchronize the circadian clock and coordinate the activity of
honey bee workers; (2) in both honey bees and bumblebees foragers have
strong circadian rhythms, whereas nurses care for brood around-the-clock
with no circadian rhythms; and (3) ant queens switch between activity
with or without circadian rhythms according to their reproductive status.
From an evolutionary perspective, these findings suggest that the evolution of sociality has shaped features of the circadian system.

I emphasize three major points in this chapter. First, there is evidence
that plasticity in the circadian system is one of the mechanisms that contributes to the temporal integration and organization of insect societies.
Second, the molecular and functional conservation in the circadian
clock, and the existence of closely related Hymenoptera species representing various levels of sociality, provides an excellent opportunity for
comparative studies that may shed light on the generality, function, and
evolution of chronobiological plasticity. Third, the circadian system provides an excellent model for research on the mechanisms of behavioral
plasticity and the social regulation of gene expression—two major issues
in contemporary sociobiology. Circadian rhythms are among the phenotypes for which the relationships between genes and behavior are best

understood. Models of the organization and function of the circadian clock in *Drosophila* and mice provide a valuable framework for studies on social insects because clock mechanisms are fairly well conserved in terms of both general organization principles and the genes involved in rhythm generation.

Circadian Rhythms

Circadian rhythms are defined as biological rhythms that meet the following three criteria: (1) they persist, or free-run, with a period of about 24 hours in the absence of external time cues; (2) they are reset, or entrained, to environmental cues, most notably daily light-dark and temperature cycles, but additional cues such as food availability and social interactions may also be important; and (3) they have a stable period length in a wide range of physiologically relevant temperatures (temperature compensations). The circadian clock influences many aspects of insect physiology and behavior such as rhythms in activity, sleep-wake, feeding, mating, oviposition, egg hatching, and pupal eclosion. The circadian clock also measures day length, and influences annual rhythms such as timing diapause and reproduction (reviewed in Dunlap, Loros, and Decoursey 2004; Saunders 2002).

The circadian clock system is commonly described as composed of three functional components. The core of the clock is the pacemaker, a cell-autonomous rhythm generator that cycles approximately, but not exactly, to a 24-hour period. The second component consists of input pathways (photic and nonphotic) that transmit environmental signals to the endogenous oscillators and keep them synchronized with external day-night cycles. The third component covers output pathways that carry temporal signals away from pacemaker cells to diverse biochemical, physiological, and behavioral processes (Dunlap, Loros, and Decoursey 2004; Saunders 2002).

Complex interactions at several levels of biological organization are involved in producing circadian rhythms in physiology and behavior. These include cellular processes such as posttranscriptional regulation, posttranslational modifications in clock proteins, chromatin remodeling, and regulation of intracellular localization. The higher-level processes include the coupling of a series of oscillators, and neuronal and humeral communication between clusters of clock cells within and between tissues

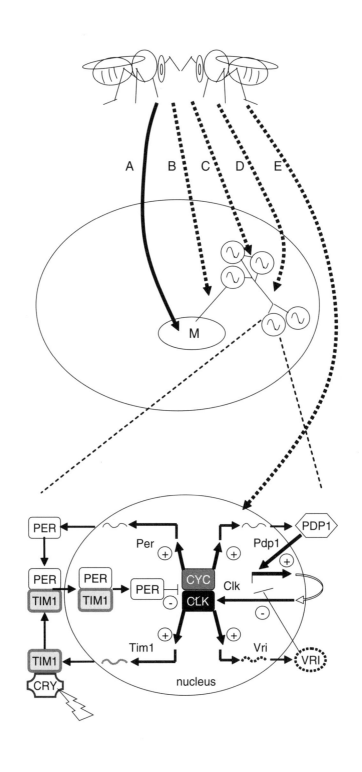

(Bell-Pedersen et al. 2005; Cermakian and Sassone-Corsi 2000; Dunlap, Loros, and Decoursey 2004; Figure 18.1). The evidence for regulation of circadian rhythms in the context of social integration reviewed below suggests there may be additional (social) levels of complexity in the multilevel regulation of circadian rhythms.

Figure 18.1. Multilevel regulation of circadian rhythms and their possible social modulation. Bottom panel—circadian rhythms are generated autonomously within pacemaker cells. The panel shows a model for the central brain pacemaker in *Drosophila* (based on Bell-Pedersen et al. 2005; Hardin 2004, 2005). Two interlocked loops are involved: a negative autoregulatory loop of Period (PER) and Timeless (TIM1) (left-hand side) that regulates their own expression by inhibiting the Clock (CLK)/ Cycle (CYC) complex. The CLK/CYC complex is also involved in a second autoregulatory loop (right-hand side) that activates *Vrille (Vri)* and *Par Domain Protein 1 (Pdp1)* transcription. VRI accumulates quickly and represses *Clk* transcription; PDP1 accumulates more slowly and activates *Clk* transcription. Light in the early morning (lightning symbol) leads to Cryptochrome (CRY)-dependent degradation of TIM1. In the absence of TIM1, PER is unstable (mediated by phosphorylation by kineases, not shown). Removal of PER releases CLK/CYC to resume transcription of *Per, Tim1, Pdp1,* and *Vri*, which restarts both loops simultaneously. The mRNA and protein for each gene is depicted by similarly tinted wavy lines and geometric shapes, respectively. Wide arrows with "+" sign delineate activation; T-end arrows with "−" sign delineate inhibition. The honey bee genome does not contain orthologs for *Tim1* and *Drosophila*-type *Cry (Cry-d)*. Therefore, it is thought that the mammalian-type *Cry (Cry-m)* takes the role of *Tim1* in the negative feedback loop of the bee (Rubin et al. 2006). Middle panel—pacemaker cells and other neuronal circuits interact to produce overt circadian rhythms. The circadian network is composed of clusters of coupled pacemaker cells (circles containing sinusoidal wave lines). The large oval shape depicts the central nervous system; M—motor controlling center. Upper panel—social interactions between bees may influence circadian rhythms in several ways. A—modulating motor controlling centers without affecting the clock ("masking," leftmost arrow); B—uncoupling the circadian system from motor controlling centers; C—modulating the interaction between pacemakers within the same cluster; D—modulating the interaction between clusters of pacemakers; E—modulating the circadian circuit within pacemaker cells.

Molecular Mechanisms Underlying Circadian
Rhythms in Insects

The basic molecular mechanisms for rhythm generation are found in organisms as diverse as cyanobacteria, plants, fruit flies, and mammals. The molecular circadian circuit in animals consists of interlocked autoregulatory transcriptional/translational feedback loops with positive and negative elements (Dunlap 1999), and is best known for the fruit fly *Drosophila melanogaster.* The current model for molecular rhythm generation in *Drosophila* involves interactions among six transcription factors: Period (PER), Timeless (TIM1), Clock (CLK), Cycle (CYC), Par Domain Protein 1 (Pdp1), and Vrille (Vri), as well as several kinases including Double-time (Dbt) and Shaggy (Sgg). Additional genes that are thought to be involved in the fly's clockwork but are less well documented will not be discussed here (for a recent review, see Bell-Pedersen et al. 2005; Dunlap, Loros, and Decoursey 2004; Hardin 2004, 2005). The products of these clock genes interact and form two interlocked feedback loops (Figure 18.1, bottom). In the first loop, PER and TIM1 negatively regulate their expression by inhibiting the transcriptional activity of CLK and CYC. The second autoregulatory loop controls the cycling levels of CLK. CLK protein inhibits its own expression in an autoregulatory loop in which VRI and PDP1 are also involved. The neuropeptide Pigment Dispersing Factor (PDF) is currently a strong candidate for an output pathway gene that influences not only locomotor activity but also the timing of eclosion. PDF also plays a crucial role in coupling the activity of pacemaker cells (Grima et al. 2004; Lin et al. 2002; Stoleru et al. 2004).

Comparison of the clock mechanism in vertebrates and flies shows that there is a high degree of conservation not only in the general design, but also in the sequence of the principal clock genes (reviewed in Bell-Pedersen et al. 2005; Dunlap 1999; Panda, Hogenesch, and Kay 2002; Young and Kay 2001). With this evidence for conservation between flies and mice, it is tempting to assume that the clocks of other insects are similar to the fly model. Insect clocks probably share many properties, but findings from various species, including the honey bee, are not easily reconciled with the *Drosophila* model (e.g., Bloch et al. 2003; Rubin et al. 2006; Sauman and Reppert 1996; Zavodska, Sauman, and Sehnal 2003).

The Circadian System of the Honey Bee

Currently, the European honey bee *Apis mellifera* is the only social insect for which the molecular- and system-level organization of the circadian clock has been described in some detail. The dearth of information on the mechanism of circadian rhythms in social insects is quite surprising, since some of the seminal discoveries in chronobiology have derived from studies on honey bees and ants. For example, studies on ants were the first to implicate the circadian clock in reproductive isolation (McCluskey 1958, 1965), and the discovery of a foraging rhythm in honey bees provided the first convincing evidence for the adaptive value of the circadian clock (Saunders 2002).

As early as 1900, von Buttel-Reepen reported that bees only visited a buckwheat field in the morning when the blossoms were secreting nectar. Forel (1910) further noticed that bees not only arrived at his table at breakfast time each day, but kept on coming on time even when reward was not available. He explained his observations by suggesting that bees have a *Zeitgedächtnis* (memory for time). Time memory was the most studied circadian behavior in bees in the years that followed. These studies established that bees can learn to arrive at a specified location at any time of the day, and can learn as many as nine time points with intervals of only 45 minutes between feeder availability, or four different feeding places at different times (Beling 1929; Koltermann 1971; von Frisch 1967; Wahl 1932, 1933). Time memory is an internal circadian rhythm—it free-runs under constant conditions, it is entrained by light—dark cycles, it can be phase shifted, and it has a narrow range of entrainment (20–26 hour cycles)—similar to other circadian oscillations (reviewed in von Frisch 1967; Moore 2001; Saunders 2002). This sophisticated time sense is thought to improve foraging efficiency as it enables bees to collect nectar and pollen at times of maximal availability, which is different for different flowers (von Frisch 1967; Willmer and Stone 2004).

Foraging bees also need a functioning circadian clock for time compensated sun-compass orientation. Bees and other social insects (e.g., Jander 1957) use the sun as a celestial compass; they orient themselves by maintaining a fixed angle to the sun (von Frisch 1967). Because of the earth's rotation, the sun moves in a predictable path that varies with latitude and season. In order to stay in the right geographical direction, the bee "consults" its circadian clock and compensates for the sun's movement during this period.

Foragers that stay for a long period inside the hive (e.g., due to bad weather) rely on their clock to correct their waggle dance in accordance with the shift in the sun's azimuth. Such "marathon dancers" repeat dances without having any view of the sky, yet they shift the direction of their dance with remarkable correlation to the sun's path in the sky (von Frisch 1967). Amazingly, dancers that were trained to two feeders, a morning feeder to the south 1 hour after sunrise and an evening feeder to the east 1 hour before sunset, danced inside the hive to the evening feeder in the early part of the night and to the morning feeder late at night (von Frisch 1967).

The activity of foragers varies considerably during the day (Moore et al. 1998). They typically forage during the day and rest inside the hive at night, showing many features of sleep found in mammals and birds (Kaiser and Steiner-Kaiser 1983; Sauer et al. 2003; Schmolz, Hoffmeister, and Lamprecht 2002). Their strong activity rhythms can be studied in the laboratory where each bee is monitored individually in a small cage (e.g., Moore and Rankin 1985; Shemesh, Cohen, and Bloch 2007; Spangler 1972; Toma et al. 2000). Under these conditions, foragers of *A. mellifera* and *A. cerana* typically have an activity cycle (free-running period) of less than 24 hours in constant darkness (DD) and of more than 24 hours in constant light (LL), a finding that is not consistent with Aschoff's rule that states that diurnal organisms have a longer free-running period in DD than in LL (Aschoff 1960). Locomotor activity rhythms are adjusted to a broad range of illumination regimes and to temperature cycles in DD (Moore and Rankin 1993; Moritz and Kryger 1994), and are influenced by social factors. Foragers also manifest circadian rhythms in physiological and biochemical processes such as metabolic activity (Hepburn et al. 1984; Moritz and Kryger 1994; Southwick and Moritz 1987; Stussi and Harmelin 1966), neuroendocrine functions (Carrington et al. 2007; Elekonich et al. 2001; Heinzeller 1976; Vogel, Heinzeller, and Renner 1977), and visual gene expression in the retina (Sasagawa et al. 2003).

Ontogeny of Circadian Rhythms in Adult Bees

Solitary insects, including solitary wasps (Fantinou, Alexandri, and Tsitsipis 1998; Fleury et al. 2000), typically eclose from the pupae at a specific time of day that is gated by the circadian clock; the newly emerged adults immediately show circadian rhythms in locomotor activity (Saunders

2002). In contrast, young honey bee (reviewed in Moore 2001) and bumblebee (*B. terrestris;* Yerushalmi, Bodenhaimer, and Bloch 2006) workers typically have no circadian rhythms in locomotor activity or metabolism for the first few days until about 2 weeks of age. Ontogeny of circadian rhythms is endogenous because it occurs under constant conditions and rhythms free-run with a period of about 24 hours. The development of overt circadian rhythms is associated with age-related changes in the expression of clock genes in the honey bee brain. Both the protein and mRNA levels for the clock gene PER are low and do not vary during the day in newly emerged bees, but are typically higher and oscillate in honey bees older than 3 weeks of age (Bloch, Toma, and Robinson 2001, Bloch et al. 2003; Bloch, Rubenstein, and Robinson 2004; Shimizu et al. 2001; Toma et al. 2000). There is evidence that environmental and seasonal variables which influence pre-adult development also affect the ontogeny and characteristics of circadian rhythms in locomotor activity of adult bees (Bloch, Shemesh, and Robinson 2006; Yerushalmi, Bodenhaimer, and Bloch 2006). However, the factors governing environmental and internal regulation of the ontogeny of behavioral rhythms and the age-related changes in PER expression of honey bees are still elusive.

Perhaps the post-eclosion ontogeny of circadian rhythms is more common in social than solitary species because individuals of social insects emerge into a protected environment. Ontogeny of rhythms may be functionally significant because it meshes with task performance in species such as the honey bee, in which the division of labor is based in part on age (Toma et al. 2000).

The Anatomical and Molecular Organization of the Circadian Clock in the Honey Bee Brain

The anatomical organization of the circadian clock has not been conclusively described for the honey bee or for any other social insect. Nevertheless, it is safe to assume that the central pacemaker controlling rhythmic behavior is located in the brain as in all other insects studied so far (Helfrich-Forster, Stengl, and Homberg 1998; Saunders 2002). There appears to be much variation in the organization of the brain circadian clock in insects. In most species the accessory medulla of the optic lobes is the site of the pacemaker, but in moths the pacemaker is apparently located in a group of PER and TIM positive cells in the central brain

(Helfrich-Forster, Stengl, and Homberg 1998; Saunders 2002; Zavodska, Sauman, and Sehnal 2003).

The initial claim that the mushroom bodies are the anatomical sites of the bee clock (Martin et al. 1978 cited in Saunders 2002; Martin and Martin 1987) was later challenged on experimental and technical grounds (Brady 1987; Moore 2001). Recent immunocytochemical studies are also not consistent with this hypothesis because the honey bee mushroom bodies were not immunostained with antibodies directed to detect the protein products of clock genes. In addition, there is no evidence implicating the mushroom bodies as the site of the pacemaker in other insects (Helfrich-Forster, Stengl, and Homberg 1998; Helfrich-Forster, Wulf, and de Belle 2002; Helfrich-Forster 2005; Saunders 2002; Zavodska, Sauman, and Sehnal 2003).

The current picture of the anatomical organization of the bee clock is largely based on immunocytochemical studies with antibodies against PER and PDF (Bloch et al. 2003; Zavodska, Sauman, and Sehnal 2003). The most consistent PER immunoreactivity (ir) was detected in the cytoplasm of about eight large cells in the lateral protocerebrum. Additional neurons in the optic lobes and other parts of the brain showed nuclear staining. PDF-ir was detected in the somata of about 20 cells located between the central brain and optic lobes in each brain hemisphere. Both the PER-ir and PDF-ir clusters are located in brain areas that are implicated in the regulation of circadian rhythms in *Drosophila* and other insects. However, by contrast to *Drosophila*, in these two studies none of the cells co-express PER-ir and PDF-ir. It is reasonable to assume that both PER and PDF are involved in the bee clock because PER is an indispensable component of the clock in both insects and vertebrates, and there is evidence implicating PDF in circadian rhythms in various insects. Thus, the circadian system of the bee seems to be differently organized than in *Drosophila*.

The brain expression pattern of clock genes in the honey bee brain fits better with the mammalian model than with that of *Drosophila*. The honey bee genome does not encode an ortholog of *Drosophila* TIM1, has only the mammalian-type *Cryptochrome* (CRY-m), and has a single ortholog for CYC, CLK, VRI, PDP1, and *Timeout (TIM2)*. In foraging honey bees expressing strong circadian rhythms, brain mRNA levels of CRY-m consistently oscillate with strong amplitude and a phase similar to PER under both light-dark and constant darkness illumination regimes. In contrast to *Drosophila*, the predicted honey bee CYC protein contains a transactivation domain and its brain transcript levels oscillate at virtually an antiphase to PER, as it does in the mouse (Rubin et al. 2006; Weinstock et al. 2006). The apparent cycling in the brain expression of PER, CRY-m, TIM2, and CYC is attenuated in around-the-clock active nurse bees (Shemesh, Cohen, and Bloch 2007). Shimizu et al. (2001) further suggested that in both *A. mellifera* and *A. cerana* there are two alternative splice forms of PER, and that their ratio varies during the day. Although there is evidence that alternative splicing in PER is functional in *Drosophila* (Majercak et al. 1999; Majercak, Chen, and Edery 2004), its role, if any, in the honey bee is yet to be determined.

The Circadian Clock and the Temporal Organization of Insect Societies

Social Synchronization of Worker Activity

Animal species differ considerably in their entrainment by social time givers (e.g., Gattermann and Weinandy 1997; Goel and Lee 1997a; Levine et al. 2002; Refinetti, Nelson, and Menaker 1992). In insect societies, social synchronization appears to be one of the mechanisms to improve coordination among individuals and social integration. For example, it is functionally important for nectar receivers and foragers to be synchronized (Crailsheim et al. 1999). There is evidence that pre-forager bees (~2–3 weeks of age) that stay inside the environmentally regulated hive already manifest synchronized circadian rhythms in their resting time several days before they perform their first foraging trip (Moore et al. 1998). There are reports that both nestmate workers and the queen function as a time-giver to the honey bee clock (Moritz and Sakofski 1991; Southwick and Moritz 1987). In what appears to be the first experimental evidence of social synchronization, Medugorac and Lindauer (1967) showed that nectar foragers who were introduced into a foreign colony continued to visit a food source at the time during which they were entrained in their source colony as well as at the feeding time of the host colony. Worker bees that are removed from time-givers in the hive have a similar phase on the first few days in isolation but later drift from the colony phase (Frisch and Koeniger 1994). Bees are synchronized even if restricted to the inner part of the hive and therefore deprived of any light and flight experience ("BigBack" bees in Figure 18.2); their brain PER mRNA levels cycled with circadian rhythms and had

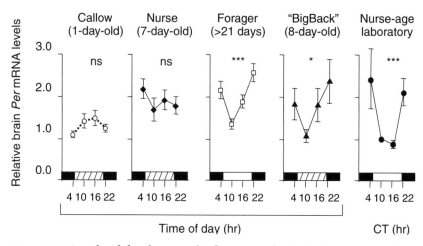

Figure 18.2. Social and developmental influences on brain clock gene expression in honey bees. The data depict *Per* mRNA levels over time (average ± SE) for workers reared in different social environments. Callows, nurses, and foragers were collected from the same typical field colony. The "BigBacks" and nurse-age (7-day-old) bees are from different experiments and source colonies. CT = circadian time, corresponds to the illumination regime during the 6 days prior to bee collection in constant darkness; CT0 and CT12 correspond to the subjective time of light on and light off, respectively. Bars at the bottom of each plot depict photoperiod: open bars = day (sunset to sunrise), dashed bars = subjective day for bees that were in constant darkness inside the hive, or were entrained to a light: dark illumination regime and collected in constant darkness (nurse-age bees in the laboratory), filled bars = night or subjective night. Brain *Per* mRNA levels differ over time in foragers, light and flight deprived ("BigBack") nurse-age bees, and nurse-age bees in the laboratory (one-way ANOVA with Fischer PLSD post hoc tests, *ns* = not significant, $^{\circ}p < 0.05$, $^{\circ\circ\circ}p < 0.001$). $N = 5$–8 bees/time point. Each experiment was repeated 2–3 times with bees from different source colonies. Based on data from Bloch et al. (2004).

higher levels at night, as is typical for foragers. When these bees were re-moved from the hive and monitored individually, they had circadian rhythms in locomotor activity with higher activity during the subjective day. These findings suggest that while in the hive, the "Big Backs" were socially entrained by nestmate workers that were exposed to the day-night fluctua-tions outside the hive (Bloch, Rubenstein, and Robinson 2004).

One emerging property of social synchronization among individuals is a colony-level circadian rhythm. Colony rhythms, like individual animals, have a stable free-running period in constant environment and their phase can be shifted by changes in environmental factors such as light, temperature, and feeding cycles (Frisch and Aschoff 1987; Frisch and Koeniger 1994; Kefuss and Nye 1970; Moore 2001).

Division of Labor-Related Plasticity

The ontogeny of circadian rhythms in honey bees is concordant with their age-related division of labor. Young bees (<10 days) typically perform activities inside the constantly dark and homeostatically regulated hive with no circadian rhythms. Foragers, which are typically older (>3 weeks of age), have strong circadian rhythms with elevated activity during the day when they forage for nectar and pollen. During the night they rest inside the hive (Crailsheim, Hrassnig, and Stabentheiner 1996; Moore et al. 1998). Variation in the environment experienced by nest bees and foragers (e.g., light and temperature) does not provide an adequate explanation for age-related plasticity in circadian rhythms because bees that are housed individually in a constant laboratory environment also show ontogeny of rhythms. Furthermore, nurses are active around-the-clock even when experiencing a light-dark illumination regime (Moore et al. 1998; Shemesh, Cohen, and Bloch 2007).

Studies on the molecular basis of task-related chronobiological plasticity in honey bees have focused mainly on the pattern of PER expression under various environmental, social, and neuroendocrine conditions. Because PER interacts directly or indirectly with all other known components of the clock (Figure 18.1), alternations in its mRNA or protein levels may reflect changes in pacemaker function. Task-associated variation in brain PER mRNA abundance includes a developmental elevation in average daily levels (the average of all time points measured during the same day), and plasticity in the temporal pattern of expression during the day (Toma et al. 2000, Bloch, Toma, and Robinson 2001; Bloch, Rubenstein, and Robinson 2004; Figure 18.2). Immunocytochemical studies with anti- (*Drosophila*) PER antibodies revealed comparable age-related changes at the protein level (Bloch et al. 2003). The functional significance of the increase in average daily PER expression is not well understood. Task-related plasticity in brain cycling has also been reported for CRY-m, TIM2, and CYC, which typically vary with circadian rhythms in foragers but are attenuated in nurses (Shemesh, Cohen, and Bloch 2007). By contrast to nurses, nurse-age bees (5–10 days of age) that grew in the laboratory without brood and a queen typically show robust oscillations in brain PER mRNA and protein levels (Toma et al. 2000; Bloch, Toma, and Robinson 2001; Bloch, et al. 2003; Bloch, Rubenstein, and Robinson 2004; Figure 18.2). These results suggest that the absence or attenuated molecular oscillation in honey bee nurses is largely linked to task, and is not merely a reflection of a circadian system that is not fully developed due to their typically young age.

This association of chronobiological plasticity with the division of labor is perhaps adaptive because it improves task specialization and colony efficiency. While around-the-clock nursing activity is necessary to meet the larvae' need for constant care, foraging behavior relies on the circadian clock. This hypothesis is supported by several lines of observations. First, nurses care for the brood around-the-clock even when kept under a light-dark illumination regime (Moore et al. 1998; Shemesh, Cohen, and Bloch 2007); second, old foragers switch back to activity with no circadian rhythms when induced to revert to brood care activity (Bloch & Robinson 2001; Bloch, Toma, and Robinson 2001); third, there is a similar task-related chronobiological plasticity in the bumblebee *B. terrestris,* in which division of labor is based primarily on size and not age as in honey bees (Yerushalmi, Bodenhaimer, and Bloch 2006).

We know very little about the external and internal modulations of plasticity in circadian rhythms and PER expression. The hypothesis that plasticity in the circadian clock is linked to division of labor predicts that factors that modulate task performance also influence the clock. Indeed, PER cycling and activity rhythms are closely associated with task performance. Bees that were induced to forage precociously showed circadian cycling in PER mRNA, whereas levels were similar throughout the day in sister nurses of similar age; PER cycling appeared attenuated in foragers reverting to around-the-clock brood care activity. These social manipulations did not affect the average daily levels of PER (Toma et al. 2000, Bloch, Toma, and Robinson 2001; Bloch, Rubenstein, and Robinson 2004). However, treating newly emerged bees with doses of juvenile hormone (JH), octopamine (OA), and cyclic GMP that advance the age of first foraging did not affect the age at onset of circadian rhythms in locomotor activity (Ben-Shahar et al. 2003; Bloch, Sullivan, and Robinson 2002; Bloch and Meshi 2007).

Mating-Related Plasticity in Circadian Rhythms in Queens

Virgin gyne and male ants time their nuptial flights to a species-specific time of the day (reviewed in McCluskey 1992; Hölldobler and Wilson 1990). This behavior is apparently adaptive as it coordinates the mating activity of males and females and facilitates reproductive isolation between species inhabiting the same region. This species-specific activity phase can be reset by changing the illumination regime and can be maintained in individually isolated ants under constant conditions in the laboratory, indicating that it is controlled by the circadian clock (McCluskey 1958, 1965, 1992; Sharma et al. 2004; Sharma, Lone, and Goel 2004). The activity profile of egg-laying queens of mature colonies is very different. Their overall activity is reduced and they have no circadian rhythms. This remarkable switch to arrhythmicity is limited to mated queens; virgin queens that were kept for similar periods, with or without wings, continued to exhibit robust circadian rhythms (McCluskey 1967; McCluskey and Carter 1969; Sharma et al. 2004; Sharma, Lone, and Goel 2004). Sharma, Lone, and Goel recently suggested that isolated mated queens return to circadian rhythmicity in locomotor activity after they cease laying eggs. This suggests that similar to the plasticity in honey bee workers, the transition from rhythmic to arrhythmic state is reversible.

Important Questions and New Methods

The study of socially mediated chronobiological plasticity integrates sociobiology and chronobiology ("sociochronobiology") and addresses two major lines of inquiry. The first relates to the mechanisms underlying chronobiological plasticity. The second is if, why, and how social evolution has shaped the clock of social insects differently from that of related solitary species. Because these lines of inquiries are in their infancy, many of the current questions are concerned with describing the phenomena and their generality, and the basic properties of the circadian clock in social insects.

Organization and Function of the Circadian Clock
of Social Insects

To be in a position to raise questions about the mechanism of plasticity in circadian rhythms and social synchronization, it is imperative to first understand the basic properties of the circadian system in social Hymenoptera. The recent finding that the molecular clockwork in the bee brain differs from the *Drosophila* model (Rubin et al. 2006; Weinstock, Robinson, Gibbs et al. 2006) strengthens the notion that although the *Drosophila* model provides a good framework, it is essential to explore the details of the clock in each focal species. These investigations need to include the input and output pathways of the clock.

Functional Significance of Plasticity in Circadian Rhythms

It is assumed that task-related plasticity in circadian rhythms of social insects is adaptive because it improves specialization and colony efficiency: however, there is only a little empirical support for this hypothesis. One approach to address this assumption is to explore the relationships between plasticity in circadian rhythms and other social functions in as many social insect species as possible (the comparative approach). The hypothesis that plasticity in worker circadian rhythms relates to task can also be tested by studying circadian rhythms in additional tasks (Moore et al. 1998). For example, in honey bees, undertakers, guards, and food storers might exhibit circadian rhythms because they need to be synchronized with day-night fluctuation, whereas activity around-the-clock is predicted to be more common in bees dedicated to in-nest activities such as thermoregulation and comb construction. Similarly, finding a mating-related switch to activity with no circadian rhythms in social insects but not in related solitary species would lend credence to the hypothesis that plasticity in circadian rhythm is associated with the evolution of sociality.

Functionality can be also investigated by manipulating a system (the experimental approach). For example, the hypothesis that around-the-clock activity is linked to brood care predicts that individuals that are induced to revert from foraging to nursing activities will switch back to activity with no circadian rhythms, and that nurses that are removed from the hive will manifest circadian rhythms. These predictions were indeed borne out for honey bees (Bloch and Robinson 2001; Shemesh, Cohen, and Bloch 2007). A complementary approach that has not yet been used, is to manipulate the circadian system (e.g., by transgenic or pharmacological means) of queens or workers and test its effect on colony fitness.

Generality of Chronobiological Plasticity in Insect Societies

It is important to test whether chronobiological plasticity underlies additional forms of temporal coordination in insect societies. There is also a need to study additional species because the value of chronobiological plasticity depends on the life history, ecology, and social organization of the species in question. For example, by contrast to honey bees and bumblebees, the nursing activity of the ant *Camponotus mus* appears to have robust circadian rhythms. Nurses of this species actively select a temperature of 30.8°C in the middle of the light phase, and 27.5°C during the dark phase, for their brood care activities. This rhythm in brood care is entrained by light and temperature cycles and is self-sustained under constant conditions, indicating that it is under the control of the endogenous circadian clock (Roces 1995; Roces and Nunez 1996).

Currently, very little can be said regarding generality because only a few species have been investigated. The studies with *C. mus*, however, suggest that brood care with no circadian rhythms is not a universal trait for all social insects. Even less is known about the generality of chronobiological plasticity in queens. Nevertheless, observations of bees suggest that it is not limited to ants. Virgin honey bee queens time their nuptial flights with species-specific characteristics (Koeniger and Koeniger 2000), whereas egg-laying queens show no diurnal periodicity in behavior (Free, Ferguson, and Simpkins 1992). Our unpublished observations suggest that there is a similar plasticity in circadian rhythms in queens of *B. terrestris* (Eban, Bellusci, and Bloch, in preparation). It is tempting to assume that social synchronization is a common feature of insect societies, yet coordination can be achieved by other means such as self-organization or social interactions, or by the clock of individuals synchronized by physical rather than social time-givers (which may be particularly effective in species that nest in the open). Some of the reasons for the shortage of comparative studies are methodological. For example, the nesting biology of many species makes it very difficult to observe activity of individuals in a natural or semi-natural social context. Also, for many species, there are no protocols for rearing isolated individuals as required for studies on properties of the circadian clock.

Social Signals Mediating Plasticity in Circadian Rhythms and Social Entrainment

The social regulation of chronobiological plasticity is currently poorly understood. We do not have answers even for such basic questions as whether interactions with adults or brood mediate plasticity in circadian rhythms or whether direct social contact is necessary. The association, and perhaps causal relationship, between chronobiological plasticity and processes such as division of labor and the regulation of fertility provides a good foundation because these processes are relatively well explored from a sociobiological standpoint. For example, in honey bees, it may be productive to test whether social factors that regulate division of labor (e.g., older workers, the queen, and the brood; Huang and Robinson 1992, 1996; Le Conte, Mohammedi, and Robinson 2001; Pankiw et al. 1998) also influence plasticity in circadian rhythms (Meshi and Bloch 2007). In the case of queens, it is still unclear whether the post-mating switch to arrhythmicity relates to her interactions with mates (including the possible involvement of agents transferred with semen; e.g., Qazi, Heifetz, and Wolfner 2003), to egg laying behavior, or to changes in her social status.

There has been some progress in studies on social entrainment. Southwick and Moritz (1987) found that bees need physical contact to achieve a synchronized group rhythm. Moritz and Kryger (1994) later reported that rhythms in temperature and oxygen consumption were partially synchronized between two groups of workers even when separated by a solid Plexiglas partition; synchronization was improved in experiments in which the Plexiglas division was punched with holes. These findings suggest that both direct contact (that may include food exchange and contact pheromones) and indirect influences (e.g., temperature and volatile pheromones) may have an impact on the clock of colony members. Moritz and Kryger proposed a self-organization model in which the activity and metabolism of all individuals produce oscillations in local ambient temperature that in turn entrains the clock of each individual. Temperature is an attractive time-giver for honey bees and other cavity or underground nesting social insects because it is effective in a constantly dark nest. The importance of temperature synchronization is questionable, however, because typical-size colonies are tightly thermoregulated (Winston 1987). Moreover, this model requires an initial entrainment by relatively subtle temperature cycles, whereas in controlled experiments minimal temperature oscillations of 6 to 10°C were needed to entrain circadian rhythms in bees (Moore and Rankin 1993; Moritz and Kryger 1994). Additional studies are needed to further test this interesting model and its possible extensions to additional products of worker activity such as comb vibrations and CO_2 concentration (Anderson and Wilkins 1989). The hypothesis that chemical signals mediate social entrainment is also attractive because pheromones modulate almost all aspects of life in insect societies (Wilson 1971; Hölldobler and Wilson 1990) and the olfactory system is implicated in the social entrainment of both rodents and flies (Goel and Lee 1997b; Levine et al. 2002).

Molecular and Physiological Underpinnings Governing Chronobiological Plasticity

Artificial manipulations, including constant light or disturbance with clock gene structure or function in model organisms such as *Drosophila* and the mouse, have indeed been able to abolish overt circadian rhythms. It is still unclear, however, what mechanisms are involved in modulating the circadian system in animals naturally switching between activities with and without circadian rhythms. Principally, behavioral plasticity can result from changes in the mechanism for rhythm generation, from uncoupling the central pacemaker from downstream mechanisms controlling behavior, or from external influences on behavior that override (mask) the influence of the internal clock. If the mechanism for rhythm generation is implicated, then modifications at several levels of clock regulation can act individually or in concert to produce modified rhythms in behavior and physiology (Figure 18.1). The finding that the cycling in brain clock gene expression is attenuated in nurses is consistent with the hypothesis that, in honey bees, task-related chronobiological plasticity is mediated by genuine clockwork reorganization (Shemesh, Cohen, and Bloch 2007).

Important New Methods

The tremendous progress in elucidating the molecular and physiological basis of behavioral rhythms in flies and mice is largely due to the availability of reliable, precise, and detailed quantitative phenotypes (e.g., locomotor activity). Behavioral data are recorded automatically and over long periods of time without any interference from a human observer. In contrast

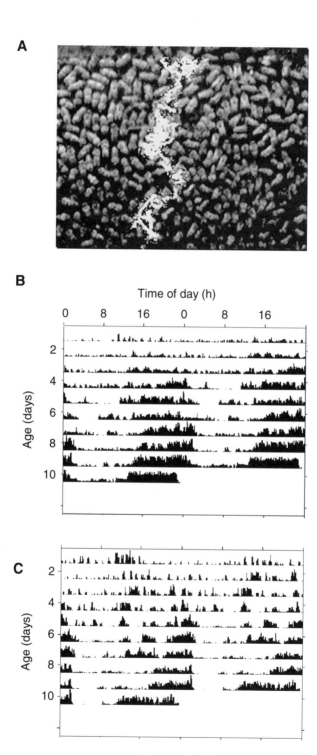

to human observers, automatic data acquisition systems have constant alertness and their recording is fully objective. However, since animals are isolated individually, these systems are limited in their applicability to studies of several interacting individuals as is necessary in the study of social insects. A possible solution to this limitation has emerged in the form of automatic systems using technologies such as pattern recognition or electric chips. For example, several commercially available systems identify and track color spots for individual identification (Meshi and Bloch 2007; Figure 18.3). Custom-made systems can be further programmed to recognize and track patterns such as geometric shapes or letters. Other monitoring systems can detect insects tagged with barcodes or distinct electric chips (e.g., Streit et al. 2003). These technologies hold great promise but are currently limited by their severely reduced efficiency in complex three-dimensional spaces. Systems that track colors are further limited by their need for broad spectrum illumination, which is not typical of cavity-dwelling species (Meshi and Bloch 2007).

New molecular techniques for measuring gene expression in individual insects are also promising. The molecular organization of the circadian pacemaker makes measurement of gene expression particularly powerful in chronobiology because the mRNA and protein products of many clock and clock-controlled genes cycle with circadian rhythms. These molecular rhythms open a window on the function of the internal pacemaker.

Figure 18.3. Automatic video tracking of activity and analyses of circadian rhythms in complex social environments. (A). The trajectory of an individually color-marked bee (outlined in white) was tracked under dim broad spectrum illumination with the EthoVision Color Pro system. Note that the system successfully tracked the marked bee despite the background of hundreds of unmarked active bees. (B). Locomotor activity over time for two bees that were tracked simultaneously from the same cage (which also contained 30 unmarked bees). Each bee was paint-marked with a different color and video tracked over 10 successive days. The distance and time data were exported to the ClockLab circadian software (Actimetrics, Evanston, IL) for chronobiological analyses (for more details see Meshi and Bloch 2007). The numbers on the Y-axis of each activity plot (actogram) depict the age of the bee (days after emergence). The height of the small bars within each day corresponds to the distance it moved in each 10 min bin. time of day (h) is depicted at the top of each actogram. Data for each day are double plotted to facilitate visual detection of rhythmicity.

Quantification of clock gene expression may also reveal alternations in basal level, for which there might be several explanations, including variation in the number of clock-gene expressing cells (Bloch, Toma, and Robinson 2001; Bloch, Rubinstein, and Robinson 2004; Toma et al. 2000; Figure 18.2). Finally, neuroanatomical methods such as immunocytochemistry are valuable for studies on the anatomical organization of the circadian system and its network properties (Helfrich-Forster 2005; Sauman et al. 2005).

Conclusions and Future Directions

The temporal organization of individual behavior and its integration into a colony-level pattern is crucial for efficient colony function. The basic properties of the circadian system make it a valuable model for this task. The circadian system integrates animal behavior and physiology with its environment, and is adjusted rapidly to changing environmental conditions. Its modulation does not require complex learning processes and may last for a period sufficient enough to produce efficient temporal coordination. However, it is important to realize that not all evidence for temporal integration among individuals in a colony should be regarded as evidence for plasticity in the circadian clock; temporal coordination can be also achieved by means of social interactions (Fewell 2003) or self organization (Camazine et al. 2003) that do not require modifications in the circadian clock.

To date, it is not clear to what extent socially mediated chronobiological plasticity is a common feature of insect societies because it has been explored in only a handful of species. My hope is that extending this socio-chronobiological approach to additional species will not only determine the generality of these phenomena but will also lead to a better understanding of their adaptive value. It is also important to continue looking for additional manifestations of socially mediated plasticity in circadian rhythms. This can be achieved by using chronobiological tools in sociobiological studies. For example, division of labor–related chronobiological plasticity was discovered by comparing diurnal and circadian rhythms of nurses and foragers, and plasticity in queens was discovered by comparing circadian rhythms before and after mating.

The evidence for socially mediated chronobiological plasticity contributes to an understanding of social evolution because it suggests that an ancient and conserved system such as the circadian clock was changed during the evolution of sociality. A multilevel comparative approach will,

I believe, draw attention to clock properties that are typical, or even unique, to social insects and set the stage for evolutionary analyses of the origin and functional significance of these "social modifications." With information on additional species it may be possible to address another crucial question; namely, are social adaptations in the clock system a prerequisite for the evolution of sociality (or eusociality), its outcome, or adaptations to specific selection pressures in some social insects?

Another goal for future studies on chronobiological plasticity will be their contribution to a better understanding of the molecular- and system-level processes underlying naturally occurring behavioral plasticity. The integration of chronobiology and insect sociobiology is promising because chronobiology provides detailed, precise, and quantitative behavioral phenotypes and comprehensive formulations of underlying molecular- and system-level mechanisms. Sociobiology contributes documentation on a set of complex behaviors that can be profoundly and precisely manipulated while maintaining a natural, ecologically relevant context. The sequencing of genomes of additional social insects and future advances in applying RNA interference (RNAi) for silencing honey bee genes (Aase et al. 2005; Beye et al. 2002) will set the stage for rigorous studies on the molecular underpinnings governing plasticity in circadian rhythms in social insects. DNA microarrays have a huge potential for studies on clock-controlled genes and their social regulation (Sato et al. 2003) and have been developed already for the honey bee.

Another important line of research is the social modulation of plasticity in circadian rhythms. We need to identify the individuals involved and the relevant social interactions. At the physiological level, it is essential to uncover the sensory modalities and specific signals conveying social information. This information in turn will allow us to question how an individual or a signal is selected from a set of possible synchronizers or relevant signals, and will set the stage for exploring the interactions between social and physical signals such as light and temperature. Perhaps social life means that there are situations where social signals override the effect of light and other prominent environmental cues. A possible example for this could be the observation that honey bee nurses with no circadian rhythms continue to care for brood around the clock under a light–dark illumination regime, whereas bees of similar age that were removed from the hive or grew in the laboratory typically manifest strong circadian rhythms in locomotor activity (Moore et al. 1998; Shemesh, Cohen, and Bloch 2007).

Acknowledgments

I would like to thank Avital Meshi, Michal Merling, and two anonymous reviewers for their comments on earlier versions of this manuscript. The author's research was supported by grants from the Israeli Science Foundation (ISF), US—Israel Binational Foundation (BSF), German—Israeli Foundation (GIF), The National Institute for Psychobiology in Israel, and the Joseph H. and Belle R. Braun Senior Lecturership in Life Sciences.

Literature Cited

Aase, A. L. T. O., G. V. Amdam, A. Hagen, and S. W. Omholt. 2005. "A new method for rearing genetically manipulated honeybee workers." *Apidologie* 36: 293–299.

Anderson, C. M., and M. B. Wilkins. 1989. "Phase resetting of the circadian-rhythm of carbon-dioxide assimilation in bryophyllum leaves in relation to their malate content following brief exposure to high- and low-temperatures, darkness and 5-percent carbon-dioxide." *Planta* 180: 61–73.

Aschoff, J. 1960. "Exogenous and endogenous components in circadian rhythms." *Cold Spring Harbor Symposia on Quantitative Biology* 25: 11–28.

Beling, I. 1929. "Über das Zeitgedächtnis der Bienen. *Zeitschrift für Vergleichende.*" *Physiologie* 9: 259–388.

Bell-Pedersen, D., V. M. Cassone, D. J. Earnest, S. S. Golden, P. E. Hardin, T. L. Thomas, and M. J. Zoran. 2005. "Circadian rhythms from multiple oscillators: Lessons from diverse organisms." *Nature Reviews Genetics* 6: 544–556.

Ben Shahar, Y., H. T. Leung, W. L. Pak, M. B. Sokolowski, and G. E. Robinson. 2003. "cGMP-dependent changes in phototaxis: A possible role for the *foraging* gene in honeybee division of labor." *Journal of Experimental Biology* 206: 2507–2515.

Beye, M., S. Hartel, A. Hagen, M. Hasselmann, and S. W. Omholt. 2002. "Specific developmental gene silencing in the honeybee using a homeobox motif." *Insect Molecular Biology* 11: 527–532.

Bloch, G., and A. Meshi. 2007. "Influences of octopamine and juvenile hormone on locomotor behavior and *period* gene expression in the honeybee, *Apis mellifera.*" *Journal of Comparative Physiology A* 193: 181–199.

Bloch, G., and G. E. Robinson. 2001. "Reversal of honeybee behavioural rhythms." *Nature* 410: 1048.

Bloch, G., C. D. Rubinstein, and G. E. Robinson. 2004. "*Period* expression in the honeybee brain is developmentally regulated and not affected by light, flight experience, or colony type." *Insect Biochemistry and Molecular Biology* 34: 879–891.

Bloch, G., Y. Shemesh, and G. E. Robinson. 2006. "Seasonal and task-related variation in free running activity rhythms in honey bees *(Apis mellifera)." Insectes Sociaux* 53: 115–118.

Bloch, G., S. M. Solomon, G. E. Robinson, and S. E. Fahrbach 2003. "Patterns of PERIOD and pigment-dispersing hormone immunoreactivity in the brain of the European honeybee *(Apis mellifera)*: Age- and time-related plasticity." *Journal of Comparative Neurology* 464: 269–284.

Bloch, G., J. P. Sullivan, and G. E. Robinson. 2002. "Juvenile hormone and circadian locomotor activity in the honeybee *Apis mellifera." Journal of Insect Physiology* 48: 1123–1131.

Bloch, G., D. P. Toma, and G. E. Robinson. 2001. "Behavioral rhythmicity, age, division of labor and *period* expression in the honeybee brain." *Journal of Biological Rhythms* 16: 444–456.

Brady, J. 1987. "Circadian rhythms—Endogenous or exogenous?" *Journal of Comparative Physiology A* 161: 711–714.

Camazine, S., J.-L. Deneubourg, N. R. Franks, J. Sneyd, G. Theraulaz, and E. Bonabeau. 2003. *Self-organization in Biological Systems* (2nd ed.). Princeton: Princeton University Press.

Carrington, E., I. C. Kokay, J. Duthie, R. Lewis, and A. R. Mercer. 2007. "Manipulating the light/dark cycle: Effects on dopamine levels in optic lobes of the honeybee *(Apis mellifera)* brain." *Journal of Comparative Physiology A* 193: 167–180.

Cermakian, N., and P. Sassone-Corsi. 2000. "Multilevel regulation of the circadian clock." *Nature Reviews Molecular Cell Biology* 1: 59–67.

Crailsheim, K., N. Hrassnigg, and A. Stabentheiner. 1996. "Diurnal behavioural differences in forager and nurse honey bees *(Apis mellifera carnica* Pollm)." *Apidologie* 27: 235–244.

Crailsheim, K., U. Riessberger, B. Blaschon, R. Nowogrodzki, and N. Hrassnigg. 1999. "Short-term effects of simulated bad weather conditions upon the behaviour of food-storer honeybees during day and night *(Apis mellifera carnica* Pollmann)." *Apidologie* 30: 299–310.

Dunlap, J. C. 1999. "Molecular bases for circadian clocks." *Cell* 96: 271–290.

Dunlap, J. C., J. J. Loros, and P. J. Decoursey. 2004. *Chronobiology: Biological timekeeping.* Sunderland: Sinauer Associates.

Elekonich, M., D. J. Schulz, G. Bloch, and G. E. Robinson. 2001. "Juvenile hormone levels in honeybee *(Apis mellifera* L.) foragers: Foraging experience and diurnal variation." *Journal of Insect Physiology* 47: 1119–1125.

Fantinou, A. A., M. P. Alexandri, and J. A. Tsitsipis. 1998. "Adult emergence rhythm of the egg-parasitoid *Telenomus busseolae*." *BioControl* 43: 141–151.

Fewell, J. H. 2003. "Social insect networks." *Science* 301: 1867–1870.

Fleury, F., R. Allemand, F. Vavre, P. Fouillet, and M. Bouletreau. 2000. "Adaptive significance of a circadian clock: Temporal segregation of activities reduces intrinsic competitive inferiority in Drosophila parasitoids." *Proceedings of the Royal Society London B* 267: 1005–1010.

Forel, A. 1910. *Das sinnesleben der insekten.* Munich: Reinhardt.

Free, J. B., A. W. Ferguson, and J. R. Simpkins. 1992. "The behavior of queen honeybees and their attendants." *Physiological Entomology* 17: 43–55.

Frisch, B., and J. Aschoff. 1987. "Circadian rhythms in honeybees: Entrainment by feeding cycles." *Physiological Entomology* 12: 41–49.

Frisch, B., and N. Koeniger. 1994. "Social synchronization of the activity rhythms of honeybees within a colony." *Behavioral Ecology and Sociobiology* 35: 91–98.

Gattermann, R., and R. Weinandy. 1997. "Lack of social entrainment of circadian activity rhythms in the solitary golden hamster and in the highly social Mongolian gerbil." *Biological Rhythm Research* 28: 85–93.

Goel, N., and T. M. Lee. 1997a. "Social cues modulate free-running circadian activity rhythms in the diurnal rodent, *Octodon degus*." *American Journal of Physiology-Regulatory Integrative and Comparative Physiology* 273: R797–R804.

———. 1997b. "Olfactory bulbectomy impedes social but not photic reentrainment of circadian rhythms in female *Octodon degus*." *Journal of Biological Rhythms* 12: 362–370.

Grima, B., E. Chelot, R. Xia, and F. Rouyer. 2004. "Morning and evening peaks of activity rely on different clock neurons of the Drosophila brain." *Nature* 431: 869–873.

Hardin, P. E. 2004. "Transcription regulation within the circadian clock: The E-box and beyond." *Journal of Biological Rhythms* 19: 348–360.

———. 2005. "The circadian timekeeping system of Drosophila." *Current Biology* 15: R714–R722.

Heinzeller, T. 1976. "Circadiane anderungen im endokrinen system der Honigbiene, *Apis mellifera*, effekt von haft und Ocellenblendung." *Journal of Insect Physiology* 22: 315–321.

Helfrich-Forster, C. 2005. "Neurobiology of the fruit fly's circadian clock." *Genes Brain and Behavior* 4: 65–76.

Helfrich-Forster, C., M. Stengl, and U. Homberg. 1998. "Organization of the circadian system in insects." *Chronobiology International* 15: 567–594.

Helfrich-Forster, C., J. Wulf, and J. S. de Belle. 2002. "Mushroom body influence on locomotor activity and circadian rhythms in *Drosophila melanogaster*." *Journal of Neurogenetics* 16: 73–109.

Hepburn, H. R., J. J. Hugo, D. Mitchell, M. J. M. Nijland, and A. G. Scrimgeour. 1984. "On the energetic costs of wax production by the African honeybee, *Apis mellifera adansonii*." *South African Journal of Science* 80: 363–368.

Hölldobler, B., and E. O. Wilson. 1990. *The ants*. Cambridge: Belknap Press.

Huang, Z. Y., and G. E. Robinson. 1992. "Honeybee colony intergration: Worker—worker interactions mediate hormonally regulated plasticity in division of labor." *Proceedings of the National Academy of Sciences of the USA* 89: 11726–11729.

———. 1996. "Regulation of honeybee division of labor by colony age demography." *Behavioral Ecology and Sociobiology* 39: 147–158.

Jander R. 1957. "Die optische Richtungsorientierung der roten Waldameise (*Formica rufa* L.)." *Zeitschrift für Vergleichende Physiologie* 40: 162–238.

Kaiser, W., and J. Steiner-Kaiser. 1983. "Neuronal correlates of sleep, wakefulness and arousal in a diurnal insect." *Nature* 301: 707–709.

Kefuss, J. A., and W. P. Nye. 1970. "The influence of photoperiod on the flight activity of honeybees." *Journal of Apicultural Research* 9: 133–139.

Koeniger, N., and G. Koeniger. 2000. "Reproductive isolation among species of the genus *Apis*." *Apidologie* 31: 313–339.

Koltermann, R. 1971. "24-Std-periodik in der langzeiterinnerung an duft- und farbsignale bei der honigbiene." *Zeitschrift für Vergleichende Physiologie* 75: 49–68.

Le Conte, Y., A. Mohammedi, and G. E. Robinson. 2001. "Primer effects of a brood pheromone on honeybee behavioural development." *Proceedings of the Royal Society B* 268: 163–168.

Levine, J. D., P. Funes, H. B. Dowse, and J. C. Hall. 2002. "Resetting the circadian clock by social experience in *Drosophila melanogaster*." *Science* 298: 2010–2012.

Lin, Y., M. Han, B. Shimada, L. Wang, T. M. Gibler, A. Amarakone, T. A. Awad, G. D. Stormo, R. N. Van Gelder, and P. H. Taghert. 2002. "Influence of the *period*-dependent circadian clock on diurnal, circadian, and aperiodic gene expression in *Drosophila melanogaster*." *Proceedings of the National Academy of Sciences USA* 99: 9562–9567.

Majercak, J., W. F. Chen, and I. Edery. 2004. "Splicing of the *period* gene 3′-terminal intron is regulated by light, circadian clock factors, and phospholipase C." *Molecular and Cellular Biology* 24: 3359–3372.

Majercak, J., D. Sidote, P. E. Hardin, and I. Edery. 1999. "How a circadian clock adapts to seasonal decreases in temperature and day length." *Neuron* 24: 219–230.

Martin, H., and U. Martin. 1987. "Transfer of a time-signal isochronous with local time in translocation experiments to the geographical longitude." *Journal of Comparative Physiology A* 160: 3–9.

McCluskey, E. S. 1958. "Daily rhythms in male harvester and Argentine ants." *Science* 128: 536–537.

———. 1965. "Circadian rhythms in male ants of five diverse species." *Science* 150: 1037–1039.

———. 1967. "Circadian rhythms in female ants, and loss after mating flight." *Comparative Biochemistry and Physiology* 23: 665–677.

———. 1992. "Periodicity and diversity in ant mating flights." *Comparative Biochemistry and Physiology* 103: 241–243.

McCluskey, E. S., and C. E. Carter. 1969. "Loss of rhythmic activity in female ants caused by mating." *Comparative Biochemistry and Physiology* 31: 217–226.

Medugorac, I., and M. Lindauer. 1967. "Der einfluss der CO_2-narkose auf das zeitgedächtnis der bienen." *Zeitschrift für Vergleichende Physiologie* 55: 450–474.

Meshi, A., and G. Bloch. 2007. "Monitoring circadian rhythms of individual honey bees in a social environment reveal social influences on post-embryonic ontogeny of activity rhythms." *Journal of Biological Rhythms:* in revision.

Moore, D. 2001. "Honeybee circadian clocks: Behavioral control from individual workers to whole-colony rhythms." *Journal of Insect Physiology* 47: 843–857.

Moore, D., J. E. Angel, I. M. Cheeseman, S. E. Fahrbach, and G. E. Robinson. 1998. "Timekeeping in the honeybee colony: Integration of circadian rhythms and division of labor." *Behavioral Ecology and Sociobiology* 43: 147–160.

Moore, D., and M. A. Rankin. 1985. "Circadian locomotor rhythms in individual honeybees, *Apis mellifera*." *Physiological Entomology* 10: 191–198.

———. 1993. "Light and temperature entrainment of a locomotor rhythm in honeybees." *Physiological Entomology* 18: 271–278.

Moritz, R. F. A., and P. Kryger. 1994. "Self-organization of circadian rhythms in groups of honeybees (*Apis mellifera* L.)." *Behavioral Ecology and Sociobiology* 34: 211–215.

Moritz, R. F. A., and F. Sakofski. 1991. "The role of the queen in circadian rhythems of honeybees (*Apis mellifera* L.)." *Behavioral Ecology and Sociobiology* 29: 361–365.

Panda, S., J. B. Hogenesch, and S. A. Kay. 2002. "Circadian rhythms from flies to human." *Nature* 417: 329–335.

Pankiw, T., Z. Y. Huang, M. L. Winston, and G. E. Robinson. 1998. "Queen mandibular gland pheromone influences worker honeybee (*Apis mellifera*) l. Foraging ontogeny and juvenile hormone titers." *Journal of Insect Physiology* 44: 685–692.

Qazi, M. C. B, Y. Heifetz, and M. F. Wolfner. 2003. "The developments between gametogenesis and fertilization: Ovulation and female sperm storage in *Drosophila melanogaster.*" *Developmental Biology* 256: 195–211.

Refinetti, R., D. E. Nelson, and M. Menaker. 1992. "Social-stimuli fail to act as entraining agents of circadian-rhythms in the golden-hamster." *Journal of Comparative Physiology A* 170: 181–187.

Roces, F. 1995. "Variable thermal sensitivity as output of a circadian clock controlling the bimodal rhythm of temperature choice in the ant *Camponotus mus.*" *Journal of Comparative Physiology A* 177: 637–643.

Roces, F., and J. A. Nunez. 1996. "A circadian rhythm of thermal preference in the ant *Camponotus mus:* Masking and entrainment by temperature cycles." *Physiological Entomology* 21: 138–142.

Rubin, E. B., Y. Shemesh, M. Cohen, S. Elgavish, H. M. Robertson, and G. Bloch. 2006. "Molecular and phylogenetic analyses reveal mammalian-like clockwork in the honeybee *(Apis mellifera)* and shed new light on the molecular evolution of the circadian clock." *Genome Research* 16: 1352–1365.

Sasagawa, H., R. Narita, Y. Kitagawa, and T. Kadowaki. 2003. "The expression of genes encoding visual components is regulated by a circadian clock, light environment and age in the honeybee *(Apis mellifera).*" *European Journal of Neuroscience* 17: 963–970.

Sato, T. K., S. Panda, S. A. Kay, and J. B. Hogenesch. 2003. "DNA arrays: Applications and implications for circadian biology." *Journal of Biological Rhythms* 18: 96–105.

Sauer, S., M. Kinkelin, E. Herrmann, and W. Kaiser. 2003. "The dynamics of sleep-like behaviour in honey bees." *Journal of Comparative Physiology A* 189: 599–607.

Sauman, I., and S. M. Reppert. 1996. "Circadian clock neurons in the silkmoth *Antheraea pernyi*—Novel mechanisms of *period* protein regulation." *Neuron* 17: 889–900.

Sauman, I., A. D. Briscoe, H. Zhu, D. Shi, O. Froy, J. Stalleicken, Q. Yuan, A. Casselman, and S. M. Reppert. 2005. "Connecting the navigational clock to sun compass input in monarch butterfly brain." *Neuron* 46: 457–467.

Saunders, D. S. 2002. *Insect clocks* (3rd ed.). Amsterdam: Elsevier Press.

Schmolz, E., D. Hoffmeister, and I. Lamprecht. 2002. "Calorimetric investigations on metabolic rates and thermoregulation of sleeping honeybees *(Apis mellifera carnica)*." *Thermochimica Acta* 382: 221–227.

Sharma, V. K., S. R. Lone, and A. Goel. 2004. "Clocks for sex: Loss of circadian rhythms in ants after mating?" *Naturwissenschaften* 91: 334–337.

Sharma, V. K., S. R. Lone, A. Goel, and M. K. Chandrashekaran. 2004. "Circadian consequences of social organization in the ant species *Camponotus compressus*." *Naturwissenschaften* 91: 386–390.

Shemesh, Y., M. Cohen, and G. Bloch. 2007. "Natural plasticity in circadian rhythms is mediated by reorganization in the molecular clockwork in honeybees." *The FASEB Journal* 21: 2304–2311

Shimizu, I., Y. Kawai, M. Taniguchi, and S. Aoki. 2001. "Circadian rhythm and cDNA cloning of the clock gene *period* in the honeybee *Apis cerana japonica*." *Zoological Science* 18: 779–789.

Southwick, E. E., and R. F. A. Moritz. 1987. "Social synchronization of circadian rhythms of metabolism in honeybees *(Apis mellifera)*." *Physiological Entomology* 12: 209–212.

Spangler, H. G. 1972. "Daily activity rhythms of individual worker and drone honeybees." *Annals of the Entomological Society of America* 65: 1073–1076.

Stoleru, D., Y, Peng, J. Agosto, and M. Rosbash. 2004. "Coupled oscillators control morning and evening locomotor behaviour of Drosophila." *Nature* 431: 862–868.

Streit, S., F. Bock, C. W. Pirk, and J. Tautz. 2003. "Automatic life-long monitoring of individual insect behaviour now possible." *Zoology* 106: 169–171.

Stussi, T., and M. L. Harmelin. 1966. "Recherche sur l'ontogenese du rythme circadien de la depense d'energie chez l'Abeille." *C R Acad Sci Hebd Seances Acad Sci D* 262: 2066–2069.

Toma, D. P., G. Bloch, D. Moore, and G. E. Robinson. 2000. "Changes in *period* mRNA levels in the brain and division of labor in honeybee colonies." *Proceedings of the National Academy of Sciences of the USA* 97: 6914–6919.

Vogel, H., T. Heinzeller, and M. Renner. 1977. "Daily protein synthesis in the pars intercerebralis and corpora allata of *Apis mellifica* L. with and without training to a foraging schedule." *Journal of Comparative Physiology A* 118: 51–60.

von Buttel-Reepen, H. B. 1900. *Sind die Bienen Reflexmaschinen?* Leipzig: Junge and John, Erlangen, 82 pp.

von Frisch, K. 1967. *The dance language and orientation of bees.* Cambridge: Harvard University Press.

Wahl, O. 1932. "Neue Untersuchungen ueber das Zeitgedaechtnis der Bienen." *Zeitschrift fuer Vergleichende Physiologie* 16: 529–589.

———. 1933. "Beitrag zur frage der biologischen bedeutung des zeitgedaechtnises der bienen." *Zeitschrift fuer Vergleichende Physiologie* 18: 709–717.

Weinstock, G. M., G. E. Robinson, R. A. Gibbs, et al. 2006. "Insights into social insects from the genome of the honeybee *Apis mellifera.*" *Nature* 443: 931–949.

Willmer, P. G., and G. N. Stone. 2004. "Behavioral, ecological, and physiological determinants of the activity patterns of bees." *Advances in the Study of Behavior* 34: 347–466.

Wilson, E. O. 1971. *The insect societies* (3rd ed.). Cambridge: Belknap Press.

Winston, M. L. 1987. *The biology of the honey bee.* Cambridge: Harvard University Press.

Yerushalmi, S., S. Bodenhaimer, and G. Bloch. 2006. "Developmentally determined attenuation in circadian rhythms links chronobiology to social organization in bees." *Journal of Experimental Biology* 209: 1044–1051.

Young, M. W. and S. A. Kay. 2001. "Time zones: a comparative genetics of circadian clocks." *Nature Review Genetics* 2: 702–715.

Zavodska, R., I. Sauman, and F. Sehnal. 2003. "Distribution of PER protein, pigment-dispersing hormone, prothoracicotropic hormone, and eclosion hormone in the cephalic nervous system of insects." *Journal of Biological Rhythms* 18: 106–122.

PART FOUR

Theoretical Perspectives on
Social Organization

JENNIFER FEWELL

WHEN I WAS a graduate student first searching for a taxon to fit my interests in behavior, foraging, and sociality, my advisor, Mike Breed, did a wonderful thing. He took me to a field covered in harvester ant nests and left me there with a paper to read. The paper was Bert Hölldobler's classic work on harvester ant sociobiology, covering 90 pages of the first issue of *Behavioral Ecology and Sociobiology*. A day of watching ants and an evening of reading that paper gave me my first real insight into the world of social insects, and I was hooked. This is the impact of Bert's work, that he inspires and informs people in all areas of social insect biology and sociobiology more generally: from those interested primarily in insects to those concerned with the human condition. His research and his perspective have helped define the field of behavioral ecology and have served as a foundation for much of the modern research in ant behavioral ecology.

This is the impact of Hölldobler's work, that he inspires and informs people in all areas of social insect biology and sociobiology more generally: from those interested primarily in insects to those concerned with the human condition. Hölldobler's impact across this volume is pervasive, but he has a particularly strong influence in the area of building theoretical frameworks. Although he is well known for his empirical contributions, his theoretical contributions have been lasting and continue today. We see his profound influence in our view of transitions in social evolution, including current characterizations of hierarchical and distributed social structures (Hölldobler and Wilson were the first to apply the concept of heterarchies

433

to social insect colonies). Hölldobler's work on communication also formed the basis of a network perspective even before the term became popular. His influence continues beyond this volume, with his recent and ongoing contributions to the discussion on levels of selection and the colony as a superorganism.

In this section, we explore current areas of expansion in social insect biology as they influence and in turn are influenced by current interests in the social insect colony as a complex system. The social insects again lead the way in providing a theoretical and empirical framework for understanding social complexity, especially in the areas of self-organization and emergence, in which we examine phenotypes and properties generated by the dynamic interactions of group members and the mechanisms by which those dynamics occur. This perspective is formally introduced in the enjoyable chapter by Nigel Franks, Anna Dornhaus, James Marshall, and Francois-Xavier Moncharmont who frame sociobiological principles in the context of mathematics. Raphaël Jeanson and Jean-Louis Deneubourg continue this perspective with their chapter illustrating how positive feedback, as the basis of self-organization, can generate complex behaviors within social groups. These mechanisms and concepts are applied to two global attributes of complex social systems in the chapters by Stephen Pratt on collective decision making and Jennifer Fewell, Shana Schmidt, and Thomas Taylor on division of labor.

Current theory in sociobiology is not just about emergence. Our theoretical contributions expand in new directions with the chapter by Gene Robinson and Andrew Barron, who apply the application of modularity modeling to behavioral development. Nina Fefferman and James Traniello also expand the utility of social insects as a model system when they integrate sociobiology and epidemiological models to examine the relationship between social complexity and disease. Our final two chapters address the question of how we view social insects as units of selection and evolution. Andrew Hamilton, Nathan Smith, and Matthew Haber present the provocative argument that cohesion and integration within the colony characterize it as an individual unit; in doing so, they open up the concept of what constitutes an individual—and an organism. In our final chapter, Jürgen Gadau and Manfred Laubichler make the case that the integrated development and evolutionary trajectories of social insect colony phenotypes are an ideal model system for the application of social behavior to evolutionary development. Emergence and modularity, levels of selection and cohesion: these were not terms bounced around in coffee conversation when Hölldobler first began looking at social insect colonies through his unique lens. However, they illuminate new directions for this field, ones on which he still leaves his mark.

CHAPTER NINETEEN

The Dawn of a Golden Age in
Mathematical Insect Sociobiology

NIGEL R. FRANKS

ANNA DORNHAUS

JAMES A. R. MARSHALL

FRANCOIS-XAVIER DECHAUME MONCHARMONT

THE TITLE OF THIS CHAPTER is a prediction. It is bold. It is also arguably overly grand and it may be illusory. History alone will judge if this is the dawn of a golden age in mathematical sociobiology. To be sure, mathematical biology has already seen a number of false dawns. It may appear, for example, that both catastrophe theory and chaos theory each have enjoyed almost all of their 15 minutes of fame. However, it is right and proper that a wave of initial excitement, or indeed, hyperbole, is followed by slower and steadier progress as a field matures.

So what justifies our unbridled optimism? The first answer is demonstrable progress. Self-organization theory and complex systems theory coupled with pioneering experiments have already revolutionized our understanding of organizational aspects of insect societies and even our own societies. Such is the gathering excitement, predictive power, and massing evidence that this endeavor has earned a new epithet: "Sociophysics" or "the physics of society" (Ball 2004; Strogatz 2004). Why sociophysics? Because the philosophy and even some of the principles of statistical mechanics are now being applied in sociobiology to great effect. The central issue in social biology to which mathematical biology is being applied is the question of how societies are organized. In shorthand, how do super-organisms work?

The second answer is that necessity is the mother of invention. The requirement for a mathematical sociobiology was not only predicted by

Wilson (1971) but he gave a clear directive that results would come from considering mass action and stochastic effects. These are key aspects of statistical mechanics. Modeling is needed because the mechanisms underlying insect societies as complex systems *are neither self-evident nor susceptible of easy proof*" (Wilson 1971; emphasis added). Thought experiments are not sufficient to predict the properties of the whole from those of the parts. Their complexity and emergent properties are such that these systems are opaque until they have been disassembled and re-assembled and interrogated with a combination of mathematical models and experiments.

In this chapter, we discuss how different modeling techniques can be used to study the functions, mechanisms, and evolution of collective behavior, and how such theoretical approaches can also feed back into empirical research, or into disciplines outside of biology. Finally, we give our—admittedly subjective—recommendations for the future of theoretical sociobiology.

A Typology of Models

Explanatory models can be split into different categories depending on their aims: teleonomic models (often called optimality models), mechanistic models, and artificial evolution models. Models of the first category aim to illuminate the function or goal of a behavior or other trait (under the assumption that the trait is the result of adaptive evolution, Cuthill 2005). Mechanistic models aim to explain the mechanisms by which a pattern of behavior or other feature is created. Finally, artificial evolution models may help in the understanding of the evolutionary process itself. We discuss each of these types in turn below, along with their benefits and certain cautionary notes.

Teleonomic Models

Teleonomic models try to portray the function of an aspect of a biological system in an evolutionary context. For example, how could we explain the biased sex-ratio in an ant colony (Bourke and Franks 1995)? The word *teleonomy* was first used by C. S. Pittendrigh (1958) in order to emphasize that the recognition and description of end-directedness does not carry a commitment to Aristotelian teleology as an efficient causal principle. In other words, an organism may evolve to an optimal state without evolution aiming for that state; a failure to recognize this is at the root of Intelligent Design criticisms of evolution. Teleonomic models do not focus on the mechanisms of decision or control of a particular behavior, but rely on the assumption that it was shaped by natural selection and thus looks as if it was optimized for some function relating to reproductive success. Some classical examples of such teleonomic approaches can be found in the optimal foraging literature, which derives optimal behavior from assumptions such as maximization of food intake (Charnov 1976).

Most, but not all, teleonomic models are mathematical in nature; the system under study is (typically) represented by differential equations, which can sometimes be solved analytically. These equations usually represent behavior at a group or population level. In this they are similar to techniques used in physics such as statistical mechanics which applies the techniques of statistics to large numbers of interacting particles to explain the macroscopic properties of populations of such particles. Statistical mechanics is needed when particles, such as molecules, are individually invisible and yet the global properties of the population to which they belong (such as a gas) need to be predicted. Furthermore, the complexity of resolving all interactions is unimaginable even if the initial state and the equations of motion are known. By contrast, classical (or Newtonian) mechanics charts the history of small numbers of visible particles that are relatively large and relatively slow. In insect sociobiology the interacting agents (e.g., workers) can often be marked individually and their movements recorded, but this is not always the case. Individual army ants almost instantly vanish among their nestmates in swarm raids and in many cases of nest construction in ants and termites the individuals disappear underground—but their mass efforts are still of considerable interest and indeed often of awesome complexity and beauty (Franks et al. 1991).

One of the founding fathers of statistical mechanics was the great Victorian physicist James Clerk Maxwell (1831–1879). Maxwell and others drew much of their inspiration for statistical mechanics from contemporary sociologists (Ball 2004). In particular, Maxwell was influenced by Henry Thomas Buckle (1821–1862), a social historian (and author of *History of Civilization in England*) who argued that societies had underlying regular characteristics notwithstanding the uniqueness and capriciousness of their individual members. For example, Maxwell stated (referring to the study of human behavior): "If we betake ourselves to the statistical method, we do so confessing that we are unable to follow the details of each case, and expecting that the effects of widespread causes, though very different in each individual, will produce an average result on the whole nation" (Maxwell 1873a).

Thus, statistical mechanics drew inspiration from studies of human societies and is now being applied to issues in our own societies, such as problems in traffic flow (Helbing and Huberman 1998; Helbing and Treiber 1998). Maxwell also tantalizingly referred to swarms of bees when he was trying to convey the idea of randomly mingling molecules: "If we wish to form a mental representation of what is going on among the molecules in calm air, we cannot do better than observe a swarm of bees, when every individual bee is flying furiously, first in one direction and then in another, while the swarm, as a whole, either remains at rest, or sails slowly through the air" (Maxwell 1873b). Maxwell attributed random movement to the bees in a swarm, likening them to random movement of air particles. Likening insects to molecules persists as a powerful metaphor (Detrain and Deneubourg 2006). However, real honey bees in swarms do not fly about at random (see for example, Couzin et al. 2005). It is a shame that it has taken so long for statistical mechanics to find its way back to one of its natural domains—insect societies.

Example 1: Partial loads in social insect foragers

Consider foraging in the honey bee. It has been observed that forager bees only partially load at a nectar food source. The optimal time spent in a flower patch of known quality may depend on the cost of flying with a heavy load (Schmid-Hempel, Kacilnek, and Houston 1985). Alternatively, it may be determined by the benefits of returning information about this new food source to the colony (Varju and Núñez 1993), or checking at the colony whether other superior food sources have been discovered (Dornhaus, Collins et al. 2006). Returning to the nest with a full load of nectar from a mediocre food source is very time consuming and could be a suboptimal strategy. Yet flying back quickly could also be a suboptimal strategy if new and better food sources are rarely discovered. Dornhaus, Collins et al. (2006) used a mathematical model to analyze the importance of information collected at the nest; because of its mathematical form, it requires just a few parameters.

The simplicity of mathematical models is both their strength and their weakness. Biological sophistication can never be captured fully in a model, but especially population-level models like the one described above cannot take individual variability into account, and in general many variables have to be aggregated in very few parameters. For some questions this is appropriate, for others it is not. However, there are other modeling techniques that allow inclusion of more biological detail. For teleonomic questions, two other approaches are also used: evolutionary modeling is used as an optimization method (as opposed to analytical derivation of an optimum as described in the example above), and individual-based models are used to quantify the benefits and costs of certain behaviors under a large number of conditions, thus also making predictions about which behavior tends to give the highest benefits in certain situations (Dornhaus, Klügl et al. 2006).

The aim of teleonomic models is to solve a problem from an engineering or strategic standpoint. For example, what decision rules should lead to the most effective nectar collection? Such an approach relies on the assumption that natural selection shaped the biological system in order to be most effective. If the experimental data deviate from the optimal predictions of the model, this will shed new light on the biological interpretation of the system.

Mechanistic Models

Mechanistic models attempt to generate and test hypotheses about proximate mechanisms creating a behavior, either at the individual or group level. Understanding the mechanisms underlying group-level social dynamics within a eusocial insect colony is difficult without abstraction tools. A colony of millions of workers, for example, is a quintessential complex system. It is here that individual-based models in particular have had their largest impact: examples of the outstanding success of such approaches in social insect studies can be found in the application of self-organization theory. Each mechanistic model assumes a particular mechanism, and predicts how the system will behave under this assumption. The predictions of several potential mechanisms can thus be compared; if any of these predict system behavior incorrectly, the corresponding mechanism can be rejected.

Example 2: The blind leading the blind and the self-organization of army ant swarm raids

Self-organization can be defined as a process that creates a pattern at the global level (e.g., the colony level) through multiple interactions among the components (e.g., the workers). The components interact through local, often simple, rules that do not explicitly code for the global pattern (see Camazine et al. 2001).

The global patterns, whose generation we will examine, are the swarm raids of the New World army ant *Eciton burchellii* (Figure 19.1). Such swarm raids are massive compared to the size of the individuals that create them. An *E. burchellii* raid can be 20 meters wide and 200 meters long and employ 200,000 individuals (Franks 1989). Furthermore, an overview

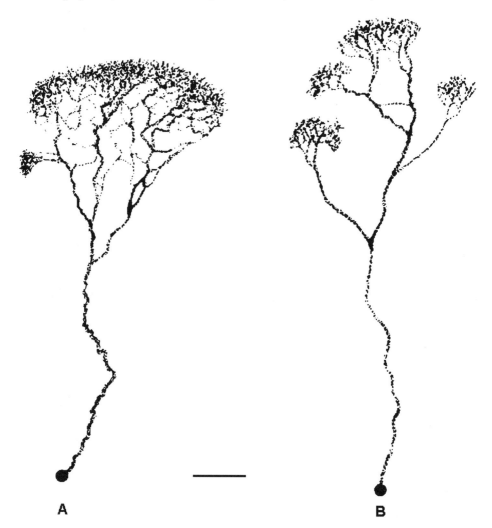

A   **B**

Figure 19.1. The typical pattern of swarm raiding by (A) *Eciton burchellii* and (B) *Eciton rapax*. The scale bar represents 5m. Redrawn from Burton and Franks, 1985, with the kind permission of Blackwell Publishing; (A) originally redrawn from Rettenmeyer (1963).

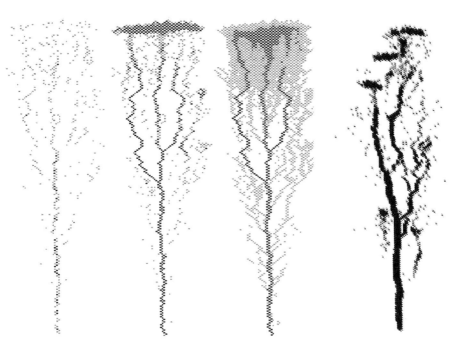

Figure 19.2. A cellular automata model of the generation of *Eciton* swarm raids. The first three illustrations are overlays of the same model raid for *Eciton burchellii*. From left to right the figure shows the density of returning ants; the density of all ants, and the density of pheromone trails. The darker the grey level the greater the density. The fourth and right-most illustration shows the density of all ants for the model re-run for a prey distribution typical of that encountered by an *Eciton rapax* raid. These illustrations were generated *de novo* by rerunning the computer simulation model described in Deneubourg et al. 1989 (see also Franks et al. 1991).

of the swarm raid pattern is not available to individual workers in *Eciton* because they have extremely rudimentary sight (Schneirla 1971; Franks 1989; Gotwald 1995). A computer simulation model of the self-organization of an army ant raid (see Figure 19.2) used the following simple set of rules (Deneubourg et al. 1989).

1. *Leading:* Each and every ant lays a followable pheromone trail wherever it goes (unless it is on a trail fully saturated with pheromone).
2. *Randomness:* If an ant is in virgin terrain it randomly goes left or right (at every bifurcation point in the computer simulation lattice).

3. *Following:* If an ant is in terrain already traversed by a nestmate it is most likely to follow the pheromone trail laid by that nestmate. (E.g., it has a higher probability of turning left than right if the previous ant turned left rather than right, or vice versa, and marked that path.) Because army ants follow one another's trails and reinforce them, trails can get stronger and stronger.

4. *Speeding:* Up to a limit, the more trail pheromone present the faster each individual will run.

5. *Crowding:* An ant will not, however, enter an area that is already over-crowded by its nestmates.

6. *Returning:* Ants only return home when they have encountered prey items. Returning ants obey the same rules for following the pheromone trail as outgoing ants, but they lay more trail pheromone than outgoing ants.

7. *Flow:* A constant number of ants leave the nest per unit time.

These seven golden rules are sufficient for the blind to lead the blind in the creation of a swarm raid of dazzling complexity and sophistication. It is imperative to note, however, that these seven qualitative rules must be executed in good quantitative agreement with the characteristics of the real army ants. The rate at which individual workers move matters; the rates of deposition and evaporation of trail pheromones matter; and the relative rates of trail-laying by outward bound and returning ants matter (Franks et al. 1991). All these quantitative variables have been established by studying experimentally the movements of *E. burchellii* workers and they have been incorporated into the model (Figure 19.2). That is, both the qualitative and quantitative assumptions of the model have been verified. Certain predictions of this modeling have also been tested (Franks et al. 1991).

The single most intriguing prediction of this model is that swarm raids have an active architecture. These are event-driven systems. The precise pattern of a raid depends on the distribution of prey encountered. Indeed, when the model is run with no prey and hence no returning ants (rule 6), the anastamosing series of columns behind the raid front is absent and a raid consists only of a broad swarm front and a principle trail. Furthermore, if prey are in large clumps that are few and far between, the model predicts that a swarm raid may break up into divergent sub-swarms. This occurs because strong return traffic flows from two (or more) directions and the outgoing ants part company. Such a raid pattern consisting of sub-swarms is seen in *E. rapax* (Burton and Franks 1985), a close relative of *E. burchellii*, which is a specialist predator of other social insect nests and thus encounters large clumps of prey that are few and far between (Figure 19.2b). This prediction has been tested by presenting the normally cohesive swarm raids of *E. burchellii* with large clumps of prey that are few and far between. The result was that *E. burchellii* adopted a swarm raid pattern similar to that of *E. rapax* (Franks et al. 1991).

Evolutionary Models and Alternative Worlds

In constructing a model of a biological system, whether an organism or a society, we need not limit ourselves to the representation of that which currently exists. Invaluable as models of existing systems are, it can be equally interesting and instructive to model that which does not exist, but may have or indeed once did. In doing so we begin to touch on a fundamental question of the contingency of evolution; if the tape of evolution were replayed, would the outcome be the same (Gould 1989)? Considering evolutionary alternatives in terms of relative effectiveness for a given task might help illuminate whether a feature evolved as it did due to a direct fitness advantage, or whether evolutionary or physical constraints may have been involved. Considering evolutionary alternatives in a sequence may also allow us to recreate plausible simulations of real evolutionary events in a species, genus, or even phylum. Of course, for some, this approach may seem problematic or even misguided due to the lack of data required for rigorous validation. A converse viewpoint would be that such models might enable us to explore the consequences of theories in the absence of empirical data by acting as extended "opaque thought experiments" in the sense of Di Paolo, Noble, and Bullock (2000).

There is, however, an even more radical viewpoint on evolutionary models: they can be alternative worlds. That is, computer models may be actual instantiations of a different, noncarbon-based form of life, rather than mere simulacra. Under this view these models act as sources of data to inform our theorizing about the nature of life and its attendant processes of ontogeny, phylogeny, and so forth. Such a viewpoint is exemplified by Ray's work with digital organisms (1994) or Bedau's "emergent thought experiment" approach to computer modeling (1999). These issues have been recently summarized more thoroughly elsewhere (Marshall and Franks

2007); here we shall limit ourselves to considering examples of modeling evolutionary alternatives in a little more detail.

More boldly, if one could measure the effectiveness of different alternative realizations of the same adaptive feature, and know which realizations were present at various points in evolutionary history, one would be faced with the tantalizing prospect of being able to reproduce a plausible evolutionary history. Some work in this area has already been carried out, such as Niklas' (1999) work on plant morphospaces. Using a simple generative plant model that can give rise to a variety of morphologies from varying a small number of parameters, and equipped with a fitness function that evaluates morphologies in terms of criteria such as light absorption, mechanical stability, and so on, it is possible to construct "adaptive walks" through the morphospace of possible plant forms via a simple local search of similar forms of higher fitness. Niklas uses such a model to consider what the effects of increasing the number of morphological optimality criteria are on the fitness landscapes and the adaptive walks over them. However, if the starting point for the adaptive walk could be determined by reference to a fossil plant for example, would it not be intriguing if such a simple model could recreate the evolution from this ancestral form to the modern day form, via various intermediate forms observed in the fossil record?

For sociobiology, reconstructing evolutionary histories may be more problematic, as behavior is somewhat difficult to fossilize, and those fossils that do result from behavior may require much interpretation (e.g. Miller, 2003). We need not despair however, as we can look to more "primitive" behaviors, in other species, that are related to our species of interest. Also, we could generate the missing origin of the adaptive walk through our own hypothesizing; for example, one plausible sequence of adaptations leading to the evolution of the collective decision-making mechanism employed by *Temnothorax albipennis* during house-hunting has been proposed by Pratt et al. (2002, 2005; Pratt, this volume). Such hypotheses may be rescued from being labeled as "just-so" stories if a suitable computer model could demonstrate plausibly the adaptive value of each link in the proposed evolutionary chain.

Classic Problems with Modeling Approaches

The two examples given above, foraging by honey bees and the self-organization of army ant raids, exemplify the value of supplementing empirical with theoretical studies. Nevertheless, models are only as good as their assumptions. Indeed, when they rely on large sets of untested assumptions, they have to be handled with care and their predictions regarded with extreme caution, or better, unrestrained skepticism. Many models, whether individual-based or even based on a modest-sized system of differential equations, cannot be studied analytically, so they have to be studied through numerical resolution. The modification of a single parameter may have incredibly dramatic consequences for the general predictions of a model. This is why theoreticians using complex mechanistic models have a duty to perform long and tedious sensitivity analyses. It is becoming clear that simulation models are a new kind of experimental system (or more accurately, the use of simulation falls somewhere between traditional tractable formulation and experimental system). Simulations are properly explored using the same experimental (experimental design) and statistical techniques (visualization, statistics, data mining) that are used to explore real-world systems. If a modeler wants to discover functional dependencies, then a barrage of trials must also be run to examine predictions across a wide range of parameters.

Another pitfall to be avoided is the illusion that such theoretical approaches can proceed without being directed by strong biological questions. Certain models, for example, claim to be open-minded and to be question-free at their inception. Hogeweg and Hesper (1990) made exactly this kind of error while reviewing strong points of individual-based models. They explicitly referred to what they call "self-structuring and non-goal directed models . . . [Individual-based models] can be non-goal-oriented models, i.e. one does not have to determine in advance what features will be studied."

Biologists should never just naïvely hope that an interesting property will emerge from their model. The hope pinned by some to individual-based models was that, once the individuals had been specified in a model, the collective level consequences would emerge naturally. Similarly, a lot of complex simulations are built without any hierarchical structure. Because there appears to be no need to think carefully about the assumptions or the parameters, one might be tempted to consider that all the parameters have the same level of accuracy or importance. Such a view is downright dangerous, not least because the effects of important parameters can be diluted among a jumble of other effects.

For example, most mechanistic models require a much more detailed understanding of variability and stochasticity, both through differences between individuals and a changing environment, than population-level models. The study of inter-individual differences has gained popularity, maybe because of that; it has revealed that individuals can differ substantially in their sensory, cognitive, and motor abilities (Scheiner, Erber and Page 1999; Scheiner and Erber, this volume), as well as differing in their behavioral responses to the same stimuli (Weidenmüller 2004; Chittka et al. 2003). This is an important aspect to consider if discussing, for example, task allocation in social insect colonies. Similarly, it is becoming clear that the physical environment, and its variability, may play a big role in creating organization. It can serve as a seed or template for structure, or it can be used as a medium not just for information transmission but also information storage (stigmergy) (Detrain and Deneubourg 2002; Dornhaus and Chittka 2005). Often the dynamics and the evolution of strategies used by social insects can only be understood by considering the type of environment and the variability to which they are exposed (Dornhaus and Chittka 2004b). The recognition of the importance of the environment should lead to new experimental studies that focus on quantifying environmental parameters and how they change, as well as the social insect behaviors themselves.

Earlier, we alluded to the term *sociophysics* and the label "the physics of society," and we think this term can be problematic. The problem arises if people imagine that societies can be reduced purely to physics. For example, Camazine et al. (2001) put extreme emphasis on the crucial point that self-organization is not an alternative explanation to natural selection for complexity in biology. Nevertheless, when their book was reviewed in *Nature,* a biophysicist repeated this error (Ben-Jacob 2002). Unfortunately, we think it is likely to be repeated again and again because some of the most dramatic recent demonstrations of self-organization in biology have experimentally accentuated the physics of these systems. Consider three recent examples from studies of ants. First, *Lasius niger* does not habitually forage on bridges (Dussutour et al. 2004). Second, *Linepithema humile* workers probably very rarely accumulate at the end of twigs and then drip-off like a chaotically leaking tap (Bonabeau et al. 1998; Theraulaz et al. 2001). Third, *Messor sancta* in nature will never encounter so many dead bodies outside their nest, in a circular arena, that they will exhibit Turing morphogenesis—producing cemeteries in regularly spaced clusters (Theraulaz 2002). Nevertheless, even though all these studies arguably excessively turn up the heat to accentuate the "signal to noise ratio," they are valuable in that they reveal the biological principles that may have an important role in these systems—only to a much more subtle degree in the natural world. As is the case with all laboratory studies of behavior, ideally they should be followed up with fieldwork whenever this is possible. Of course, the purpose of some of these experiments may not have been the study of (natural) ant behavior, but the study of collective behavior of a complex system where one tests, for example, how different patterns can be created. This can be a worthy goal in itself, especially if the resulting insights can be used successfully to design artificial systems that employ logically similar processes to generate useful patterns and procedures. It is important to note that none of these findings show that biology reduces to physics. Biology is unique among the sciences because it is the domain of evolution by natural selection, thus it is underpinned by physics and chemistry but does not reduce to them.

New Topics in (Empirical) Insect Sociobiology

The use of a variety of modeling approaches has made sociobiology a truly interdisciplinary effort. Once such connections between disciplines are established, methods and ideas can be exchanged in both directions. This has inspired and helped the empirical study of social insects as well. Biologists are now exploring new questions experimentally, inspired by theoretical advances in other fields. Networks theory, with the now famous "small world effects," was established in mathematics and sociology, but is now being applied to the pattern of interactions in social groups of animals (Fewell 2003; Couzin et al. 2005) as well as that of ecological interactions (Memmott 1999; Memmott, Waser, and Price 2004). The realization in mathematics and physics that positive feedback systems can lead to pattern formation has prompted biologists to examine feedback loops, for example, in recruitment (Beekman, Sumpter, and Ratnieks 2001), traffic flow (Couzin and Franks 2002), aggregation behavior (Jeanson et al. 2004; Jeanson and Deneubourg, this volume), and spatial structure (Theraulaz et al. 2002; Bonabeau et al. 1998). Researchers in economics and epidemiology have developed methods of analyzing the spread of ideas and diseases, looking for patterns of information flow in heterogeneous populations (Britton et al. 2002; Feffermann and Traniello this volume) and discovering how some disappear while others spread rapidly. Similarly, social insect researchers have realized that information is not evenly distributed in an insect colony and have started to study ways of information flow (e.g., Seeley 1998; Dornhaus and Chittka 2004a). They, in turn, found bifurcation and collective decision making involving minimal colony sizes and quorum thresholds reminiscent of critical masses (Franks et al. 2002; Beckers et al. 1989; Anderson and McShea 2001; Seeley and Visscher 2004). Others have demonstrated how decision making in a social insect colony may function in a similar manner to neural circuits in the primate brain, and thus similarly achieve optimal decision making (Marshall et al., submitted). More parallels are sure to be found between such systems, and rather than reinventing wheels separately, researchers would do well to explore the use of techniques and results already established for similar systems in other fields.

Nevertheless, interdisciplinary research has its difficulties, not just because one has to admit ignorance and be prepared to learn about a new system. It also often means adjusting to the different traditions of research and communication in another field. It is therefore not surprising that a certain reluctance has to be overcome before a common language is found and ideas are fruitfully exchanged between disciplines. To some degree, the complexity of the models used in theoretical sociobiology has forced biologists to interact with mathematicians and computer scientists, and the complexity of many man-made systems has led computer scientists and engineers to look for problem-solving strategies in (equally complex) biological systems. The emergence of pattern from collective activity—self-organization—has thus become a buzz-word in several disciplines. Social insect research, neuroscience, physics, sociology, computer science, and other fields have discovered the similarities in the processes underlying pattern formation in their systems, which was only possible after they had started creating a common language to talk about them (Camazine et al. 2001; Ball 1999). This interaction has introduced not only new ideas to biology, and particularly social insect science, but also new tools, such as individual-based simulations and other mathematical and computational techniques. A growing number of studies now look at mechanisms of how collective pattern is created (e.g., Bonabeau et al. 1998; Franks and Deneubourg 1997; Camazine et al. 2001; Millor et al. 1999; Roces 2002; Sendova-Franks and Franks 1999; Watmough and Camazine 1995), in addition to studying why animals collaborate in the first place. The latter involves weighing benefits against costs and thus often mathematical models; the former, however, is usually too complex to be solved analytically and is only possible by the rise in the power of available computational tools.

New modeling techniques can open up new areas to be studied, and thereby inspire new experiments. In fact, the more interesting results of modeling studies are not those that confirm old hypotheses; they are those results that show us that we do not actually understand why and how things happen the way they do. This, in an ideal world, should lead to new and experimental ways of approaching the system under study, but the temptation is great to bend and "correct" the model in such a way that it conforms to previous hypotheses. It is vital that we let our views of how social insect colonies work be challenged by results from new modeling studies; at the same time, all models should suggest ways to test any new hypotheses in the real world.

Attempting to build a model of collective behavior usually leads to the realization of how much we still do not know. Too often, the interaction is a one-way street: models use experimentally collected data, but predictions of models are then not tested on a real-world system. We think that this is one of the big opportunities, but also challenges, opened up by using models: to close the gap, to help identify and lead to the collection of missing data, and to test new hypotheses and predictions from modeling studies experimentally. However, it is also important to realize that models can lead not only to new experiments but also to important new questions.

Repaying the Favor: Feedback from Biology into Other Disciplines

Thus far, we have sought to demonstrate the contribution that mathematical models and computers have made to the development of insect sociobiology. In return, the study of social insects has made contributions of its own to computer science, mathematics, and related disciplines. Computer scientists are increasingly realizing that social insects have evolved solutions to some difficult problems, such as foraging; problems that are similar in many respects to those encountered by computer scientists. In the twenty-first century the ubiquity of computers and networks such as the Internet has created a host of problems in which a global, accurate picture of the entire problem is not available for the planning of a solution. Rather, approaches to these problems must try to optimize a solution using only local and uncertain information. It is precisely these constraints that social insects work under; consider the example of foraging behavior in social insects, in which individual insects only have access to local information (the food sources they discover or are recruited to by other scouts) which is uncertain (food sources can fluctuate in quality, appear or disappear, etc.). For such problems the collective behaviors of social insects may provide inspiration for robust and efficient solutions. One example is the application of an algorithm based on pheromone-based foraging in ants to the routing of connections in a telecoms network (Dorigo and Stützle 2004). In the ants, shorter paths to a food source receive more ant traffic and hence more pheromone, and positive feedback for those shorter paths occurs. However, if new and better paths become available or old high-use ones are blocked, the colony is often able to adapt. Other social insect behaviors may be applicable to engineering problems; the house-hunting behavior of *Temnothorax albipennis*, for example, has been studied with reference to decentralized control problems such as process migration in computer networks (Marshall et al. 2006).

Even in cases where a global view of the problem is available, there is a significant class of problems in computer science and mathematics that are combinatorial; that is, there are too many possible solutions to evaluate all the alternatives exhaustively. One famous example is a kind of shortest path problem known as the Traveling Salesman Problem, in which the shortest possible tour visiting all the cities in a country exactly once must be discovered. Heuristic approaches for solving such problems are required, and here again the pheromone-based foraging behavior of ants has provided inspiration. By allowing a simulated ant colony to "forage" repeatedly for the shortest tour around the cities, important components of the tour are found to be those that repeatedly are heavily marked with virtual pheromone, and these components may be combined into a single good quality solution (Bonabeau, Dorigo, and Theraulaz 1999). Social insect-inspired engineering solutions have been derived from spatial sorting (Lumer 1994) or task allocation (Bonabeau, Dorigo, and Theraulaz 1999), and are used in job-shop scheduling (Cicirello 2004), software 'agents' (Parunak 1997; Weiss 1999), optimization of communication networks (DiCaro and Dorigo 1998), and collective robotics (e.g., for planetary exploration; Brooks and Flynn 1989; Krieger, Billeter, and Keller 2000). Moreover, computer science and mathematics are not the only disciplines to benefit from insights into the collective behavior of social insects; such disparate fields as corporate organization and sociology have also taken inspiration from social insect organization (Costa 2002; Parunak 1997). Of course, interest in what lessons social insects might hold for the organization of human affairs has a long and venerable history: "It is true that certain living creatures, as bees and ants, live sociably one with another (which are therefore by Aristotle numbered amongst political creatures) . . . and man may desire to know why mankind cannot do the same." (Hobbes 1651).

Recommendations for the New Generation Mathematical Sociobiologist

As the French mathematician Henri Poincaré (1905) said, "Science is built of facts as a house is built of stones; but an accumulation of facts is no more science than a heap of stones is a house." If biology is to be more than just "stamp collecting" (or stone heaping), and more like physics, then theory is essential. In fact, recent advances in the mathematical and computational sciences have brought theory to the fore in biology. However, Jacob (1970) was rather overstating the matter when he said, "One doesn't study life in laboratories these days." Rather, we feel that while modeling has tremendous contributions to make to the development of biological understanding, biological experimentation and validation should always be the final arbiter (see also Bray 2001; May 2004).

We believe that the quality of a modeling study is directly related to how clearly the questions, the assumptions, and the hypotheses are laid out, and how well the method used can distinguish between these hypotheses. Progress is only made when questions are answered. Merely achieving a similarity of certain model results with empirically observed ones does not guarantee that the underlying mechanisms are the same (Bonabeau and Theraulaz 1994). The scientific approach of devising hypotheses and attempting to falsify them is bread and butter to any empirical biologist, but is not necessarily part of a mathematician's or computer scientist's daily work. Particularly with stochastic models, the same techniques of multiple sampling and statistical analysis have to be used as with an empirical study. This also is needed to check that any model results apply to the biologically relevant parameter values.

Simple models are usually more illuminating than complex ones. Estimating parameter values used in numerical models always entails the necessity of performing a sensitivity analysis. If many parameters have to be estimated, a sensitivity analysis can become cumbersome—and this is one of the most important reasons for preferring a simple model to a complex one. When a model is designed, the feasibility of analyzing it should be considered at the outset. Further, in complex models the underlying assumptions are often hard to specify. However, understanding how certain assumptions lead to the observed results is key to the explanatory value of a model.

Lastly, in order to make an impact on a field that is mostly empirical, modelers should aim to communicate their results to empiricists. This can be helped by clearly stating how model results follow from particular assumptions. Models should make testable predictions and such tests should be spelled out explicitly in a modeling study.

What are the hallmarks of good modeling studies? (1) They should answer a biologically relevant question by spelling out hypotheses and disproving some of them. (2) They should show that the results apply to biologically relevant parameter values, and are independent of some variation in parameter estimates. (3) They should clearly indicate which assumptions led to the results. (4) Last, but not least, they should suggest empirical ways of testing the conclusions of the model. Models that meet this "gold-standard" should not fail to make a substantial impact in this field.

In sum, as insect sociobiologists we have an unrivaled opportunity to observe our study organizations part and parcel. We can then employ recent developments in statistical physics, its sister disciplines of complexity theory and self-organization theory, and the new realm of individual-based modeling to generate testable hypotheses. And we can evaluate and test these ideas through close-coupled iterated loops of progressive modeling and experimentation. The future is indeed bright for insect sociobiology.

Acknowledgments

We wish to thank BBSRC (EF19832) and EPSRC (GR/S78674/01) and the DFG (German Science Foundation, Emmy Nöther Fellowship to A.D.) for funding.

Literature Cited

Anderson, C., and D. W. McShea. 2001. "Individual vs. social complexity, with particular reference to ant colonies." *Biological Reviews* 76: 211–237.

Ball, P. 1999. *The self-made tapestry.* Oxford: Oxford University Press.

———. 2004. *Critical mass: How one thing leads to another.* London: William Heinemann.

Beckers, R., S. Goss, J. L. Deneubourg, and J. M. Pasteels. 1989. "Colony size, communication and ant foraging strategy." *Psyche* 96: 239–256

Bedau, M. A. 1999. "Can unrealistic computer models illuminate theoretical biology?" In A. S. Wu, ed., *Proceedings of the 1999 genetic and evolutionary computation conference workshop programme,* 20–23. San Francisco: Morgan Kaufmann.

Beekman, M., D. Sumpter, and F. Ratnieks. 2001. "Phase transition between disordered and ordered foraging in Pharaoh's ants." *Proceedings of the National Academy of Sciences USA* 98: 9703–9704.

Ben-Jacob, E. 2002. "When order comes naturally." *Nature* 415: 370.

Bonabeau, E., M. Dorigo, and G. Theraulaz. 1999. *Swarm intelligence: From natural to artificial systems.* New York: Oxford University Press.

Bonabeau, E., G. Theraulaz, J.-L. Deneubourg, N. R. Franks, O. Rafelsberger, J. L. Joly, and S. Blanco. 1998. "A model for the emergence of pillars, walls and royal chambers in termite nests." *Philosophical Transactions of the Royal Society of London: Biological Sciences* 353: 1561–1576.

Bonabeau, E., and G. Theraulaz. 1994. "Why do we need artificial life?" *Artificial Life* 1: 303–325.

Bonabeau, E., G. Theraulaz, J.-L. Deneubourg, A. Lioni, F. Libert, C. Sauwens, and L. Passera. 1998. "Dripping faucet with ants." *Physical Review E Volume* 57: 5904–5907.

Bourke, A. F. G., and N. R. Franks. 1995. *Social evolution in ants.* Princeton: Princeton University Press.

Bray, D. 2001. "Reasoning for results." *Nature* 412: 863.

Brooks, R. A., and A. M. Flynn. 1989. "Fast, cheap, and out of control: A robot invasion of the solar system. *Journal of the British Interplanetary Society* 20: 478–485.

Britton, N. F., N. R. Franks, S. C. Pratt, and T. D. Seeley. 2002. "Deciding on a new home: How do honeybees agree?" *Proceedings of the Royal Society B* 269: 1383–1388.

Burton, J. L., and N. R. Franks. 1985. "The foraging ecology of the army ant *Eciton rapax:* An ergonomic enigma?" *Ecological Entomology* 10: 131–141.

Camazinc, S., J.-L. Deneubourg, N. R. Franks, J. Sneyd, G. Theraulaz, and E. Bonabeau. 2001. *Self-organization in biological systems.* Princeton: Princeton University Press.

Charnov, E. L. 1976. "Optimal foraging, the marginal value theorem." *Theoretical Population Biology* 9: 129–136.

Chittka, L., A. G. Dyer, F. Bock, and A. Dornhaus. 2003. "Bees trade off foraging speed for accuracy." *Nature* 424: 388.

Cicirello, S. and S. F. Smith 2004. "Wasp-like agents for distributed factory coordination." *Journal of Autonomous Agents and Multi-Agent Systems* 8: 237–266.

Costa, J. T. 2002. "Scale models? What insect societies teach us about ourselves." *Proceedings of the American Philosophical Society* 146: 170–180.

Couzin, I. D., and N. R. Franks. 2002. "Self-organized lane formation and optimized traffic flow in army ants." *Proceedings of the Royal Society B* 270: 139–146.

Couzin, I. D., J. Krause, N. R. Franks, and S. A. Levin. 2005. "Effective leadership and decision-making in animal groups on the move." *Nature* 433: 513–516.

Cuthill, I. C. 2005. "The study of function in behavioural ecology. *Animal Biology* 55: 399–417.

Deneubourg, J. L., S. Goss, N. R. Franks, and J. M. Pasteels. 1989. "The blind leading the blind: Modeling chemically mediated army ant raid patterns." *Journal of Insect Behavior* 2: 719–725.

Detrain, C., and J.-L. Deneubourg. 2002. "Complexity of environment and parsimony of decision rules in insect societies." *The Biological Bulletin* 202: 268–274.

———. 2006. "Self-organized structures in a superorganism: Do ants "behave" like molecules?" *Physics of Life Review* 3: 162–187.

DiCaro, G., and M. Dorigo. 1998. "AntNet: Distributed stigmergic control for communications networks." *Journal of Artificial Intelligence Research* 9: 317–365.

Di Paolo, E. A., J. Noble, and S. Bullock. 2000. "Simulation models as opaque thought experiments." In M. A. Bedau, Bedau MA.; Snyder E and N. H. Packard, eds., *Proceedings of the Seventh International Conference on Artificial Life,* 497–506. Cambridge: MIT Press.

Dorigo, M., and T. Stützle. 2004. *Ant colony optimization.* Cambridge: MIT Press.

Dornhaus, A., and L. Chittka. 2004a. "Information flow and regulation of foraging activity in bumble bees." *Apidologie* 35: 183–192.

———. 2004b. "Why do honey bees dance?" *Behavioral Ecology and Sociobiology* 55: 395–401.

———. 2005. "Bumble bees (*Bombus terrestris*) store both food and information in honeypots." *Behavioral Ecology* Behavioral Ecology 16: 661–666.

Dornhaus, A., E. J. Collins, F.-X. Dechaume-Moncharmont, A. Houston, N. R. Franks, and J. McNamara. 2006. "Paying for information: partial loads in central place foragers." *Behavioral Ecology and Sociobiology* 61: 151–161.

Dornhaus, A., F. Klügl, C. Oechslein, F. Puppe, and L. Chittka. 2006. Benefits of recruitment in honey bees: ecology and colony size." *Behavioral Ecology* 17: 336–344.

Dussutour, A., V. Fourcassié, D. Helbing, and J.-L. Deneubourg. 2004. "Optimal traffic organization in ants under crowded conditions." *Nature* 428: 70–73.

Fewell J. H. 2003. "Social insect networks." *Science* 301: 1867–1870.

Franks, N. R. 1989. "Army ants: A collective intelligence." *American Scientist* 77: 138–145.

Franks, N. R., and J.-L. Deneubourg. 1997. "Self-organizing nest construction in ants: individual worker behaviour and the nest's dynamics." *Animal Behaviour* 54: 779–796.

Franks, N. R., S. C. Pratt, E. Mallon, N. Britton, and D. Sumpter. 2002. "Information flow, opinion-polling and collective intelligence in house-hunting social insects." *Philosophical Transactionsof the Royal Society: Biological Sciences* 357: 1567–1583.

Franks, N. R., N. Gomez, S. Goss, and J.-L. Deneubourg. 1991. "The blind leading the blind in army ant raid patterns: Testing a model of self-organization." *Journal of Insect Behavior* 4: 583–607.

Gotwald, W. H. 1995. *Army ants: The biology of social predation.* Ithaca: Cornell University Press.

Gould, S. J. 1989. *Wonderful life: The Burgess shale and the nature of history.* New York: Norton.

Helbing, D., and B. A. Huberman. 1998. "Coherent moving states in highway traffic." *Nature* 396: 738–740.

Helbing, D., and M. Treiber. 1998. "Traffic theory—Jams, waves and clusters." *Science* 282: 2001–2003.

Hobbes, T. 1651/1985. *Leviathan.* London: Penguin.

Hogeweg, P., and B. Hesper. 1990. "Individual-oriented modelling in ecology." *Mathematical and Computer Modelling* 13: 83–90.

Jacob, F. 1970. *La logique du vivant.* Paris: Gallimard.

Jeanson, R., J.-L. Deneubourg, A. Grimal, and G. Theraulaz. 2004. "Modulation of individual behavior and collective decision-making during aggregation site selection by the ant *Messor barbarus.*" *Behavioral Ecology and Sociobiology* 55: 388–394.

Krieger, M. J. B., J.-B. Billeter, and L. Keller. 2000. "Ant-like task allocation and recruitment in cooperative robots." *Nature* 406: 992–995.

Lumer, F. 1994. "Diversity and adaptation in populations of clustering ants." *From Animals to Animats 3: Proceedings of the 3rd International Conference on Simulation of Adaptive Behavior,* 501–508.

Marshall, J. A. R., A. Dornhaus, N. R. Franks, and T. Kovacs. 2006. "Noise, cost and speed-accuracy trade-offs: Decision making in a decentralized system." *Journal of the Royal Society: Interface* 3: 243–254.

Marshall, J. A. R., and N. R. Franks. 2007. "Whys and wherefores of computer modelling in behavioural biology." In Laubichler MD Müller, G.B. eds., *Modeling biology—Structures, behaviors, evolution.* Cambridge: MIT Press.

Maxwell, J. C. 1873a. "Science and free will." In L. Campbell and W. Garnett, eds., *The life of James Clerk Maxwell*, 438–439. London: Macmillan.

———. 1873b. "Molecules [from *Nature*, Vol. 8]." In W. D. Niven, ed., *The scientific papers of James Clerk Maxwell, Vol II*. Cambridge: Cambridge University Press.

May, R. M. 2004. "Uses and abuses of mathematics in biology." *Science* 303: 790–793.

Memmott, J. 1999. "The structure of a plant-pollinator food web." *Ecology Letters* 2: 276–280.

Memmott, J., N. Waser, and M. Price. 2004. "Tolerance of pollination networks to species extinctions." *Proceedings of the Royal Society B* 271: 2605–2611.

Miller, W. III. (ed). 2003. "New interpretations of complex trace fossils." *Palaeogeography, Palaeoclimatology, Palaeoecology* (Special Issue): 192.

Millor, J., M. Pham-Delegue, J.-L. Deneubourg, and S. Camazine. 1999. "Self-organized defensive behavior in honeybees." *Proceedings of the National Academy of Sciences USA* 96: 12611–12615.

Niklas, K. 1999. "Evolutionary walks through a land plant morphospace." *Journal of Experimental Botany* 50: 39–52.

Parunak, H. D. 1997. " 'Go to the ant': Engineering principles from natural multi-agent systems." *Annals of Operations Research* 75: 69–101.

Pittendrigh, C. S. 1958. "Adaptation, natural selection and behavior." In A. Roe and G. G. Simpson, eds., *Behavior and evolution*. New Haven: Yale University Press.

Poincaré, H. 1905. *La science et l'hypothèse*. Paris: Flammarion.

Pratt, S. C., E. B. Mallon, D. J. T. Sumpter, and N. R. Franks. 2002. "Quorum sensing, recruitment, and collective decision-making during colony emigration by the ant *Leptothorax albipennis*." *Behavioral Ecology and Sociobiology* 52: 117–127.

Pratt, S. C., D. J. T. Sumpter, E. B. Mallon, and N. R. Franks. 2005. "An agent-based model of collective nest choice by the ant *Temnothorax albipennis*." *Animal Behaviour* 70: 1023–1036.

Ray, T. S. 1994. "An evolutionary approach to synthetic biology: Zen and the art of creating life." *Artificial Life* 1: 179–210.

Roces, F. 2002. "Individual complexity and self-organization in foraging by leaf-cutting ants." *Biological Bulletin* 202: 306–313.

Scheiner, R., J. Erber, and R. E. Page. 1999. "Tactile learning and the individual evaluation of the reward in honey bees (Apis mellifera L.)." *Journal of Comparative Physiology A* 185: 1–10.

Schmid-Hempel, P., A. Kacelnik, and A. L. Houston. 1985. "Honeybees maximize efficiency by not filling their crop." *Behavioral Ecology and Sociobiology* 17: 61–66.

Schneirla, T. C. 1971. *Army ants: A study in social organization.* San Francisco: Freeman.

Seeley, T. D. 1998. "Thoughts on information and integration in honey bee colonies." *Apidologie* 29: 67–80

Seeley, T. D., and P. K. Visscher. 2004. "Quorum sensing during nest-site selection by honeybee swarms." *Behavioral Ecology and Sociobiology* 56: 594–601.

Sendova Franks, A. B., and N. R. Franks. 1999. "Self-assembly, self-organization and division of labour." *Philosophical Transactions of the Royal Society of London: Biological Science* 354: 1395–1405.

Strogatz, S. 2004. "The physics of crowds." *Nature* 428: 367–368.

Theraulaz, G., E. Bonabeau, S. C. Nicolis, R. V. Sole, V. Fourcassie, S. Blanco, R. Fournier, J. L. Joly, P. Fernandez, A. Grimal, P. Dalle, and J.-L. Deneubourg. 2002. "Spatial patterns in ant colonies." *Proceedings of the National Academy of Sciences USA* 99: 9645–9649.

Theraulaz, G., E. Bonabeau, C. Sauwens, J.-L. Deneubourg, A. Lioni, F. Libert, L. Passera, and R. Solé. 2001. "Model of droplet dynamics in the argentine ant *Linepithema humile* (Mayr)." *Bulletin of Mathematical Biology* 63: 1079–1093.

Varju, D., and J. Núñez. 1993. "Energy balance versus information exchange in foraging honeybees." *Journal of Comparative Physiology* 172: 257–261.

Watmough, J., and S. Camazine. 1995. "Self-organized thermoregulation of honeybee clusters." *Journal of Theoretical Biology* 176: 391–402.

Weidenmüller, A. 2004. "The control of nest climate in bumblebee *(Bombus terrestris)* colonies: Interindividual variability and self reinforcement in fanning response." *Behavioral Ecology* 15: 120–128.

Weiss, G. 1999. "Multiagent systems: A modern approach to distributed artificial intelligence." Cambridge: MIT Press.

Wilson, E. O. 1971. *The insect societies.* Cambridge: Belknap Press.

CHAPTER TWENTY

Positive Feedback, Convergent Collective Patterns, and Social Transitions in Arthropods

RAPHAËL JEANSON

JEAN-LOUIS DENEUBOURG

THE ORIGIN OF SOCIAL LIFE represents a major evolutionary transition which has occurred repeatedly across many lineages (Maynard-Smith and Szathmary 1999). The expression of sociality encompasses the production of a multiplicity of structures ranging from undifferentiated assemblages of individuals to highly integrated societies. In this chapter, we address the question of whether a set of universal principles can account for two seemingly contradictory aspects of sociality: the large diversity in social organizations and the existence of convergent collective patterns across taxa.

Many similarities in collective activities and structures have been reported in distinct lineages. For instance, collective scent trails are used for finding and exploiting resources (e.g., food patches and nests) in termites (Reinhard and Kaib 2001), ants (Hölldobler and Wilson 1990), caterpillars (Fitzgerald 1995), bumblebees (Cameron and Whitfield 1996), and stingless bees (Nieh et al. 2004). Synchronization of individual behaviors occurs during cooperative hunting in social spiders (Krafft and Pasquet 1991), acoustic signaling in katydids (Greenfield and Roizen 1993), waving display in crabs (Aizawa 1998), foraging in caterpillars (Fitzgerald and Visscher 1996), flashing in fireflies (Buck 1988), and production of activity cycles in social insect colonies (Cole and Trampus 1999; Bloch, this volume). Task specialization in group living has been reported in many taxa, including caterpillars (Underwood and Shapiro 1999), shrimps (Duffy 1996), thrips (Crespi 1992), aphids (Rhoden and Foster 2002), as well as the Hymenoptera (Hölldobler and Wilson 1990;

Seeley 1995). Such analogous collective behaviors lead us to examine whether generic principles can be invoked to account for these convergent patterns.

At the same time, it is evident that the complexity and degree of cooperation achieved by social groups strongly differ within and between taxa. Several classifications have been proposed with the aim of ordering the multiple social structures. The traditional classification in insects was coined by Michener (1969) and used three criteria to classify social systems: overlap of generations, cooperative brood care, and reproductive division of labor. The co-occurrence of all three characteristics defines *eusociality*. Several authors (Gadagkar 1994; Crespi and Yanega 1995; Sherman et al. 1995) have proposed to revise the scope of eusociality. Proceeding by a top-down approach, these classifications catalogue societies initially by the presence or absence of the required attributes and subsequently label systems lacking, singly or in combination, these features (Costa and Fitzgerald 1996). However, this continued emphasis on differences in degree of cooperation may hide the existence of common principles of organization across social levels. A central concern that remains is whether the contribution of fundamentally new rules should necessarily parallel transitions toward more complex or derived forms of cooperation and organization, or whether invariant mechanisms can account for the production of a wide range of social structures.

Analysis of the dynamic among societal subunits is essential to understanding the interplay between social organization and evolution (Fewell 2003; Fewell, Schmidt, and Taylor, this volume). In this chapter, our objective is to examine how one aspect of dynamic systems, positive feedback (an inherent ingredient of social life), might contribute to the production of similar collective patterns in independent lineages, and could have driven social transitions from simple forms of sociality, such as undifferentiated groups, to highly integrated societies, such as eusocial colonies. We explore the hypothesis that a system of behavioral rules involving positive feedback governs aggregation in group-living arthropods regardless of the sophistication of their means of communication and levels of sociality. To illustrate that generic rules based on amplification processes can generate similar collective behaviors, we also review examples of collective decision making in diverse species of arthropods. Finally, we discuss how positive feedback can promote the emergence of further complex collective patterns and transition to more advanced forms of cooperation.

Positive Feedback

From gene expression (Novick and Wiener 1957) to ecosystem shifts (Scheffer and Carpenter 2003), biological processes are regulated by networks of feedback loops involving the mutual coupling of different variables. Negative feedback loops elicit reactions counteracting perturbations imposed on a system, returning the system toward the previous point, as in homeostasis. In contrast, positive feedback reinforces changes in the direction of the initial deviation (DeAngelis, Post, and Portis 1986), so that the system moves farther in that direction each time reinforcement occurs. Positive feedback can arise from the combination of positive interactions among components, so that the behavior of one component (or individual) reinforces performance of the same behavior in the next (Thomas 1998). For example, a pheromone trail laid by one ant may increase the probability that other ants follow that path and reinforce it (Figure 20.1). Negative interactions also may promote positive feedback; for instance, the probability of leaving a group might decrease with group size and consequently favor large group formation.

In animal behavior and physiology, positive feedback loops can be implemented in two ways. First, they can be involved at the individual level through positive or negative self-reinforcement processes. For example, learning is an amplification mechanism where past experience conditions the probability of returning to a food patch and can induce the specialization of workers in the exploitation of foraging areas such as in the ant *Neoponera apicalis* (Deneubourg et al. 1987). Similarly, the individual probability of performing a task may increase (decrease) as the individual repeatedly does (not) tackle the task. For example, workers of the bumblebee *Bombus terrestris* control nest climate by fanning their wings and in-

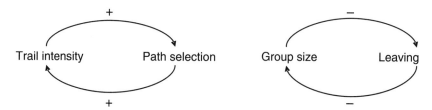

Figure 20.1. Examples of positive feedback loops, "+" and "−" signs represent positive and negative influences, respectively.

dividual response thresholds have been shown to change with experience (Weidenmüller 2004). Second, positive feedback can be implemented at the collective level through a multiplicity of interactions among individuals (Bonabeau et al. 1997; Camazine et al. 2001). The likelihood of an individual adopting a behavior may depend on the number of nestmates already engaged in that behavior, such as during aggregation where the probability for an individual to cluster increases with the size of the group or during synchronous flashing among fireflies (Camazine et al. 2001).

Aggregation

Aggregation is a fundamental attribute of sociality and could be defined, in its broadest sense, as any assemblage of individuals with a higher density than in the surroundings areas (Camazine et al. 2001). Many studies have been devoted to the costs and benefits of group life (Krause and Ruxton 2002), but little attention has been paid to the proximate causes of aggregation, particularly the dynamic aspects of clustering. Aggregation can result either from individual responses to external or environmental heterogeneities that act as a template specifying the patterns of aggregation (Fraenkel and Gunn 1961), or from social interactions (Parrish and Hamner 1997; Parrish and Edelstein-Keshet 1999).

Understanding the role of aggregation in the evolution of sociality involves three important points. First, aggregation constitutes the most fundamental expression of social life; it has been reported in a wide range of taxa from bacteria to mammals (Allee 1931; Parrish and Edelstein-Keshet 1999). Second, evolutionary transitions in social structure involve changes in the processes of aggregation and dispersion. Indeed, two classic routes of sociality have been proposed for insects (Lin and Michener 1972) and other arthropods (Shear 1970): the clustering of solitary individuals and the retention of juveniles within the natal nest. Both of these scenarios involve aggregation as the essential prerequisite for socialization.

Finally, within highly integrated societies, aggregation processes are involved in the regulation of various collective activities and in the spatio-temporal organization of colonies. For instance, aggregation contributes to bivouac formation in army ants (Gotwald 1995), and thermoregulation (Heinrich 1981) and cooperative defense in honey bees (Ono et al. 1995). The same categories of mechanisms may account for the clustering of items such as corpses (Theraulaz et al. 2002), brood (Franks and Sendova-Franks

1992), and leaves (Hart and Ratnieks 2000) in ants and comb formation in honey bees (Camazine 1991).

The mechanisms of aggregation contribute at many scales in social organization. To illustrate this, consider the individual and collective behaviors in the German cockroach, *Blattella germanica*. As in many other cockroach species, they form mixed clusters of males and females with overlapping generations. Cockroaches aggregate readily in an experimental situation. After introduction into a homogeneous and circular arena, they initially distribute themselves uniformly around the periphery. The emergence of aggregates benefits from several sources of positive feedback: the greater a cluster, the more cockroaches are attracted. Aggregation also relies on the modulation of individual behaviors depending on the presence of congeners in their immediate vicinity. Thus, the probability of an individual stopping and its mean resting time in a group are greater when the number of cockroaches is higher (Jeanson, Rivault, et al. 2005; Figure 20.2a). The formation of aggregates relies on inter-attraction between individuals following simple rules based on local information without reference to the global emerging pattern.

Although cockroaches and eusocial insects strongly differ in their social organization, the same behavioral rules described for cockroaches (Figure 20.2b) promote aggregation in ants. Indeed, aggregation of workers of *Lasius niger* in a homogeneous environment results from amplification mechanisms where the time spent in a cluster increases with its size (Depickère, Fresneau, and Deneubourg 2004). In these examples, clustering requires close contact between individuals to initiate. However, aggregation can also be influenced via long-range attraction based on the diffusion of aggregation pheromones (Wertheim et al. 2005). For example, larvae of the bark beetle *Dendroctonus micans* cluster in intracortical chambers of spruce trees in response to the production of volatile chemicals probably originating from the oxidation of monoterpenes produced by host-trees (Deneubourg, Grégoire, and Le Fort 1990). Clustering based on the diffusion of aggregation pheromones shares common processes with cooperative defense through recruitment via alarm pheromones, where amplification processes also exert a key role (Millor et al. 1999; Hölldobler and Wilson 1990).

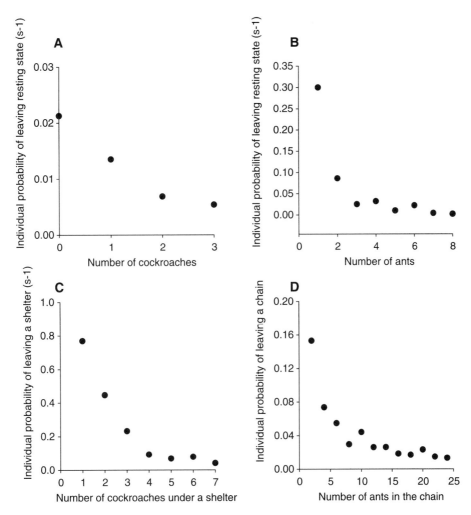

Figure 20.2. Modulation of individual behaviors depending on the presence of conspecifics. (A) *Blattella germanica* (from Jeanson et al. 2005); (B) *Lasius niger* (from Depickere et al. 2004); (C) *Periplaneta americana* (Sempo and Deneubourg, unpublished data); (D) *Oecophylla smaragdina* (from Lioni and Deneubourg 2004).

Collective Decision Making

In the examples reported above, aggregation results from amplification processes associated with the modulation of individual behaviors' depending on the presence of conspecifics in their vicinity. In this section, we examine how the presence of environmental heterogeneities in association with positive feedbacks generated by social interactions can support collective decision making in various species of arthropods.

Organisms living in patchy environments have to make many decisions to exploit their environment effectively. Such choices may include the selection of nest sites, food sources, and other resources. In such circumstances, social arthropods face a double challenge: the selection of the most suitable or profitable resources and, simultaneously, the maintenance of cohesion among group members. Group members with pertinent information about a resource or location can act as permanent or temporary leaders, and exert more weight on the outcome of the decision (Conradt and Roper 2005; Couzin et al. 2005). For instance, group cohesion can be ensured by a leader guiding conspecifics to particular locations. For example, a mother lace bug, *Gargaphia solani*, prevents her offspring dividing by blocking all but one path at the leaf axil (Tallamy and Wood 1986). And, during group recruitment in ants, successful foragers often guide nestmates to food sources (Hölldobler and Wilson 1990).

Decision processes also can be evenly distributed among *identical* individuals through multiple iterations of interactions (Camazine et al. 2001). In self-organized systems, collective decisions can then arise as a result of competition between different sources of information that are amplified in a social context through positive feedback (Deneubourg, Lioni, and Detrain 2002). For instance, colonies of the ant *Lasius niger* generally exploit one of two identical food sources (Beckers et al. 1990; Camazine et al. 2001). After an initially equal exploitation of both sources, a sudden transition occurs with ants focusing their foraging activity toward a unique source. This shift arises spontaneously after one trail becomes, by chance, slightly stronger than the other and is further reinforced by the ants. Collective decision making could then be defined as the process whereby a group selects, through interactions among individuals, one solution (or one course of action) among different alternatives.

From an experimental perspective, one simple approach to examine the underlying mechanisms of decision making is through recourse to binary choices. Binary choices with identical alternatives are initially required to distinguish choices arising in response to environmental heterogeneities or through social interactions. Indeed, a skewed distribution of individuals between identical resources evidences the contribution of amplifications through social interactions. In contrast, a symmetrical repartition of individuals suggests that the selection is not governed by inter-attraction but likely by the individual response to environmental heterogeneities. Once the role of social interactions has been evidenced, choices between the alternatives can contribute to the understanding of how individuals modulate their behaviors and/or the information transmitted to congeners, depending on the characteristics of the resources.

Below, we provide case studies of collective decision making relative to the selection of an aggregation or nest site, because both individual mechanisms and collective dynamics have been precisely characterized in these contexts.

Spiders

Coordinated migrations have been reported in social spiders after the destruction of their web (*Anelosimus eximius*, Vollrath 1982), and during the foundation of a new colony (*Acharearanea wau*, Lubin and Robinson 1982). Sociality in spiders ranges from transient gregariousness after hatching from the maternal cocoon to permanent social structures with overlapping generations. Although quasi solitary as adults (Avilès 1997), all spider species undergo at least a transient gregarious phase after their emergence from the maternal cocoon (Krafft 1979). During this stage, spiderlings might face collective displacement after the accidental dissociation of the group or to cluster in a more favorable environment. Studies were performed both in a social species, *Anelosimus eximius* (Saffre et al. 1999), and with the spiderlings of a solitary species, *Larinioides cornutus* (Jeanson, Deneubourg, and Theraulaz 2004). According to D'Andrea's classification (1987), these species are at the opposite ends of the sociality spectrum in spiders.

In the studies, spiders were offered a choice between two identical branches in a Y-shaped set-up. During their displacement, individuals produced dragline silk as security thread. This was attached to the substrate in a discrete pattern. Once the first spider accessing the set-up reached the bifurcation, it randomly selected one branch and formed, incidentally, a shortcut between the stem and the chosen branch. This silk shortcut was more likely to be followed by the next individual to reach the branch. Through consecutive passages, spiders selected one site of aggregation. In both species, the maintenance of the cohesion in the migrating group relied on amplification processes due to reinforcement of the silk road, which was based on a discrete pattern of attachment of successive stands of silk. In spiders, a simple mechanism relying on the incident deposition of silk can thus lead to the collective selection of an aggregation site.

Cockroaches

During their resting period in dark places, urban cockroaches (e.g., *Periplaneta americana* and *Blattella germanica*) form mixed clusters of males and females with generation overlap. Shelters scaled to the size of larvae and adults (males) were offered to pure groups of larvae, males, and to mixed groups (Jeanson and Deneubourg, 2007). Groups of males or larvae tended to form a unique cluster, respectively, under a high or a low shelter. When mixed groups were offered a binary choice between a low and a high shelter, the cockroaches mainly aggregated under the high shelter. Thus, the preferences of males for a high shelter were consistent across treatments but the preferences of larvae shifted depending on the social context. The presence of males overrided the affinity of larvae for scaled-size shelters. This final repartition is hypothesized to originate from a differential attraction between larvae and males as a function of their relative body size and the larger production of pheromones of aggregation by males. Cockroaches thus collectively select a unique shelter through the incident attraction toward large individuals without any explicit coding.

Chain Formation in *Oecophylla*

Ants of the genus *Oecophylla* hang onto one other to form chains in order to bridge empty spaces between branches (Hölldobler and Wilson 1990). When faced with a choice between two potential sites for chain formation, workers favor one branch and discard the other one. The presence of a small visual cue (mimicking a branch) at one potential site of chain formation led to the selection of one particular site (Lioni and Deneubourg 2004).

In both cockroaches and ants, choice relies on an interplay of behavioral modulation, which depends on the presence of environmental cues (e.g., greater individual resting time under a dark rather than a light shelter) and the presence of conspecifics locally perceived in the chain or the shelter. The probability of leaving (or entering) an aggregate decreases (or increases) with the number of conspecifics (Figure 20.2c, d).

Nest Site Selection in *Temnothorax albipennis*

The ants *Temnothorax albipennis* form small colonies within cracks in rocks. In the laboratory, after migrations were experimentally induced by removing their original nest roof, scouts hunted for other potential nest sites (Franks et al. 2002; Pratt, this volume). Depending on the quality of the nest sites discovered, scouts modulated their behaviors. The latency between the first entry in a nest and the first recruitment to this site was shorter when scouts visited a superior nest site rather than a mediocre one (Mallon, Pratt, and Franks 2001). This modulation led to a more rapid initiation of recruitment toward the superior nest. The first stage of re-cruitment was achieved by tandem running, in which a scout leads a single nestmate from the old to the new nest (Franks and Richardson 2006). Once the nest population reached a threshold (i.e., quorum), scouts switched from slow tandem running to transport, a faster mode of recruit-ment in which an ant carries a nestmate (Pratt et al. 2002; Pratt 2005). The initial difference in recruitment latency between superior and mediocre sites was amplified by the detection of a quorum leading to a more rapid increase of the population in the superior site and then to its selection (Figure 20.3a).

Nest Site Selection in Ants Using Mass Recruitment

The harvester ant *Messor barbarus* forms colonies of several thousand in-dividuals laying chemical trails on the substrate to recruit nestmates for the exploitation of resources. In ants using mass communication, informa-tion relative to the qualities of potential aggregation and nest sites can be conveyed as a signal through the chemical trail that serves as a repository of information on collective choice. Groups of *M. barbarus* were pre-sented experimentally with a binary choice between potential aggregation sites (Jeanson et al. 2004). At the beginning of the migration, both sites were equally visited by ants, then a dramatic increase in the exploitation of one site occured while the other was discarded (Figure 20.3b). Scouts used pheromones trails to recruit nestmates toward the discovered site. They actively modulated the information conveyed to nestmates depending on the qualities of the sites: both the probability to lay a trail and the intensity of trail-laying were influenced by the characteristics of the aggregation sites. Thus the selection of a site relies on the modulation of trail-laying, which acts in synergy with the aggregation processes taking place within sites to decrease the probability of an ant leaving as the population in-creases.

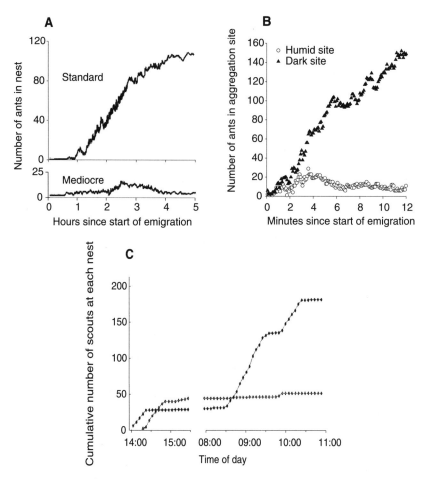

Figure 20.3. (A) dynamic of nest emigration in *Leptothorax albipennis* between a standard and mediocre nest (from Mallon et al. 2001); (B) dynamic of aggregation site selection between two different sites in *Messor barbarus* (from Jeanson et al. 2004a); (C): buildup of scouts between two potential nest sites in honey bees (from Camazine et al. 1999).

Nest Site Selection in Honeybees

In honey bees, as a part of the life cycle to ensure colony propagation, a portion of the colony and the old queen move out of the old nest (Seeley 1995). After the swarm settles, scouts begin hunting for potential nest sites. When scouts dance to recruit nestmates to new nest sites, they modulate their behavior by tuning the duration of the dance and the waggle run rate in relation to nest site quality (Seeley and Buhrman 2001). This induces a stronger buildup of bees visiting and recruiting for a good site, leading eventually to its selection (Figure 20.3c; Camazine et al. 1999; Seeley and Buhrman 2001). A difference worth noting is in the processes of decision making between bees and the other examples above. In bees, the decision precedes action, because the swarm lifts off to the selected location only once unanimity has been reached (Seeley and Visscher 2004). In the other examples, the decision arises during the course of action.

We have restricted our examples to nest and aggregation site selection, but positive feedback regulates multiple activities in arthropod societies, including foraging (Camazine et al. 2001) and nest defense (Hölldobler 1981). The selection of richer or more plentiful food sources by many ants and by central-place foraging caterpillars relies on positive feedback associated with a modulation of trail-laying behaviors (Beckers, Deneubourg, and Goss 1993; Fitzgerald 1995). Honey bee foragers tune the intensity of their dance to recruit nestmates toward more profitable food sources (Seeley, Camazine, and Sneyd 1991; Seeley, Mikheyev, and Pagano 2000). In treehoppers, nymphs attract conspecifics to feeding sites with vibrational signals and convey information about patch quality with signaling rate (Cocroft and Rodriguez 2005).

In all the examples reported above, the key principle supporting collective decision making relies on the modulation of individual behaviors in association with positive feedback. Small differences in individual preferences for environmental heterogeneities can be amplified in a social context to support the selection or exploitation of one resource. From the conclusion that the underlying mechanisms of aggregation constitute an efficient source of amplification, we speculate that all organisms clustering through social interactions should display the ability to achieve collective decision making relative to the selection of an aggregation site. The existence of sophisticated modes of communication, such as trail pheromones, does not appear necessary to induce collective choices, but they can supplement the basic mechanisms inherent in social life: positive feedback resulting from the multiplicity of direct or indirect interactions among group members.

Tuning of Positive Feedback

A frequent characteristic of systems involving positive feedback is the existence of a critical value of parameters beyond which the system's behavior changes suddenly (Glansdorff and Prigogine 1971; DeAngelis, Post, and Portis 1986). We can identify two distinct levels where small changes can take place to drive the system toward different states: implementation of behavioral rules at the individual level and group size.

The implementation of behavioral rules at the individual level can vary depending on context or species. When considering positive feedbacks arising from the multiplicity of social interactions, one critical parameter is the sensibility of an individual to conspecifics. Small variations in the nature of social interactions can influence the production of collective patterns.

In bees, the probability of following a dance to a nectar source i depends on the relative number of bees dancing for that source (Camazine and Sneyd 1991). In ants (e.g., *Lasius niger*), the probability of following branch i leading to a food source depends on the relative pheromone concentration on branch i (Beckers, Deneubourg, and Goss 1993):

$$P_i = \frac{(k + Y_i^\eta)}{(k + Y_1^\eta) + (k + Y_2^\eta)} \qquad i = 1, 2$$

For bees, Y_i ($i = 1, 2$) represents the number of bees dancing for source i and $\eta = 1$. For ants, Y_i represents the pheromone concentration on branch i and $\eta = 2$. The noticeable difference between both choice functions is the value of the parameter η, which has been determined from experimental data in ants and bees (see Camazine et al. 2001). As predicted theoretically and validated empirically, ants generally exploit only one source between two identical sources whereas bees distribute their foraging effort evenly between identical nectar sources. Thus, the stronger the sensibility (here η) to conspecifics or their signals, the higher the probability for a group to select one resource among different alternatives. In both species, the same choice function regulates foraging but variation in one parameter can drastically affect the collective patterns.

In self-organized systems, variations in external parameters can also incidentally affect the implementation of positive feedback to produce a wide diversity of collective patterns without any explicit change at the individual level. For instance, abiotic environmental factors can play a decisive role. The nature of the substrate can affect the persistence of pheromones that might favor or, in contrast, impede the formation of efficient foraging trails (Detrain, Natan, and Deneubourg 2001; Jeanson, Ratnieks, and Deneubourg 2003). Thus, small variations in the nature of the ecological niche occupied by group-living species can have important repercussions for their organization.

Group size is another critical factor regulating the production of collective patterns. The formation of efficient foraging or exploration trails in ants is affected by colony size via its influence on traffic flow and, consequently, trail persistence (Beekman, Sumpter, and Ratnieks 2001; Detrain et al. 1991). In locusts, the transition from disordered movement to aligned collective motion depends on the density of individuals within groups (Buhl et al. 2006). In termites, pillars are built with soil pellets impregnated with pheromones and their construction includes two phases: the random deposition of soil pellets followed by the emergence of pillars. The transition between these uncoordinated and coordinated phases requires the presence of many individuals to overcome the disappearance of pheromones between successive trips by individual workers (Bruinsma 1979; Grassé 1984; Camazine et al. 2001). In each of these examples—sensitivity, substrate persistence, and group size—no behavioral changes at the individual level are required to account for the production of collective patterns.

We can speculate that minor changes in behavioral rules through social evolution might have been shaped by natural selection to produce diverse collective patterns through the central contribution of positive feedbacks, without changing the essential nature of the underlying rules. Similarly, the incidental implementation of positive feedback without any explicit modification at the individual level may have significantly contributed to the expression of more complex forms of social organizations.

Cascade of Complexity

Group-living species share at least one common attribute: aggregation based on inter-attraction. Such assemblages of individuals might constitute an effective source of positive feedback that will favor cascades of further social patterns. As shown above, simple mechanisms of aggregation can lead to collective decisions and patterns that may themselves promote new social or spatial organizations, such as synchronization, quorum sensing, and cooperative defense, once a critical number of individuals or interactions has been reached. From initially undifferentiated aggregates, at the incipient stages of sociality, small differences at the individual level can be amplified in a social context to promote the expression of new social patterns without requiring the contribution of fundamentally new sets of rules (Camazine et al. 2001). For instance, forced associations of solitary bee or ant foundresses can lead to the emergence of division of labor by exploiting interindividual variability in task performance (Helms-Cahan and Fewell 2004; Jeanson, Kukuk, and Fewell 2005). Similarly, amplification through self-reinforcement can induce the formation of stable hierarchies in bumblebees (Hogeweg and Hesper 1983), crayfishes (Issa, Adamson, Edwards 1999), wasps (Theraulaz, Bonabeau, and Deneubourg 1995), and crayfishes (Goessmann, Hemelrijk, and Huber 2000), and favors the monopolization of resources leading to asymmetrical opportunities of reproduction in social spiders (Rypstra 1993).

Theoretical models (Gautrais et al. 2002; Jeanson et al. 2007), validated by empirical studies in wasps (Karsai and Wenzel 1998) and ants (Thomas and Elgar 2003), show that the degree of behavioral specialization increases with group size. In relation to group size, density promotes soldier differentiation in aphids (Shibao, Kutsukake, and Fukatsu 2004). As it had been suggested for multicellular organisms (Bell and Mooers 1997), Bourke (1999) argued that colony size is a major determinant of social complexity in insects and showed that group size is associated with many traits that together define the level of complexity. As group size increases, colonies face new challenges that may favor the evolution of further behaviors. For instance, group size might shape the communication system to convey information over greater distance and favor the transitions from individual to collective foraging (Beckers et al. 1989; Anderson and McShea 2001). The trend toward an increase in social complexity with group size might rely on a growing number of opportunities to launch positive feedback, with each new state more sensitive to new perturbations and thus further transitions (DeAngelis, Post, and Portis 1986).

In association with the capacity to exploit new ecological opportunities in larger colonies, group size could influence the dynamics of organization of social groups through changes in the rate of interactions among group members (Pacala, Gordon, and Godfray 1996) or modifications of the range of environmental variations perceived by individuals within the aggregate (Anderson and McShea 2001). An increase in colony size might then favor the emergence of new social features that feedback and influence the colony size itself (Bourke 1999). For instance, positive feedback related to group size might explain the loss of social totipotency and the differentiation between reproductive and workers castes (Jeon and Choe 2003).

Conclusion

In this chapter, we have examined how positive feedback can contribute to the production of convergent collective patterns in arthropods and might account for the wide diversity of social organization. Positive feedback is a key mechanism shared by group-living species promoting the emergence of convergent collective patterns across taxa through the amplification of social information. Positive feedback may have also played a critical role in driving the multiple transitions from simple to complex societies which have arisen repeatedly within independent lineages (Bourke 1999; Crespi 2004). Within this perspective, social structures could be considered as networks of positive feedback loops. Simple forms of sociality such as gregariousness involve one or a few amplification processes underlying group formation based on inter-attraction; whereas multiple interconnected loops support the organization of complex and highly cooperative societies. Hölldobler and Wilson (2005) emphasized that common decision rules might be at work at the developmental and evolutionary time scales; networks of positive feedback are likely to be one of these general principles of organization.

Literature Cited

Aizawa, N. 1998. "Synchronous waving in an ocypodid crab, *Ilyoplax pusilla:* Analyses of response patterns to video and real crabs." *Marine Biology* 131: 523–532.

Allee, W. C. 1931. *Animal aggregations: A study in general sociology.* Chicago: University of Chicago Press.

Anderson, C., and D. McShea. 2001. "Individual versus social complexity, with particular reference to ant colonies." *Biological Review* 76: 211–237.

Avilès, L. 1997. "Causes and consequences of cooperation and permanent-sociality in spiders." In J. C. Choe and B. J. Crespi, eds., *The evolution of social behavior in insects and arachnids*, 476–498. Cambridge: Cambridge University Press.

Beckers, R., J.-L. Deneubourg, and S. Goss. 1993. "Modulation of trail laying in the ant *Lasius niger* (Hymenoptera: Formicidae) and its role in the collective selection of a food source." *Journal of Insect Behavior* 6: 751–759.

Beckers, R., J.-L. Deneubourg, S. Goss, and J. M. Pasteels. 1990. "Collective decision making through food recruitment." *Insectes Sociaux* 37: 258–267.

Beckers, R., S. Goss, J. L. Deneubourg, and J. M. Pasteels. 1989. "Colony size, communication and ant foraging strategy." *Psyche* 96: 239–256.

Beekman, M., D. J. T. Sumpter, and F. L. W. Ratnieks. 2001. "Phase transition between disordered and ordered foraging in Pharaoh's ants." *Proceedings of the National Academy of Sciences USA* 98: 9703–9706.

Bell, G., and A. O. Mooers. 1997. "Size and complexity among multicellular organisms." *Biological Journal of the Linnean Society* 60: 345–363.

Bonabeau, E., G. Theraulaz, J.-L. Deneubourg, S. Aron, and S. Camazine. 1997. "Self-organization in social insects." *Trends in Ecology and Evolution* 12: 188–193.

Bourke, A. F. G. 1999. "Colony size, social complexity and reproductive conflict in social insects." *Journal of Evolutionary Biology* 12: 245–257.

Buck, J. 1988. "Synchronous rhythmic flashing of fireflies. II." *Quarterly Review of Biology* 63: 265–289.

Bruinsma, O. H. 1979. "An analysis of building behaviour of the termite *Macrotermes subhyalinus* (Rambur)." Ph.D. diss., Landbouwhogeschool, Wageningen.

Buhl, J., D. J. T. Sumpter, I. D. Couzin, J. J. Hale, E. Despland, E. R. Miller, and S. J. Simpson. 2006. "From disorder to order in marching locusts." *Science* 312: 1402–1406.

Camazine, S. 1991. "Self-organizing pattern formation on the combs of honey bee colonies." *Behavioral Ecology and Sociobiology* 28: 61–76.

Camazine, S., J. L. Deneubourg, N. R. Franks, J. Sneyd, G. Theraulaz, and E. Bonabeau. 2001. *Self-organization in biological systems.* Princeton: Princeton University Press.

Camazine, S., and J. Sneyd. 1991. "A model of collective nectar source selection by honey bees: Self-organization through simple rules." *Journal of Theoretical Biology* 149: 547–571.

Camazine, S., P. K. Visscher, J. Finley, and R. S. Vetter. 1999. "House-hunting by honey bee swarms: Collective decisions and individual behaviors." *Insectes Sociaux* 46: 348–360.

Cameron, S. A., and J. B. Whitfield. 1996. "Use of walking trails by bees."
    Nature 379: 125.
Cocroft, R. B., and R. L. Rodriguez. 2005. "The behavioral ecology of insect
    vibrational communication." *Bioscience* 55: 323–334.
Cole, B. J., and F. I. Trampus. 1999. "Activity cycles in ant colonies: Worker
    interactions and decentralized control." In C. Detrain, J. L. Deneubourg,
    and J. M. Pasteels, eds., *Information processing in social insects,* 289–307.
    Basel: Birkhäuser Verlag.
Conradt, L., and T. J. Roper. 2005. "Consensus decision making in animals."
    Trends in Ecology and Evolution 20: 449–456.
Costa, J. T., and T. D. Fitzgerald. 1996. "Developments in social terminology:
    Semantic battles in a conceptual war." *Trends in Ecology and Evolution* 11:
    285–289.
Couzin, I. D., J. Krause, N. R. Franks, and S. A. Levin. 2005. "Effective lead-
    ership and decision-making in animal groups on the move." *Nature* 433:
    513–516.
Crespi, B. J. 1992. "Eusociality in Australian gall thrips." *Nature* 359: 724–726.
———. 2004. "Vicious circles: Positive feedback in major evolutionary and
    ecological transitions." *Trends in Ecology and Evolution* 19: 627–633.
Crespi, B. J., and D. Yanega. 1995. "The definition of eusociality." *Behavioral
    Ecology* 6: 109–115.
D'Andrea, M. 1987. "Social behaviour in spiders (Arachnida, Araneae)."
    Monitore Zoologico Italiano (Nuova Serie), Monographia 3: 1–156.
DeAngelis, D. L., W. M. Post, and C. C. Portis. 1986. *Positive feedback in nat-
    ural systems.* New York: Springer-Verlag.
Deneubourg, J. L., S. Goss, J. M. Pasteels, D. Fresneau, and J. P. Lachaud.
    1987. "Self-organization mechanisms in ant societies (II): Learning in for-
    aging and division of labor." In J. M. Pasteels and J.-L. Deneubourg, eds.,
    From individual to collective behavior in social insects. Basel: Birkhäuser
    Verlag.
Deneubourg, J.-L., J.-C. Grégoire, and E. Le Fort. 1990. "Kinetics of larval
    gregarious behavior in the bark beetle *Dendroctonus micans* (Coeloptera:
    Scolytidae)." *Journal of Insect Behavior* 3: 169–182.
Deneubourg, J.-L., A. Lioni, and C. Detrain. 2002. "Dynamics of
    aggregation and emergence of cooperation." *Biological Bulletin* 202:
    262–267.
Depickere, S., D. Fresneau, and J.-L. Deneubourg. 2004. "A basis for spatial
    and social patterns in ant species: Dynamics and mechanisms of aggrega-
    tion." *Journal of Insect Behavior* 17: 81–97.
Detrain, C., J.-L. Deneubourg, S. Goss, and Y. Quinet. 1991. "Dynamics of
    collective exploration in the ant *Pheidole pallidula.*" *Psyche* 98, 21–31.

Detrain, C., C. Natan, and J.-L. Deneubourg. 2001. "The influence of the physical environment on the self-organised patterns of ants." *Naturwissenschaften* 88: 171–174.

Duffy, J. E. 1996. "Eusociality in a coral-reef shrimp." *Nature* 381: 512–514.

Fewell, J. H. 2003. "Social insect networks." *Science* 301: 1867–1870.

Fitzgerald, T. D. 1995. *The tent caterpillars.* Ithaca: Cornell University Press.

Fitzgerald, T. D., and C. R. Visscher. 1996. "Foraging behavior and growth of isolated larvae of a social caterpillar, Malacosoma americanum." *Entomologia Experimentalis Et Applicata* 81: 293–299.

Fraenkel, G. S., and D. L. Gunn. 1961. *The orientation of animals: Kineses, taxes and compass reactions.* New York: Dover Publications.

Franks, N. R., and A. B. Sendova-Franks. 1992. "Brood sorting by ants: Distributing the workload over the work-surface." *Behavioral Ecology and Sociobiology* 30: 109–123.

Franks, N. R., S. C. Pratt, E. B. Mallon, N. F. Britton, and D. J. T. Sumpter. 2002. "Information flow, opinion polling and collective intelligence in house-hunting social insects." *Philosophical Proceedings of the Royal Society London B* 357: 1567–1583.

Franks, N. R., and T. Richardson. 2006. "Teaching in tandem-running ants." *Nature* 439: 153.

Gadagkar, R. 1994. "Why the definition of eusociality is not helpful to understand its evolution and what should we do about it." *Oikos* 70: 485–488.

Gautrais, J., G. Theraulaz, J.-L., Deneubourg, and G. Theraulaz. 2002. "Emergent polyethism as a consequence of increased colony size in insect societies." *Journal of Theoretical Biology* 215: 363–373.

Glansdorff, P., and I. Prigogine. 1971. *Thermodynamic theory of structure, stability and fluctuations.* New York: J. Wiley & Sons.

Goessmann, C., C. Hemelrijk, and R. Huber. 2000. "The formation and maintenance of crayfish hierarchies: Behavioral and self-structuring properties." *Behavioral Ecology and Sociobiology* 48: 418–428.

Gotwald, W. H. J. 1995. *Army ants: The biology of social predation.* Ithaca: Cornell University Press.

Grassé, P.-P. 1984. *Termitologia, Tome II. Fondation des Sociétés. Construction.* Paris: Masson.

Greenfield, M. D., and I. Roizen. 1993. "Katydid synchronous chorusing is an evolutionarily stable outcome of female choice." *Nature* 364: 618–620.

Hart, A., and F. Ratnieks. 2000. "Leaf caching in Atta leafcutting ants: Discrete cache formation through positive feedback." *Animal Behaviour* 59: 587–591.

Heinrich, B. 1981. "Energetics of honeybee swarm thermoregulation." *Science,* 212: 565–566.

Helms Cahan, S., and J. H. Fewell. 2004. "Division of labor and the evolution of task sharing in queen associations of the harvester ant *Pogonomyrmex californicus.*" *Behavioral Ecology and Sociobiology* 56: 9–17.

Hogeweg, P., and B. Hesper. 1983. "The ontogeny of the interaction structure in bumble bee colonies: A MIRROR model." *Behavioral Ecology and Sociobiology* 12: 271–283.

Hölldobler, B. 1981. "Foraging and spatiotemporal territories in the honey ant *Myrmecocystus mimicus* Wheeler (Hymenoptera: Formicidae)." *Behavioral Ecology and Sociobiology* 9: 301–314.

Hölldobler, B., and E. O. Wilson. 1990. *The ants.* Cambridge: Belknap Press of Harvard University Press.

Hölldobler, B., and E. O. Wilson. 2005. "Eusociality: Origin and conse-quences." *Proceedings of the National Academy of Sciences of the USA* 102: 13367–13371.

Issa, F. A., D. J. Adamson, and D. H. Edwards. 1999. "Dominance hierarchy formation in juvenile crayfish *Procambarus clarkii.*" *Journal of Experimental Biology* 202: 3497–3506.

Jeanson, R., J.-L. Deneubourg, and G. Theraulaz. 2004. "Discrete dragline attachment induces aggregation in spiderlings of a solitary species." *Animal Behaviour* 67: 531–537.

Jeanson, R., J.-L. Deneubourg, R. Grimal, and G. Theraulaz. 2004. "Modulation of individual behavior and collective decision-making during aggregation site selection by the ant *Messor barbarus.*" *Behavioral Ecology and Sociobiology* 55: 388–394.

Jeanson, R., J. H. Fewell, R. Gorelick, and S. M. Bertram. 2007. "Emergence of division of labor as a function of group size." *Behavioral Ecology and Sociobiology* 62: 289–298.

Jeanson, R., F. L. W. Ratnieks, and J.-L. Deneubourg. 2003. "Pheromone trail decay rates on different substrates in the Pharaoh's ant, *Monomorium pharaonis.*" *Physiological Entomology* 28: 192–198.

Jeanson, R., and J.-L. Deneubourg. 2007. "Conspecific attraction and shelter selection in gregarious insects." *American Naturalist.* 170: 47–58.

Jeanson, R., P. K. Kukuk, and J. H. Fewell. 2005. "Emergence of division of labor in halictine bees: Contributions of social interactions and behavioral variance." *Animal Behaviour* 70: 1183–1193.

Jeanson, R., C. Rivault, J.-L. Deneubourg, S. Blanco, R. Fournier, C. Jost, and G. Theraulaz. 2005. "Self-organized aggregation in cockroaches." *Animal Behaviour* 69: 169–180.

Jeon, J., and J. C. Choe. 2003. "Reproductive skew and the origin of sterile castes." *American Naturalist* 161: 206–224.

Karsai, I., and J. W. Wenzel. 1998. "Productivity, individual-level and colony-level flexibility, and organization of work as consequences of colony size." *Proceedings of the National Academy of Sciences USA* 95: 8665–8669.

Krafft, B. 1979. "Organisation et évolution des sociétés d'araignées." *Journal de Psychologie* 1: 13–51.

Krafft, B., and A. Pasquet. 1991. "Synchronized and rhythmical activity during the prey capture in the social spider *Anelosimus eximius* (Araneae, Theridiidae)." *Insectes Sociaux* 38: 83–90.

Krause, J., and G. D. Ruxton. 2002. *Living in groups*. Oxford: Oxford University Press.

Lin, N., and C. D. Michener. 1972. "Evolution of sociality in insects." *The Quarterly Review of Biology* 47: 131–159.

Lioni, A., and J.-L. Deneubourg. 2004. "Collective decision through self-assembling." *Naturwissenschaften* 91: 237–241.

Lubin, Y. D., and M. H. Robinson. 1982. "Dispersal by swarming in a social spider." *Science* 216: 319–321.

Mallon, E. B., S. C. Pratt, and N. R. Franks. 2001. "Individual and collective decision-making during nest site selection by the ant *Leptothorax albipennis*." *Behavioral Ecology and Sociobiology* 50: 352–359.

Maynard-Smith, J., and E. Szathmary. 1999. *The origins of life: From the birth of life to the origin of language*. Oxford: Oxford University Press.

Michener, C. D. 1969. "Comparative social behavior of bees." *Annual Review of Entomology* 14: 299–342.

Millor, J., M. Pham-Delegue, J.-L. Deneubourg, and S. Camazine. 1999. "Self-organized defensive behavior in honeybees." *Proceedings of the National Academy of Sciences of the USA* 96: 12611–12615.

Nieh, J. C., F. A. L. Contrera, R. R. Yoon, L. S. Barreto, and V. L. Imperatriz-Fonseca. 2004. "Polarized short odor-trail recruitment communication by a stingless bee, *Trigona spinipes*." *Behavioral Ecology and Sociobiology* 56: 435–448.

Novick, A., and M. Weiner. 1957. "Enzyme induction as an all-or-none phenomenon." *Proceedings of the National Academy of Sciences of the USA* 43: 553–566.

Ono, M., T. Igarashi, E. Ohno, and M. Sasaki. 1995. "Unusual thermal defense by a honeybee against mass attack by hornets." *Nature* 377: 334–336.

Pacala, S. W., D. M. Gordon, and H. C. J. Godfray. 1996. "Effects of social group size on information transfer and task allocation." *Evolutionary Ecology* 10: 127–165.

Parrish, J. K., and L. Edelstein-Keshet. 1999. "Complexity, pattern, and evolutionary trade-offs in animals aggregation." *Science* 284: 99–101.

Parrish, J. K., and W. H. Hamner, eds. 1997. *Animal groups in three dimensions*. Cambridge: Cambridge University Press.

Pratt, S. C. 2005. "Quorum sensing by encounter rates in the ant *Temnothorax albipennis.*" *Behavioral Ecology* 16: 488–496.

Pratt, S. C., E. B. Mallon, D. J. T. Sumpter, and N. R. Franks. 2002. "Quorum sensing, recruitment, and collective decision-making during colony emigration by the ant *Leptothorax albipennis.*" *Behavioral Ecology and Sociobiology* 52: 117–127.

Reinhard, J., and M. Kaib. 2001. "Trail communication during foraging and recruitment in the subterranean termite Reticulitermes santonensis De Feytaud (Isoptera, Rhinotermitidae)." *Journal of Insect Behavior* 14: 157–171.

Rhoden, P. K., and W. A. Foster. 2002. "Soldier behaviour and division of labour in the aphid genus Pemphigus (Hemiptera, Aphididae)." *Insectes Sociaux* 49: 257–263.

Rypstra, A. L. 1993. "Prey size, social competition, and the development of reproductive division of labor in social spider groups." *American Naturalist* 142: 868–880.

Saffre, F., R. Furey, B. Krafft, and J. L. Deneubourg. 1999. "Collective decision-making in social spiders: Dragline-mediated amplification process acts as a recruitment mechanism." *Journal of Theoretical Biology* 198: 507–517.

Scheffer, M., and S. R. Carpenter. 2003. "Catastrophic regime shifts in ecosystems: Linking theory to observation." *Trends in Ecology and Evolution* 18: 648–656.

Seeley, T. D. 1995. *The wisdom of the hive: The social physiology of honey bee.* Cambridge: Harvard University Press.

Seeley, T. D., and S. C. Buhrman. 2001. "Nest-site selection in honey bees: How well do swarms implement the "best-of-N" decision rule?" *Behavioral Ecology and Sociobiology* 49: 146–427.

Seeley, T. D., S. Camazine, and J. Sneyd. 1991. "Collective decision-making in honey bees: How colonies choose among nectar sources." *Behavioral Ecology and Sociobiology* 28: 277–290.

Seeley, T. D., A. S. Mikheyev, and G. J. Pagano. 2000. "Dancing bees tune both duration and rate of waggle-run production in relation to nectar-source profitability." *Journal of Comparative Physiology* A 186: 813–819.

Seeley, T. D., and P. K. Visscher. 2004. "Group decision making in nest-site selection by honey bees." *Apidologie* 35: 101–116.

Shear, W. A. 1970. "The evolution of social phenomena in spiders." *Bulletin of the British Arachnological Society* 1: 65–76.

Sherman, P. W., E. A. Lacey, H. K. Reeve, and L. Keller. 1995. "The eusociality continuum." *Behavioral Ecology* 6: 102–108.

Shibao, H., M. Kutsukake, and T. Fukatsu. 2004. "Density triggers soldier production in a social aphid." *Proceedings of the Royal Society of London Series B-Biological Sciences* 271: S71–S74.

Tallamy, D. W., and T. K. Wood. 1986. "Convergence patterns in subsocial insects." *Annual Review of Entomology* 31: 369–390.

Theraulaz, G., E. Bonabeau, and J.-L. Deneubourg. 1995. "Self-organization of hierarchies in animal societies—The case of the primitively eusocial wasp *Polistes dominulus* Christ." *Journal of Theoretical Biology* 174: 313–323.

Theraulaz, G., E. Bonabeau, S. C. Nicolis, R. V. Solé, V. Fourcassié, S. Blanco, R. Fournier, J. L. Joly, P. Fernandez, A. Grimal, P. Dalle, and J.-L. Deneubourg. 2002. "Spatial patterns in ant colonies." *Proceedings of the National Academy of Sciences USA* 99: 9645–9649.

Thomas, M. L., and M. A. Elgar. 2003. "Colony size affects division of labour in the ponerine ant *Rhytidoponera metallica*." *Naturwissenchaften* 90: 88–92.

Thomas, R. 1998. "Laws for the dynamics of regulatory networks." *International Journal of Developmental Biology* 42: 479–485.

Underwood, D. L. A., and A. M. Shapiro. 1999. "Evidence for division of labor in the social caterpillar *Eucheira socialis* (Lepidoptera: Pieridae)." *Behavioral Ecology and Sociobiology* 46: 228–236.

Vollrath, F. 1982. "Colony foundation in a social spider." *Zeitschrift für Tierpsychologie* 60: 313–324.

Weidenmüller, A. 2004. "The control of nest climate in bumblebee *(Bombus terrestris)* colonies: Interindividual variability and self reinforcement in fanning response." *Behavioral Ecology* 15: 120–128.

Wertheim, B., E. A. van Baalen, M. Dicke, and L. E. M. Vet. 2005. "Pheromone-mediated aggregation in nonsocial arthropods: An evolutionary ecological perspective." *Annual Review of Entomology* 50: 321–346.

CHAPTER TWENTY-ONE

Division of Labor in the Context
of Complexity

JENNIFER FEWELL

SHANA K. SCHMIDT

THOMAS TAYLOR

SOCIOBIOLOGY IS UNDERGOING a shift in its theoretical framework toward the paradigm that societies are complex and dynamical systems rather than amalgamated groups of individuals (Bonabeau 1998; Camazine et al. 2001; Fewell 2003; Detrain and Deneubourg 2006). Although components of this perspective have been embedded in social insect biology for a long time, the tools and theory inherent in complexity science have added new contributions to our conceptualization of insect societies as dynamic and cohesive units. This framework considers social groups inclusively from the perspective of group members' individual phenotypes, societal-level phenotypes that extend beyond simple summation of individual behaviors, and the interactions that link these levels. Such a paradigm shift is epistemologically significant. Viewing the social group as a dynamic system requires a different experimental approach than dissecting components or removing elements to determine function (Woese 2004). Viewing social groups as dynamic and thus cohesive units also has a major impact on our understanding of how selection shapes social evolution (Wilson and Sober 1989; Moore, Brodie, and Wolf 1997; Hamilton, Smith, and Haber, this volume).

Some of the elements of this paradigm are seen in the foundational literature of insect sociobiology. Wheeler's (1928) concept of the superorganism emphasized that eusocial insect colonies function as an interactive whole. And Wilson and Hölldobler's (1988) characterization of colonies as heterarchies provided the framework for viewing the colony as a decentralized and densely interactive network (Fewell 2003). Over the past few decades, there has been a large increase in the pervasiveness of complexity theory across disciplines, allowing us to tap into a growing theoretical framework (Bechtel and Richardson 1993; Holland 1995; Auyang 1998; Levin 1998). Simultaneously, the number of theoretical and empirical studies on social dynamics, and especially self-organization in social insects, has dramatically increased (Camazine et al. 2001; Fewell 2003; Franks et al., Jeanson and Deneubourg, and Pratt, this volume).

In this chapter we provide one perspective on how complexity thinking has been applied to division of labor. We explore how division of labor fits within (and expands on) current thinking on social complexity. We also examine how task organization within social groups can be shaped by the interplay of emergent phenotypes and natural selection, from the simple dynamics within communal systems to the more derived and sophisticated systems in eusocial colonies.

Defining and Measuring Division of Labor

We define *division of labor* as the degree to which different individuals within a social group specialize on different tasks, following Michener's (1974) characterization of division of labor "to include any behavioral pattern that results in some individuals in a colony performing different functions from others, even if only temporarily." As nicely explained by Michener, division of labor is a statistical, rather than an absolute, designation that can be measured across different time scales (e.g., from specialization by foragers on a specific food source to a worker's task ontogeny over her lifetime).

This conceptualization allows us to explore division of labor across diverse social systems, but such explorations have been limited in part because division of labor is often equated with discrete morphological castes (Costa and Pierce 1997; Clutton Brock, Russell, and Sharpe 2003). From Michener's definition, however, behavioral task differentiation among group members appears pervasively across social taxa, including shrimp (Duffy, Morrison, and Macdonald 2002), burying beetles (Trumbo 2006), caterpillars (Underwood and Shapiro 1999), spiders (Aviles, Madison, and Agnarsson 2006), lions (Stander 1992), dolphins (Gazda et al. 2005), monkeys (Hattori and Kuroshima 2005), rats (Grasmuck and Desor 2002), multiple bird species (Arnold, Owens, and Goldizen 2004; Bartlett, Mock, and Schwagmeyer 2005; Komdeur 2006), and, of course, humans.

Embedded in the concept of division of labor is that of task specialization, which has been characterized variously as (1) disproportionate performance of a task by one or a few individuals relative to others in the group (Oster and Wilson 1978), or (2) a bias by individuals toward increased performance of one task relative to available tasks (Kolmes 1985; Beshers and Fewell 2001). Although task specialization is often measured by how frequently different individuals perform a specific task (Fewell and Page 1999; Weidenmüller 2004), this metric is not sufficient to differentiate specialization from elitism, where some individuals simply work harder, performing multiple tasks at higher rates (Oster and Wilson 1978; Gorelick et al. 2004).

For division of labor to be high, two things must happen: first, individuals must be biased toward a subset of tasks relative to all those available, and second, individuals must be distributed across tasks so that different individuals perform different tasks. In contrast, division of labor is low or absent if one individual is simply more active than others, or if all individuals specialize on the same task. We can capture this dynamic using Shannon's mutual entropy, derived from information theory (Shannon 1948; Gorelick et al. 2004). Shannon's mutual entropy measures information flow from signaler to receiver, such that entropy is lowest when given a signaler, one can predict the receiver, and given the receiver, one can predict the signaler. The division of labor (DOL) measure (Gorelick et al. 2004; Gorelick and Bertram 2007) applies this metric to division of labor. DOL is highest when, knowing an individual worker, we have a high probability of predicting the task, and conversely, given a specific task, we have a high certainty of knowing which worker or group of workers is performing it. This metric provides a starting point to ask questions about how division of labor changes over different scales, from changes within colonies through ontogeny to evolutionary transitions from primitively social groups to highly eusocial colonies.

The ease of application of information theory metrics to division of labor illustrates that the basic mechanistic algorithms generating complex patterns are broadly applicable across systems and scales. Division of labor could be considered most basically as a system of behavioral differentiation, as individuals separate across different tasks, structures, or niches. Parallel differentiation occurs across biological scales, from tissue differentiation within an organism to the differentiation of species within communities (Gorelick and Bertram 2007). Given that we can measure these processes using the same metric, we should also ask whether similar algorithms or self-organizing principles apply across scales.

Complexity and Self-Organization

Although the term *complexity* is used pervasively, there is significant debate on what characteristics actually define systems as complex. According to Rosen (1991), systems can be characterized as simple if they can be completely (or almost so) simulated via simple algorithms, and complex if they cannot. This is not a completely satisfying description. Many aspects of social organization within a social insect colony can, theoretically, be explained by rather simple rule sets, but the colony's behavioral outcomes are by no means simple (Camazine et al. 2001). Algorithms or rule sets describing social interactions are not enough to describe the colony-level outcome. These must be coupled with information about the participating individuals themselves, and the ways by which individuals are connected (the social network).

Another perhaps more useful approach is to consider system complexity within the context of the paired attributes of decomposition and localization (Bechtel and Richardson 1993). Decomposition allows that complex systems can be divided into subcomponents, but also recognizes that those subunits are never completely independent of each other (nearly decomposable systems, Simon 1962). To this we can add the expectation that the subcomponents of more complex systems are generally more heterogeneous and often more numerous (Simon 1962; Levin 1998). Thus, the behavior of a social group as a complex system would be described in part by the behaviors of the individual group members, but it could not be completely described without also considering the interactions among them.

Localization is the degree to which individual components are responsible for some subset or range of phenomena—their contribution to the dynamics of the system. If a social system has multiple clusters of individuals with localized interactions, and if they are each connected somehow to the group as a whole (and thus contribute to the behavior of the whole group), the social group becomes a complex system. Thus, complex systems are by definition distributed, because the behavior of the collective whole results from multiple local interactions rather than being directed externally or from a central source. These local interactions collectively produce group-level phenotypes, or emergent properties, at the larger scale that cannot be described or explained simply by measuring the behaviors of the individual group members alone. The process of local dynamics generating emergent effects is called self-organization, and the ubiquity of this effect has led to the suggestion that the presence of self-organization could be considered the defining characteristic of complex systems.

Self-Organization and Division of Labor

Self-organization of social behavior is produced primarily by positive feedback loops in which a behavioral effect is amplified over successive iterations, either by a cascading increase in the number of individuals performing the behavior or by increasing the probability that it will be repeated by any one individual (Bonabeau, Theraulaz, and Deneubourg 1998; Theraulaz, Bonabeau, and Deneubourg 1998; Jeanson and Deneubourg, this volume). The addition of negative feedback loops (in which performance of a behavior reduces its subsequent likelihood or intensity) enables social systems to reach behavioral equilibrium or homeostasis (Camazine et al. 2001; Sumpter 2006). Self-organizational mechanisms have been linked to a diversity of social insect colony phenotypes, including nest construction, house hunting, recruitment to food or defense, trail-laying, and dominance (Goss et al. 1989; Camazine et al. 2001; Britton et al. 2002; Hogeweg and Hesper 1983; Hemelrijk 2002; Franks et al., Jeanson and Deneubourg, and Pratt, this volume). However, of all the social phenotypes characterized with self-organization, division of labor is perhaps the most global and ubiquitous. It occurs at the level of the entire society and is present continuously over the life of the colony, rather than associated with specific events or tasks.

How can division of labor self-organize? The dominant current paradigm for its emergence is the response threshold model (Robinson and Page 1989, Bonabeau, Theraulaz, and Deneubourg 1998; Beshers and Fewell 2001). In this model the assumption is made that any individual has a threshold for responding to stimuli for any given task. An individual begins performing the task if stimulus levels reach their threshold, and if they are not already engaged. Individuals within a group initially vary in task thresholds either intrinsically or from differences in experience and/or development. As stimulus levels for a task rise, those group members with lower thresholds perform it first. When they do so, they lower the probability that other group members will perform it also. Thus, they become, by default, the specialists for that task. Because individual thresholds vary across tasks, different group members specialize on different tasks.

The response threshold model provides a simple but powerful framework for understanding how division of labor can be generated via social dynamics, and has gained increasing theoretical support. Variations of the model have demonstrated a positive relationship between inter-individual variation in thresholds and differentiation in task performance (Bonabeau, Theraulaz, and Deneubourg 1998; Page and Mitchell 1998), as well as its contribution to colony stability and homeostasis in task performance (Myerscough and Oldroyd 2004). Recent models have also used the response threshold concept as a starting point to explore the effect of group size on division of labor (Gautrais et al. 2002; Jeanson et al. 2007).

There is increasing empirical support for the mechanisms by which individual variation in task thresholds (generally measured as responsiveness to a task; see Scheiner and Erber, this volume) contribute to task specialization and division of labor. Contexts in which variation in individual thresholds has been tested include: fanning in honey bees and bumblebees (Weidenmueller 2004; Jones et al. 2004), undertaking (Robinson and Page 1988) and foraging in honey bees (Scheiner and Erber, Page, Linksvayer, and Amdam, this volume); as well as nest excavation (Fewell and Page 1999; Helms Cahan and Fewell 2004), foraging (Stuart and Page 1991), and prey capture (Theraulaz et al. 2002) in ants. In some eusocial species, task specialization and thus differentiation is also intensified by self-reinforcement, in which performance of a task increases the likelihood that a worker will perform it again (Theraulaz, Bonabeau, and Deneubourg 1998; Ravary et al. 2007)

It is important to keep in mind that despite their utility in colony-level descriptions of division of labor, response threshold models do provide an overly simplistic representation of individual response. Individual responses to task stimuli are not simple on–off switches. In response to temperature change, individual bumblebees vary in their response threshold (the temperature at which they begin fanning), but also in their probability of fanning, and frequency and amount of time spent fanning (Weidenmueller 2004; Gardner, Foster, and O'Donnell 2007). Workers within social insect colonies may also change thresholds (measured as probability of task performance) with experience, a process called self-reinforcement (Theraulaz, Bonabeau, and Deneubourg 1998; Weidenmueller 2004; Ravary et al. 2007).

The emergence of division of labor could also occur via mechanisms associated with spatial differentiation in addition to, or alternately to, intrinsically based variation. For example, the foraging for work model hypothesizes that task differentiation can be generated by spatial heterogeneity (Franks and Tofts 1994). Individuals without work move away from filled tasks toward those that are available, correspondingly moving from the nest center where they emerge to the nest periphery and associated tasks such as foraging. Repeated performance of individual tasks (increasing specialization) can also theoretically be influenced by spatial fidelity, in which individuals tend to return to an area in which they were previously located (Sendova-Franks and Franks 1994).

Evolutionary Transitions in Division of Labor

Division of labor is generally considered the premier adaptation of eusociality (Wilson 1971), and as such, it is essentially viewed as a derived trait, one that evolved and was selected for within social groups. However, self-organizational models allow for the possibility that it could in fact emerge spontaneously within simple social groups, and even at the origins of sociality (Page and Mitchell 1998; Fewell and Page 1999). There is empirical support for this assertion. Normally solitary *Ceratina* bees show division of labor when forced to nest together, with one bee guarding the nest while the other forages (Sakagami and Maeta 1987). Similarly, solitary Halictid bees *(Lasioglossum [Ctenonomia] NDA-1)* forced together into observation nests show differentiation for nest excavation and guarding the entrance (Jeanson, Kukuk, and Fewell 2005).

Harvester ant queens, *Pogonomyrmex barbatus,* who normally found nests alone (haplometrosis) also show division of labor when forced into associations (Fewell and Page 1999). Which ant takes over the task of excavation can be predicted from variation in their individual excavation behavior while alone, or from their behavior in previous associations. Also consistent with the emergence hypothesis, specialization appears even when two individuals are paired who previously performed the task at high levels. One takes over nest excavation, while the other significantly reduces her excavation behavior and spends more time tending brood (Fewell and Page 1999; Fewell, unpublished data).

The data on harvester ants suggest that intrinsic differences in task propensity drive the emergence of division of labor in these associations (Fewell and Page 1999; Helms Cahan and Fewell 2004). However, in

Halictid bees, task differentiation is also a product of aggression and consequent spatial dynamics. These ground-nesting bees dig a narrow central nest tunnel; traveling from the bottom of the nest (where excavation occurs) to the top (for guarding) often requires that individuals pass each other. Although bees of the communal species *L. (Chilalictus) hemichalceum* are tolerant of each other, females of a related solitary species, *L. (Ctenonomia) NDA-1*, often prevent each other from passing. These species show higher levels of division of labor than their communal cousins, and the increase is accounted for in large part by the spatial separation in the nest that results from lower tolerance (Jeanson, Kukuk, and Fewell 2005). These data demonstrate that, although the basic algorithms remain similar, behavioral divergence can be generated via multiple biological mechanisms which vary across species and social contexts.

Transitions in Division of Labor from Solitary to Communal

We can use the wide range of social systems in the social Hymenoptera to explore how selection and self-organization interact to shape division of labor over social evolution. One of the most common social transitions is from solitary life to communal and/or quasisocial societies, in which often unrelated adults cooperate in nest building, nest defense, and (in quasisocial systems) brood care (Michener 1974; Wcislo and Tierney, this volume). Surprisingly, in comparisons of division of labor between associations of normally solitary and normally communal harvester ant and sweat bees, the normally solitary queens, when forced into associations, consistently show higher levels of task specialization than do the communal groups (Fewell and Page 1999; Helms Cahan and Fewell 2004; Jeanson, Kukuk, and Fewell 2005; Figure 21.1). Thus, the transition from solitary to communal seems to involve a reduction in division of labor (Helms Cahan and Fewell 2004), counter to our traditional understanding of division of labor as a social adaptation.

Why would division of labor be reduced in species that exhibit a higher level of social organization? To answer this question, we must consider how self-organization and selection interact. Individual tasks vary in risk and physiological cost. For example, in harvester ant associations, nest excavation is associated with cuticular abrasion and water loss (Johnson 2000). Queens that become the excavation specialist often have higher mortality rates than queens who tend brood (and often consume eggs as a part of that task;

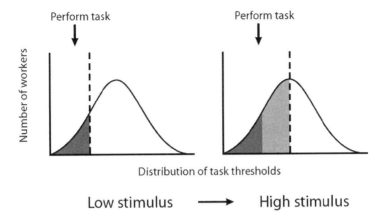

Figure 21.1. Diagram of the relationship between diversity in task thresholds and flexible task performance in a social insect colony. The graphs show a colony of workers with a normal distribution of task thresholds. Only a narrow subset of individuals performs the task under low stimulus conditions. These are the task specialists. As conditions change so that stimulus levels for the task increase, a wider range of thresholds are met, and more workers perform the task.

Fewell and Page 1999; Helms Cahan and Fewell 2004). The probability of a given queen becoming the excavator (and dying) or the brood specialist (and inheriting the nest) is essentially dependent on the social context, specifically the intrinsic response thresholds of the other queens in her group. Thus, the emergence of division of labor is coupled with the emergence of cost disparities that could potentially block the evolution of cooperative associations. This point may contribute to the "sociality paradox"—that social species are often highly successful, but sociality itself is rare.

If division of labor presents a barrier to the evolution of cooperation, the leap from solitary to social requires groups to reduce specialization for tasks that can generate significant fitness disparities among members. This evolutionary hurdle is consistent with the general observation that division of labor tends to be much lower in communal than eusocial taxa. Variable task costs likely also influence differential helping behavior in birds (Arnold, Owens, and Goldizen 2004) and mammals (Heinsohn and Packer 1995).

The evolutionary transition from emergent task specialization to more evenly distributed sharing of tasks (Helms Cahan and Fewell 2004) requires a shift in social roles to minimize cost disparities. This implicitly assumes that while some individuals gain in fitness, others must reduce fitness advantages accordingly. The implication of this transition is that selection on communal or cooperatively breeding groups involves the interplay of individual and group effects. In other words, multilevel selection already plays a role in social evolution within communal systems.

Transitions to Division of Labor in Eusocial Species

The transition to eusociality provides a different selective context for division of labor. By definition, in eusocial colonies the reproductive division of labor is coupled with high intragroup relatedness, so that selection acts predominantly at the group level. Thus, individual cost disparities become less important as long as individual task specialization contributes to group function. The increased size of highly eusocial colonies can itself have emergent effects on division of labor. Larger group size is widely associated with increased division of labor, both phylogenetically and ontogenetically (Gautrais et al. 2002; Jeanson et al. 2007). For example, small colonies of independent-founding Polistinae wasps typically exhibit less task specialization than swarm-founding wasps characterized by larger colonies (Jeanne 1991; Karsai and Wenzel 1998).

Changes in division of labor with increased size are influenced in part by changes in task number and demand (Jeanson et al. 2007). Demand, defined as the need for tasks to be performed relative to the availability of workers to perform them, is expected to scale negatively with group size. This effect can be viewed as an economy of scale; the additional workload imposed by adding workers to the colony should not increase as fast as the capacities of those individuals to do work. Empirical evidence generally supports this assertion; in larger colonies the number of inactive or reserve workers increases, while work performed per individual generally decreases with group size (Wilson 1986; Lachaud and Fresneau 1987; Mailleux, Deneubourg, and Detrain 2003; but see also Schmid Hempel 1990). According to the model, at low demand levels there are enough workers available that all work is likely to be taken care of maintaining low stimulus levels for each task; in this case only those individuals with lower response thresholds are likely to repeatedly engage in a given task. When demand is high, not all tasks are performed completely and their corresponding stimulus levels build up. This, in turn, causes many individual thresholds to be met for these tasks, and generates random task performance rather than division of labor.

Increased group size can also theoretically affect division of labor via increased task number. Simple eusocial societies such as the ponerine ants *Amblyopone pallipes* tend to perform 5 to 10 tasks (Traniello 1978), while more complex social groups such as the Myrmicinae or honey bees perform 20 to 40 tasks (Oster and Wilson 1978; Kolmes 1985). Colonies also increase the number of tasks ontogenetically. Nest construction in the paper wasp requires foraging for paper materials, foraging for water, and chewing paper and water together into pulp. In early stages of colony growth when colony population is low, each worker performs all of the tasks in sequence. As the colony expands, workers begin partitioning the task, setting up a network of workers, some of whom forage for materials while others process them for nest construction (Karsai and Wenzel 1998).

Noise, Complexity, and Genetic Task Specialization

Social insect workers are theoretically pleuripotent; they are capable of performing any task with the exception of reproduction. However, the evolution of eusociality is associated with the evolution of mechanisms that increasingly predispose individuals to perform only a subset of tasks. In the more highly derived eusocial species these mechanisms include age polyethism and morphological castes. Although temporal and morphological polyethism are often viewed in terms of individual developmental as opposed to social dynamical processes, they involve network interactions among genetics, development and social context (Robinson 1992; Page, Linksvayer, and Amdam, this volume). As one example, the task repertoire of workers in the ant *Pheidole dentata* follows a developmental progression in which repertoire size increases with age, but task performance varies with colony task need. Individual task performance is flexible, yet older workers show more plasticity than younger ants (Seid and Traniello 2006).

Another mechanism for task differentiation in highly eusocial systems is intrinsically or genetically based task specialization, where individual workers are biased toward performing a subset of available tasks (Oldroyd and Fewell 2007). Genetic task specialization is widespread across social insect taxa (Oldroyd and Fewell 2007). The literature on genetically based task specialization in honey bees is particularly large, and includes demonstrations of behavioral differences based on genotype as well as explorations into the specific genetic and developmental architectures generating task specialization (Oldroyd and Thompson 2007; Scheiner and Erber, Page, Linksvayer, and Amdam this volume).

What benefit does genetic task specialization bring to eusocial colonies? The generally proposed answer is increased individual efficiency of task performance, usually measured by performance time or rate (Robinson 1992). There is some evidence that some task specialists do perform tasks faster. Undertaker specialists in ants remove dead workers much more quickly than other workers (Julian and Cahan 1999), and honey bees with hygienic genotypes are much better at stopping infection by sealing contaminated corpses (Arathi and Spivak 2001). However, there are also multiple studies documenting no specific changes in individual performance rate or efficiency from task specialization (Oldroyd and Fewell 2007).

A different potential benefit of genetic task specialization is that it could filter the multiple signals associated with task regulation (Gronenberg and Riveros, Kleineidam and Rössler, this volume). As colony size and complexity increase, the associated signal "noise" within the colony can generate an information load problem for individual workers. Signaling modalities in the social insects are especially rich and complex (Hölldobler and Wilson 1991), with a huge number of signals and cues being transmitted at any time within a large colony and thus creating a potential cacophony of competing signals. This is what we describe as "the problem of the Borg," where all individuals instantaneously communicate with all others. Such a system might quickly overload the capacity of any one individual to process and respond to information. Therefore, the strongest efficiency benefit of genetic or intrinsic task specialization may occur at the point of task choice, if individual workers "notice" only the signals for tasks to which they are predisposed.

Genetic Diversity and Colony Resiliency

Although individual task specialization may increase individual productivity, it creates a potential problem at the colony level when worker flexibility is reduced (Oster and Wilson 1978; Page et al. 1995; Oldroyd and Fewell 2007). The colony must retain sufficient flexibility to allocate tasks across a constantly changing environment. There is increasing theoretical and empirical evidence that genotypic variation in highly eusocial colonies contributes to division of labor (Oldroyd and Fewell 2007). In particular, variation in response thresholds enhances colony flexibility and resiliency because of the ability to recover more quickly or completely from perturbation. For example, genetically more diverse honey bee colonies consistently maintain more uniform temperatures in their brood nests, an indication that they are more resilient to external temperature fluctuations (Jones et al. 2004).

Genetic subgroups within honey bee colonies vary in their propensity for pollen collection (Calderone and Page 1988; Fewell and Page 1993; 2000). Under low stimulus conditions, as when extra pollen stores are added, only one or a few subgroups forage for pollen. When stores are removed, additional genetic subgroups are recruited to pollen foraging, and total colony effort increases accordingly (Fewell and Page 1993; Figure 21.2). In selection experiments for pollen hoarding, Page et al. (1995) effectively narrowed genetic variation for pollen foraging, and the resulting

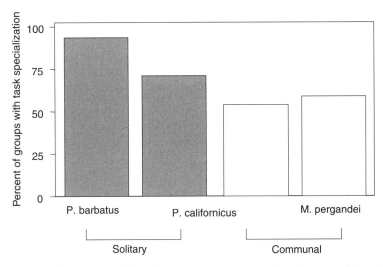

Figure 21.2. The percent of foundress pairs in three species (four populations) of harvester ants (*Pogonomyrmex* and *Messor*) showing task specialization for excavation behavior. Data include both haplometrotic (solitary founding) and pleometrotic (communal founding) populations of *P. californicus* (compiled from Fewell and Page 1999; Helms Cahan and Fewell 2004). Foundresses were collected from the field just prior to nest excavation and placed together into nests filled with soil. Pairs were observed over a 1- or 2-day period and the number of observed excavation bouts (moving soil to the surface) were recorded for each individual. Significant differences in excavation behavior were determined for each pair using α^2, with an acceptable alpha value of $p < 0.05$.

colonies were far more productive in their pollen foraging. However, the colonies did not respond appropriately either to manipulated changes in pollen stores (Fewell and Page 2000) or to naturally occurring seasonal changes in pollen. In that case, colonies of genetic specialists for the task were less flexible or resilient than wild-type colonies (Page et al. 1995).

New Explorations: The Colony as a Complex Adaptive System

Division of labor within eusocial colonies shows an amazing ability to function consistently and homeostatically within the face of a dynamic and constantly changing external environment. The properties of emergence and resiliency outlined above lead to the consideration of social insects as complex adaptive systems (CAS). The concept of CAS was developed primarily to understand how some systems of interacting agents can develop and maintain a group-level structure in the absence of a central organizing source (Rosen 1991; Holland 1995; Auyang 1998; Levin 1998). It originated in the observation that groups of interacting units, from economies to ecosystems, seem to share certain properties in common (Holling 1973; Levin 1998).

Subset's of complex systems are also able to change behavior in ways that increase performance or success; this provides the definition of *adaptive* within complex adaptive systems. In biological terms, this is closer to the concept of adaptable rather than adaptive; in other words, CAS have the capacity to change and learn from experience (Holling 1973; Rosen 1991; Auyang 1998). These attributes provide an excellent fit with the dynamics of eusocial insect colonies, especially in the context of division of labor. Eusocial colonies in particular share the traits of a dense but distributed network (Wilson and Hölldobler 1988; Bonabeau 1998; Fewell 2003). The network structures of other social systems, such as communal and primitively eusocial societies, are less well described, as are their behaviors in terms of resiliency and adaptability. In order to diagram the evolution of insect societies as complex systems, it would be informative to explore how well these systems meet or deviate from a CAS organizational structure.

Recent explorations in complexity theory help reframe how we evaluate colony function. Rather than placing features of CAS within an optimality context (which tends to emphasize one parameter, usually efficiency), the more recent focus has been on system robustness, or the maintenance of functionality in the face of change. More robust systems are expected to operate in ways that balance efficiency with the ability to adapt faster and respond more flexibly and resiliently to environmental change.

Social insect colonies do more than maintain homeostasis, however. They grow and reproduce; in their developmental trajectories they react appropriately to the external environment, but they also show developmental stability beyond what we typically associate with robustness (Wilson 1985; Hölldobler and Wilson 1991; Bonabeau 1998). Eusocial colonies are also more closed than other social systems, in that individuals are added or removed from the colony via birth and death but not via immigration and emigration. Are these attributes consistent with CAS, or do they move eusocial colonies even further, to the realm of organismal characteristics? Theoretical considerations of CAS are only beginning to explore these possibilities.

Finally, the exploration of social groups as complex systems provides important insights into how selection acts on social organization from the individual to the group level (Auyang 1998; Fewell 2003). Self-organization and emergence generate a bi-directional pathway between the individual and the group; the route from the bottom up determines collective behavior, and the route from the top down defines how the aggregate environment influences and constrains the behaviors of the group members (Moore, Brodie, and Wolf 1997; Fewell 2003). This lesson is critical to our understanding of social insect evolution in the context of selection. As soon as self-organizational dynamics come into play, the group becomes an entity with effects on individual phenotype and fitness. For example, as described above, a harvester ant queen is a winner or a loser in the fitness lottery depending on the foundress association in which she randomly joins after mating (Fewell and Page 1999; Helms Cahan and Fewell 2004). The group cohesion generated by complexity dynamics cements the fitness connection between individual and group, and emphasizes that social evolution cannot be completely understood without the consideration of how selection and complexity dynamics interact.

Literature Cited

Arathi, H. S., and M. Spivak. 2001. "Influence of colony genotypic composition on the performance of hygienic behaviour in the honeybee *Apis mellifera* L." *Animal Behavior* 62: 57–66.

Arnold, K. E., I. P. F. Owens, and A. W. Goldizen. 2005. "Division of labour within cooperatively breeding groups." *Behaviour* 142: 1577–1590.

Auyang, S. Y. 1998. *Foundations of complex-system theories in economics, evolutionary biology, and statistical physics.* New York: Cambridge University Press.

Avilés, L., Maddison, W. P., and Agnarsson, I. 2006. "A new independently derived social spider with explosive colony proliferation and a female size dimorphism." *Biotropica* 38: 743–753.

Bartlett, T. L., D. W. Mock, and P. L. Schwagmeyer. 2005. "Division of labor: Incubation and biparental care in house sparrows *(Passer domesticus)." The Auk* 122: 835–842.

Bechtel W., and R. C. Richardson. 1993. *Discovering complexity: Decomposition and localization as strategies in scientific research.* Princeton: Princeton University Press.

Bonabeau, E. 1998. "Social insect colonies as complex adaptive systems." *Ecosystems* 1: 437–443.

Bonabeau, E., G. Theraulaz, and J.-L. Deneubourg. 1998. "Fixed response thresholds and the regulation of division of labor in insect societies." *Bulletin of Mathematical Biology* 60: 753–807.

Britton, N. F., S. C. Pratt, N. R. Franks, and T. D. Seely. 2002. "Deciding on a new home: How do honey bees agree?" *Proceedings of the Royal Society of London* 269: 1383–1388.

Calderone, N. W., and R. E. Page. 1988. "Genotypic variability in age polyethism and task specialization in the honey bee, *Apis mellifera* (Hymenoptera: Apidae)." Behavioral Ecology and Sociobiology 22: 17–25.

Camazine, S., J.-L. Deneubourg, N. Franks, J. Sneyd, G. Theraulaz, and E. Bonabeau. 2001. *Self-organization in biological systems.* Princeton: Princeton University Press.

Clutton-Brock, T. H., A. F. Russell, and L. L. Sharpe. 2003. "Meerkat helpers do not specialize in particular activities." *Animal Behaviour* 66: 531–540.

Costa, J. T., and N. E. Pierce. 1997. "Social evolution in the Lepidopera: Ecological context and communication in larval societies." In J. C. Choe and B. J. Crespi, eds., *The evolution of social behavior in insects and arachnids,* 407–442. Cambridge: Cambridge University Press.

Detrain, C., and J.-L. Deneubourg. 2006. "Self-organized structures in a superorganism: Do ants "behave" like molecules?" *Physics of Life Reviews* 3: 162–187.

Duffy, J. E., Morrison, C. L., and K. S. Macdonald. 2002. "Colony defense and behavioural differentiation in eusocial shrimp, *Synalpheus regalis." Behavioral Ecology and Sociobiology* 51: 488–495.

Fewell, J. H. 2003. "Social insect networks." *Science* 301: 1867–1870.

Fewell, J. H., and R. E. Page. 1993. "Genotypic variation in foraging responses to environmental stimuli by honey bees, *Apis mellifera." Experientia* 49: 1106–1112.

————. 1999. "The emergence of division of labour in forced associations of normally solitary ant queens." *Evolutionary Ecology Research* 1: 537–548.

————. 2000. "Colony-level selection effects on individual and colony foraging task performance in honeybees, *Apis mellifera* L." *Behavioral Ecology and Sociobiology* 48: 173–181.

Franks, N. R., and C. Tofts. 1994. "Foraging for work: How tasks allocate workers." *Animal Behaviour* 48: 470–472.

Gardner, K. E., R. L. Foster, and S. O'Donnell. 2007. "Experimental analysis of worker division of labor in bumblebee nest thermoregularion (*Bombus huntii*, Hymenoptera: Apidae). *Behavioral Ecology and Sociobiology* 61: 783–792.

Gautrais, J., G. Theraulaz, J.-L. Deneubourg, and C. Anderson. 2002. "Emergent polyethism as a consequence of increased colony size in insect societies." *Journal of Theoretical Biology* 215: 363–373.

Gazda, S. K., R. C. Connor, R. K. Edgar, and F. Cox. 2005. "A division of labour with role specialization in group-hunting bottlenose dolphins (*Tursiops truncatus*) off Cedar Key, Florida." *Proceedings of the Royal Society of London Series B-Biological Sciences* 272: 135–140.

Gorelick, R., and S. M. Bertram. 2007. "Quantifying division of labor: Borrowing tools from sociology, sociobiology, information theory, landscape ecology, and biogeography." *Insectes Sociaux* 54: 105–112.

Gorelick, R., S. M. Bertram, P. Killeen, and J. H. Fewell. 2004. "Normalized mutual entropy in biology: Quantifying division of labor." *American.Naturalist* 164: 677–682.

Goss, S., S. Aron, J.-L. Deneubourg, and J. M. Pasteels. 1989. "Self-organized shortcuts in the Argentine ant." *Naturwissenschaften* 76: 579–581.

Grasmuck, V., and D. Desor. 2002. "Behavioural differentiation of rats confronted to a complex diving-for-food situation." *Behavioural Processes* 58: 67–77.

Hattori, Y., and H. Kuroshima. 2005. "Cooperative problem solving by tufted capuchin monkeys (*Cebus apella*): Spontaneous division of labor, communication, and reciprocal altruism." *Journal of Comparative Psychology* 119: 335–342.

Heinsohn, R., and C. Packer. 1995. "Complex cooperative strategies in group-territorial African lions." *Science* 269: 1260–1262.

Helms Cahan, S., and J. H. Fewell. 2004. "Division of labor and the evolution of task sharing in queen associations of the harvester ant *Pogonomyrmex califonicus*." *Behavioral Ecology and Sociobiology* 56: 9–17.

Hemelrijk, C. K. 2002. "Self-organization and natural selection in the evolution of complex despotic societies." *Biological Bulletin* 202: 283–288.

Hogeweg, P., and B. Hesper. 1983. "The ontogeny of the interaction structure in bumble bee colonies: A MIRROR model." *Behavioural Ecology and Sociobiology* 12: 271–283.

Holland, J. H. 1995. *Hidden order: How adaptation builds complexity.* Reading: Addison Wesley.

Hölldobler, B., and E. O. Wilson. 1991. *The ants.* Cambridge: The Belknap Press of Harvard University Press.

Holling, C. S. 1973. "Resilience and stability of ecological systems." *Annual Review of Ecology and Systematics* 4: 1–23.

Jeanne R. L. 1991. "The swarm-founding Polistinae." In K. G. Ross and R. W. Matthews, eds., *The social biology of wasps,* 191–231. Ithaca: Comstock Publishing Associates, Cornell University Press.

Jeanson, R., J. H. Fewell, R. Gorelick, and S. M. Bertram. 2007. "Emergence of increased division of labor as a function of group size." *Behavioral Ecology and Sociobiology* 62: 289–298.

Jeanson, R., P. F. Kukuk, and J. H. Fewell. 2005. "Emergence of division of labour in halictine bees: Contributions of social interactions and behavioural variance." *Animal Behaviour* 70: 1183–1193.

Johnson, R. A. 2000. "Water loss in desert ants: Caste variation and the effect of cuticle abrasion." *Physiological Entomology* 25: 48–53.

Jones, J. C., M. R. Myerscough, S. Graham, and B. P. Oldroyd. 2004. "Honey bee nest thermoregulation: Diversity promotes stability." *Science* 305: 402–404.

Julian, G. E., and S. Cahan. 1999. "Undertaking specialization in the desert leaf-cutter ant, *Acromyrmex versicolor.*" *Animal Behavior* 58: 437–442.

Karsai, I., and J. W. Wenzel. 1998. "Productivity, individual-level and colony-level flexibility, and organization of work as consequences of colony size." *Proceedings of the National Academy of Sciences USA* 95: 8665–8669.

Kolmes, S. A. 1985. "An information-theory analysis of task specialization among worker honey bees performing hive duties." *Animal Behaviour* 33: 181–187.

Komdeur, J. 2006. "Variation in individual investment strategies among social animals." *Ethology* 112: 729–747.

Lachaud, J. P., and D. Fresneau. 1987. "Social regulation in ponerine ants." In J. M. Pasteels and J.-L. Deneubourg, eds., *From individual to collective behaviour in social insects,* 197–217. Basel: Birkhäuser.

Levin, S. A. 1998. "Ecosystems and the biosphere as complex adaptive systems." *Ecosystems* 1: 31–36.

Mailleux, A. C., J.-L. Deneubourg, and C. Detrain. 2003. "How does colony growth influence communication in ants?" *Insectes Sociaux* 50: 24–31.

Michener, C. D. 1974. *The social behavior of the bees: A comparative study.* Cambridge: Harvard University Press.

Moore, A. J., E. D. Brodie, and J. B. Wolf. 1997. "Interacting phenotypes and the evolutionary process: I. Direct and indirect genetic effects of social interactions." *Evolution* 51: 1352–1362.

Myerscough, M., and B. P. Oldroyd. 2004. "Simulation models of the role of genetic variability in social insect task allocation." *Insectes Sociaux* 51: 146–152.

Oldroyd, B. P., and J. H. Fewell. 2007. "Genetic diversity promotes homeostasis in social insect colonies." *Trends in Ecology and Evolution* 22: 408–413.

Oldroyd, B. P., and G. J. Thompson. 2007. "Behavioural genetics of the honey bee, *Apis mellifera*." *Advances in Insect Physiology* 33: 1–49.

Oster, G. F., and E. O. Wilson. 1978. *Caste and ecology in the social insects.* Princeton: Princeton University Press.

Page, R. E., and S. D. Mitchell. 1998. "Self-organization and the evolution of division of labor." *Apidologie* 29: 171–190.

Page, R. E., G. E. Robinson, M. K. Fondrk, and M. E. Nasr. 1995. "Effects of worker genotypic diversity on honey bee colony development and behavior (*Apis mellifera*)." *Behavioral Ecology and Sociobiology* 36: 387–396.

Ravary, F., E. Lecoutey, G. Kaminski, N. Chaline, and P. Jaisson. 2007. "Individual experience alone can generate lasting division of labor in ants." *Current Biology* 17: 1308–1312.

Robinson, G. E. 1992. "Regulation of division of labor in insect societies." *Annual Review of Entomology* 37: 637–665.

Robinson, G. E., and R. E. Page. 1989. "Genetic basis for division of labor in an insect society." In M. D. Breed and R. E. Page, eds., *The genetics of social evolution*, 61–68. Boulder: Westview Press.

Rosen, R. 1991. *Life Itself.* New York: Columbia University Press.

Sakagami, S. F., and Y. Maeta. 1987. "Sociality, induced and/or natural, in the basically solitary small carpenter bees *(Ceratina)*." In Y. Ito, J. L. Brown, and J. Kikkawa, eds., *Animal societies: Theories and facts*, 1–16. Tokyo: Japan Scientific Societies Press.

Schmid-Hempel, P. 1990. "Reproductive competition and the evolution of work load in social insects." *American Naturalist* 135: 501.

Seid, M. A., and J. F. A. Traniello. 2006. "Age-related repertoire expansion and division of labor in *Pheidole dentata* (Hymenoptera: Formicidae): A new perspective on temporal polyethism and behavioral plasticity in ants." *Behavioral Ecology and Sociobiology* 60: 631–644.

Sendova-Franks, A. B., and N. R. Franks. 1994. "Social resilience in individual worker ants and its role in division of labor." *Proceedings of the Royal Society of London B* 256: 305–309.

Shannon, C. E. 1948. "A mathematical theory of communication." *Bell System Technical Journal* 27: 379–423, 623–656.

Simon, H. A. 1962. "The architecture of complexity." *Proceedings of the American Philosophical Society* 106: 467–482.

Stander, P. E. 1992. "Cooperative hunting in lions: The role of the individual." *Behavioral Ecology and Sociobiology* 29: 445–454.

Stuart, R. J., and R. E. Page. 1991. "Genetic component to division of labor among workers of a leptothoracine ant." *Naturwissenschaften* 78: 375–377.

Sumpter, D. 2006. "The principles of collective animal behaviour." *Philosophical Transactions of the Royal Society B* 361: 5–22.

Theraulaz, G., E. Bonabeau, and J.-L. Deneubourg. 1998. "Response thresholds reinforcement and division of labor in insect societies." *Proceedings of the Royal Society of London Series B-Biological Sciences* 265: 327–332.

Theraulaz, G., E. Bonabeau, R. V. Solé, B. Schatz, and J.-L. Deneubourg. 2002. "Task partitioning in a ponerine ant." *Journal of Theoretical Biology* 215: 481–489.

Traniello, J. F. A. 1978. "Caste in a primitive ant: Absence of age polyethism in *Amblyopone*." *Science* 202: 770–772.

Trumbo, S. T. 2006. "Infanticide, sexual selection and task specialization in a biparental burying beetle." *Animal Behaviour* 72: 1159–1167.

Underwood, D. L. A., and A. M. Shapiro. 1999. "Evidence for division of labor in the social caterpillar *Eucheira socialis* (Pieridae: Lepidoptera)." *Behavioral Ecology and Sociobiology* 46: 221–227.

Waldschmidt, A. M., L. A. O. Campos, and P. J. De Marco. 1997. "Genetic variability of behavior in *Melipona quadrifasciata* (Hymenoptera: Meliponinae)." *Brazilian Journal of Genetics* : Vol 20. No 4.

Weidenmüller, A. 2004. "The control of nest climate in bumblebee *(Bombus terrestris)* colonies: Interindividual variability and self reinforcement in fanning response." *Behavioral Ecology* 15: 120–128.

Wheeler, W. M. 1928. *The social insects: Their origin and evolution.* New York: Harcourt, Brace & Co.

Wilson, D. S., and E. Sober. 1989. "Reviving the superorganism." *Journal of Theoretical Biology* 136: 337–356.

Wilson, E. O. 1971. *The insect societies.* Cambridge: Harvard University Press.

———. 1985. "The sociogenesis of insect colonies." *Science* 228: 1489–1495.

———. 1986. "Caste and division of labor in *Erebomyrma*, a genus of dimorphic ants (Hymenoptera: Formicidae: Myrmicinae)." *Insectes Sociaux* 33: 59–69.

Wilson, E. O., and B. Hölldobler. 1988. "Dense heterarchies and mass communication as the basis of organization in ant colonies." *Trends in Ecology and Evolution* 3: 65–67.

Woese, C. R. 2004. "A new biology for a new century." *Microbiology and Molecular Biology Reviews* 68: 173–186.

CHAPTER TWENTY-TWO

Insect Societies as Models for
Collective Decision Making

STEPHEN C. PRATT

SOCIAL INSECTS ARE OF SPECIAL INTEREST for the study of decision making because they make choices not only as individuals, but also as groups. When a honey bee forager decides between feeding at a particular blossom or searching for something better, she behaves much like any solitary animal, and her decision process can be studied with the same concepts applicable to a foraging jay or shrew. This approach need not go beyond the information available to the bee, the workings of her nervous system, and the effect of her decision on her future behavior and condition. When the same bee decides whether to recruit nestmates to a patch of flowers, however, her choice makes sense only in terms of the colony-level decision of which it forms but a small part. At this larger scale, the colony as a whole decides how to allocate thousands of foragers across scores of potential food sources over tens of square kilometers. No individual bee will ever consider every source and decide how many foragers should visit each. Instead, each bee will make narrow decisions based on local information—whether, for example, to search for new food sources or to follow a recruitment signal to a patch already discovered by a nestmate. A coherent group decision emerges from many such individual choices, guided by appropriate behavioral rules and coordinated via specialized signals and cues. To comprehend this process, we must analyze the bee as part of a complex and highly integrated network.

These decision-making networks are not unique to social insects. Group-living vertebrates make consensus choices about when and where to travel (Conradt and Roper 2005). Fish schools and bird flocks make spectacular acrobatic movements requiring highly coordinated decisions (Couzin and Krause 2003). At a smaller scale, colonial bacteria jointly decide when to activate virulence factors or perform other acts that require the cooperation of many cells (Miller and Bassler 2001). Even the decision making of a single brain can be described usefully as collective choice by a "society" of neurons. Despite their diversity, these and similar systems share common themes, including coordination without a central controller, the emergence of global order from local interactions, and the generation of precise group behaviors from the contributions of many imprecise individuals. This commonality means that understanding the collective decision making of an ant or bee colony can expand to knowledge well beyond the social insects.

In recent years, ants and bees have provided some of the best-described mechanisms of collective decision making, thanks in large part to the relative simplicity, openness, and experimental tractability of their societies. Individual workers—the rough analogs of neurons in a decision-making brain—can be uniquely identified and tracked, and colonies can be experimentally taken apart and re-assembled to rigorously test specific hypotheses of colony function. Further, the rich data sets that result can aid the development of realistic mathematical models and computer simulations.

Table 22.1. Summary of collective abilities and underlying behavioral mechanisms addressed in this chapter

---

I. Decision-making abilities of colonies
   Selecting the best of several options
   Comparing options that vary in multiple attributes
   Adjusting preferences according to environmental context and
      colony state
   Trading off the speed and accuracy of decision-making

II. Behavioral mechanisms underlying colony abilities
   Positive feedback generated by nestmate recruitment
   Quorum rules that mark key decision points and filter out
      individual errors
   Individual assessment rules based on simple but informative local cues
   Behavioral algorithms that can be tuned to decision context
   Independent decision-making by large numbers of individuals
   Abandonment of options by individual workers, to avoid stalemates

---

Such models are especially important in the study of complex collectives, where unaided intuition has difficulty connecting individual and group behavior (Sumpter and Pratt, 2003; Franks, et al. this volume).

In this chapter, I review the current understanding of collective decision-making by insect societies. The capacities of colonies as a whole are considered first, followed by an examination of underlying behaviors and social interactions (Table 22.1). The goals of this review are both to highlight recent discoveries about the remarkable collective abilities of these societies, and to suggest the advantages of insect societies as experimental model systems for the general phenomenon of collective decision making.

Decision-Making Abilities of Insect Societies

Many of the best-studied cases of collective decision making concern foraging. Colonies of the honey bee *Apis mellifera,* for example, use their waggle dance recruitment signal to direct foraging efforts to clumped resources. When given a choice between two artificial feeders differing in the richness of their sucrose solution, colonies make a clear choice, concentrating most of their foragers on the better site. When foraging on natural sources, they can track the changing quality of multiple sources over a large region. Many ant species use pheromone trails to recruit nestmates to rich foraging sites, and they too will selectively exploit the better of two feeders (Beckers et al. 1990, Sumpter and Beekman 2003) (Figure 22.1). Trail-following ants can also pick the better of two routes to a food source—given a choice between a long and a short bridge leading to a single feeder, *Lasius niger* workers begin by using both routes, but soon direct all traffic to the shorter one (Beckers, Deneubourg, and Goss 1992).

Much has also been learned about how colonies choose among nest sites. These decisions have significant fitness consequences, because the structure and location of the nest determine whether proper conditions can be met for brood development, food storage, and effective defense from predators and parasites (Visscher 2007). Colonies of *Temnothorax* ants, for example, typically nest in natural cavities such as hollow stems, rock crevices, or nut shells. If their home is damaged they emigrate to a new one, guided by an active minority of workers who find the new nest and carry the rest of the colony to it (Mallon et al. 2001, Pratt et al. 2002). When more than one site is available, the colony chooses among them on the basis of several

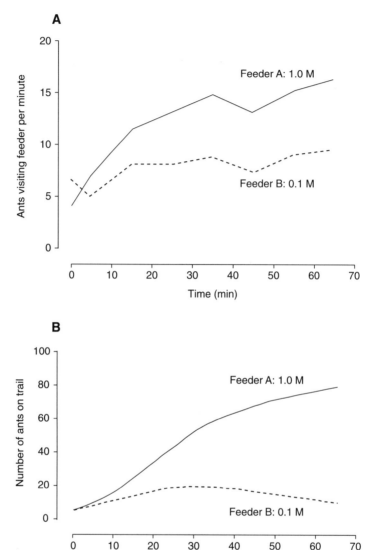

Figure 22.1. Collective decision making by colonies of the ant *Monomorium pharaonis* presented with two feeders containing sucrose solution of different concentration. (A) The ants form a stronger recruitment trail to the better feeder. (B) The ants' behavior can be fit with a simple mathematical model that assumes: (1) the rate at which ants join a trail is proportional to the amount of pheromone already deposited on the trail; (2) the rate at which ants reinforce the trail with pheromone is proportional to feeder quality; and (3) the rate at which ants lose the trail is a declining nonlinear function of the number of ants on the trail. Adapted from Sumpter & Beekman (2003).

attributes, including entrance size, cavity volume, light level, nest hygiene, and nearness to conspecific colonies (Visscher 2007; Franks et al. 2007). Honey bee swarms, also cavity-dwellers, show similarly complex discrimination when choosing a site (Seeley, Visscher, and Passino 2006; Visscher 2007). Species that live in cavities may place an especially strong emphasis on choosing a site well, because they will have less opportunity to improve a deficient site compared with societies that excavate their nests or build them from scratch. Nonetheless, nest site selection is a general problem for all insect societies, and noncavity dwellers can also collectively choose among nest sites that vary in quality (Jeanson et al. 2004).

Choosing among Options with Multiple Attributes

Whether choosing among food sources or nest sites, a colony's first decision-making challenge is the complex nature of its options. Candidate nest sites, for example, vary in multiple attributes measured along very different scales. This places a high cognitive burden on colonies, as they must assess each available option for several distinct properties and somehow integrate this information to determine the best alternative. Moreover, they must deal with the likely circumstance that no single site is the best choice for every attribute. Human decision makers cope with analogous problems by using simplifying rules of thumb, and animals likely take a similar approach (Hutchinson and Gigerenzer 2005). Some of these rules reduce the problem's complexity by evaluating only a subset of the alternatives—under the satisficing strategy, for instance, options are considered singly until one is encountered that exceeds a threshold value for each attribute (Simon 1990). Other rules, such as the lexicographic strategy, consider all options, but attend only to a subset of their features. Each option is evaluated only for the most important attribute and the highest ranked is chosen (Payne, Bettman, and Johnson 1993). Further attributes are considered only if necessary to break a tie.

   A more taxing rule, the weighted-additive strategy, always considers every attribute of every option (Payne, Bettman, and Johnson 1993). Alternatives are ranked according to a weighted sum of their separate attribute scores, and the highest scorer is chosen. *Temnothorax* colonies appear to use this rule when choosing a nest site (Franks, Mallon, et al. 2003). They reliably choose a dark nest with an undesirably low ceiling over a bright one with the preferred ceiling height, showing that they weight darkness more heavily than height. Likewise, they choose a thick nest with a big entrance over a thin one with a small entrance, showing that thickness is weighted more heavily than entrance size. More tellingly, they can be convinced to choose a bright nest over a dark one only if the bright nest is superior in both of the other attributes. This implies that colonies evaluate and weight all attributes, not only the most important ones, such that a sufficiently strong showing in lesser attributes can make up for a poor score in the top feature.

Adaptive Tuning of Preferences

A colony's preferences are not static; they depend both on colony state and on environmental context. For example, colonies tune their preferences to match the quality range of available options. A honey bee colony will ignore artificial nectar feeders offering less than a threshold sucrose solution if better natural or artificial sources are available. If foraging conditions are poor, however, colonies will eagerly accept these dilute sources (Seeley 1995). Likewise, emigrating colonies of *Temnothorax* ants consistently reject a mediocre nest if a good one is also present, but they will readily move into the same mediocre site if it is the best on hand (Mallon, Pratt, and Franks 2001; Pratt 2005). This adjustment is not a mere loss of discriminatory power under difficult conditions. Colonies continue to make choices, because they reject still worse nests made available at the same time as the mediocre one. With only relatively low-quality sites to choose from, however, colonies shift their discrimination range downward, making distinctions among options that would all be rejected if a better candidate were present. This flexibility and context-dependence are important, because they allow a colony to take advantage of excellent options when possible, but prevent paralyzing indecision when all alternatives are poor.

In addition to their sensitivity to the external environment, colonies change their preferences according to their own state. For example, honey bees derive their protein from pollen, and they actively regulate the amount stored in their combs at about 1 kg, taking in more pollen when their internal stores are low than when they are high (e.g., Fewell and Winston 1992; Weidenmüller and Tautz 2002; Camazine 1993). They also adjust pollen foraging according to the protein demand created by developing brood, with greater pollen-foraging effort at higher brood amounts (e.g., Dreller, Page, and Fondrk 1999). Ants and stingless bees make similar adjustments based on the relative amounts of nectar and pollen currently stored (Hofstede and Sommeijer 2006; Biesmeijer et al. 1999; Judd 2005; Stein, Thorrilson, and Johnson 1990). Colonies also take internal state into account in decision contexts other than foraging. Honey bees monitor both the size of their stock of empty comb and the relative quantity of worker and drone comb when deciding on the timing and type of new construction (Pratt 2004). In general, colony state will be as important to adaptive decision making as the quality of the options themselves.

Unanimity versus Balance

Many decisions require consensus on a single choice. An emigrating colony, for example, generally needs to move cohesively to one new home. Honey bee swarms are especially adept at reaching unanimity, perhaps because of the high cost paid by those that split. A swarm consists of a single queen and several thousand workers (Seeley, Visscher, and Passino 2006). If the swarm should divide and move to two separate nests, the queenless portion will be doomed to a quick death with no real hope of reproduction. The queenright portion, meanwhile, will lose the considerable investment and immediate work capacity of the missing bees. Splitting would appear to be a real danger, because a swarm's scouts find and assess large numbers of candidate sites. The swarm is nonetheless able to winnow these down to a single choice before flying off to the new home (Seeley, Visscher, and Passino 2006). In rare cases, swarms have been observed to lift off and engage in an aerial tug-of-war, apparently because multiple groups of scouts were attempting to lead the swarm to different sites (Lindauer 1967). The disagreements generally ended with the swarm settling down to resume its deliberations until a single site was chosen. Swarms combine this cohesiveness with accurate decision making, consistently selecting a high-quality artificial nest provided among four mediocre alternatives (Seeley and Buhrman 2001). In nature, swarms commonly consider a dozen or more sites, so this test likely does not show the limits of a swarm's abilities (Lindauer 1967; Seeley, Visscher, and Passino 2006; Visscher 2007). *Temnothorax* ants also avoid splits, although not as effectively as honey bees (Pratt et al. 2005). Like the bees, ant colonies can consistently choose a single good nest among four inferior ones (Franks, Mallon, et al. 2003).

Unlike nest site selection, foraging decisions are often not unanimous. When presented with two feeders containing sucrose solution of different concentrations, a honey bee colony directs most foragers to the richer one, but it sends a small proportion to the lesser one as well. If the feeders are identical, it divides its foraging force roughly equally between them. In contrast, colonies of *Lasius niger* faced with similar situations always set up a foraging trail to only one site: the better one if the feeders differ, and a randomly selected one if they are identical (Beckers et al. 1990; Portha, Deneubourg, and Detrain 2002). In nature, these ants often choose among patches of honeydew-producing aphids. These are rich and long-lived resources, and colonies may benefit from concentrating their foraging forces at one or a few patches in order to defend them effectively from competitors. Consistent with this explanation, colonies show a much more homogeneous distribution of foragers when presented with two identical protein baits (Portha, Deneubourg, and Detrain 2002). Proteinaceous prey items are more likely to be scattered and ephemeral and thus to reward a more even dispersion of foragers. The diffuse foraging strategy of honey bees may enable rapid response to temporal changes in source quality. Experimental reversal of feeder richness is followed within a few hours by a complete reversal in the distribution of foragers across the two sources (Seeley 1995). *L. niger* colonies apparently lack this flexibility; once they have set up a strong trail to a mediocre source, they cannot switch their attention to a better source made available later (Beckers et al. 1990).

Speed/Accuracy Trade-Offs

A colony will benefit not only from picking the best option, but also from making its decision quickly. These priorities are likely to be in opposition, because gathering the information needed for an accurate decision takes time. Human decision makers must also choose between speed and accuracy, and they can do so adaptively, emphasizing one or the other as the situation demands (Ratcliff and Rouder 1998). Similar abilities should be of great value to an insect society. When a colony's nest has been destroyed or is menaced by predators, or when it faces intense competition for ephemeral food sources, it may prize a rapid choice more highly than a good one. In other circumstances, time pressure may be less and a colony should invest time in improving its choice. *Temnothorax* ants can dramatically alter their speed/accuracy trade-off when choosing between good and mediocre nest sites in conditions of either high or low urgency (Franks,

Dornhaus, et al. 2003; Pratt and Sumpter 2006). If their old nest is destroyed they typically complete emigration within 3 hours, but they often move into the lesser site. If their old nest is simply of poor quality but nonetheless habitable, they take much longer to move and nearly always pick the good site.

Collective Decision-Making Mechanisms

Group decision making poses something of a paradox. Collectives should be better decision makers than the individuals that compose them because they have the benefit of multiple information gatherers and processors. At the same time, the individuals within the collective are at a disadvantage because they know much less about the decision at hand than would a solitary animal. An ant on a foraging trail, for example, typically knows only the food source to which the trail leads, and nothing of the other sources her nestmates are exploiting elsewhere. Comparison of multiple sources must therefore be carried out by the colony as a whole, rather than independently by many well-informed individuals. For this reason, collective decision making depends crucially on the specialized behavioral rules that these individuals apply to limited information and on the communication pathways that allow each worker to adapt her behavior to that of her nestmates. These rules have been described in detail for a number of species solving different decision problems. Although the details vary, a number of common themes shine through.

Recruitment Communication and Positive Feedback

At the heart of most collective decision-making systems is some process of positive feedback (Jeanson, and Deneubourg this volume). For many ants, feedback arises from nestmate recruitment via pheromone trails. *Lasius niger* provides a particularly well-studied example. Foragers that find a sucrose bait lay a pheromone recruitment trail that attracts and guides nestmates to the site. These recruits may in turn reinforce the trail, increasing its ability to bring still more recruits. This process can rapidly concentrate a large force at a profitable location, boosting the colony's foraging rate. When more than one source is present, positive feedback plays a crucial role in deciding among them. Individual ants condition the intensity with which they deposit pheromone on the richness of the food source (e.g., Beckers, Deneubourg, and Goss 1993). Better sites thus experience stronger trail reinforcement. This effect will be compounded by the successive decisions of each recruit and can lead to runaway growth of the trail to the better site, and a withering away of recruitment at the lesser site (Figure 22.1). This process does not depend on any ant visiting both sites and comparing them. It emerges from the competition among sites to attract the same pool of potential recruits. The fastest-growing site will monopolize this resource and thus prevent lesser sites from attracting and retaining foragers. The positive feedback of trail-laying also underlies the choice of an optimal route. Because the shorter of two routes enjoys a higher rate of passages, and thus trail reinforcement, it rapidly becomes the favored path for all ants (Beckers, Deneubourg, and Goss 1992).

Besides trail pheromones, any behavioral interaction creating positive feedback on the numbers of individuals choosing an option can produce a collective decision. This includes waggle dance recruitment by honey bees and tandem run recruitment by *Temnothorax* ants. The necessary positive feedback is possible even without explicit recruitment. In many social arthropods, aggregation decisions rely on simple rules that make joining a group more likely (and leaving it less likely) as group size increases (Jeanson, and Deneubourg this volume). Positive feedback can also operate through imitative behavior, as when members of a group align their movement vectors with those of their neighbors. Simulations suggest that a flying swarm of bees following this strategy can direct itself toward a goal known only to a small minority of informed scouts (Beekman, Fathke, and Seeley 2006; Couzin et al. 2005). If the scouts are divided between adherents to two different goals, the imitative behavior can amplify even a slight majority of the scouts into a collective decision for their favored heading (Couzin et al. 2005).

While all of these systems are based on positive feedback, they show important differences in collective performance, as noted above. Much of this variation arises from differences in the response of an individual insect to the increasing strength of the cues that stimulate her selection. In some cases, such as trail-laying ants, this function is highly nonlinear. That is, the likelihood of an ant choosing a given trail increases in a nearly step-like fashion as the pheromone concentration increases. In other cases, such as the waggle dances of honey bees and the tandem runs of *Temnothorax*, the response of individuals to recruitment cues is more linear. A doubling of the number of bees dancing for a site will roughly double the number being recruited. A doubling of pheromone trail concentration by *Lasius niger*, on the other hand, leads to far more than a doubling of its effectiveness in bringing recruits to a food source. This nonlinearity leads to highly discrete decision making, with complete dedication of the colony to a single choice even when options do not vary in quality (Jeanson, and Deneubourg in this volume). Honey bees, with their more linear recruitment system, evenly divide their workforce between similar options.

Nonlinearity can also influence the response to changing conditions. As mentioned, *Lasius niger* ants that have already set up a trail to a mediocre food source cannot switch their efforts to a better source made available later (Beckers et al. 1990). It appears that strong positive feedback traps foragers in the already established trail. *Tetramorium caespitum* colonies, in contrast, can abandon a trail for a new food source in favor of a later-discovered but better one (Beckers et al. 1990). This ability depends on *T. caespitum*'s mixed foraging strategy, in which ants are recruited either by strong pheromone trails or by scouts who directly lead a small group of recruits. These leaders can ignore the pull of a powerful trail and set up a new foraging effort at a better site. *Myrmica sabuleti* can likewise switch to a better source, at least for large enough differences in quality (de Biseau, Deneubourg, and Pasteels 1991). Their ability does not depend on group recruitment, but instead may be explained by stronger modulation of trail-laying behavior in response to food quality. As long as the new source is sufficiently superior to the old one, the new trail can self-amplify with enough strength to divert the colony's foragers from the old trail. This kind of modulation may explain why other species with strong pheromone trail feedback are more flexible than *L. niger* (Beckers et al. 1990; Sumpter and Beekman, 2003).

Quorums and Thresholds

Honey bees and tandem-running ants must sometimes make unanimous decisions, as when they decide among nest sites. How do they do so if the linearity of their recruitment system disperses workers over multiple options? Interestingly, nonlinearity appears to re-enter the system at a higher level of organization. Both honey bees and *Temnothorax* rely on a quorum rule when choosing among nests (Pratt et al. 2002; Seeley and

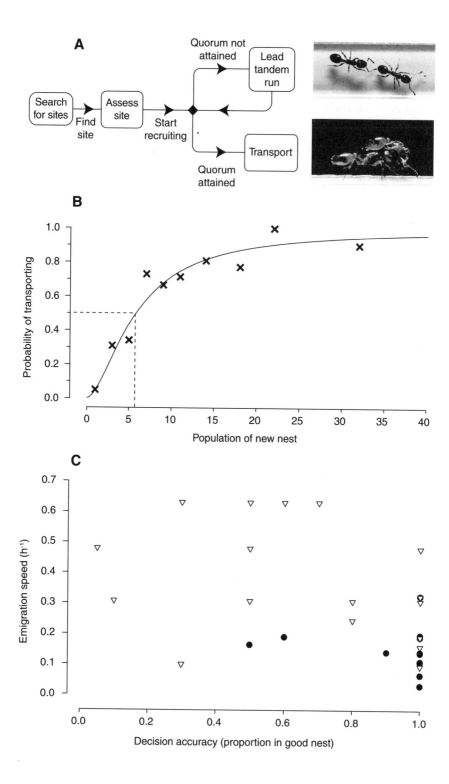

Visscher 2004). In each case, scouts search for sites and recruit fellow scouts to promising finds. The bees recruit by waggle dances which communicate the distance and direction of the new site, while the ants directly guide recruits to the new site using tandem runs. In both cases scouts vary the strength of their recruitment depending on the quality of the site they have found. Honey bees perform a larger number of waggle runs for better sites (Seeley and Buhrman 2001) and ants initiate tandem runs sooner to better sites (Mallon, Pratt, and Franks 2001). Recruitment of scouts continues only until a threshold number have been summoned to the new site. The scouts themselves sense the attainment of this quorum and dramatically change their behavior, making a full commitment to the site as the colony's new home (Figure 22.2). For honey bees, this means returning to the swarm and inducing it to prepare for liftoff and travel to the new site (Seeley and Visscher 2004). For the ants it means abandoning tandem runs in favor of transports that more rapidly carry the nonscouting majority of the colony to their new home (Pratt et al. 2002). Models show that this step-like change increases the colony's likelihood of unanimously moving into the best nest (Pratt et al. 2005; Passino and Seeley 2006).

Quorum rules are a recent discovery in collective decision making, but they may prove to be a general phenomenon. They offer a simple and accessible cue that allows each worker to coordinate her own efforts with

Figure 22.2. Collective decision making by emigrating colonies of the ant *Temnothorax curvispinosus*. (A) Decision algorithm used by active scouts. Search consists of exploratory journeys from the old nest and gives way to assessment once the ant finds a site. Assessment ceases at a rate that depends on site quality, and the next behavior is chosen on the basis of a quorum rule. (B) At low site population, the ant canvasses her fellow active ants by leading them to the site in slow tandem runs. Once she judges the population to have reached the quorum, the ant fully commits to the site and uses speedy transports to bring the passive adults and brood that constitute the majority of the colony. (C) Speed/accuracy trade-offs for colonies choosing between a good and a mediocre nest. Ants can tune their algorithm to move fast in times of high urgency, but at the cost of lower accuracy, measured as the proportion of the colony ending up in the good nest (open triangles). If urgency is low, the ants instead take more time and make a more accurate choice (filled circles). Adapted from Pratt & Sumpter (2006).

those of her nestmates. In effect, an option must pass two tests to be chosen: it must receive the positive judgment of the scout herself, and it must also attract sufficient attention from other scouts to push it above the quorum. This stringent testing accentuates the colony's discrimination between options of different quality. House-hunting scouts of *Temnothorax*, for example, initiate recruitment sooner to better sites, ensuring faster growth of the scout population at those sites. The quorum rule amplifies this difference by significantly accelerating recruitment at the first site to attain a quorum, which will most likely be of relatively high quality. More generally, a quorum rule provides an error check against individual workers who recruit too soon or too eagerly to a relatively poor option. Any nonlinear system is endangered by such errors, because positive feedback can rapidly exaggerate them. The quorum requirement filters out the effects of individual errors by demanding that several insects "vote" for an option before the group fully commits to it.

Assessment Rules

Effective decision making requires each worker to assess an option and tune her recruitment effectiveness to the option's value for the colony. Assessment can sometimes reduce to evaluation of a single attribute. For example, foragers of the ant *Pheidole pallidula* adjust their probability of recruiting nestmates to a prey item according to the difficulty they have in dragging it—only items too large to be shifted by a single ant will experience strong recruitment of nestmates, who can then help to cut the prey into manageable pieces (Detrain and Deneubourg, 1997). This simple rule allows a colony to match its foraging effort to the demands posed by a food source. Likewise, a *Lasius niger* worker foraging at a patch of honeydew-secreting aphids will lay a recruitment trail only if she can fill her crop to a desired threshold volume (Mailleux, Jean-Louis, and Detrain 2000). Simulations show that colonies following this rule can effectively match the number of foragers exploiting a site to its total honeydew production (Mailleux, Deneubourg and Detrain 2003).

Assessment of an option may also require integration of data about the colony itself. Honey bee foragers attend to the concentration of sugar in a nectar source, adjusting the number of waggle runs in their recruitment dance according to the strength of the advertised source, thus allowing the colony to match foraging effort to food quality (Seeley 1995). This simple rule is further modulated by a separate evaluation each bee makes at the nest—she lowers her dance effort as she experiences a greater delay in finding a nestmate to receive her nectar load (Seeley 1995). This delay measures the imbalance between the colony's current nectar intake and its capacity for more nectar storage, and thus allows the forager to calibrate her judgment of site quality to the colony's current need. If the delays are sufficiently large, the forager abandons waggle dancing altogether and performs a tremble dance, which serves to recruit more nectar receivers and thus alleviate the processing bottleneck (Seeley 1995). In addition to these temporal cues, foragers may learn about the colony's total nectar influx from the number of food receivers that unload them, with larger numbers indicating lower influx (De Marco 2006).

Adaptive Tuning of Individual Behavior

As external conditions change, so can a colony's decision priorities; therefore, the colony must be able to alter its decision behavior accordingly. One way to do so is to rely on specialized choice mechanisms for each context, but the emigration behavior of *Temnothorax* suggests a different strategy: quantitative tuning of a single, flexible algorithm. The decision rules used by scout ants comprise a steplike series of increasing commitment to a candidate site. By adjusting key parameters governing their advancement through these stages, ants can influence how the colony as a whole balances the speed and accuracy of decision-making (Pratt et al. 2005) (Figure 22.2). By raising the rate at which they leave the old nest to search for a new home, and the rate at which they initiate recruitment to a site of given quality, scouts can sharply increase emigration speed, but at the cost of reducing decision accuracy (Pratt and Sumpter 2006). Thus, the ants can use a single complex algorithm to deal with two very different problems—abandonment of their old nest in a crisis versus careful choice of a new home when better ones become available. This is possible because the algorithm's quorum-based design allows quantitative tuning of key parameters to produce major functional changes.

Independent Individual Decisions

It is important to note that individual workers are not driven mindlessly by the overwhelming power of recruitment mechanisms. Effective decision making requires that each worker act as an independent information processor. Recruitment brings a worker to an option, but she then decides on her own whether to become a recruiter herself. Theoretically, integrating the independent choices of many decision makers leads to an overall choice that is more precise than that made by any individual (Simons 2004). This benefit would be lost if each worker slavishly imitated the behavior of its nestmates.

The Importance of Forgetting

Another important aspect of decision making is an individual's capacity to abandon an earlier choice. If colonies are to respond to changes in their environment or to new knowledge, they must redirect attention from formerly promising options. For example, honey bee nest site scouts that have started to recruit to a candidate nest site eventually cease doing so, even if the site remains attractive and the colony has not yet made a decision (Seeley 2003). Retirement occurs later for better sites, but it occurs eventually for every site, even the one that is ultimately chosen by the swarm. One potential advantage of retirement is to avoid deadlock, with competing groups of scouts at multiple sites persistently advertising their candidates. Simulation of the swarm's decision making shows that too-fast retirement greatly slows decision making by making it difficult for any site to reach a quorum, while too-slow retirement yields frequent split decisions in which a quorum is reached at more than one site (Passino and Seeley 2006). A retirement rate near that seen in actual colonies makes splitting rare, but ensures a decision after a few days of deliberation.

Comparisons by Individual Insects

The preceding discussion emphasized highly local decisions by individual insects, each lacking full knowledge of all options under consideration. This is the typical situation in which workers act, but this does not mean that individuals lack the ability to directly compare options and choose the better one. Evidence suggests, for example, that scouts of *Temnothorax albipennis* that have visited two candidate nest sites are fully capable of selecting the better one (Mallon, Pratt, and Franks 2001). These well-informed ants were far more likely to recruit nestmates only to the better site than would be expected if they chose a recruitment target randomly. It should not be surprising that an insect capable of assigning a quality score to a complex, multi-attribute option can compare such scores for two options. It is well-known that solitary insects carry out similar or more complicated acts of comparison, and there is no reason to suppose that social individuals have inferior cognitive abilities. For insect societies the key point is the rarity of opportunities to make such comparisons, given the large array of options likely to come under scrutiny of at least one colony member. The difficulty in the organization of collective decision making is not the cognitive limitations of the colony's members, but rather their ignorance. Insect societies generally lack information-sharing systems that grant each colony member access to all knowledge gathered by her nestmates, or that channel this knowledge to a select cadre of centralized decision makers (Seeley 2002). Instead, insects make use of carefully tailored decision rules and local information-sharing pathways that allow an adaptive decision to emerge in an entirely decentralized way.

Conclusions

Do discoveries about insect societies hold lessons for the study of other collective decision makers? One reason to think that they do is the increasing attention to similar mechanisms in social vertebrates. Because of their cognitive complexity, it might be assumed that vertebrate groups make decisions through well-informed leaders who process all relevant data and direct the rest of the group to the best option. For many groups, however, consensus emerges instead from simple local interactions among individuals, without direction by a central controller (Conradt and Roper 2005). Vertebrate societies may have many reasons to rely on such decentralized mechanisms. For one, just as in social insects, the scale of large groups may deny any single animal a synoptic view of the problem at hand. Another reason is that a consensus built with input from all may better allow negotiation of different preferences among group members (Conradt and Roper 2005). Even when all group members have the same preferences, a decentralized approach may produce a superior choice by smoothing out the errors made by individual animals (Simons 2004).

The collective mechanisms of social insects also find parallels at lower levels of biological organization, such as the highly decentralized organization of nervous systems. Cognition arises from a massive number of local interactions among neurons, in which threshold effects and positive feedback play key roles. Recent work on the neural mechanisms of decision making suggests that competing populations of neurons accumulate bits of data in favor of each option, until a threshold activity level is reached for one of them (Bogacz 2007). This model finds a remarkable echo in the quorum-dependent house-hunting algorithms of ants and bees. A more obvious example of quorum-sensing at the cellular level is seen in many kinds of colonial bacteria that monitor the local density of cells and initiate certain behaviors only when a threshold density is exceeded (Miller and Bassler 2001). The quorum-detection mechanism and goals are very different from those of insects, but the tactic for group coordination is fundamentally similar.

Ultimately, understanding bacterial coordination, nervous systems, or vertebrate social structure requires detailed study of those systems, yet social insects can offer basic insights that may inform these studies. Understanding any collective system requires an algorithmic description of its behavior—a detailed accounting of the behavioral rules followed by group members and the contextual cues and interactions that guide them. This goal is especially attainable for insect societies owing to their small size, intermediate complexity, and experimental tractability. The resulting descriptions must be paired with the testing of precisely-stated hypotheses for how adaptive group behavior emerges from the interactions of many insects following these algorithms. In general, the less-than-obvious implications of individual actions for group behavior mean that mathematical models and computer simulations are a vital tool. Even in the social insects, this approach is still only partly developed, but a few well-described cases, described in this review, can now be seen as model systems. Further development of these systems and other well-chosen examples promises insights not only into the organization of colonies of ants and bees, but of collective decision-making systems in general.

Literature Cited

Beckers, R., J. L. Deneubourg, and S. Goss. 1992. "Trails and U-turns in the selection of a path by the ant *Lasius niger.*" *Journal of Theoretical Biology* 159: 397–415.

———— 1993. "Modulation of trail laying in the ant *Lasius niger* (Hymenoptera, Formicidae) and its role in the collective selection of a food source." *Journal of Insect Behavior* 6: 751–759.

Beckers, R., J. L. Deneubourg, S. Goss, and J. M. Pasteels. 1990. "Collective decision-making through food recruitment." *Insectes Sociaux* 37: 258–267.

Beekman, M., R. L. Fathke, and T. D. Seeley. 2006. "How does an informed minority of scouts guide a honeybee swarm as it flies to its new home?" *Animal Behaviour* 71: 161–171.

Biesmeijer, J. C., M. Born, S. Lukacs, and M. J. Sommeijer. 1999. "The response of the stingless bee *Melipona beecheii* to experimental pollen stress, worker loss and different levels of information input." *Journal of Apicultural Research* 38: 33–41.

Bogacz, R. 2007. "Optimal decision-making theories: Linking neurobiology with behaviour." *Trends in Cognitive Sciences,* 11: 118–125.

Camazine, S. 1993. "The regulation of pollen foraging by honey bees: How foragers assess the colony need for pollen." *Behavioral Ecology and Sociobiology* 32: 265–272.

Conradt, L., and T. J. Roper. 2005. "Consensus decision making in animals." *Trends in Ecology & Evolution* 20: 449–456.

Couzin, I. D., and J. Krause. 2003. "Self-organization and collective behavior in vertebrates." *Advances in the Study of Behavior* 32: 1–75.

Couzin, I. D., J. Krause, N. R. Franks, and S. A. Levin, 2005. "Effective leadership and decision-making in animal groups on the move." *Nature* 433: 513–516.

de Biseau, J. C., J. L. Deneubourg, and J. M. Pasteels. 1991. "Collective flexibility during mass recruitment in the ant *Myrmica sabuleti* (Hymenoptera: Formicidae)." *Psyche* 98: 323–336.

de Marco, R. J. 2006. "How bees tune their dancing according to their colony's nectar influx: Re-examining the role of the food-receivers' 'eagerness'." *Journal of Experimental Biology* 209: 421–432.

Detrain, C., and J. L. Deneubourg. 1997. "Scavenging by *Pheidole pallidula*: A key for understanding decision-making systems in ants." *Animal Behaviour* 53: 537–547.

Dreller, C., R. E. Page, and M. K. Fondrk. 1999. "Regulation of pollen foraging in honeybee colonies: Effects of young brood, stored pollen, and empty space. *Behavioral Ecology and Sociobiology* 45: 227–233.

Fewell, J. H., and M. L. Winston. 1992. "Colony state and regulation of pollen foraging in the honey bee, *Apis mellifera* L." *Behavioral Ecology and Sociobiology* 30: 387–393.

Franks, N. R., A. Dornhaus, J. P. Fitzsimmons, and M. Stevens. 2003. "Speed versus accuracy in collective decision making." *Proceedings of the Royal Society of London B Biological Sciences* 270: 2457–2463.

Franks, N. R., A. Dornhaus, G. Hitchcock, R. Guillem, J. Hooper, and C. Webb. 2007. "Avoidance of conspecific colonies during nest choice by ants." *Animal Behaviour* 73: 525–534.

Franks, N. R., E. B. Mallon, H. E. Bray, M. J. Hamilton, and T. C. Mischler. 2003. "Strategies for choosing between alternatives with different attributes: Exemplified by house-hunting ants." *Animal Behaviour* 65: 215–223.

Hofstede, F. E., and M. J. Sommeijer. 2006. "Influence of environmental and colony factors on the initial commodity choice of foragers of the stingless bee *Plebeia tobagoensis* (Hymenoptera, Meliponini)." *Insectes Sociaux* 53: 258–264.

Hutchinson, J. M. C., and G. Gigerenzer. 2005. "Simple heuristics and rules of thumb: Where psychologists and behavioural biologists might meet." *Behavioural Processes* 69: 97–124.

Jeanson, R., J. L. Deneubourg, A. Grimal, and G. Theraulaz. 2004. "Modulation of individual behavior and collective decision-making during aggregation site selection by the ant *Messor barbarus*." *Behavioral Ecology and Sociobiology* 55: 388–394.

Judd, T. M. 2005. "The effects of water, season, and colony composition on foraging preferences of *Pheidole ceres* (Hymenoptera: Formicidae)." *Journal of Insect Behavior* 18: 781–803.

Lindauer, M. 1967. *Communication among social bees*. New York: Atheneum.

Mailleux, A. C., J. L. Deneubourg, and C. Detrain. 2003. "Regulation of ants' foraging to resource productivity." *Proceedings of the Royal Society of London B-Biological Sciences* 270: 1609–1616.

Mailleux, A. C., D. Jean-Louis, and C. Detrain. 2000. "How do ants assess food volume?" *Animal Behaviour* 59: 1061–1069.

Mallon, E. B., S. C. Pratt, and N. R. Franks. 2001. "Individual and collective decision-making during nest site selection by the ant *Leptothorax albipennis*." *Behavioral Ecology and Sociobiology* 50: 352–359.

Miller, M. B., and B. L. Bassler. 2001. "Quorum sensing in bacteria." *Annual Review of Microbiology* 55: 165–199.

Passino, K. M. and T. D. Seeley. 2006. "Modeling and analysis of nest-site selection by honeybee swarms: The speed and accuracy trade-off." *Behavioral Ecology and Sociobiology* 59: 427–442.

Payne, J. W., J. R. Bettman, and E. J. Johnson. 1993. *The adaptive decision maker*. Cambridge: Cambridge University Press.

Portha, S., J. L. Deneubourg, and C. Detrain. 2002. "Self-organized asymmetries in ant foraging: A functional response to food type and colony needs." *Behavioral Ecology* 13: 776–781.

Pratt, S. C. 2004. "Collective control of the timing and type of comb construction by honey bees (*Apis mellifera*)." *Apidologie* 35: 193–205.

——— 2005. "Behavioral mechanisms of collective nest-site choice by the ant *Temnothorax curvispinosus*." *Insectes Sociaux* 52: 383–392.

Pratt, S. C., E. B. Mallon, D. J. T. Sumpter, & N. R. Franks. 2002. "Quorum sensing, recruitment, and collective decision-making during colony emigration by the ant *Leptothorax albipennis.*" *Behavioral Ecology and Sociobiology* 52: 117–127.

Pratt, S. C., and D. J. T. Sumpter. 2006. "A tunable algorithm for collective decision-making." *Proceedings of the National Academy of Sciences of the United States of America* 103: 15906–15910.

Pratt, S. C., D. J. T. Sumpter, E. B. Mallon, and N. R. Franks. 2005. "An agent-based model of collective nest choice by the ant *Temnothorax albipennis.*" *Animal Behaviour* 70: 1023–1036.

Ratcliff, R., and J. N. Rouder. 1998. "Modeling response times for two-choice decisions." *Psychological Science* 9: 347–356.

Seeley, T. D. 1995. *The wisdom of the hive.* Cambridge: Belknap Press of Harvard University Press.

———. 2002. "When is self-organization used in biological systems?" *Biological Bulletin* 202: 314–318.

———. 2003. "Consensus building during nest-site selection in honey bee swarms: The expiration of dissent." *Behavioral Ecology and Sociobiology* 53: 417–424.

Seeley, T. D., and S. C. Buhrman, 2001. "Nest-site selection in honey bees: How well do swarms implement the "best-of-N" decision rule?" *Behavioral Ecology and Sociobiology* 49: 416–427.

Seeley, T. D., and P. K. Visscher. 2004. "Quorum sensing during nest-site selection by honeybee swarms." *Behavioral Ecology and Sociobiology* 56: 594–601.

Seeley, T. D., P. K. Visscher, and K. M. Passino. 2006. "Group decision making in honey bee swarms." *American Scientist* 94: 220–229.

Simon, H. A. 1990. "Invariants of human behavior." *Annual Review of Psychology* 41: 1–19.

Simons, A. M. 2004. "Many wrongs: The advantage of group navigation." *Trends in Ecology & Evolution* 19: 453–455.

Stein, M. B., H. G. Thorvilson, and J. W. Johnson. 1990. "Seasonal changes in bait preference by red imported fire ant *Solenopsis invicta* (Hymenoptera, Formicidae)." *Florida Entomologist* 73: 117–123.

Sumpter, D. J. T., and M. Beekman. 2003. "From nonlinearity to optimality: Pheromone trail foraging by ants." *Animal Behaviour* 66: 273–280.

Sumpter, D. J. T., and S. C. Pratt. 2003. "A modelling framework for understanding social insect foraging." *Behavioral Ecology and Sociobiology* 53: 131–144.

Visscher, P. K. 2007. "Group decision making in nest-site selection among social insects." *Annual Review of Entomology* 52: 255–275.

Weidenmüller, A., and J. Tautz. 2002. "In-hive behavior of pollen foragers (*Apis mellifera*) in honey bee colonies under conditions of high and low pollen need." *Ethology* 108: 205–221.

CHAPTER TWENTY-THREE

From Social Behavior to Molecules:
Models and Modules in the Middle

ANDREW B. BARRON

GENE E. ROBINSON

SOCIOGENOMICS (molecular and genomic studies of social behavior) is still young, but the field is advancing rapidly (Robinson, Grozinger, and Whitfield 2005) due largely to the explosive growth of genomic resources. Thanks to new genomic databases and technologies it is increasingly feasible to identify genes or molecular pathways associated with aspects of social behavior. The annotated honey bee genome, for example, can assist positional cloning efforts building on studies that have identified quantitative trait loci for social traits (Hunt et al. 1995; Page, Gadau, and Beye 2002; Rueppell et al. 2004; Rueppell, Pankiw, and Page 2004). It facilitates testing candidate genes suggested from comparative analyses with other organisms (Ben-Shahar et al. 2002; Kucharski et al. 2000; Toma et al. 2000), and enables genome-wide microarray analyses (Grozinger et al. 2003; Whitfield, Cziko, and Robinson 2003). Identifying molecular elements associated with social behavior is one challenge; analyzing how genes or molecules function to influence social behavior is a different and perhaps more difficult question.

The path linking genes and behavior is invariably complicated (Hall 2003), but social behavior adds an additional tier of complexity because it depends on interaction and communication between individuals. For the advanced insect societies, a society contains tens or even hundreds of thousands of individuals. Social interactions are multimodal and complex (Wilson 1971), and many aspects of behavior are regulated by social feedback

(Leoncini et al. 2004). The colony-level traits of particular interest to socio-biologists, such as division of labor and nest construction, are in part self-organizing (Bonabeau and Theraulaz 1999; Bonabeau et al. 1997; Franks et al., Fewell, Schmidt and Taylor et al., this volume) and therefore particularly difficult to relate to molecular events occurring within an individual.

In this chapter we discuss how molecular analyses of social behavior can be aided by dissecting social behavior into simpler behavioral modules that are easier to relate to molecular or neural processes. The concept of modularity has broad utility in fields of biology examining complex interacting systems, particularly evolutionary and developmental biology. Although the concept of modularity is somewhat intuitive, modules of a complex system can be difficult to define precisely. In developmental biology, *modules* are defined as distinct organizational or functional units within the body (Carroll 2005). We borrow that perspective here to describe a behavioral module as a distinct organizational or functional unit in the expression of complex behavior.

Modules in developmental biology are somewhat obvious since most animal body plans are constructed from serially repeated structures or segments (Carroll 2005). Vertebrae and the associated segmental musculature are an example. Modification of the basic module generates structural and functional variation within an animal (e.g., the lumbar, thoracic, and cervical vertebrae), and between-species evolution has generated variation in body plans by varying the design and number of basic modules (Carroll 2005). Evolutionary developmental biology has made enormous strides toward understanding the genetic control of development and evolution of body plans by focusing attention on how genes build modules and how modules have been adapted both within and between species (Carroll 2005; Laubichler and Gadau, this volume).

Behavioral modules are more abstract than the segmental modules of developmental biology since behavioral modules are based on function rather than structure. For example, rhythm is considered a functional module in the expression of many forms of behavior, and behavioral genetics has made enormous strides toward understanding the genetic basis of behavior by focusing on rhythm. In *Drosophila* this led to the discovery of the molecular clock (Weiner 2000). The clock module crops up repeatedly in *Drosophila* as a regulator of very different forms of behavior cycling over very different periods, including diurnal activity and courtship songs (Hall 1998; Weiner 2000); hence, it seems that within a species the same basic clock module has been repeatedly adopted for different functions. Focusing on how the clock module has been modified has revealed how evolution has generated variation in courtship songs between species (Weiner 2000; Gleason 2005).

There are many approaches for dissecting social behavior into behavioral modules; these can be grouped into two broad categories. The first involves modeling social behavior to suggest relevant behavioral modules and to propose hypotheses for gene function. The second approach involves borrowing behavioral modules from a different (usually solitary) context and exploring their role in social behavior. Something is usually known of the molecular or neural mechanisms underlying the borrowed modules, and consequently this approach also leads to hypotheses for mechanistic analyses of social behavior. Because of the diversity of social behavior, the most appropriate strategy for behavioral dissection varies case by case.

In this chapter, we highlight cases that illustrate how taking a modular approach to social behavior can help explain the underlying molecular mechanisms. First, we discuss how the response threshold model and state space model have facilitated molecular analyses of division of labor in honey bees and the molecular basis of aggression in lobsters. We then consider how borrowing a reinforcement module known in the context of individual learning has helped understand pair bonding in voles and division of labor in honey bees. We do not champion one approach over another; rather, we use these examples to illustrate the successful synergy among modeling, modular, and molecular analyses of social behavior.

The Response Threshold Model

The response threshold model is one of the oldest and simplest models of animal behavior. The model states that responses to various stimuli by an individual are based on stimulus thresholds; stimulus levels below an individual's threshold result in no response whereas super-threshold stimuli elicit a reaction. The response threshold model was originally developed to aid in predicting individual behavior (Loeb 1918; Manning 1967), and it proposes a strategy for dissecting behavior in a way that is easily related to neuronal properties. An individual neuron only responds when a stimulus sufficiently depolarizes the membrane potential, eliciting either graded or all-or-nothing responses (depending on the type of neuron). The response threshold model suggests a behavioral dissection into modules associated with stimuli involved in the behavior. For example, nest guarding in honey bees is a complex social interaction involving tolerance of nestmates, rebuttal of non-nestmates, and the ability to distinguish between the two. Applying the stimulus response model simplifies guarding by focusing on identifying stimuli that increase the likelihood of an aggressive response by the guard toward an arriving bee (Breed et al. 1995).

The Response Threshold Model and Division of Labor in Insect Colonies

The response threshold model has been very successfully elaborated to model division of labor in social insect colonies (Beshers, Robinson, and

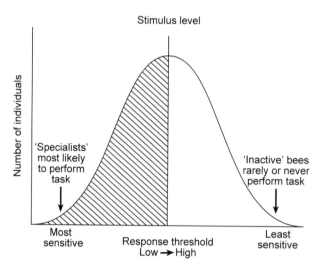

Figure 23.1. Hypothetical response threshold model to explain inter-individual differences in task specialization in an insect colony. Shown is a hypothetical distribution describing the variation in response thresholds to a task-related stimulus between individuals in a colony. Individuals with response thresholds lower than the current stimulus level (shaded area) will be actively engaged in the associated task. Individuals at the margins of the distribution will show 'extreme' behavior. Those with very high response thresholds will rarely or never perform the behavior, but bees with very low response thresholds will be most likely to perform the task, and may be classed as 'specialists'. Adapted from Page and Robinson (1991).

Mittenthal 1999; Bonabeau and Theraulaz 1999; Page and Erber 2002; Robinson 1987; Wilson 1984). In a social insect colony, individual workers typically specialize on different tasks and thereby increase the efficiency of the colony as a whole (Beshers, Robinson, and Mittenthal 1999; Page and Mitchell 1998). The response threshold model can be applied to groups to suggest an elegant and simple process by which division of labor can occur (Figure 23.1). If there is variation in stimulus thresholds between individuals in a group, then there will be variation in the likelihood of different individuals performing tasks associated with those stimuli. Variation in response thresholds can create specialists, since the workers with the lowest response threshold for a stimulus will be the first to respond and the longest to act (Beshers, Robinson, and Mittenthal 1999; Page and Mitchell 1998; Page and Robinson 1991; Weidenmuller 2004).

Applying the response threshold model to groups in this way has aided molecular analysis of division of labor in honey bees. The model proposes that division of labor, a group dynamic, can be understood in terms of individual variation in responsiveness to task-related stimuli (Beshers, Robinson, and Mittenthal 1999; Beshers and Fewell 2001; Scheiner and Erber, this volume), leading to modules at the individual level that can be more easily related to molecular pathways. In honey bees, division of labor unfolds as a process of behavioral development, with worker bees changing tasks as they age (Winston 1987). Adult workers perform in-hive tasks such as brood care ("nursing") when young, and shift to foraging when they are about 2 to 3 weeks old (Winston 1987). To identify genes associated with foraging behavior in honey bees, Ben-Shahar et al. (2002) focused their search on honey bee homologs of candidate genes (Fitzpatrick et al. 2005) known to be involved in foraging in other species. The *foraging* gene (*for*) in *Drosophila melanogaster* encodes a guanosine 3',5'-monphosphate (cGMP)-dependent protein kinase (PKG) (Osborne et al. 1997). Allelic variation in *for* influences variation in larval feeding behavior in *Drosophila* causing some to roam widely while foraging (rovers) and others to explore just the immediate area (sitters) (Sokolowski, Pereira, and Hughes 1997). There is an appealing analogy between rover and sitter *Drosophila* larvae and the exploratory behavior of honey bee foragers versus the stay-at-home behavior of nurses. Forager honey bees, like rover flies, have higher expression of the orthologous *Amfor* in the brain relative to nurses (Ben-Shahar et al. 2002), suggesting there may be greater PKG activity in the the brains of foragers versus nurses. Feeding bees cGMP pharmacologically increases brain PKG activity, and applying this treatment to young bees resulted in a precocious onset of foraging (Ben-Shahar et al. 2002).

These two lines of evidence suggest the *foraging* gene (*for*) is involved in the expression of foraging behavior in honey bees, but a more difficult question remains: How does elevated PKG activity in the bee brain cause the onset of foraging behavior? Reasoning from the response threshold model, Ben-Shahar et al. (2003) proposed that PKG modulated individual response thresholds to one or more foraging-related stimulus thereby creating variation within the colony in individual responses to foraging-related stimuli. Foraging is stimulated by an array of olfactory, mechanical, and visual stimuli (Page and Erber 2002; Page and Mitchell 1998; Winston 1987), but in honey bees for is strongly expressed in the visual system, particularly the optic lobe lamina and regions of the mushroom bodies known to receive visual input (Ben-Shahar et al. 2003; Ehmer and Gronenberg 2002; Gronenberg and Riveros 2002, this volume). Therefore, Ben-Shahar et al. (2003) examined the effects of PKG on individual responsiveness to light. Honey bees experience a major change in exposure to light when they shift from working in the dark hive to foraging outside. Foragers are positively phototactic, much more so than nurses (Ben-Shahar et al. 2003; Menzel and Greggers 1985). In a laboratory bioassay, cGMP treatment increased the phototactic response of young bees (Ben-Shahar et al. 2003). Positively phototactic bees were attracted to the nest entrance, or even to leave the hive. Once in the vicinity of the nest entrance, bees were more likely to encounter other stimuli known to stimulate foraging, such as dances and incoming nectar (Seeley 1995), hence increased phototaxis could be part of the mechanism driving the onset of foraging.

Electroretinogram analysis indicated that the cGMP-induced increase in positive phototaxis was not based on effects of sensitivity to light *per se*. Perhaps PKG is involved in modifying the function of neuronal circuits in the lamina and/or mushroom bodies via phosphorylation of some component molecules, which is similar to the affect of PKG on olfaction in mammals (Kroner et al. 1996). *Amfor* expression also has been shown to be high in the brains of undertaker bees, but not similarly aged bees engaged in tasks performed solely in the hive (Ben-Shahar et al. 2003). Undertakers are behaviorally specialized in removing corpses from the colony. They leave the hive for short flights but do not forage. This suggests that the upregulation of *Amfor* may be more generally related with working outside the hive than with foraging specifically. In this example, the candidate gene approach identified *Amfor* as a gene associated with honey bee foraging, but the inspiration for experiments exploring how *Amfor* influenced foraging and division of labor came from the modular dissection of division of labor suggested by the response threshold model.

State-Space Models of Behavior

Models of behavior must sometimes take into account a variety of possible responses that an animal may make to a given stimulus, or that the form of behavior expressed can depend on interactions between multiple stimuli. State-space models are a multivariate elaboration of the philosophy of the response threshold model to simultaneously consider multiple stimuli and/or behavioral alternatives (McFarland and Houston 1981; McFarland and Sibly 1975).

State-space models postulate that the total behavioral repertoire is controlled by a set of causal factors (McFarland and Sibly 1975). External factors, such as the perception of food stimuli, and internal "motivational" factors, such as the size of fat stores, circadian rhythms, or reproductive condition, are all considered causal factors and treated similarly in the model. The state of these causal factors can be represented as a Euclidean space, the axes of which represent the causal factors. This causal factor space is multidimensional, with the number of dimensions depending on the number of causal factors relevant to the animal's behavior.

McFarland and Sibly (1975) assumed that an animal's behavioral repertoire can be classified into mutually exclusive categories (activities) such that actions belonging to one activity are incompatible with actions belonging in another. For instance, all the actions associated with fighting can be considered one activity and all the actions associated with mating another, since an animal cannot fight and mate at the same time. The modeling process, summarized by McFarland and Houston (1981), maps behavioral tendencies onto the causal factor space such that every point in the causal factor space is associated with a tendency of performing a specific action. A consequence of this model is that there can be more than one point in causal factor space with the same behavioral tendency. For example, assuming for simplicity that feeding depends solely on the two causal factors food availability and hunger, the same feeding tendency might be observed in a situation where there is high food availability and low hunger as when there is low food availability and high hunger. Points in causal factor space sharing the same behavioral tendencies are linked by motivational isoclines (Figure 23.2).

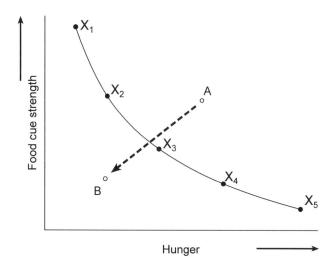

Figure 23.2. The state-space model: hypothethical causal factor space for feeding. This model assumes that feeding tendency is influenced by hunger and food cues only. The feeding tendency is the same at points $x_1 \ldots \ldots x_5$ and the solid line joining these points is a motivational isocline. Point A represents the causal factor state of an animal before feeding, and point B after feeding (assuming that feeding reduces both hunger and local food availability). The dashed arrow indicates the trajectory in causal factor space that results from feeding. Adapted from McFarland and Houston (1981).

The causal factor state is a point in causal factor space that describes the animal's current state. The causal factor state is constantly changing due to shifts in environmental conditions, internal factors, and the outcomes of the animal's own behavior, and this change can be represented as a trajectory in causal factor space. As the trajectory crosses motivational isoclines there will be a shift in the relative balance of behavioral tendencies. If the causal factor state moves to an area of causal factor space where a different behavioral tendency is dominant, a change in behavior will be observed (Figure 23.2). The state-space approach, therefore, makes it possible to predict when a change in conditions will result in a change in behavior.

State-space models consider how both internal and external factors influence behavior, and how the various factors interact in an attempt to describe a program for control of behavior dependent on all causal factors. Such a model aids mechanistic analyses because once a molecular pathway has been found to be associated with behavior, the results of experiments manipulating the molecular pathway can be added to the model as a new causal factor. The result is that information on the basic effects of an experimental treatment become integrated into a program for behavioral control, which can reveal much more detail of how molecules contribute to complex behavior, as illustrated by the following example.

State-Space Models and the Basis of Decapod Aggression

Aggression tends to dominate intraspecific encounters between decapod crustaceans. With the exception of mating and courtship, most intraspecific encounters in lobsters and crayfish result in escalating fights until one individual withdraws. The biogenic amine neuromodulators octopamine and serotonin were found to be involved in agonistic behavior based on observations that injection of serotonin and octopamine into freely moving lobsters (*Homarus americanus*) generated postures resembling those seen in dominant (serotonin) and subordinate (octopamine) animals when they encounter one another (Huber et al. 1997). Later studies demonstrated that infusion of serotonin in crayfish (*Astacus astacus*) caused treated individuals to engage larger opponents in prolonged bouts of fighting, even in instances that carried substantial risk of injury (Kravitz and Huber 2003). These analyses identified a neurochemical pathway involved in decapod aggression, but did not explain how changing serotonergic signaling impacted on decapod behavior to make the animals more aggressive.

Serotonin did not simply make decapods indiscriminately more aggressive (Huber et al. 1997). Fights between decapods are complex affairs due to the many alternative behavioral strategies employed by each combatant (Huber and Delago 1998; Huber et al. 1997). The expression of different fighting strategies varies with internal causal factors such as hunger states and previous agonistic success, and external causal factors such as proximity to a shelter or food resource (Kravitz and Huber 2003). The progress of the fight and its eventual outcome are determined both by an animal's prior actions in the fight and its responses to the opponent (Kravitz and Huber 2003). Winners are much more likely to dominate future encounters, whereas losers are much more likely to retreat. Consequently, fights are influenced by a web of interdependent causal factors with any action affecting subsequent actions.

Huber and Delago (1998) used detailed ethological dissection and a multivariate Principal Components Analysis to determine the effects of serotonin on the progression of fights. This analysis embraced a state-space approach by considering how the many causal factors that influence a fight sequence were interrelated. Their results showed that the dominant effect of serotonin was to change the decision to retreat, decreasing the likelihood that subordinates would withdraw from the attacks of opponents (Huber and Delago 1998; Huber et al. 1997). In the terms of the state-space model, experimentally elevating serotonin moved lobsters to an area of causal factor space where the tendency to retreat was very low.

Molecular components of the serotonin pathway that appear to be involved in modulating the decision to retreat include two serotonin receptor subtypes, 5HT1 and 5HT2 (Yeh, Fricke, and Edwards 1996; Yeh, Musolf, and Edwards 1997). 5HT1 and 5HT2 receptors affect the excitability of peripheral lateral giant (LG) neurons, which are known to influence the expression of behaviors related to both dominance and retreat. Dominant individuals exhibit more excitable LG neurons, apparently due to an increase in 5HT2-mediated signaling, and they are less likely to retreat during an encounter. Subordinate individuals show an inhibited LG neuron response and an increase in 5HT1-mediated signaling, and they are more likely to retreat. Expression of the 5HT1 and 5HT2 receptors is affected by prior success in fights, but it is not yet known whether these socially mediated changes are due to transcriptional or post-transcriptional mechanisms.

Together, the findings from the molecular and behavioral analyses suggest that the decision to retreat may depend on the balance between 5HT1 and 5HT2 signaling on the excitability of the LG neurons, which itself is socially regulated. Prior success in fights influences the expression of 5HT1 and 5HT2, so this molecular mechanism can provide an explanation as to why previous winners are less likely to retreat in later encounters. The next step in unraveling serotonin's role in aggression is to determine what stimuli release serotonin to activate the 5HT1 and 5HT2 receptors. Behavioral analyses suggest that serotonin is released by threatening stimuli (Huber and Delago 1998; Huber et al. 1997), but the nature of these stimuli have yet to be explored. In this example, the state-space approach enabled a dissection of decapod fight behavior, which led to hypotheses for serotonin's role in fighting behavior, and focused later molecular analyses on pathways that mediated assessment of threat and retreat.

Extrapolating Behavioral Modules from Solitary to Social Behavior

New forms of social behavior do not arise *de novo*. It is increasingly clear that many aspects of social behavior have evolved by adopting and adapting preexisting modules from aspects of solitary behavior (Amdam et al. 2006; Robinson 2007; West-Eberhard 1996). A consequence is that underlying molecular mechanisms have also been adopted and adapted for new roles in social behavior (Robinson and Ben-Shahar 2002). Sometimes molecular pathways associated with a form of social behavior are suggestive of an association with a behavioral module known in a solitary context, and this in turn generates new hypotheses for how the molecular pathway might be acting to shape social behavior. This was the case in molecular analyses of pair bond formation in voles.

Reinforcement and Pair Bonding in Voles

Social attachment is essential for the formation of many stable social groups, and recent studies with voles have established a novel approach to explore the molecular basis of social cohesion. Two related species of vole display strongly contrasting social organizations. Prairie voles *(Microtus ochrogaster)* are usually monogamous, but the montane vole *(Microtus montanus)* is nonmonogamous (Insel, Preston, and Winslow 1995). Most of the differences in pair bonding between these two closely related species appear to be caused by differences in the expression of two genes that encode receptors for the neuropeptides oxytocin and arginine (vasopressin): OTR and V1aR. Oxytocin and vasopressin are involved in social recognition and bonding between mates or parents and offspring in many mammalian species: oxytocin influences female behavior and vasopressin influences male behavior (Insel 1992; Insel, Preston, and Winslow 1995). The monogamous prairie vole expressed the oxytocin and vaspressin receptor genes OTR and V1aR in female and male brains at strikingly high densities in regions of the brain normally associated with reward processing; a pattern not seen in the nonmonogamous vole species (Young et al. 2001; Young and Wang 2004). Further, viral-driven over expression of V1aR in the reward processing brain regions of the nonmonogamous male meadow vole enhanced pair bonding in this species (Lim et al. 2004).

The differences in the expression of the OTR and V1aR genes in females and males are somehow involved in the differences in social affiliation between the monogamous and nonmonogamous vole species. These differences are most pronounced in brain regions better known for their involvement in reinforcement of reward learning (Schultz 2001; Schultz, Apicella, and Ljungberg 1993); an anatomical insight suggesting reinforcement could be a behavioral module relevant to pair bond formation.

Reinforcement has been well studied, both theoretically and physiologically, as a module of learning. Reinforcement-learning theory models how animals learn to achieve a desired state and avoid undesirable ones as efficiently as possible. This outcome is achieved under the guidance of reinforcement signals, which serve to "criticize" actions with respect to how well they serve obtaining the desired state (Doya 2002; Montague, Hyman, and Cohen 2004). There is now strong evidence that dopamine modulates a reinforcement signal in the mammalian brain (Schultz, Apicella, and Ljungberg 1993; Hollerman and Schultz 1998; Schultz 2001, 2002). Robinson (1987) argued that dopamine release from neurons in the mid and fore brain mediates the attachment of reward value to an action or an object to reinforce future behavioral choices.

Insel and Young (2001) hypothesized that reinforcement could be a behavioral module "reused" in pair bond formation. They proposed that for monogamous species, features of the partner could be strongly associated with an innately rewarding stimulus such as mating, so that partners become motivated to remain together, whereas for the nonmonogamous species, the identity of the partner is less reinforcing.

Molecular studies support this interpretation of pair bonding. Both oxytocin and vasopressin are involved in the neural processing of sensory cues involved in social learning, especially learning the olfactory signatures of conspecifics (Ferguson et al. 2000). Mating causes a release of vasopressin or oxytocin into the prefrontal cortex, nucleus accumbens, and ventral pallidum (Young and Wang 2004). Mating in both males and females also causes dopamine release into the same brain regions (Young and Wang 2004). Concurrent activation of oxytocin/vasopressin *and* the dopamine pathway is necessary for pair bond formation (Figure 23.3). In nonmonogamous species the lower density of OTR and V1aR in these brain regions would suggest that the dopamine system and ocytocin/vasopressin systems are only weakly coupled, so the smell of a partner might not be intimately associated with a rewarding mating experience (Young and Wang 2004). In this example molecular analyses first suggested a link between pair bond-

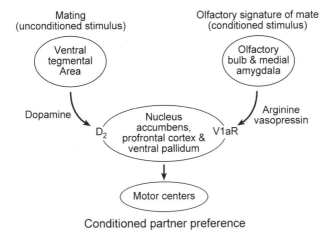

Conditioned partner preference

Figure 23.3. Neurobiological model for pair bonding in male monogamous prairie voles based on reward theory. The act of mating stimulates the ventral tegmental area, which releases dopamine into the nucleus accumbens and prefrontal cortex. Olfactory cues from the female are processed by the olfactory bulb and the medial amygdala, which are critical for social recognition. The medial amygdala releases arginine vasopressin to the ventral pallidum. Concurrent activation of Dopamine D2 receptors in the nucleus accumbens and vasopressin V1aR receptors in the ventral pallidum of males results in a conditioned pair bond.

ing and reinforcement. Having made that bridge, knowledge of the neural and molecular mechanisms of the reinforcement module generated new, testable hypotheses for exploring how the genes OTR and V1aR cause pair bonding.

Reinforcement as a Module in Insect Social Behavior

Reinforcement is a concept that has been developed and explored almost exclusively in mammalian model systems, but the module can be also applied to insect behavior. It is clear that many insects, especially honey bees, rapidly learn features of rewarding situations (Giurfa 2003; Menzel and Giurfa 2001); however, whereas dopamine modulates reinforcement in mammals, evidence suggests octopamine modulates the reinforcement signal in insects. In bees, local injection of octopamine into the antennal lobes or mushroom body calyces can substitute for reward in an associative learning paradigm (Hammer and Menzel 1998). Similar findings have been reported for *Drosophila;* octopamine pathways in their mushroom bodies modulated learning with positive reinforcement and, intriguingly, dopamine pathways mediated learning with negative reinforcement (Schwaerzel et al. 2003). Octopamine stimulates feeding and food seeking in blow flies (Long and Murdock 1983), and social foraging in honey bees (Barron et al. 2002; Schulz, Barron, and Robinson 2002).

Considering how reinforcement may act as a behavioral module can help to dissect some of the most complex aspects of social behavior. Recent investigations into the neural systems underlying the honey bee dance language provide an example. Honey bees use symbolic dances to communicate the location and value of resources to their colony (Frisch 1967). The dance language is one of the few well-characterized examples of symbolic communication in animals (Frisch 1967), but after nearly 50 years of intensive study relatively little is known about the neural mechanisms underlying the dance. This is primarily a consequence of the complex and integrative nature of dance behavior, which makes it difficult to know where and how to begin mechanistic analyses.

Treatment with octopamine makes forager bees more likely to dance on return to the colony (Barron et al. 2007), but why does octopamine make bees dance? The function of the dance is to recruit additional foragers to valuable resources; consequently, returning foragers will only dance for valuable or underexploited resources needed by the colony (Seeley 1995). Elements of the dance (vigor and duration) represent the value of the resource to inform recruitment (Seeley 1995). Given octopamine's role in reinforcement of reward learning, it was hypothesized that the role of octopamine in dance behavior could be as a modulator of the reinforcement signal involved in the computation of resource value. Dances are complex multicomponent signals communicating both the location and value of resources through different aspects of the dance movement. Therefore, to test this hypothesis, Barron et al. (2007) dissected the dance structure to examine the effect of octopamine treatment on the representation of location and quality. Octopamine treatments caused dose-dependent changes in dance aspects related to resource value, but did not change the representation of location. This study supports the idea that octopamine's function in the context of dance is as part of the reinforcement module calculating resource value, and is yet another example of how a modular approach can help dissect complex social behavior.

Conclusion and Future Directions

In this chapter we have discussed cases that illustrate how dissecting complex behavior to simpler behavioral modules can help work out how genes and molecular pathways contribute to social behavior. Various strategies exist for identifying behavioral modules and no approach is necessarily better than another; rather, different strategies are useful in different contexts. Models can both dissect complex traits into simpler modules and suggest hypotheses for testing how identified molecular pathways might influence behavior. This was illustrated by the application of the response threshold model to division of labor in bees to explore the role of the *foraging* gene in honey bee division of labor. Modules can also be borrowed from aspects of solitary behavior and used to generate hypotheses for the organization of social behavior. This was illustrated by the case of pair bonding in voles where initial molecular analyses implied a link between pair bond formation and reinforcement systems.

We have entered an exciting phase in insect sociology. Thanks to a strong tradition of behavioral and phylogenetic analysis the social insects have been recognized as behavioral innovators and models for the evolution of new social systems. Social insect societies have been famously described as a major evolutionary transition to a new level of biological complexity (Maynard-Smith and Szathmáry 1985), and by combining evolutionary, molecular, and genomic analyses of behavior we can now identify social innovations and examine how they evolved. When examining how molecular pathways influence social behavior, taking a modular approach to complex behavior is likely to be most effective.

Literature Cited

Amdam, G. V., A. Csondes, M. K. Fondrk, and R. E. Page. 2006. "Complex social behaviour derived from maternal reproductive traits." *Nature* 439: 76–78.

Barron, A. B., R. Maleszka, R. K. V. Meer, and G. E. Robinson. 2007. "Octopamine modulates honey bee dance behavior." *Proceedings of the National Academy of Science USA* 104: 1703–1707.

Barron, A. B., D. J. Schulz, and G. E. Robinson. 2002. "Octopamine modulates responsiveness to foraging-related stimuli in honey bees *(Apis mellifera)*." *Journal of Comparative Physiology A* 188: 603–610.

Ben-Shahar, Y., H. Leung, W. Pak, M. Sokolowski, and G. Robinson. 2003. "cGMP-dependent changes in phototaxis: A possible role for the foraging gene in honey bee division of labor." *Journal of Experimental Biology* 59: 269–278.

Ben-Shahar, Y., A. Robichon, M. B. Sokolowski, and G. E. Robinson. 2002. "Influence of gene action across different time scales on behavior." *Science* 296: 741–744.

Beshers, S. N., and J. H. Fewell. 2001. "Models of division of labor in social insect colonies." *Annual Review of Entomology* 46: 413–440.

Beshers, S. N., G. E. Robinson, and J. E. Mittenthal. 1999. "Response thresholds and division of labor in insect colonies." In C. Detrain, J. L. Deneubourg, and J. M. Pasteels, eds., *Information processing in social insects,* 115–139. Basel: Birkhauser Verlag.

Bonabeau, E., and G. Theraulaz. 1999. "Role and variability of response thresholds in the regulation of division of labor in insect societies." In C. Detrain, J. L. Deneubourg, and J. M. Pasteels, eds., *Information processing in social insects,* 141–164. Basel: Birkhäuser.

Bonabeau, E., G. Theraulez, J.-L. Deneubourg, S. Aron, and S. Camazine. 1997. "Self-organization in social insects." *Trends in Ecology and Evolution* 12: 188–193.

Breed, M. D., M. F. Garry, A. N. Pearce, B. E. Hibbard, L. B. Bjostad, and R. E. Page. 1995. "The role of wax comb in honey-bee nestmate recognition." *Animal Behaviour* 50: 489–496.

Carroll, S. B. 2005. *Endless forms most beautiful: The new science of evo devo.* New York: W.W. Norton.

Doya, K. 2002. "Metalearning and neuromodulation." *Neural Networks* 15: 495–506.

Ehmer, B., and W. Gronenberg. 2002. "Segregation of visual input to the mushroom bodies in the honeybee (Apis mellifera)." *Journal of Comparative Neurology* 451: 362–373.

Ferguson, J. N., L. J. Young, E. F. Hearn, T. R. Insel, and J. T. Winslow. 2000. "Social amnesia in mice lacking the oxytocin gene." *Nature Genetics* 25: 284–288.

Fitzpatrick, M. J., Y. Ben-Shahar, H. M. Smid, L. E. M. Vet, G. E. Robinson, and M. B. Sokolowski. 2005. "Candidate genes for behavioural ecology." *Trends in Ecology and Evolution* 20: 96–104.

von Frisch, K. 1967. *The dance language and orientation of honeybees.* Cambridge: Harvard University Press.

Giurfa, M. 2003. "Cognitive neuroethology: Dissecting non-elemental learning in a honeybee brain." *Current Opinion in Neurobiology* 13: 726–735.

Gleason, J. M. 2005. "Mutations and natural genetic variation in the courtship song of *Drosophila*." *Behavior Genetics* 35: 265–277.

Grozinger, C. M., N. M. Sharabash, C. W. Whitfield, and G. E. Robinson. 2003. "Pheromone-mediated gene expression in the honey bee brain." *Proceedings of the National Academy of Sciences of the United States of America* 100(Suppl. 2): 14519–14525.

Hall, J. C. 1998. "Genetics of biological rhythms in *Drosophila*." *Advances in Genetics* 38: 135–184.

———. 2003. "A neurogeneticist's manifesto." *Journal of Neurogenetics* 17: 1–90.

Hammer, M., and R. Menzel. 1998. "Multiple sites of associative odor learning as revealed by local brain microinjections of octopamine in honeybees." *Learning & Memory* 5: 146–156.

Hollerman, J. R., and W. Schultz. 1998. "Dopamine neurons report an error in the temporal prediction of reward during learning." *Nature Neuroscience* 1: 304–309.

Huber, R., and A. Delago. 1998. "Serotonin alters decisions to withdraw in fighting crayfish, *Astacus astacus:* The motivational concept revisited." *Journal of Comparative Physiology* A 182: 573–583.

Huber, R., K. Smith, A. Delago, K. Isaksson, and E. A. Kravitz. 1997. "Serotonin and aggressive motivation in crustaceans: Altering the decision to retreat." *Proceedings of the National Academy of Science USA* 94: 5939–5942.

Hunt, G. J., R. E. Page, M. K. Fondrk, and C. J. Dullum. 1995. "Major quantitative loci affecting honey bee foraging behaviour." *Genetics* 141: 1537–1545.

Insel, T. R. 1992. "Oxytocin—a Neuropeptide for Affiliation—Evidence from behavioral, receptor autoradiographic, and comparative studies." *Psychoneuroendocrinology* 17: 3–35.

Insel, T. R., S. Preston, and J. T. Winslow. 1995. "Mating in the monogamous male: Behavioral consequences." *Physiology & Behavior* 57: 615–627.

Insel, T. R., and L. J. Young. 2001. "The neurobiology of attachment." *Nature Reviews Neuroscience* 2: 129–136.

Kravitz, E. A., and R. Huber. 2003. "Aggression in invertebrates." *Current Opinion in Neurobiology* 13: 736–743.

Kroner, C., I. Boekhoff, S. M. Lohmann, H. G. Genieser, and H. Breer. 1996. "Regulation of olfactory signalling via cGMP-dependent protein kinase." *European Journal of Biochemistry,* 236: 632–637.

Kucharski, R., E. E. Ball, D. C. Hayward, and R. Maleszka. 2000. "Molecular cloning and expression analysis of a cDNA encoding a glutamate transporter in the honeybee brain." *Gene,* 242: 399–405.

Leoncini, I., D. Crauser, G. E. Robinson, and Y. Le Conte. 2004. "Worker-worker inhibition of honey bee behavioural development independent of queen and brood." *Insectes Sociaux* 51: 392–394.

Lim, M. M., Z. X. Wang, D. E. Olazabal, X. H. Ren, E. F. Terwilliger, and L. J. Young. 2004. "Enhanced partner preference in a promiscuous species by manipulating the expression of a single gene." *Nature* 429: 754–757.

Loeb, J. 1918. *Forced movements, tropisms, and animal conduct.* New York: Dover Publications.

Long, T. F., and L. L. Murdock. 1983. "Stimulation of blowfly feeding behavior by octopamiergic drugs." *Proceedings of the National Academy of Sciences of the United States of America,* 80: 4159–4163.

Manning, A. 1967. *An introduction to animal behavior.* Reading, MA: Addison-Wesley.

Maynard-Smith, J., and E. Szathmáry. 1985. *The major transitions in evolution.* Oxford: Oxford University Press.

McFarland, D., and A. Houston. 1981. *Quantitative ethology.* Boston: Pitman Advanced Publishing Program.

McFarland, D. J., and R. M. Sibly. 1975. "The behavioural final common path." *Philosophical Transactions of the Royal Society of London B* 270: 265–293.

Menzel, R., and M. Giurfa. 2001. "Cognitive architecture of a mini-brain: The honeybee." *Trends in Cognitive Science* 5: 62–71.

Menzel, C. R., and U. Greggers. 1985. "Natural phototaxis and its relationship to color vision in honey bees." *Journal of Comparative Physiology A* 157: 311–322.

Montague, P. R., S. E. Hyman, and J. D. Cohen. 2004. "Computational roles for dopamine in behavioural control." *Nature* 431: 760–767.

Osborne, K. A., A. Robichon, E. Burgess, S. Butland, R. A. Shaw, A. Coulthard, H. S. Pereira, R. J. Greenspan, and M. B. Sokolowski. 1997. "Natural behavior polymorphism due to a cGMP-dependent protein kinase of *Drosophila.*" *Science* 277: 834–836.

Page, R. E., and J. Erber. 2002. "Levels of behavioral organization and the evolution of division of labor." *Naturwissenschaften* 89: 91–106.

Page, R. E., J. Gadau, and M. Beye. 2002. "The emergence of hymenopteran genetics." *Genetics* 160: 375–379.

Page, R. E., and S. D. Mitchell. 1998. "Self-organisation and the evolution of division of labor." *Apidologie* 29: 171–190.

Page, R. E., and G. E. Robinson. 1991. "The genetics of division of labour in honey bee colonies." *Advances in Insect Physiology* 23: 117–169.

Robinson, G. E. 1987. "Regulation of honey bee age polyethism by juvenile hormone." *Behavioral Ecology and Sociobiology* 20: 329–338.

Robinson, G. E., and Y. Ben-Shahar. 2002. "Social behavior and comparative genomics: new genes or new gene regulation?" *Genes Brain and Behavior* 1: 197–203.

Robinson, G. E., C. M. Grozinger, and C. W. Whitfield. 2005. "Sociogenomics: Social life in molecular terms." *Nature Reviews Genetics* 6: 257–270.

Rueppell, O., T. Pankiw, D. I. Nielsen, M. K. Fondrk, M. Beye, and R. E. Page. 2004. "The genetic architecture of the behavioral ontogeny of foraging in honeybee workers." *Genetics* 167: 1767–1779.

Rueppell, O., T. Pankiw, and R. E. Page. 2004. "Pleiotropy, epistasis and new QTL: The genetic architecture of honey bee foraging behavior." *Journal of Heredity* 95: 481–491.

Schultz, W. 2001. "Reward signaling by dopamine neurons." *Neuroscientist* 7: 293–302.

———. 2002. "Getting formal with dopamine and reward." *Neuron* 36: 241–263.

Schultz, W., P. Apicella, and T. Ljungberg. 1993. "Responses of monkey dopamine neurons to reward and conditioned stimuli during successive steps of learning a delayed response task." *Journal of Neuroscience* 13: 900–913.

Schulz, D. J., A. B. Barron, and G. E. Robinson. 2002. "A role for octopamine in honey bee division of labor." *Brain Behavior and Evolution* 60: 350–359.

Schwaerzel, M., M. Monasterioti, H. Scholz, F. Friggi-Grelin, S. Birman, and M. Heisenberg. 2003. "Dopamine and octopamine differentiate between aversive and appetitive olfactory memories in *Drosophila*." *Journal of Neuroscience* 23: 10495–10502.

Seeley, T. D. 1995. *The wisdom of the hive.* Cambridge: Harvard University Press.

Sokolowski, M. B., H. S. Pereira, and K. Hughes. 1997. "Evolution of foraging behavior in *Drosophila* by density-dependent selection." *Proceedings of the National Academy of Sciences of the United States of America* 94: 7373–7377.

Toma, D. P., G. Bloch, D. Moore, and G. E. Robinson. 2000. "Changes in *period* mRNA levels in the brain and division of labor in honey bee colonies." *Proceedings of the National Academy of Science USA* 97: 6914–6919.

Weidenmuller, A. 2004. "The control of nest climate in bumblebee *(Bombus terrestris)* colonies: Interindividual variability and self reinforcement in fanning response." *Behavioral Ecology,* 15: 120–128.

Weiner, J. 2000. *Time, love, memory: A great biologist and his quest for the origins of behavior.* London: Faber & Faber.

West-Eberhard, M. J. 1996. Wasp societies as microcosms for the study of development and evolution. In S. Turillazzi and M. J. West-Eberhard, eds., *Natural history and evolution of paper wasps,* 290–317. Oxford: Oxford University Press.

Whitfield, C. W., A.-M. Cziko, and G. E. Robinson. 2003. "Gene expression profiles in the brain predict behavior in individual honey bees." *Science* 302: 296–299.

Wilson, E. O. 1971. *The insect societies.* Cambridge: Harvard University Press.

———. 1984. "The relation between caste ratios and division of labor in the ant genus *Pheidole* (Hymenoptera: formicidae)." *Behavioral Ecology and Sociobiology* 16: 89–98.

Winston, M. L. 1987. *The biology of the honey bee.* Cambridge: Harvard University Press.

Yeh, S. R., R. A. Fricke, and D. H. Edwards. 1996. "The effects of social experience on serotonergic modulation of the escape circuit of crayfish." *Science* 271: 366–369.

Yeh, S. R., B. E. Musolf, and D. H. Edwards. 1997. "Neuronal adaptations to changes in the social dominance status of crayfish." *Journal of Neuroscience* 17: 697–708.

Young, L. J., M. M. Lim, B. Gingrich, and T. R. Insel. 2001. Cellular mechanisms of social attachment. *Hormones and Behavior* 40: 133–138.

Young, L. J., and Z. X. Wang. 2004. "The neurobiology of pair bonding." *Nature Neuroscience* 7: 1048–1054.

CHAPTER TWENTY-FOUR

Social Insects as Models in Epidemiology: Establishing the Foundation for an Interdisciplinary Approach to Disease and Sociality

NINA H. FEFFERMAN

JAMES F. A. TRANIELLO

SOCIAL INSECTS have an abundance and diversity that, with the exception of beetles, is unrivaled among animal taxa. Two groups, the ants and termites, are ecologically dominant in tropical rainforests (Hölldobler and Wilson 1990; More than 12,000 species of ants and 2,650 species of termites have been identified so far (Bolton et al. 2006), and their local diversity and abundance can be extraordinary. The amazing diversity of ants is exemplified by Floren and Linsenmair's (2000) discovery of 239 species on just 19 individual trees in Borneo, with as many as 61 species found on a single tree. In the Southern Cameroon, densities of 10,000 termites per square meter have been recorded (Eggleton et al. 1996). The soil and ground litter of tropical forests is the environment of adaptation and diversification of these two groups (Hölldobler and Wilson 1990; Eggleton 2000; Moreau et al. 2006). These environments are also inhabited by what is potentially an equally or even more diverse community of microbes that can be abundant and pathogenic.

Together with the diversification of the angiosperms and the litter of tropical angiosperm forests (Moreau et al. 2006), adapting to the disease risks inherent in soil, litter, and decaying wood environments exploited so well by ants and termites very likely constituted a major event in the diversification of ants and termites. Indeed, Bulmer and Crozier (2004, 2005) described selection on immune genes associated with the transition from grass feeding to decayed wood feeding in Australian termites. Bees and wasps are also diverse and ecologically significant social insect clades whose colonies are impacted by arthropod, fungal, bacterial and viral parasites (cf. Bailey and Ball 1991; Rose, Harris, and Glare 1999). What role did sociality play in the evolution of disease resistance? Did the high density of individuals in social insect colonies render them more susceptible to infection through enhanced pathogen transmission? Or did group living offer new, cooperative mechanisms to lower disease risk? How can we begin to approach these and other questions of historical significance in social insect evolution?

We believe that concepts from human epidemiology have heuristic value for insect sociobiology and ecological immunology. Concepts derived from epidemiological theory can be tailored to questions that might at first appear to be outside the domain of studies of human disease dynamics. Extending the application of these theories beyond human disease to a novel system that might appear to hold little in common, presents a significant opportunity to deepen the understanding of the underlying principles of infectious disease spread. Human infectious disease epidemiology is based on a set of standard, disease-related statistical measures (cf. Nelson, Williams, and Graham 2001). Implicit in these measures is the underlying factor of exposure to disease. Human epidemiologists have generally assumed that the probability of exposure in a population is uniform for particular sets of susceptible individuals at a given time. The probability of exposure is integrated within the probability of becoming infected, developing the signs and symptoms of a disease, infecting others, and then either recovering or dying.

Exposure to disease is also a significant factor in evolutionary pathobiology. In group-living animals, especially those with complex social organization, probabilities governing the exposure of individuals to disease within a society can show strong variance (Andersson 1997). Behavior plays a critically important role in infection control: the removal, quarantine, or exile of infected individuals can greatly reduce the exposure of a society's population once disease is present (Clancy 1996). Additionally, behaviors such as decreasing activity after infection, avoiding infected individuals, nursing infected group members, soliciting care, and other social processes of infection control greatly impact the probability of exposure to infectious disease. All facets of social behavior directly impact the exposure and transmission risks of infectious diseases within a population, and therefore greatly influence the selective pressures on individuals that live in societies.

Just as exposure is implicit in principle epidemiological concepts, so is the concept of resistance to disease. Resistance can be defined narrowly as protection arising solely from immunity in the molecular and cellular sense, but when integrating the perspectives of epidemiology, immunology, evolutionary ecology, and insect sociobiology a more encompassing concept would offer greater applicability, allowing us to consider an array of defense mechanisms extending beyond purely physiological adaptations to social characteristics that protect against disease. Owens and Wilson (1999) proposed that immunocompetence be defined as "a measure of the ability of an organism to minimize the fitness cost of an infection *via any means* [emphasis added], after controlling for previous exposure to appropriate antigens." Thus, social mechanisms of disease resistance such as herd immunity—the insulation of susceptible individuals from exposure by surrounding them with immune individuals that are incapable of transmitting infection—would be included in this definition. Immunocompetence, in this holistic sense, can also circumscribe colony-level mechanisms of infection control (Hart, Bot, and Brown 2002) and the social enhancement of resistance to infection (Traniello, Rosengaus, and Savoie 2002). An expanded view of immunocompetence would also involve prophylaxis and prevention from initial infection, as well as survival subsequent to the contraction of a disease. In this way, social groups can be said to resist disease, and resistance can be measured in terms of their persistence despite pathogen presence and the challenges of infection. Social behavior can affect exposure risk and, thus, positively impact survival, playing a crucial role in immunocompetence overall.

The study of the significance of behavior to exposure and immunocompetence can be unified from the perspectives of epidemiology (following paths of disease spread) and evolution (describing the survival and fitness consequences of transmission) to create a deeper understanding of disease dynamics in social groups (cf. Bonds et al. 2005). Manipulations can be difficult in non-human populations and in field studies of non-human social species. Theoretical modeling, however, can supplement empirical research to investigate the impact of behavior and social organization on exposure and the induction and maintenance of immunocompetence. Progress in achieving a more complete view of how individual-, group-, and population-level disease dynamics lead to local pathogen spread, affect survival and fitness, and ultimately select for resistance mechanisms could be accelerated by identifying and developing model systems for an integrative approach. A system lending itself to empirical and theoretical research would allow hypothesis testing to occur in a manner that would ensure a realistic representation of the effect of each studied mechanism of immunocompetence on disease dynamics at the level of the society. In this chapter, we argue that social insects are an ideal model system for an integrated sociobiological and epidemiological approach.

An Accelerated Arms Race: Social Insects and the Selective Pressures of Infectious Pathogens

Disease can seriously impact survival and reproductive success and, as a result, pathogens are potent agents of selection. Pathbreaking studies of the coevolution and epidemiology of animal and plant parasites demonstrate that parasites have exerted significant effects on the evolution of their hosts (Anderson and May, 1982; Ewald 1991, 1994; Andersson 1994; Briggs and Godfray 1995; Rothman and Myers 1996). This is especially true of infectious diseases in group-living species in which interactions among individuals within a dense social structure can lead to increased exposure and transmission relative to solitary animals (Grenfell and Dobson 1995). Recent research has focused on host defenses and life-history traits and the potential fitness costs associated with infection resistance (Schmid-Hempel 1998; Moret and Schmid-Hempel 2000; Hasselquist, Wasson and Winkler 2001; Norris and Evans 2000; Calleri, 2006) using solitary, gregarious, and eusocial insects as models in ecological immunology (Cotter et al. 2004; Pie et al. 2005; Schmid-Hempel 2005).

The focus on immunological and ecological factors in the cyclic nature of epidemics to examine disease dynamics can be adapted to sociobiological research problems by considering the scale of operation of these factors. From this perspective, immunological factors are attributes of the host/parasite association that manifest in individuals, matri/patrilines, demographic subgroups, colonies, and populations of colonies, potentially leading to patterns of recurrent epidemics. All inducible colony defenses play important roles in the prevention of establishment and transmission of disease and the induction of resistance. Ecological factors, such as foraging, or competitive or predatory actions that ultimately could lead to the introduction of infected individuals from neighboring colonies of the same or different social insect species, are paramount to understanding exposure, although very little is known about them.

In the same way that disease resistance and its underlying mechanisms (such as physiological immunity) can be thought of as having been shaped over evolutionary time by a diverse array of disease challenges, social behavior and colony structure have also formed under a set of selective pressures from pathogens. Considering insect immune defenses as "evolved traits" (Schmid-Hempel 2005), and focusing on the interface of insect immunology and sociobiology, we can identify the selective forces that have favored the variety of resistance mechanisms that characterize different species (Schmid-Hempel and Ebert 2003; Pie et al. 2005; Fefferman et al. 2007). By considering the impact of disease on the evolution of social structure and behavior, colony-level mechanisms of immunocompetence can be considered as adaptive responses to the selective pressure of infectious disease.

We predict that the individual and social disease-resistance mechanisms that comprise colony-level immunocompetence will reflect variation in microbial loads associated with nesting and feeding ecology (Cruse 1998; Rosengaus et al. 2003; Bulmer and Crozier 2004, 2005). In eusocial insects, disease transmission risks that are inherent in their nesting ecologies are likely to be high due to the level of relatedness of workers, leading to increased susceptibility to infectious agents due to the lower genetic heterozygosity of nestmates (e.g., Shykoff and Schmid-Hempel 1991). At the same time, the contributions of the individual through selection for adaptive patterns of division of labor, especially when considered in terms of group size and structure, are critical to the survival and reproductive success of the colony. This leads to a trade-off in selection for genetic variation to enhance colony immunocompetence or colony efficiency. The relationships among these variables, however, are far from clear (e.g., Rosset, Keller, and Chapuisat 2005).

Social Insect Ecology, Behavior, and Disease Resistance

The nesting habits of social insects highly recommend them for epidemiological study: they form multigenerational families in environments that are particularly challenging in terms of acute and chronic exposure to pathogens and parasites. Ants and termites—two diverse and abundant groups—nest and feed in soil and decayed-wood environments in which a diverse and abundant microbial community flourishes (Rosengaus et al. 2003). Social insects often live in densely populated colonies and mortality risks may be compounded through inter-individual transmission of infection (Rosengaus et al. 1998; Schmid-Hempel 1998 and references therein).

Adapting to disease has long been considered to have been a major event in the evolution of sociality and diversification of social insects (e.g., Wilson 1971), but only recently has the evolutionary significance of social insect pathobiology been the focus of empirical and theoretical investigation. If mechanisms of disease resistance compromise the energetics of colony operations, we expect they will be selected against, and that mechanisms that afford protective benefits at low cost will be highly unlikely due to the evolutionary responses of the pathogens in question (Figure 24.1). A benefit/cost analysis involving humans is instructive here: to control infectious disease, simple hygiene such as hand washing, a low-cost behavior, can provide significant protective benefits against infection (Daniels and Rees 1999). Analogously, through self-grooming, ants can physically remove microbes and spread the antibiotic secretions of the metapleural gland, but they have been shown to be metabolically costly to produce (Poulsen et al. 2002). Humans have antibiotic dermal peptides (e.g., Harder et al. 1997); their production costs are unknown. Inducible physiological responses such as fever help combat infection, but can also have moderately high costs (e.g., energy loss or febrile seizures; Nesse and Williams 1994). Evolution has also favored the maintenance of a trait with enormous protective benefits against malaria in heterozygotes, but at the high cost of sickle-cell anemia (a disease with a high associated mortality of

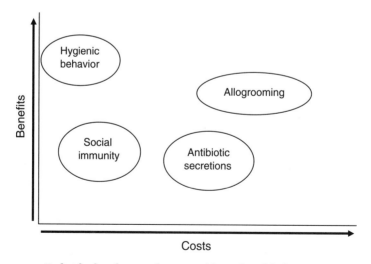

Figure 24.1. Individual and societal costs and benefits of defensive mechanisms.

its own, Platt et al. 1994) in homozygotic individuals. Allogrooming can both decrease the likelihood of illness after exposure for the individual contacted, but can also spread infection to otherwise unexposed individuals (Rosengaus and Traniello 1997). Due to heightened levels of interaction in the course of colony function, hygienic behavior can provide significant protective benefits at what would appear to be a relatively low cost. This example provides rich opportunities to study exceptions to theoretical expectations.

In group-living animals, mating system, group size, population density, and parasite transmission are significant to infection risk and control (Freeland 1976, 1979; Hamilton 1987; Reeson et al. 1998; Nunn, Gittleman, and Antonovics 2000; Read and Allen 2000; Cotter et al. 2004). Social insects face similar challenges (e.g., Hamilton, Axelrod, and Tanese 1990; Keller 1995; O'Donnell 1997; Thorne and Traniello 2003; O'Donnell and Beshers 2004). There is ample support for the hypothesis that the presence of pathogens has selected for a variety of disease-protection mechanisms in phylogenetically different groups of social insects. In the haplodiploid social Hymenoptera, there is evidence that behavior, physiology, and biochemistry (as well as colony size and organization, colony and population genetics, and mating system) have been influenced by pathogens (Hölldobler and Wilson 1990; Baer and Schmid-Hempel 1999; Rosengaus et al. 1998; Schmid-Hempel 1998, 2005; Schmid-Hempel and Crozier 1999; Starks, Blackie, and Seeley 2000; Tarpy and Seeley 2006). The biology of the diploid termites (Order Isoptera) reflects individual physiological, life history, and social adaptations, again indicating a prominent role for pathogens (Rosengaus et al. 1998; Rosengaus, Traniello, et al. 1999; Rosengaus, Jordan, et al. 1999; Rosengaus and Traniello 2001; Thorne and Traniello 2003; Traniello et al. 2002; Calleri et al. 2005; Calleri, Rosengaus, and Traniello 2006).

The risks of disease and its subsequent impact on colony fitness are affected by social behavior. Crowding can lead to increased transmission, but if behavior can improve the recovery of infected nestmates or if individual defenses can be summed over a colony's population to lower susceptibility, then overall mortality can be decreased even if transmission rates are increased. Moreover, the removal, quarantine, or isolation of diseased individuals may counteract the increased exposure risk incurred by group living. Other forms of social interaction, such as allogrooming, which can prove more efficacious than self-grooming (Rosengaus et al. 1998), can be highly effective in the prevention of disease. Allogrooming is an inducible defense that can be increased through signaling in the dampwood termites *Zootermopsis angusticollis*. Nymphs communicate information about exposure risk (Rosengaus, Jordan, et al. 1999); when individuals encounter lethal concentrations of fungal conidia, they remain in place and vibrate, transmitting substrate-borne signals that cause nearby nestmates to abscond and then increase allogrooming. The signal activates behaviors that are equivalent to the induction of the molecular cascades that govern individual immune response. These signals serve as public service announcements—sending cues to nestmates to up-regulate hygienic behavior and preemptively defend against disease as well as avoid infection.

Social Insect Immunocompetence

Insects, in general, have nonspecific and short-lived cellular immune responses such as encapsulation, nodule formation, and phagocytosis, which characterize innate immunity (Hoffmann et al. 1999; Russell and Dunn 1996). For example, some dictyopteran ancestors of termites defend against disease through acquired immunity (Faulhauber and Karp 1992). Humoral immune responses are characterized by specificity and memory (e.g., Hultmark 1994; Hoffman et al. 1999). Specificity in insect immunity has been reported: cercopins, diptericin, drosocin, mucin, and attacin have antibacterial activity; drosomycin appears to be fungicidal (Lemaitre et al. 1996; Lemaitre, Reichhart, and Hoffman 1997). Although Hamilton (1987) once noted that in insect societies "nothing like the immune system . . . is known to exist," recent research has found social insects to have adaptive immunity and produce antimicrobial peptides (Rosengaus, Traniello, et al. 1999, 2007; Lamberty et al. 2001; Bulmer and Crozier 2004; Evans 2004). Social insect disease defenses are likely to have evolved in part from traits present in ancestral solitary and/or presocial species. Because social insects live in groups in ecological circumstances characterized by variation in disease risk (Rosengaus et al. 2003), we expect that adaptive immune system variation will match the qualitative and quantitative nature of pathogen load. That said, acquired immunity is poorly understood in ants, bees, wasps, termites, and other gregarious, presocial, colonial, and eusocial insects.

Immune defenses can be considered on both a colony-level and individual scale (Rosengaus et al. 1999b; Traniello, Rosengaus, and Savoie 2002).

Individual immunity can confer protection against pathogen infection, either by innate or adaptive defense, and the transfer of immunity to nestmates can also confer protection to a naïve individual (Traniello, Rosengaus, and Savoie 2002). Allogrooming reduces the pathogen load of an individual and thus overall exposure risk. Conversely, high levels of immunity in a colony may buffer susceptible individuals from exposure to infection. Studies of herd immunity suggest that substantial protective benefits accrue only after a large percentage of the total population becomes immune (Castillo-Chavez et al. 2002; but see Fletcher 2003).

In addition to providing protection from disease, immunity has associated costs at the individual and colony level. For the individual, mounting an immune response entails metabolic/energetic costs which have been estimated in a variety of systems (cf. Nesse and Williams 1994). There can be additional costs to the individual: physiological responses can have adverse effects. For example, while fever is beneficial, extreme elevation of body temperature can be dangerous (Kluger 1979). Additionally, the actual protective function of an immune system can itself result in autoimmune disorders, which exact severe costs to the health and fitness of an individual (Nesse and Williams 1994). Defense mechanisms at the colony level have costs too. For example, hygienic behavior can result in the culling of particularly susceptible individuals once disease is present in the colony (Rosengaus and Traniello 2001). While this may prove effective in limiting disease spread in the short term (Fefferman et al. 2007), it can negatively impact long-term colony growth.

Social Organization and Infection Control

Colony organization may enhance disease resistance, and colony demography—a chief feature of the social architecture of an insect colony—can impact disease susceptibility. When pseudomutant colonies of Z. angusticollis were exposed to conidia of the entomopathogenic fungus *Metarhizium anisopliae*, groups with a mixed-instar demographic distribution had significantly higher survivorship than colonies composed only of individuals of a single instar. Same-age groups (10 individuals of the same instar) had more than double the hazard ratio of death than mixed-age groups. Relatively young individuals were the most susceptible to infection (Rosengaus and Traniello 2001). These studies indicate that age-structure can serve as a mechanism of defense against disease and age- and caste-distribution structures could in part have evolved under the selective pressures of parasitism (Schmid-Hempel and Schmid-Hempel 1993). Indeed, in social insects with worker task specialization, there can be colony-wide up-regulation of hygienic behaviors, which offer greater protection to the colony as a whole (e.g., Hart, Bot, and Brown 2002). Middens workers are more likely to come into contact with pathogenic fungi, which are thus restricted from freely moving to other areas of the nest.

The life histories of basal termites are characterized by plasticity in the reproductive options of individuals and variation in colony social organization that can impact disease resistance (Thorne and Traniello 2003). In these species, cycles of outbreeding and inbreeding associated with the reproductive plasticity can alter genetic heterozygosity and affect innate immunity as well as social mechanisms of infection control. For example, low levels of genetic variation in the offspring of inbred supplementary reproductives could result in greater susceptibility to infection. In *Zootermopsis angusticollis*, a single generation of inbreeding significantly reduced heterozygosity and allelic diversity (Calleri et al. 2006). This lower genetic variation in turn affected the resistance of isolated and grouped individuals depending on the mode of transmission of the pathogen used in experimental exposure. Inbred and outbred, isolated and grouped termites inoculated internally with the bacterium *Pseudomonas aeruginosa*, exposed to a low dose of the fungal pathogen *Metarhizium anisopliae* on the cuticle, or challenged with a pseudopathogen (an implanted nylon monofilament) did not differ in immunocompetence. This is because internal infection by bacteria apparently cannot be controlled by social interactions such as mutual grooming and low concentrations of conidia do not impact survival. Inbred termites housed in groups and exposed to a relatively high concentration of fungal conidia, however, had significantly greater mortality than outbred grouped termites, presumably because of a genetic effect on allogrooming or another group mode of disease control. Inbred termites also had significantly higher cuticular microbial loads, likely due to less effective grooming by nestmates. Decreased genetic heterozygosity associated with inbreeding thus appeared to increase disease susceptibility by influencing the efficacy of social behavior rather than physiological immunity.

Finally, living in groups can augment individual immunity in social insects. The ability of *Z. angusticollis* nymphs to resist a fungal pathogen is enhanced in naïve nymphs (i.e., termites having no prior exposure to a challenge pathogen) allowed to associate with immunized nestmates (Traniello,

Rosengaus and Savoie 2002). Increased immunocompetence at the colony level can thus be improved through the social interactions of nestmates that vary in their immune status and ability to resist infection. Although social immunization in termites would appear to be most closely related to vaccination in humans, social immunization involves the up-regulation of immune response in naïve individuals by an immune nestmate(s). Social immunization may, therefore, most accurately be approximated in the realm of human epidemiology by the vertical transmission of immunity from mother to infant through antibodies passed in the colostrum (the nutrient and immune-factor rich secretions of human mammary glands within the first few days after childbirth and prior to the onset of milk production). In termites this transmission is not limited to parent-offspring interaction. This difference in the mechanistic basis by which groups can achieve greater collective immunity shows an instance where, by limiting the definition of an epidemiological concept to human physiological capability, the underlying theory itself is narrowed. By incorporating a broader (insect-based) concept of individual-to-individual transmission, both vertical and horizontal, of immunity, it may be that human epidemiologists may arrive at more efficient algorithms for socially based inoculation strategies (e.g., a "new and improved chicken pox party").

Modeling Disease Resistance in Social Insects

Attaining a balance between accuracy and simplicity is required for successful modeling in biology (Levin 1992). In identifying useful model systems, researchers attempt to identify commonalities between the system being studied and a more accessible system which serves as a model. The more attributes the two systems hold in common, the greater the applicability and credibility of the model, and the greater its heuristic value. The nature and complexities of some pairs of systems—social insects and humans and how they cope with disease—can initially appear highly divergent. We anticipate that some researchers will balk at the idea of bridging human epidemiology and insect sociobiology, using mathematical modeling as the foundation of the interdisciplinary linkage, but we believe such an approach is timely, appropriate, and fruitful. Social insects can be viewed as an epidemiological system that is simple relative to humans and many group-living vertebrates, and by creating mathematical models based on insect sociobiology and human epidemiology, it will be possible to contribute new insights into traditional problems in both disciplines.

Models can focus at the level of individual action or emergent system-wide properties, so the analysis of social insect behavior and colony epidemiology through computational modeling is a natural extension of more traditional empirical experimentation. The wealth of research in human epidemiology has identified parameters of disease prevention and transmission that are fundamental to the analysis of mechanisms of disease control in social insects, but many of these factors have been neglected or only infrequently examined or explicitly considered in theoretical or empirical studies (Table 24.1).

Even among factors considered in the context of disease, very few have been examined in combination to determine plausible synergistic effects within the scope of human epidemiology and none have, as yet, been examined in social insects. Indeed, as Table 24.1 illustrates, important aspects of the social phenotype of insect colonies that may have been studied in other contexts have yet to be examined in relation to infection. For example, polyethism in social insects is well studied; it has long been known that workers perform tasks (e.g., foraging, nest maintenance, and brood care) in an organized manner to allow the colony to efficiently function. The probability of performing a particular task can be determined by the individual's developmental or physiological state, or can be dependent on the requirements of the colony and/or social interactions (Robinson 1992; Seid and Traniello 2006). The performance of each task may have an associated risk to the individual and cost to the colony (e.g., being eaten by a predator while foraging outside the nest). Task performance can also carry associated disease-related costs and benefits that are rarely studied. Foraging might increase exposure to pathogens, for example, and subsequent contact with an infected nestmate can reduce individual pathogen loads (through allogrooming) and/or spread infection. Processes of task allocation can be described with a set of minimum threshold values, probability distributions, and net contributions to colony function, accurately comprising a detailed mathematical representation of the organization of colony labor. The importance of these factors for withstanding a pathogen challenge can then be examined.

Recently, a few models have been developed using some parameters generally considered within an epidemiological framework to examine disease spread in social insect colonies, specifically showing how nest

Table 24.1. Parameters considered in models of infectious disease prevention and transmission as developed in human epidemiology and insect sociobiology.

Effect	Epidemiology	Insect sociobiology
Single factor effects		
Antibiotics/prophylaxis	Yes°	Yes°
Distribution of antibiotics	Yes	No
Heritable variation in resistance	Yes	No
Vaccination	Yes°	No
Quarantine	Yes	Yes°
Hygiene	Yes	Yes°
Exposure	Yes	Yes°
Transmission	Yes°	No
Infectivity	Yes°	Yes°
Attack rate	Yes°	No
Pathogenicity	Yes°	No
Virulence	Yes°	No
Immunity	Yes°	Yes°
Immunogenicity	Yes°	No
Herd effect and social immunity	Yes	Yes°
Population density	Yes°	Yes°
Age structured susceptibility	Yes°	Yes°
Division of labor	No	Yes
Activity level	Yes	Yes
Environment structure	Yes	Yes°
Additive multifactor effects		
Antibiotics and quarantine	Yes	No
Antibiotics and immunity	Yes	No
Antibiotics and vaccination and virulence	Yes	No

Sources: Based on Antia and Lipsitch 1997; Brisson et al. 2000; Brooker et al. 2002; Brookmeyer et al. 2003; Davison and Nair 2005; Eames and Keeling 2002; Fefferman et al. 2007; Hupert et al. 2002; Kaplan 1990; Longini et al. 2004; Meltzer et al. 2001; Michaels et al. 2004; Naug and Camazine 2002; Pie et al. 2004, 2005; Reilly et al. 2004; Starr and Campbell 2001; Temime et al. 2004; and Wendelboe et al. 2005

Note: Although vaccination, antibiotic protection and prophylaxis are distinct concepts in the context of social insects, in human epidemiology there may be some scenarios in which they operate as the same factor.

°Indicates that the effect of these metrics on disease spread has been measured empirically.

architecture, behavior, social immunity, and nest hygienic behavior can affect disease transmission (Naug and Camazine 2002; Pie, Rosengaus, and Traniello 2004; Fefferman et al. 2007). Many aspects of social organization pertinent to disease dynamics within the population of a colony can be empirically determined, such as demography, task allocation, and nest structure. In addition, some of the standard epidemiological metrics employed in human disease studies (e.g., infectivity) have been measured (Table 24.1). The mechanisms that likely govern exposure are more complex, but in the case of social insects, where mechanisms of exposure can be clearly defined and observed, it is possible to measure epidemiological factors to validate models and thus increase our understanding of exposure and the influences of individual behavior and social structure on infection control.

To illustrate the ways in which these standard epidemiological rates rely on exposure and the types of predictive models that can be developed, consider the following mathematical expressions for etiologic- and exposure-based parameters:

$E_{X, T}$ = The probability of exposure, for each worker, in a group of individuals of caste X at time T.

I = The probability of becoming infected from exposure (subscript denotes time since previous infection, with some measure of immunity conferred that influences a current probability of infection).

S_T = The probability of becoming symptomatic at time T, given infection at time $T=0$.

C_T = The probability of becoming contagious at time T, given infection at time $T=0$.

M_T = The probability of death at time T, given the onset of symptoms at time $T=0$.

While it is still possible to consider each of these variables as population-wide transition averages (as is done in traditional susceptible-infected-recovered epidemiological models, also called "mean field approximation" models), this method brings the successive conditional probabilities into the definitions. Standard epidemiological rates can then be defined

mathematically: pathogenicity as $\sum_{T=0}^{n} S_T$, virulence as $\sum_{T=0}^{n} M_T$, and infectivity as $I \cdot C_T$, where n = the duration of disease until either death or recovery and infection beginning on day 0. Similarly, attack rate may be defined as $I^* \sum_{T=0}^{n} S_T$, where n = the end of the window for disease expression. Expressed mathematically, these standard concepts in epidemiology are conditional probabilities, building on each other, and each fundamentally relying on the underlying probability of exposure (see Figure 24.2 for a visualization of how these concepts follow directly from exposure risks).

In social insect research, $E_{X,T}$ can be experimentally manipulated and S_T, C_T, and M_T can all be measured. None of this is possible in human

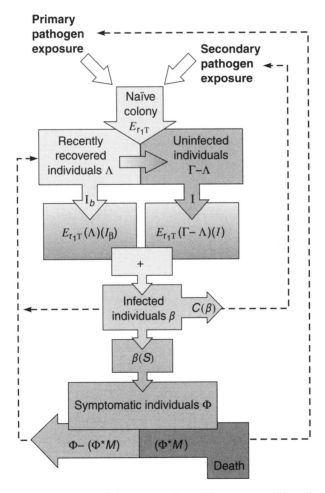

Figure 24.2. A visualization of the progression of disease spread based on underlying exposure.

studies. Pathogenicity, virulence, infectivity, and attack rate can all then be computed, thereby providing known parameter values for use in previously purely theoretical modeling. This allows the testing of predictive models of outbreak scenarios (including the traditional SIR (S = susceptible; I = infections; R = recovered) models)—something which has not been possible in epidemiological modeling on such a scale. Furthermore, the ability to validate models in this way opens the possibility for the theoretical investigation of the efficacy of specific interventions or defenses to combat disease spread by influencing the underlying exposure rates.

By dissecting the system of disease dynamics in this way, we can then examine the roles of individual aspects of social phenotype. By manipulating hypothesized mechanisms of exposure and calculating their effect on dependent metrics, we are able to tease apart the effects of individual behavior and colony structure on disease spread. The demographic distribution of a population and the associated age-specific etiologies of a given pathogen can be recorded. Modes of infection can be isolated and rates of transmission among nestmates and colonies can be determined. The induction and subsequent efficacy of immunity can be tested. All of the standard epidemiological parameters can be investigated and models based on these parameters can provide insight into how each influences patterns of disease in a population. However, social insect colonies provide an opportunity to investigate the effect on disease dynamics of other factors often assumed pertinent, but rarely amenable to experimental manipulation in humans, such as behavioral responses to pathogen presence (e.g., allogrooming, alarm signaling, contact rates between susceptible and infected individuals). Computational models are being developed to examine the effects of such influences in cases where direct empirical manipulation, even in social insects, is difficult (Fefferman et al. 2007).

Infectious disease epidemiology in social insect colonies has begun to examine the effects of social behavior and colony organization through modeling. Naug and Camazine (2002) created a cellular automata model to examine the effects of division of labor and interaction networks among nestmates and colony demography, focusing on transmission rates among colony subgroups as defined by colony size, and using an underlying spatially explicit nest structure. We altered the density of infected workers at the outset of each model run within castes and varied the inter-caste probabilities of transmission to represent differences in their rates of interaction. We then examined the trajectory of disease in a closed population, using the number of infected individuals at any given time as their metric of comparison. Our results indicate that although the presence of distinct subgroups within the nest was by itself ineffective as a protective measure, when combined with the altered caste-specific replacement rates the two properties were successful at reducing the number of infected individuals. This implies that division of labor within a colony can act in concert with demographic change to provide some measure of protection against disease risks.

Pie, Rosengaus, and Traniello (2004) dealt more explicitly with exposure by constructing both a mean field approximation and cellular automata models of disease in a single termite colony. Here, the effects of disease on colony-wide survivorship were compared, rather than the number of individuals infected at any given time. In the mean field approximation model, contact was based on the density of infected individuals, which was found to have a significant effect on the likelihood of epidemics within the colony. In the cellular automata model, transmission occurred between infected and susceptible individuals if they co-occupied the same cell according to a uniform probability. Nest architecture was found to have increasing protective benefit against colony-wide spread of disease as the nest became increasingly spatially segregated, and thus subdivided. By focusing the cellular automata models on the restrictions of movement in a spatially explicit environment and holding the probabilities of transmission constant subsequent to exposure, these models allowed the examination of the effects of nest architecture and worker activity level on exposure risk.

Fefferman et al. (2007) developed a series of cellular automata models to analyze the relative efficacy of allogrooming, immunity, and nest hygiene in combating pathogen challenges hypothesized to be associated with the transition from solitary to presocial to eusocial life. Nest hygiene was found to provide an immediate survival benefit, and immunity lowered overall disease susceptibility under constant and periodic disease exposure scenarios. Allogrooming increased survivorship in chronically pathogen-challenged colonies, but also increased transmission rates when exposure was recurrent. Colonies having demographies biased toward either young or old individuals had slightly higher mortality than those with evenly distributed demographies. These models indicated that nest hygiene and immunity function on different temporal scales and can interact with demography to lower disease risks and with the age distribution of individuals to lower, to varying extents, pathogen-related mortality risks in social insects. Indeed, nest hygiene had an immediate and significant impact on improved survivorship. Allogrooming benefited overall survival in the short term, but the advantage was outweighed over time by the loss of protection conferred via socially transmitted immunity. The equilibrium point of the trade-off between the costs and benefits of allogrooming is therefore likely influenced by natural patterns of exposure. The results suggest that infection control systems in social species were built upon the inducible immune defenses and nest hygienic behaviors of solitary and presocial ancestors and served as important preadaptations to manage disease transmission in colonies of incipiently eusocial species.

While social insects are an appealing model system for the study of disease and sociality in all group-living animals due to generalities of the social phenotype, there are a number of equally appealing differences that make them more accessible to study than many other social species. Human behavior is too complex and diverse to model effectively; examining how individual and group behavior can provide infection-control benefits is difficult (if not unethical) through empirical study. The inability to manipulate contributing factors, the lack of replication, and the lack of control/understanding of possible confounding factors (e.g., genetics or environmental conditions) create the need for a simpler model system. Therefore, we have demonstrated that social insects can provide a simplified study system where social structure, behavioral decisions, and their epidemiological impact can be examined empirically on both an individual and colony-wide level, and modeled accordingly.

The use of a simplified biological system that retains crucial but otherwise impenetrable aspects of group living is a powerful tool, allowing the external validation of theoretical models of disease spread that incorporate the substantial effects of behavior and social structure in ways that have not previously been possible. As a model system, social insects offer significant and diverse opportunities for analysis within colonies, between colonies, and/or across populations of colonies. Contact rates among foragers of different colonies can be easily made to represent different populations that vary in rates of emigration and immigration. Additionally, the focus can be on an individual rather than a population.

Socioecoimmunology can help us to understand the selective pressures and constraints that accompanied adaptations to pathogen exposure as insects evolved socially and diversified ecologically, and to identify the infection-reducing benefits associated with colonial life. As we have argued, the ability of social insects to control infection results from the behavioral and physiological attributes of individuals as well as the social groups in which they reside. The nature of social interactions among nestmates, as well as the mechanisms of disease transmission and pathogen defense, may directly parallel those of other group-living animals. Social insects exhibit both innate and adaptive immune responses to pathogen challenge and some species have social mechanisms of generating immunity (Traniello, Rosengaus, and Savoie 2002) that can be thought of as similar to human vaccination. They face both primary and secondary exposure to disease and are vulnerable to a variety of micro- and macro-parasites via many different modes of transmission. Social insect behavior can be observed with relative ease, and colony subpopulations that perform specific tasks with different associated risks of exposure to pathogens can be monitored. Workers are capable of modifying their behaviors in response to the threat of pathogen exposure, including quarantining possibly infected individuals and sending alarm signals that cause nestmates to avoid contaminated areas of the nest. They can generate effective immune responses (e.g., Mackintosh et al. 1998; Rosengaus Traniello, et al. 1999) and regulate pathogen spread with hygienic behavior (Spivak and Gilliam 1998a, b). These behaviors parallel human behavioral interactions (e.g., parents being more frequently exposed to childhood diseases than adults without children, healthcare workers being exposed to a greater diversity of pathogens, etc.) and public health efforts (e.g., Food & Drug Administration recalls of contaminated food, or flags to indicate pollution along public beaches), albeit as the result of local rather than group decision making (e.g., Rosengaus, Jordan, et al. 1999). Finally, as in many group-living animals, social insects exhibit differential susceptibility to some diseases based on the developmental stage of colony members (Rosengaus and Traniello 2001; Evans and Lopez 2004).

Traditionally, modeling social insect biology has been somewhat limited in scope, but the properties of colonies make them ideal biological systems for the study of a wide array of aspects of infectious disease epidemiology. The fact that accurate representation of this biological system in computational models may be verified by empirical investigation presents a unique research opportunity. In no other system is it possible to manipulate as many of the factors directly governing exposure, infection, and disease defense as is possible with social insects. By using empirical studies to test the predictions of models, we can have greater confidence in the accuracy of results of more complicated, purely theoretical investigations than has previously been possible for epidemiologists to achieve. Investigations in insect socioecoimmunology can advance our understanding of the role of sociality in disease risk and control in both the study of social insects and the field of infectious disease epidemiology in ways that each discipline might not be able to accomplish separately.

Over the last few years, models have begun to set the groundwork for deeper investigation into the epidemiology and socioecoimmunology of insect colonies and provide methods to examine how individual behaviors can have potentially synergistic contributions to colony health and survival. They allow the examination of how colony structure and organization can directly affect exposure, as well as illuminating how disease resistance traits of the solitary ancestors of eusocial species were adaptively expressed within the context of group life (Fefferman et al. 2007).

An entirely new scope of questions about the nature of disease dynamics governed by the social phenotype and influencing its evolution can now begin to be examined: To what extent is disease regulation built on emergent properties of a colony? At what point will compromising colony efficiency lower disease risk and maximize fitness? How does selection for disease resistance at multiple levels affect the ergonomic organization of a colony? How do immunological systems change as a consequence of social evolution and how do they interact with social behavior to control disease? Each of these questions (among many others) are of direct interest, not only as significant and timely questions in sociobiology, but also as a first step toward addressing analogous questions about the structure of human society and disease.

Literature Cited

Anderson, R. M., and R. M. May. 1982. "Coevolution of hosts and parasites." *Parasitology* 85: 411–426.

Andersson, H. 1997. "Epidemics in a population with social structures." *Mathmatical Biosciences* 140: 79–84.

Andersson, M. 1994. *Sexual selection.* Princeton: Princeton University Press.

Antia, R., and M. Lipsitch. 1997. "Mathematical models of parasite responses to host immune defences." *Parasitology* 115: 155–167.

Baer, B., and P. Schmid-Hempel. 1999. "Experimental variation in polyandry affects parasite loads and fitness in a bumblebee." *Nature* 397: 151–154.

Bailey, L., and B. V. Ball. 1991. *Honey bee pathology.* Kent: Harcourt Brace Jovanovich.

Bolton B., G. Alpert, P. S. Ward, and P. Naskrecki. 2006. *Bolton's Catalogue of Ants of the World.* Harvard University Press: Cambridge, Mass.

Bonds, M. H., Keenan, D. C., Leidner, A. J., and P. Rohani. 2005. "Higher disease prevalence can induce greater sociality: A game theoretic coevolutionary model." *Evolution* 59: 1859–1866.

Briggs, C. J., and H. C. J. Godfray. 1995. "The dynamics of insect-pathogen interactions in stage-structured populations." *American Naturalist* 145: 855–887.

Brisson, M., W. J. Edmunds, N. J. Gay, B. Law, and G. De Serres. 2000. "Modelling the impact of immunization on the epidemiology of *Varicella zoster* virus." *Epidemiology and Infection* 125: 651–669.

Brooker, S., S. I. Hay, and D. A. P. Bundy. 2002. "Tools from ecology: useful for evaluating infection risk models?" *Trends in Parasitology* 18(2): 70–74.

Brookmeyer, R., E. Johnson, and R. Bollinger. 2003. "Modeling the optimum duration of antibiotic prophylaxis in an anthrax outbreak." *Proceedings of the National Academy of Science USA* 100(17): 10129–10132.

Bulmer, M. S., and R. H. Crozier. 2004. "Duplication and diversifying selection among termite antifungal peptides." *Molecular Biology and Evolution* 21: 2256–2264.

Bulmer, M. S., and R. H. Crozier. 2006. "Variation in positive selection in termite GNBPs and Relish." *Molecular Biology and Evolution* 23: 317–326.

Calleri, D. V., E. M. Reid, R. B. Rosengaus, E. L. Vargo, and J. F. A. Traniello. 2006. "Inbreeding and disease resistance in a social insect: Effects of heterozygosity on immunocompetence in the termite *Zootermopsis angusticollis.*" *Proceedings of the Royal Society of London B* 273: 2633–2640.

Calleri, D. V., R. B. Rosengaus, and J. F. A. Traniello. 2005. "Disease and colony foundation in the dampwood termite *Zootermopsis angusticollis:* The survival advantage of nestmate pairs." *Naturwissenschaften* 92: 300–304.

Calleri, D. V., R. B. Rosengaus, and J. F. A. Traniello. 2006. "Trade-offs between immunity and reproduction in the dampwood termite *Zootermopsis angusticollis.*" *Insectes Sociaux* 53: 204–211.

Castillo-Chavez, C., S. Blower, P. van den Driessche, D. Kirschner, and A. Yakubu. 2002. *Mathematical approaches for emerging and reemerging infectious diseases: An introduction.* New York: Springer-Verlag.

Clancy, D. 1996. "Carrier-borne epidemic models incorporating population mobility." *Mathmatical Biosciences* 132: 185–204.

Cotter, S. C., R. S. Hails, J. S. Cory, and K. Wilson. 2004. "Density-dependent prophylaxis and condition-dependent immune function in Lepidopteran larvae: A multivariate approach." *Journal of Animal Ecology* 73: 283–293.

Cruse, A. 1998. "Termite defences against microbial pathogens." Ph.D. diss., Macquarie University, Australia.

Daniels, I. R., and B. I. Rees. 1999. "Handwashing: Simple, but effective." *Annals of the Royal College of Surgeons of England* 81: 117–118.

Davison, F., and V. Nair. 2005. "Use of Mareks disease vaccines: could they be driving the virus to increasing virulence?" *Expert Review of Vaccines* 4(1): 77–88.

Eames, K. T. D., and M. J. Keeling. 2002. "Modeling dynamic and network heterogeneities in the spread of sexually transmitted diseases." *Proceedings of the National Academy of Science USA* 99(20): 13330–13335.

Eggleton, P., D. E. Bignell, W. A. Sands, N. A. Mawdsley, J. H. Lawton, T. G. Wood, and N. C. Bignell. 1996. "The diversity, abundance and biomass of termites under differing levels of disturbance in the Mbalmayo Forest Reserve, southern Cameroon." *Philosophical Transactions of the Royal Society London B* 351: 51–68.

Eggleton, P. 2000. "Global diversity patterns." In T. Abe, D. E. Bignell, and M. Higashi, eds., *Termites: Evolution, Sociality, Symbiosis, Ecology*, 25–51, Dordrecht: Kluwer Academic Publishers.

Evans, J. D. 2004. "Transcriptional immune responses by honey bee larvae during invasion by the bacterial pathogen, *Paenibacillus larvae*." *Journal of Invertebrate Pathology* 85: 105–11.

Evans, J. D., and D. L. Lopez. 2004. "Bacterial probiotics induce an immune response in the honey bee *(Hymenoptera: Apidae)*." *Journal of Economic Entomology* 97: 752–756.

Ewald, P. W. 1991. "Waterborne transmission and the evolution of virulence among gastrointestinal bacteria." *Epidemiology and Infection* 106: 83–119.

———. 1994. *Evolution of infectious disease.* Oxford: Oxford University Press.

Faulhauber L. M., and R. D. Karp. 1992. "A diphasic immune response against bacteria in the American cockroach." *Immunology* 75: 378–381.

Fefferman, N. H., J. F. A. Traniello, R. B. Rosengaus, and D. V. Calleri. 2007. "Evolution of disease prevention and resistance in social insects: Modeling the survival consequences of immunity, hygienic behavior and colony organization." *Behavioral Ecology and Sociobiology* 61: 565–577.

Fletcher, S. 2003. "Creating immunity." *Canadian Medical Association Journal* 169: 282–283.

Floren, A., and K. E. Linsenmair. 2000. "Do ant mosaics exist in pristine lowland rain forests?" *Oecologia* 123: 129–137.

Freeland, W. J. 1976. "Pathogens and the evolution of primate sociality." *Biotropica* 8: 12–24

———. 1979. "Primate social groups as biological islands." *Ecology* 60: 719–728.

Grenfell, B. T. and A. P. Dobson, eds. 1995. *Ecology of Infectious Diseases in Natural Populations.* Cambridge: Cambridge University Press.

Hamilton, W. D. 1987. "Kinship, recognition, disease, and intelligence:

Constraints of social evolution." In Y. Ito, J. L. Brown, and J. Kikkawa, eds., *Animal societies: Theory and facts.* Tokyo: Japanese Scientific Society.

Hamilton, W. D., R. Axelrod, and R. Tanese. 1990. "Sexual reproduction as an adaptation to resist parasites (a review)." *Proceedings of the National Academy of Science USA* 87: 3566–3573.

Harder, J., J. Bartels, E. Christophers, and J. M. Schroder. 1997. "A peptide antibiotic from human skin." *Nature* 388: 416.

Hart, A. G., A. Bot, and M. J. Brown. 2002. "A colony-level response to disease control in a leaf-cutting ant." *Naturwissenschaften* 89: 257–277.

Hasselquist, D., M. F. Wasson, and D. W. Winkler. 2001. "Humoral immuno-competence correlates with date of egglaying and reflects work load in female tree swallows." *Behavioral Ecology* 12: 93–97.

Hoffmann, J. A., F. C. Kafatos, C. A. Janeway, Jr, and R. A. B. Ezekowitz. 1999. "Phylogenetic perspectives in innate immunity." *Science* 284: 1313–1318.

Hölldobler, B., and E. O. Wilson. 1990. *The ants.* Cambridge: Harvard University Press.

Hultmark, D. 1994. "Ancient relationships." *Nature* 367: 116–117.

Hupert, N., A. I. Mushlin, and M. A. Callahan. 2002. "Modeling the public health response to bioterrorism: using discrete event simulation to design antibiotic distribution centers." *Medical Decision Making* 22(1): S17–S25.

Kaplan, E. H. 1990. "Modeling HIV infectivity: must sex acts be counted?" *Journal of Acquired Immune Deficiency Syndrome* 3(1): 55–61.

Keller, L. 1995. "Parasites, worker polymorphism, and queen number in social insects." *American Naturalist* 145: 842–847.

Kluger, M. J. 1979. *Fever, its biology, evolution, and function.* Princeton: Princeton University Press.

Lamberty, M., D. Zachary, R. Lanot, C. Bordereau, A. Roberts, J. A. Hoffmann, and P. Bulet. 2001. "Insect immunity—Constitutive expression of a cysteine-rich antifungal and a linear antibacterial peptide in a termite." *Journal of Biological Chemistry* 276: 4085–4092.

Lemaitre, B., E. Nicolas, L. Michaut, J. Reichhart, and J. A. Hoffmann. 1996. "The dorsoventral regulatory gene cassette spatzel/Toll/cactus controls the potent antifungal response in *Drosophila* adults." *Cell* 86: 973–983.

Lemaitre, B., J. M. Reichhart, and J. A. Hoffmann. 1997. "Drosophila host defense: Differential induction of antimicrobial peptide genes after infection by various classes of microorganisms." *Proceedings of the National Academy of Science USA* 94: 14614–14619.

Levin, S. A., ed. 1992. *Mathematics and biology: The interface.* Berkeley: Lawrence Berkeley Laboratory, University of California.

Longini Jr., I. M., M. E. Halloran, A. Nizam, and Y. Yang. 2004. "Containing pandemic influenza with antiviral agents." *American Journal of Epidemiology* 159: 623–633.

Mackintosh, J. A., D. A. Veal, A. J. Beattie, and A. A. Gooley. 1998. "Isolation from an ant Myrmecia gulosa of two inducible O-glycosylated proline rich antibacterial peptides." *Journal of Biological Chemistry* 273: 6139–6143.

Meltzer, M.I., I. Damon, J. W. LeDuc, and J. D. Millar. 2001. "Modeling potential responses to smallpox as a bioterrorist weapon." *Emerging Infectious Disease* 7(6): 959–969.

Michaels, B., C. Keller, M. Blevins, G. Paoli, T. Ruthman, E. Todd, and C. J. Griffith. 2004. "Prevention of food worker transmission of foodborne pathogens: risk assessment and evaluation of effective hygiene intervention strategies." *Food Service Technology* 4(1): 31–49.

Moreau, C. S., C. D. Bell, R. Vila, S. B. Archibald, and N. Pierce. 2006. "Phylogeny of ants: diversification in the age of angiosperms." *Science* 312: 101–104.

Moret, Y., and P. Schmid-Hempel. 2000. "Survival for immunity: The price of immune system activation for bumblebee workers." *Science* 290: 1166–1168.

Naug, D., and S. Camazine. 2002. "The role of colony organization on pathogen transmission in social insects." *Journal of Theoretical Biology* 215: 427–439.

Nelson, K. E., C. M. Williams, and N. M. H. Graham. 2001. *Infectious diseases epidemiology, theory and practice.* Maryland: Aspen Publishers.

Nesse, R., and G. Williams. 1994. *Why we get sick.* New York: Random House.

Norris, K., and M. Evans. 2000. "Ecological immunology: Life-history trade-offs and immune defense in birds." *Behavioral Ecology* 11: 19–26.

Nunn, C. L., J. L. Gittleman, and J. Antonovics. 2000. "Promiscuity and the primate immune system." *Science* 290: 1168–1170.

O'Donnell, S. 1997. "How parasites can promote the expression of social behaviour in their hosts." *Proceedings of the Royal Society of London B* 264: 689–694.

O'Donnell, S., and S. N. Beshers. 2004. "The role of male disease susceptibility in the evolution of haplodiploid insect societies." *Proceedings of the Royal Society of London B* 271: 979–984.

Owens, I. P. F. and K. Wilson. 1999. "Immunocompetence: neglected life-history trait or conspicuous red herring?" *Trends in Ecology and Evolution* 14, 170–172.

Pie, M. R., R. B. Rosengaus, and J. F. A. Traniello. 2004. "Nest architecture, activity pattern, worker density and the dynamics of disease transmission in social insects." *Journal of Theoretical Biology* 226: 45–51.

Pie, M. R., R. B. Rosengaus, D. V. Calleri II, and J. F. A. Traniello. 2005. "Density and disease resistance in group-living insects: Do eusocial species exhibit density-dependent prophylaxis?" *Ethology, Ecology & Evolution* 17: 41–50.

Platt, O. S., D. J. Brambilla, W. F. Rosse, P. F. Milner, O. Castro, M. H. Steinberg, and P. P. Klug. 1994. "Mortality in Sickle Cell Disease—Life expectancy and risk factors for early death." *New England Journal of Medicine* 330: 1639–1644.

Poulsen, M., A. N. Bot, M. G. Nielsen, and J. J. Boomsma. 2002. "Experimental evidence for the costs and hygienic significance of the antibiotic metapleural gland secretion in leaf-cutting ants." *Behavioral Ecology and Sociobiology* 52: 151–157.

Read, A. F., and J. A. Allen. 2000. "The economics of immunity." *Science* 290: 1104–1105.

Reeson, A. F., K. Wilson, A. Gunn, R. S. Hails, and D. Goulson. 1998. "Baculovirus resistance in the noctuid *Spodoptera exempta* is phenotypically plastic and responds to population density." *Proceedings of the Royal Society of London B* 265: 1787–1791.

Reilly, M., A. Salim, E. Lawlor, O. Smith, I. Temperley, and Y. Pawitan. 2004. "Modelling infectious disease transmission with complex exposure pattern and sparse outcome data." *Statistics in Medicine* 23(19): 3013–3032.

Robinson, G. E. 1992. "Regulation of division of labor in insect societies." Annual Review of Entomology 37: 637–665.

Rose, E. A. F., R. J. Harris, and T. R. Glare. 1999. "Possible pathogens of social wasps *(Hypenoptera: Vespidae)* and their potential as biological control agents." *New Zealand Journal of Zoology* 26: 179–190.

Rosengaus, R. B., T. Cornelisse, K. Guschanski, and J. F. A. Traniello. 2007. "Inducible immune proteins in the dampwood termite Zootermopsis angusticollis." *Naturwissenschaften* 94: 25–33.

Rosengaus, R. B., C. Jordan, M. L. Lefebvre, and J. F. A. Traniello. 1999. "Pathogen alarm behavior in a termite: A new form of communication in social insects." *Naturwissenschaften* 86: 544–548.

Rosengaus, R. B., A. B. Maxmen, L. E. Coates, and J. F. A. Traniello. 1998. "Disease resistance: A benefit of sociality in the dampwood termite *Zootermopsis angusticollis* (Isoptera: Termopsidae)." *Behavioral Ecology and Sociobiology* 44: 125–134.

Rosengaus, R. B., J. E. Moustakas, D. V. Calleri, and J. F. A. Traniello. 2003. "Nesting ecology and cuticular microbial loads in dampwood *(Zootermopsis angusticollis)* and drywood termites (Incisitermes minor, I. schwarzi, *Cryptotermes cavifrons)." Journal of Insect Science* 3: 31.

Rosengaus, R. B., and J. F. A. Traniello. 1997. "Pathobiology and disease transmission in dampwood termites *Zootermopsis angusticollis* (Isoptera: Termopsidae) infected with the fungus *Metarhizium anisopliae* (Deuteromycotina: Hypomycetes)." *Sociobiology* 30: 185–195.

Rosengaus, R. B., and J. F. A. Traniello. 2001. "Disease susceptibility and the adaptive nature of colony demography in the dampwood termite *Zootermopsis angusticollis.*" *Behavioral Ecology and Sociobiology* 50: 546–556.

Rosengaus, R. B., J. F. A. Traniello, T. Chen, J. J. Brown, and R. D. Karp. 1999. "Immunity in a social insect." *Naturwissenschaften* 86: 588–591.

Rosset, H., L. Keller, and M. Chapuisat. 2005. "Experimental manipulation of colony genetic diversity had no effect on short-term task efficiency in the Argentine ant *Linepithema humile.*" *Behavioral Ecology and Sociobiology* 58: 87–98.

Rothman, L., and J. Myers. 1996. "Debilitating effects of viral diseases on host Lepidoptera." *Journal of Invertebrate Pathology* 67: 1–10.

Russell, V., and P. E. Dunn. 1996. "Antibacterial proteins in the midgut of *Manduca sexta* during metamorphosis." *Journal of Insect Physiology* 42: 65–71.

Schmid-Hempel, P. 1998. *Parasites in social insects.* Princeton: Princeton University Press.

———. 2005. "Evolutionary ecology of insect immune defenses." *Annual Review of Entomology* 50: 529–551.

Schmid-Hempel, P., and R. H. Crozier. 1999. "Polyandry versus polygyny versus parasites." *Philosophical Transactions of the Royal Society Biological Sciences* 354: 507–515.

Schmid-Hempel, P., and D. Ebert. 2003. "On the evolutionary ecology of specific immune defence." *Trends in Ecology & Evolution* 18: 27–32.

Schmid-Hempel, P., and R. Schmid-Hempel. 1993. "Transmission of a pathogen in *Bombus terrestris,* with a note on division of labour in social insects." *Behavioral Ecology and Sociobiology* 33: 319–327.

Seid, M., and J. F. A. Traniello. 2006. "Age-related repertoire expansion and division of labor in *Pheidole dentata* (Hymenoptera: Formicidae): A new perspective on temporal polyethism and behavioral plasticity in ants." *Behavioral Ecology and Sociobiology* 60: 631–644.

Shykoff, J. A., and P. Schmid-Hempel. 1991. "Genetic relatedness and eusociality: Parasite-mediated selection on the genetic composition of groups." *Behavioral Ecology and Sociobiology* 28: 371–376.

Spivak, M. and M. Gilliam. 1998a. "Hygienic behaviour of honey bees and its application for control of brood diseases and varroa. Part I. Hygienic behaviour and resistance to American foulbrood." *Bee World* 79: 124–134.

Spivak, M. and M. Gilliam. 1998b. "Hygienic behaviour of honey bees and its application for control of brood diseases and varroa. Part II. Studies on hygienic behaviour since the Rothenbuhler Era." *Bee World* 79: 169–186.

Starks, P. T., C. A. Blackie, and T. D. Seeley. 2000. "Fever in honeybee colonies." *Naturwissenschaften* 87: 229–231.

Starr, J. M., and A. Campbell. 2001. "Mathematical modeling of *Clostridium difficile* infection." *Clinical Microbiology and Infection* 7(8): 432.

Tarpy, D. R., and T. D. Seeley. 2006. "Lower disease infections in honeybee (*Apis mellifera*) colonies headed by polyandrous vs. monandrous queens." *Naturwissenschaften* 93: 195–199.

Temime, L., D. Guillemot, and P. Y. Boëlle. 2004. "Short- and long-term effects of pneumococcal conjugate vaccination of children on penicillin resistance." *Antimicrobial Agents and Chemotherapy* 48(6): 2206–2213.

Thorne, B. L., and J. F. A. Traniello. 2003. "Comparative social biology of basal taxa of ants and termites." *Annual Review of Entomology* 48: 283–306.

Traniello, J. F. A., R. B. Rosengaus, and K. Savoie. 2002. "The development of immunity in a social insect: Evidence for the group facilitation of disease resistance." *Proceedings of the National Academy of Science USA* 99: 6838–6842.

Wendelboe, A. M., A. Van Rie, S. Salmaso, and J. A. Englund. 2005. "Duration of immunity against pertussis after natural infection or vaccination." *Pediatric Infectious Disease Journal: The Global Pertussis Initiative* 24(5): S58–S61.

Wilson, E. O. 1971. *The insect societies.* Cambridge: Harvard University Press.

CHAPTER TWENTY-FIVE

Social Insects and the Individuality Thesis: Cohesion and the Colony as a Selectable Individual

ANDREW HAMILTON

NATHAN R. SMITH

MATTHEW H. HABER

Social Insects, Cohesion, and the Units-of-Selection Problem

Evolutionary theory is general in a way that is often not appreciated (Okasha 2006), partly because of the long-standing focus at the organism and gene levels. In the abstract, however, there is nothing special from an evolutionary perspective about any particular level of biological organization. One concrete problem for researchers is determining which levels are special because of the causal or historical circumstances of evolutionary change. John Maynard Smith captured the issue with characteristic pith in 1988: "Any population of entities with the properties of multiplication (one entity can give rise to many), variation (entities are not all alike, and some kinds are more likely to survive and multiply than others), and heredity (like begets like) will evolve: A major problem for current evolutionary theory is to identify the relevant entities."

Our task in this chapter is to point to a new way to frame this problem as it pertains to social insect colonies and to colony-level selection. We argue below that there are two general superorganism approaches: one focused on similarities between organisms and colonies that has its roots in the developmental and organicist traditions followed by William Morton Wheeler (Wheeler 1911, 1928; Seeley 1995; Moritz and Southwick 1992;

Moritz and Fuchs 1998), and one that emphasizes the colony as a unit of selection and has its roots in kin and group selection theory (Wilson and Hölldobler 2005; Reeve and Hölldobler, 2007). The similarity approach is very widely used, but we think it obscures important issues about evolution. The selection approach does attend to evolutionary subtleties, but it largely ignores development.

Here we offer an alternative conceptualization of colonies in terms of the individuality thesis of Ghiselin (1969, 1974) and Hull (1976, 1978) that brings together the best features of both superorganism approaches while avoiding the shortcomings of each. The individuality thesis says that complex or higher-level biological objects *are* individuals, rather than that they are *like* organisms. While the individuality thesis was originally articulated to address a set of issues around the reality and nature of species, we argue that it applies well to colonies and that it frames an important set of questions about colony-level multiplication, variation, and heredity, thus throwing light on the colony as a unit of selection. Most importantly, it helps reopen a discussion about development at the colony level.

Understanding what colonies are and how they function from an evolutionary perspective turns out to be very similar to understanding species and how they function. We draw out this parallel below to illustrate the individuality thesis in detail, and argue that all real biological taxa are concrete, spatio-temporally located individuals rather than abstract classes or sets, and that the individual, rather than the organism, is the paradigm unit. We also begin to apply the individuality thesis to colonies. In particular, we situate our arguments within an evolutionary framework by sketching a picture of reproduction at the colony level—a *sine qua non* of colony-level selection. In the final section we point to some work that remains in applying the individuality thesis to colonies.

Individuals, Not Superorganisms

The central idea in this chapter is that colonies are individuals. This thesis is worth arguing for two reasons. The first is that it frames the discussion of colony-level selection in a way that can be obscured by thinking of colonies as superorganisms. We think the similarity approach to superorganisms is metaphorical in a manner that leads away from the most interesting questions about social insects because it relies on the brittle notion that colonies and organisms are similar.

The second reason that attention to the individuality thesis is worthwhile is that it frames a more general set of questions about what it means to be an evolutionary unit and reframes disagreements over what it means to be a superorganism within a selectionist approach. As we shall see, the selection approach to superorganisms that is being developed by Hölldobler and colleagues is a shift away from, and an improvement on, the similarity approach and toward an account given in terms of the evolutionary consequences of sociality, but it still has some shortcomings that can be addressed by thinking about colonies as individuals. One important reason the individuality approach is superior is that it connects the selection approach to a developmental account of social insect colonies.

The Similarity Approach to Superorganisms

One problem with the similarity approach to superorganisms is that similarity is a notoriously difficult relation to capture meaningfully (Goodman 1974), for the reason that, as Sterelny and Griffiths (1999) have put it, "similarity without theory is empty." There are any number of similarities (and dissimilarities) between any two biological entities, and the similarity approach to superorganisms gives very little guidance about what the *relevant* similarities are or how to capture them. Put another way, the superorganism metaphor has had the effect of hiding rather than emphasizing the theory that is needed to put flesh on the bones of similarity claims.

This objection can be put more concretely. In their defense of the superorganism metaphor, Wilson and Sober (1989) argue that a superorganism is "a collection of single creatures that together possess the functional organization implicit in the formal definition of an organism." There are two strains to Wilson and Sober's argument. On the one hand, they sometimes speak of superorganisms as being real entities. On the other hand, they sometimes argue that colonies are relevantly *like* organisms because both are functionally organized. We are sympathetic to the former notion but wish to raise some concerns about their reliance on metaphor in the latter. Colonies are not individuals because they are functionally organized, but are functionally organized because they are individuals. Wilson and Sober argue in favor of the superorganism concept partly, at least, to advance their thesis that natural selection operates at multiple levels of biological organization. We are quite sympathetic to this thesis for reasons that will be clear in the next section, but superorganism talk of this sort is not illuminating precisely because it provokes commentators to ask after the closeness of the similarity, rather than directly about the quality of the causal claims being made. Moreover, what counts as similar *enough* varies with one's theoretical perspective in a way that renders the matter virtually impossible to settle.

Take, for instance, the objections to Wilson and Sober raised by Mitchell and Page (1992). Mitchell and Page are also proponents of multi-level selection, but argue that Wilson and Sober's defense of the superorganism metaphor "obscures our vision" because there are important respects in which functional organization at the colony level varies across species of social insects and, thus, that colonies are unlike organisms. Again, colonies and organisms are both similar and dissimilar in different respects and to different degrees, and refining the superorganism metaphor has not led researchers to a better understanding of or agreement about selection at the colony level.

We hope to do better. Approaching colonies in terms of individuality allows one to ask what it means for them to participate in relevant evolutionary and ecological processes, rather than what similarities there are (or are not) between colonies and organisms. From our perspective, it is because colonies are individuals that participate in various biological processes that they are relevantly similar to individual organisms and are functionally organized (as opposed to Wilson and Sober, who argue that it is *because* colonies are functionally organized that they are superorganisms). Despite this talk of similarity, however, the individuality approach shifts the discussion substantially. The new emphasis is on the particular causal relations that hold between parts of a whole such that they form a cohesive individual, as well as on what it means for that individual to participate in evolutionary processes.

The Selection Approach to Superorganisms

The selection approach to superorganisms, exemplified in the strain of Wilson and Sober's (1989) argument that takes superorganisms to be real things, and in the work of Hölldobler and collaborators, places participation in evolutionary processes at the fore of defining what a superorganism is. From this approach, to be a superorganism is to be a colony in which within-group competition is nearly nonexistent, while between-group competition is high (Reeve and Hölldobler 2007). This approach is explicitly about natural selection (and implicitly about the causal relations that hold between the organisms in a colony such that they form a "selectable" unit) because in a scenario in which cooperation within groups and competition between groups are both pronounced, the case is made that selection is operating at the level of the group.

This approach has the advantage that it puts evolutionary processes in the foreground and is clearly in line with the individuality thesis, but it has the disadvantage that it does not obviously have the resources to address colony-level multiplication, variation, and heredity directly. It also leads to some disagreement about how social a colony must be before it is properly called a superorganism. Should a colony only be called a superorganism when within-group competition is nearly nonexistent as Hölldobler argues, or whenever within-group competition is a less powerful evolutionary force than between-group competition, as E. O. Wilson (1975) apparently argues (Keim 2007)? This disagreement is not necessarily over ontology, but may be usefully recast (and redirected) by placing it in the context of the individuality thesis. As will be seen, this amounts to prioritizing the question of what it means to be a colony, the result of which is to resurrect talk of development and shift the important foci of the debate away from superorganism talk. The advantages of understanding colonies this way can be made clearer by analyzing what it means to say that colonies and other kinds of biological entities are individuals.

Individuality and Cohesion: Two Parallel Cases

The thesis that species are biological individuals has been much discussed and, we think, widely misapplied. Misapplications result in part from thinking that the thesis gives particular advice about what sorts of individuals species are (e.g., Ghiselin 1997) or about what processes drive macroevolution (e.g., Eldredge 1985; Cracraft 1987). As we read it, all the thesis says is that (i) species are defined by ancestry, not by possession of any properties or characters, and (ii) that species are spatio-temporally located biological wholes constituted by parts (as opposed to having members).

What determines whether something is a part of a biological whole are the relations between it and other parts of that whole, as opposed to possession of any particular (set of) property(ies). That is, biological taxa are not sets, classes, or any other kind of abstract entity (at least not as these are traditionally understood). The thesis denies essentialism of the kind decried by Mayr (1959) and others (Cain 1958; Simpson 1961; Hull 1965; see also Winsor 2006), but taken alone, says nothing at all about what particular relations do or must obtain for parts to be unified into a given biological whole or how cohesive an individual must be in order to be a unit of selection. In other words, the individuality thesis makes a very general ontological cut, and leaves a fair bit of work to be done in specific cases.

Our contention, developed below, is that there are many causal relations by which parts cohere into biological wholes—many cohesion-generating relations (CGRs)—and that the most familiar one is not always the one at work in a given case (Haber and Hamilton 2005; Hamilton and Haber 2006). The individuality thesis says that physically scattered entities, like colonies and species, are no less individuals than are more familiar ones that are bounded by membranes or skin. The cohesion-generating glue that binds the parts into a unified whole is somewhat different with colonies than with species or organisms, but this difference is not ontologically relevant. More importantly, this difference is not visible to natural selection, provided that other conditions are met.

Given this reading of the individuality thesis, its usefulness lies in framing a discussion about multiplication, fitness, variation, and heredity at the colony level, rather than in giving information about how we should understand particular entities or the relationships between their constituent parts. To say that something is an individual is to say something incomplete. One wants more information: What unifies the parts such that they form a single entity? Among the various kinds of relations—gene flow (Mayr 1963; Ehrlich and Raven 1969), phylogeny (Mishler and Theriot 2000; Wheeler and Platnik 2000), shared evolutionary fates (Wiley 1978; Wiley and Mayden 2000), and so on—that generate cohesion, which are salient in particular cases? Under what conditions do particular CGRs break down? In particular cases of CGR disruption, what happens? When and why do new relationships obtain and how do they causally partition the world into parts and wholes? We attempt to orient the reader first by framing answers to these questions for species. We then move on to discuss CGRs for social insects, with a focus on sociality.

Species as Individuals

As biological individuals, species are both made up of biological parts and are themselves parts of larger biological entities (wholes). What these various parts and wholes are is, famously, controversial. Furthermore, it is not enough to simply be composed of biological parts and to be a part of a larger biological entity. To be a species is to be made up of parts that stand in some appropriate CGR, and to stand in relevant relationships to other species (e.g., phylogenetic relationships). What these relations are is also famously controversial. These controversies are tightly linked to, and often simply map onto, debates over species concepts. So, for example, one theory of what it means for something to be a species is that it is composed of individual organisms that interbreed, and is itself a part of a lineage of populations. Species, unlike (most) individual organisms, are not bound by membranes; instead their parts cohere in other ways.

Were the boundaries of species easily discernable, then debates over the CGRs that are the glue of species may have been more easily resolvable (though membranes hardly settle the matter for organisms). Unfortunately, the beginning and end of a species (both spatially and temporally) rarely presents itself in any obvious manner. The matter is not simply an epistemic one; data alone will not be sufficient to determine the boundaries of a species. Which data are salient depends on the theoretical and conceptual framework in which they are implemented. Researchers working with different species concepts may agree on the data, but disagree over what constitutes a species boundary, or even over which data count as evidence for a that boundary. This situation may be resolved in many ways. One option is to advocate a pluralistic approach to species and species boundaries (Ereshefsky 2001). Another is to argue that a particular species concept is the only or best one. Species also may be more or less cohesive, and this complicates matters. The degree of cohesiveness necessary for a group of organisms to count as a species will be specified by particular species concepts. Again, this is an arena about which there is much controversy, and tracks very closely to the debates over how to delimit the boundaries of a species.

Far from settling the question of which species concept is superior or even the question of whether one ought to be a pluralist or monist about species concepts, the individuality thesis helps to demarcate the contours of the debate. When two researchers advocate different concepts, it is often because of deeper commitments; namely, they disagree about which CGRs are most salient. Take, for instance, Mayr's (2000) criticism of the evolutionary species concept of Wiley and Mayden (2000). Mayr argued that "the capacity for evolving is not the crucial biological criterion of a species; that would be the protection of its gene pool." Mayr's objection, essentially, is that he understands gene flow to be the most important causal process at the species level, whereas Wiley and Mayden take it that the suite of causal processes that render particular populations unique in their evolutionary trajectories are most important. Here we have a disagreement over species concepts that is driven by differing understandings of which CGRs are most important; that is, the dispute is over what kinds of biological individuals species are.

Colonies as Individuals

Now that we have seen how the individuality thesis applies to species, we can explore in detail what it means for colonies of social insects to be individuals. All by itself, the thesis carries no information about the features of units of selection that have interested Maynard Smith and other theoreticians. That is, the individuality thesis applied to social insects tells us little about colonies that reproduce differentially, vary, and have heritable traits. To fill in these details, it will be necessary to discover what *kind* of biological individuals colonies of social insects are by specifying which CGRs unify them. These details have to be worked out if the case is to be made that social insect colonies are both individuals and units of selection.

Colonies of social insects are individuals in the sense that they, like all other biological individuals, are defined by ancestry and are concrete rather than abstract (i.e., are spatio-temporally located). This line of thought can be fleshed out by anticipating an objection about dissimilarities between colonies and organisms. We take it that the latter are paradigm individuals for most people, and that some will not want to countenance colonies as individuals for the reason that colonies (and species for that matter) are not physically integrated in the same way that organisms are. Organisms seem to have relatively clear boundaries set by physical membranes that enclose the parts of the organisms. Colonies are not like this. A social insect colony is composed of many discrete parts—the individual insects—which can be spread over space in a way that the parts of organisms generally can not.

This objection confuses what it means to be an individual with what it means to be an organism, and argues against the superorganism metaphor once again on the grounds that colonies are not relevantly like organisms. Not all individuals, however, are organisms (Wilson and Sober 1989). As

Ghiselin (1974) and Hull (1976, 1978) pointed out in their original articulation of the individuality thesis about species, being spatially spread out in a way that most organisms are not is no reason to discount individuality. All sorts of scattered objects are rightly regarded as individuals (e.g., universities, corporations, and solar systems).

Physical integration is just one kind of cohesion that unifies parts into wholes, and it comes in degrees. Sociality is another kind of cohesion (Queller 2000), and it also comes in degrees. Relatedness (Hamilton 1964; Gadau and Laubichler 2006) and functional integration (Wilson and Sober 1989) are other, relevant CGRs for colonies of social insects. Whether or not colonies are rightly countenanced as units of selection depends on whether they participate in causal processes *qua* unified whole rather than only by means of the interactions between their parts. The relevant colony-level causal processes for natural selection at every level are multiplication, variation, and the passing on of heritable, fitness-relevant traits. What these processes look like at the colony level is the topic of the next section.

Colonies are certainly *logical* individuals, in that they are constituted by parts; the relevant question is whether they are rightly counted as *biological* individuals. To decide this issue, it is necessary to resolve whether these parts stand in some biologically interesting CGR, whether colonies stand in relevant relations to other colonies, and whether colonies are themselves parts of larger biological wholes. Answering these questions will address whether colonies cohere enough to be "seen" as a selectable object from the "point of view" of natural selection, and whether they cohere enough and in the right way to be selectable. At a minimum, this will mean that colonies multiply, vary, and have heritable traits. Notice that the selection approach focuses on the cohesion that arises from within-group cooperation, allowing the confrontation of problems about how colonies participate in evolutionary processes from a particular theoretical perspective. What happens when we prioritize the ontology of development over any particular CGR is explored below.

Colonies as Selectable Individuals: Multiplication, Reproduction, and Development

Now that we have a clearer understanding of the individuality thesis as it applies to colonies, we can ask what it means for colonies to be the kind of individual on which natural selection operates. In other words, if colonies are to be selectable, it must be the case that colonies reproduce differentially in a way that leads to a fitness-relevant modification of a population of colonies. Notice that, despite our focus on evolution, this is a substantially different way of thinking about colonies than what is found in the selection approach to superorganisms. Our approach prioritizes discussion about what it means for colonies to reproduce other colonies rather than pointing at a particular CGR as *the* relation that marks superorganisms off from other kinds of biological individuals. Instead, relevant CGR are determined by fall out of empirical research and conceptual framing of colony reproduction.

Space considerations do not allow for elaboration of all aspects of colony-level processes here, so we focus on multiplication/reproduction, which we take to be the hardest case. In order for colonies to be selectable individuals, they must reproduce other colonies. Though loose talk of colony reproduction may be common, here we consider the details of colony behavior against the background of individuality and a general theory of biological reproduction, and it becomes clear that it is the colonies—not just their constituent organisms—that reproduce.

In order to make the case that colonies, rather than only organisms, reproduce, it will be necessary first to have a look at a general account of reproduction. The most highly articulated account that we know of is by Griesemer (2000), who argued that biological reproduction has two components: progeneration and development. Progeneration is a special kind of multiplication on which material progenerants overlap across generations. The material overlap requirement is meant to distinguish reproduction from mere multiplication or copying, and thus to mark off the biological processes that result in an increase in the number of entities from other processes that have the same result; that is, there is no reproduction at a distance in biology.

According to this definition, photocopying is multiplication because it increases the number of entities of the same kind, but it isn't reproduction, partly because there is no material overlap. Because reproducers come in various shapes and sizes, the relevant generation-spanning material will vary by taxon and level of organization. In our own species, the relevant material is gametes and the subsequent fertilized egg with its complete diploid complement of genes. In prokaryotic cells that reproduce by binary fission, various cellular materials are shared between generations. With honey bees, a new colony is born when a mated foundress and a contingent of workers leave their former hive and establish a new one. In this case, the overlapping material is the colony sub-unit.

A foundress and a contingent of workers (i.e., a colony propagule) does not, of course, make a colony any more than a gamete by itself constitutes a human being. This brings us to development, the second component of Griesemer's (2000) account of reproduction. Griesemer understood development to be the acquisition of the capacity to reproduce. Reproduction, then, is the progeneration by material overlap of entities or material that have the capacity to develop in such a way that the process is repeated. This way of thinking about reproduction is iterative, since developing into a reproducer is built in, but the account avoids circularity by bottoming out at null development. Not all reproducers need to acquire the capacity to reproduce; it is at least possible that some have this capacity at progeneration.

With this brief sketch in mind, we can proceed to discuss colonies in more detail. The most familiar case of what Michener (1974) calls "colony multiplication" is, perhaps, the swarm behavior of eusocial bees. In swarming, the colony splits fairly abruptly, and a new colony site is located and communicated by forager bees turned scouts. A colony propagule then departs for the new location and establishes a colony. Swarm behavior is interesting in the context of our argument for two reasons. First, it is fairly easy to see that we have a case of material overlap, even if the material is not at the level of organization that we are used to thinking about with respect to reproduction: here the colony is the individual and the overlapping material is the colony propagule that make up the nascent daughter colony. Second, it is also clear that we do not have a case of organism-level reproduction by another name. The foundress will rear new workers quickly, but this, we suggest, is best understood as part of the development of the new colony. After all, it is not the reproduction of any particular organism that counts as the production of a new colony. Only the coordinated reproduction and development of individual workers will tell the whole story of the establishment of a new colony. This is a function of what it means to be an individual colony.

Highly eusocial species like some of those in the apid subfamily Apinae are characterized by a high degree of task specialization. The gyne lacks the physical structure for pollen foraging, and thus cannot survive apart from the colony. Similarly, the workers generally do not reproduce (Visscher 1996). The set of tight functional relationships that obtain at the colony level are intrinsic features of the individual colony, though notice that these supervene on the extrinsic relational properties (CGRs) that obtain between the parts of that colony (i.e., the foundress and workers). To track individual colony reproduction is to mark the relevant CGRs of the parts of a colony, and to identify and discover the relevant patterns of CGR disruption and formation. The shape and nature of *these* patterns will determine whether colonies are units of evolution or selection.

A new colony cannot, of course, be the source of a second immediate swarm: in *Apis,* about two hundred bees are necessary (Michener 1974). Even if the relevant (non-density dependent) stimuli were present, the incipient colony would have to forage for pollen and nectar stores, scout a new location, and rear a new gyne. Two of these three tasks will usually require the building of comb, and this task also requires a minimum number of bees (Darchen 1957). This is just to say that the incipient colony has the capacity to acquire the capacity to reproduce: incipient colonies have the capacity to build a hive (development) that confers the capacity to be the parent colony for a swarm. This is development at the colony level. In becoming fully functional, the colony acquires the capacity to be the source of a swarm.

It should now be easy to see how to proceed along these lines by way of taking up the problem framed by Maynard Smith (1988); with a plausible account of reproduction at the colony level in place for highly eusocial bees, accounts of variation and fitness of colony-level traits will follow without much difficulty given existing work on the evolution of eusocial colonies (e.g., the selection approach account). There will surely be competition for resources among proximate colonies, and the colonies are often more or less isolated reproductively from sister colonies. Indeed, there is already a large body of literature on these topics (Wade 1978; Owen and Harder 1995; Page and Fondrk 1995; Moore, Brodie, and Wolf 1997; Wilson and Dugatkin 1997; Sober and Wilson 1998; Fewell 2003; Tarpy, Gilley, and Seeley 2004). There is another, prior step to understanding selection and heritability, as they may or may not apply to colonies. The need for more conceptual work is illustrated by paying attention to variation in the kinds of sociality—the kinds and degrees of CGRs—among social insects. With eusocial species, the reproducer case is relatively easy to make because of the tight functional integration among the various task specialists. But what of colonies that are less social? Does it make sense to think that they are reproducers as well?

Take, for instance, the extreme opposite case of sleeping clusters, in which a relatively small number of mostly male bees (or wasps) gather overnight. Michener (1974) reported that bees of all families except the Apidae form such clusters, and he further argued that sleeping clusters are not colonies because "the bees . . . in a cluster are not inhabiting a nest, rearing young, and the like." According to Michener, allodapine species—whose organization ranges from solitary to primitively eusocial—do form colonies, but these colonies are often or always founded by a lone female. Because all the young are reared together with care only from the mother, the colony is not social at this stage. Some allodapine species go on to behave cooperatively. For instance, in some species, some of the adult daughters will become workers while the mother becomes (temporarily, at least) a queen.

The point of this look at different organizational structures is that the account of colony reproduction we gave in terms of highly eusocial species is a special case, and that it is unclear where the line should be drawn between colony-level reproduction and organismal reproduction for lesser degrees of sociality. Seeking a threshold on the continuum of sociality that marks the relevant degree for this particular CGR is probably a fool's errand. Whether or not it is relevant that there is material overlap in the case of allodapine bees (the mated foundress leaves one nest to establish another), and that in some cases this new nest will come to house a group that has varying degrees of sociality, will depend in turn on one's concept of a colony. In like manner, whether or not allodapine nest establishment means that we have a new entity that should properly be called a daughter colony established by colony-level reproduction will also depend on one's colony concept. None of this is worrying in an individuality context, as it is expected that individuality comes in degrees (e.g., of physical proximity).

Colony Concepts and Superorganisms

In the last section we sketched an account of colony-level reproduction, and ended up addressing colony concepts. Even where the relevant CGR is clear, there is still the matter of *degree* of cohesion, thus taking us back to disagreements stemming from Hölldobler's concept of superorganism. How much cohesion will we require before we mark something as a unit? How much disruption of this cohesion will we accept before we want to mark a unit as having broken into its parts? These two questions require clarity about colony concepts. Notice, however, that they are precisely the questions that systematists ask of species. The cases are parallel because they are doing the same work: they are asking where we might locate the boundaries of scattered individuals and acknowledging that the task is a conceptual one that turns on giving principled reasons for some CGR and for the degree to which that CGR should hold.

The benefit of this approach to colony-level selection is that it makes it possible to understand the two remaining issues related to selection. Accounts of colony-level variation, and colony-level heredity of those variations against a fitness environment, will depend on having an appropriate colony concept. Because there are several relevant CGRs, however, there is no reason at the outset to think that there is a privileged colony concept given in terms of some CGR. Again, this is exactly the case with species.

Given the mishmash of debate about concepts surrounding species, what productive work is the individuality thesis doing? And why is it applicable to thinking about colonies? We have argued that thinking about colonies as individuals will generate the right kinds of questions about colonies, facilitate useful theoretical and conceptual debates about the nature of social insect colonies, and, perhaps most important, provide a powerful explanatory and research framework for how colonies may (or may not) participate in various evolutionary processes.

For example, thinking of colonies as individuals provides new traction for debates over levels of selection. Given that biological individuals are parts of other biological individuals, and are themselves (often) constituted by biological individuals, we should not be surprised to find advocates of individual-level selection arguing that selection may act on individuals simultaneously at multiple levels (Sober and Wilson 1999); or, alternatively, that distinctions between kinds of group selections must be drawn (Damuth and Heisler 1988; Michod 2005; Okasha 2006).

Like species, colonies may be more or less cohesive. This is just to say that social insects, like species and every other kind of biological individual, exhibit a range of kinds and degrees of CGRs. This is why the argument over whether only eusocial colonies are the only true superorganisms seems to us to be wide of the mark: the interesting concerns are not over what constitutes a superorganism, but what kinds of cohesions generate evolutionary individuals. What degree of cohesiveness is necessary for a colony to count as an evolutionary individual is a matter that will not be settled by empirical facts alone. Conceptual and theoretical frameworks of being-a-colony are needed to determine which facts are salient; that is, which data are evidence one way or another. Likewise, what will count as colony-level variation or heritability will, in large part, be determined by the colony concept being pressed into service. Sociality and the functional integration that comes with it, clearly is a CGR that will play a central role in any theory of being a colony, but there is a great more conceptual work to be done than most applications of the superorganism concept suggest.

Literature Cited

Cain, A. J. 1958. "Logic and memory in Linnaeus's system of taxonomy." *Proceedings of the Linnaean Society of London* 169: 144–163.

Cracraft, J. 1987. "Species concepts and the ontology of evolution." *Biology and Philosophy* 2: 329–346.

Damuth, J., and I. L. Heisler. 1988. "Alternative foundations of multi-level selection." *Biology and Philosophy* 3: 407–430.

Darchen, J. 1957. "La eeine d'*Apis mellifica*, les ouvrières pondeuses et les constructions cirières." *Insectes Sociaux* 4: 321–325

Ehrlich, P., and P. Raven. 1969. "Differentiation of populations." *Science* 165: 1228–1232.

Eldredge, N. 1985. *Unfinished synthesis.* New York: Oxford University Press.

Ereshefsky, M. 2001. *The poverty of the Linnaean hierarchy.* New York: Cambridge University Press.

Fewell, J. H. 2003. "Social insect networks." *Science* 301: 1867–1870.

Gadau, J., and M. D. Laubichler. 2006. "Relatedness: Capturing cohesion in biological systems." *Biological Theory* 414: 417.

Ghiselin, M. T. 1969. "The evolution of hermaphroditism among animals." *The Quarterly Review of Biology* 44: 189–208.

———. 1974. "A radical solution to the species problem." *Systematic Zoology* 23: 536–544.

———. 1997. *Metaphysics and the origin of species.* Albany: State University of New York Press.

Goodman, N. 1974. "Seven strictures on similarity." In N. Goodman, ed., *Problems and projects,* 437–447. New York: Bobbs-Merrill.

Griesemer, J. 2000. "The units of evolutionary transition." *Selection* 1: 67–80.

Haber, M., and A. Hamilton. 2005. "Coherence, consistency, and cohesion: Clade selection in Okasha and beyond." *Philosophy of Science* 72: 1026–1040.

Hamilton, W.D. 1964. "The Genetical Evolution of Social Behaviour. I." *Journal of Theoretical Biology* 7: 1–16.

Hamilton, W.D. 1964. "The Genetical Evolution of Social Behaviour. II." *Journal of Theoretical Biology* 7: 17–52.

Hamilton, A., and M. Haber. 2006. "Clades are reproducers." *Biological Theory* 1: 381–391.

Hull, D. L. 1965. "The effects of essentialism on taxonomy: Two thousand years of stasis." *The British Journal for the Philosophy of Science* 15: 314–326; 16: 1–18.

Hull, D. L. 1976. "Are species really individuals?" *Systematic Zoology* 25: 174–191.

———. 1978. "A matter of individuality." *Philosophy of Science* 45: 335–360.

Keim, 2007. "A brief history of the superorganism." *Wired Science,* Retrieved from http://blog.wired.com/wiredscience/2007/07/a-brief-history.html.

Reeve, H. K., and B. Hölldobler. 2007. "The emergence of a superorganism through intergroup competition." *Proceedings of the National Academy of Sciences* 104: 9736–9740.

Maynard Smith, J. 1988. "Evolutionary progress and the levels of selection." In M. H. Nitecki, ed., *Evolutionary progress,* 219–236. Chicago: University of Chicago Press.

Mayr, E. 1959. "The emergence of evolutionary novelties." In S. Tax, ed., *The evolution of life, Vol. 1: Evolution after Darwin,* 349–380. Chicago: University of Chicago Press.

———. 1963. *Populations, species, and evolution.* Cambridge: Harvard University Press.

———. 2000. "A critique from the biological species concept perspective: What is a species, and what is not?" In Q. D. Wheeler and R. Meier, eds., *Species concepts and phylogenetic theory,* 70–89. New York: Columbia University Press.

Michener, C. D. 1974. *The social behavior of the bees.* Cambridge: The Belknap Press of Harvard University Press.

Michod, R. E. 2005. "On the transfer of fitness from cell to organism." *Biology and Philosophy* 20: 967–987.

Mishler, B. D., and E. C. Theriot. 2000. "The phylogenetic concept (*sensu* Mishler and Theriot): Monophyly, apomorphy, and phylogenetic species concepts." In Q. D. Wheeler and R. Meier, eds., *Species concepts and phylogenetic theory,* 44–54. New York: Columbia University Press.

Mitchell, S .D., and R. E. Page. 1992. "Idiosyncratic paradigms and the revival of the superorganism." *Report NR. 26/92 of the Research Group on Biological Foundations of Human Culture.* Bielefeld: Germany.

Moore, A. J., E. D. Brodie, and J. B. Wolf. 1997. "Interacting phenotypes and the evolutionary process: I. Direct and indirect genetic effects of social interactions." *Evolution* 51: 1352–1362.

Motitz, R. F. A. and E. E. Southwick. 1992. *Bees as Superorganisms: An Evolutionary Reality.* Springer: New York.

Moritz, R. F. A. and S. Fuchs. 1998. "Organization of honeybee colonies: characteristics and consequences of a superorganism concept." *Apidologie* 29: 7–21.

Okasha, S. 2006. *Evolution and the levels of selection.* New York: Oxford University Press.

Owen, R. E., and D. L. Harder. 1995. "Heritable allometric variation in bumble bees: Opportunities for colony-level selection of foraging ability." *Journal of Evolutionary Biology* 8: 725–738.

Page, R. E., and M. K. Fondrk. 1995. "The effects of colony-level selection on the social organization of honey bee *(Apis mellifera L.)* colonies: Colony-level components of pollen hoarding." *Behavioral Ecology and Sociobiology* 36: 135–144.

Queller, D. C. 2000. "Relatedness and the fraternal major transition." *Philosophical Transactions: Biological Sciences* 355: 1647–1655.

Reeve, H. K. and B. Hölldobler. 2007. "The emergence of a superorganism through intergroup competition." *Proceedings of the National Academy of Sciences U.S.A.* 104: 9736–9740.

Seeley, T. D. 1995. *The Wisdom of the Hive.* Harvard University Press: Cambridge, Mass.

Simpson, G. J. 1961. *Principles of animal taxonomy.* New York: Columbia University Press.

Sterelny, K., and P. E. Griffiths. 1999. *Sex and death.* Chicago: University of Chicago Press.

Sober, E., and D. S. Wilson. 1998. *Unto others.* Cambridge: Harvard University Press.

Tarpy, D. R., D. C. Gilley, and T. D. Seeley. 2004. "Levels of selection in a social insect: A review of conflict and cooperation during honey bee *(Apis mellifera)* queen replacement." *Behavioral Ecology and Sociobiology* 55: 513–523.

Visscher, P. K. 1996. "Reproductive conflict in honey bees: A stalemate of worker egg-laying and policing." *Behavioral Ecology and Sociobiology* 39: 237–244.

Wade, M. J. 1978. "A critical review of the models of group selection." *The Quarterly Review of Biology* 53: 101–114.

Wheeler, W. M. 1911. "The ant-colony as an organism." *Journal of Morphology* 22: 307–325.

Wheeler, W. M. 1928. *The Social Insects*. Harcourt, Brace & Co.: New York.

Wheeler, Q. D., and N. I. Platnik. 2000. "The phylogenetic species concept (*sensu* Wheeler and Platnik." In Q. D. Wheeler and R. Meier, eds., *Species concepts and phylogenetic theory*, 55–69. New York: Columbia University Press.

Wiley, E. O. 1978. "The evolutionary species concept reconsidered." *Systematic* Zoology 24: 233–243.

Wiley, E. O., and R. L. Mayden. 2000. "The evolutionary species concept." In Q. D. Wheeler and R. Meier, eds., *Species concepts and phylogenetic theory*, 70–89. New York: Columbia University Press.

Wilson, E. O. 1975. *Sociobiology*. Cambridge: The Belknap Press of Harvard University Press.

Wilson, D. S., and L. A. Dugatkin. 1997. "Group selection and assortative interactions." *American Naturalist* 149: 336–351.

Wilson, E. O., and B. Hölldobler. 2005. "Eusociality: Origin and consequences." *Proceedings of the National Academy of Sciences USA* 102: 13367–13371.

Wilson, D. S., and E. Sober. 1989. "Reviving the superorganism." *Journal of Theoretical Biology* 136: 337–356.

Winsor, M. P. 2006. "The creation of the essentialism story: An exercise in metahistory." *History and Philosophy of the Life Sciences* 28: 149–174.

CHAPTER TWENTY-SIX

Social Insects, Evo-Devo, and the
Novelty Problem: The Advantage of
"Natural Experiments" Sensu Boveri

MANFRED LAUBICHLER

JÜRGEN GADAU

ONE OF THE CONTINUING CHALLENGES of evolutionary biology is to mechanistically explain the origin of complex novel structures and behaviors, a problem Darwin already struggled with in the *Origin of Species*. Recent advances in comparative genomics have undermined any idea of a simple correlation between new phenotypes and new genes, as it is now clear that both protein-coding and regulatory genes are conserved across a wide range of different species (Gehring 1998; Carroll, Grenier, and Weather bee 2005). With this straightforward solution no longer an option, how do we account for the obvious phenotypic differences between and within groups of organisms and for the emergence of novel structures and behaviors in the course of evolution? The short answer is that changes in the developmental systems of these organisms and, more specifically, changes in gene regulatory networks are responsible for these differences in observable phenotypes (Wilkins 2002; Davidson 2006; Davidson and Erwin 2006). This conclusion makes intuitive sense as all morphological and behavioral differences of adults first emerge during the development and life history of individual organisms.

Variation in developmental processes will thus always be the immediate, or proximate, cause of phenotypic variation. But the devil is, as always, in the details: Exactly how do developmental mechanisms contribute to phenotypic variation? How can explanations of developmental mechanisms be integrated into the theoretical framework of evolutionary biology, with its emphasis on speciation and adaptation through the differential fixation of alleles by natural selection? And, more practically, what are the best model systems to study these questions experimentally and comparatively? We are suggesting social insects as new models for such investigations because, as Boveri (1906) noted more than 100 years ago, nature has already performed many evolutionary experiments, leaving the results for us to interpret. Darwin's ingenious book on speciation was almost exclusively based on observations and interpretations of nature's experiments in evolution rather than detailed experimental studies. Focusing on ants alone leaves us the tantalizing number of at least 20,000 natural experiments, many with novel phenotypic and behavioral features if we consider each species as an independent experiment.

Almost 150 years after the publication of *Origin,* studies of evolutionary novelties still face many difficulties. In part, these can be attributed to the choice of model systems, which tend to reflect major morphological transformations such as the fin-limb transition in early tetrapod evolution (Hinchliffe 2002). Most of these transitions happened tens of millions of years ago, thus rendering a detailed mechanistic and stepwise reconstruction of the evolutionary events responsible for these transformations almost impossible. To be more explicit, it is not clear whether the currently observed regulatory mechanisms or the genetic architecture generating a limb in vertebrates are identical to the ones originally involved in the generation of the first limb, a problem that has been identified as the inference gap in explanations of evolutionary novelties (Wagner 2001). Here, we argue that expanding the focus of evolutionary developmental biology (Evo-Devo) to the evolution and development of behavioral phenotypes not only broadens the perspective of Evo-Devo, it also helps to overcome the inference gap.

Social insects are particularly well-suited model systems for studying the mechanisms underlying the origin of evolutionary novelties experimentally as well as theoretically. Social insects display a remarkable diversity in behavior and morphological structures, within species and between closely related species, allowing the repeated and direct study of the evolution of novelties on different evolutionary scales, from populations to orders. Novelties in social insects include morphological, physiological, and behavioral innovations on both individual and colony levels; for example, worker or queen polymorphism, worker-queen dimorphism, the bee dance, queen numbers, ratio of major to minor workers in a colony, and division of labor (e.g., Hölldobler and Wilson 1990; West-Eberhard 2003). Social insects are well studied and their phylogenetic relationships have recently been resolved for many taxa (e.g., bees: Danforth, Conway, and Ji 2003; see also http://www.entomology .cornell.edu/BeePhylogeny/; ants: Brady et al. 2006; Moreau et al. 2006; bumblebees: Cameron, Hines, and Williams 2007; wasps: Hines et al. 2007). In addition, many details of their genetic architecture and developmental mechanisms of important behavioral traits are known.

Social insects can be manipulated on both individual and colony levels, in several cases actually inducing novel types of social behavior *de novo* (e.g., Fewell and Page 1999, Fewell, Schmidt and Taylor, this volume). In 2006, the honey bee became the first social insect with a fully annotated genome, concurrently with the tools to manipulate expression levels of specific genes. Social insects are unique in that they represent a system in which individuals can express phenotypes that, had it been solitary, are detrimental or maladaptive for the individual (e.g., altruism) but are highly adaptive in the context of a social lifestyle. Explanations of these particular features of social insects have contributed enormously to the further development of evolutionary theory over the last few decades (e.g., through kin selection and multi-level selection models), which has led to a more complex and sophisticated understanding of evolutionary dynamics. In addition, the expression of certain phenotypes of social insects is heavily context dependent, as can, for instance, be seen in the different morphologies and behaviors of queens and workers. Social insects are, therefore, ideal model systems for the study of evolutionary novelties and the interacting roles of genes, developmental processes including epigenetic effects, and environmental factors during the evolution of a novel trait. In this chapter, we review several research questions related to evolutionary novelties in social insects and how they can be addressed using an Evo-Devo framework.

Evo-Devo and Evolutionary Novelties

Addressing the problem of evolutionary novelties and innovations requires first clearly defining *novelties*; second, developing a set of causal hypotheses about the developmental changes involved in the emergence of any particular evolutionary novelty; and, third, formulating a model describing the evolutionary dynamics of these traits. Müller and Wagner (1991) defined morphological novelty as "a structure that is neither homologous to any structure in the ancestral species, nor homonomous to any other structure of the same organism." While this rather general definition still leaves open many details, it does have one practical implication. The problem of identifying novelties is squarely placed within a comparative framework, as their recognition depends on both a good phylogeny and a detailed assessment of homology or homonomy. As we have reasonably good phylogenies for social insects, this pre-condition for the study of evolutionary novelties is clearly met.

Setting aside, for the moment, many of the practical problems connected with actually identifying novelties, which are similar to the problems of assessing homology (Wagner 2001a), we can recognize some of the steps required for establishing a causal hypothesis about the origin of evolutionary novelties within the context of Evo-Devo. The first question that needs to be addressed is: What specific developmental mechanisms are responsible for a new derived character as distinct from a character state that has been identified as an evolutionary novelty? Answering this question requires the detailed analysis of the developmental mechanisms that generate specific phenotypic characters. Thus, it is part of the experimental program of Evo-Devo that studies developmental and developmental genetic mechanisms in a comparative framework. In the case of social insects, we already have a good general appreciation about the mechanisms of phenotypic determination, as well as a set of well-developed molecular, genetic, and physiological tools to study these developmental mechanisms experimentally.

The next couple of question's build on this analysis of developmental mechanisms and ask: Did the developmental mechanisms that are currently responsible for the derived novel character originate at the same time as this character? More specifically, what are the exact developmental mechanisms responsible for the initial changes leading to a novel character? These are difficult inqueries that require us to compare the developmental mechanisms of ancestral and derived novel characters. In many cases this will not be feasible, as it is impossible to reconstruct the exact ancestral condition of developmental processes, especially if the transformation in question happened hundreds of millions of years ago, like fin-limb transition. Model organisms to answer these questions would ideally have sister taxa where one has retained the ancestral state and the other has evolved the novel state. Thus the social insects again serve as an ideal model system as there are many instances of sister taxa showing ancestral and derived novel phenotypes, especially with regard to social behavior (Abouheif and Wray 2002; West-Eberhard 2003).

Similar problems apply to the next concern: Are the observed *genetic* differences between these two developmental systems sufficient to account for the observed *phenotypic* differences? This last question focuses on the mechanistic details of the changes in the developmental system and to what extent observable genetic changes provide a complete explanation of evolutionary transformations (Wagner et al. 2001a). To answer it properly, we need to be able to experimentally manipulate the developmental system, something that can be done, at least on some organizational levels, with social insects.

Which evolutionary novelties do we want to study using social insects? The Evo-Devo literature is mostly focused on morphological characters— not the least because these have the potential to leave traces in the fossil record, and their developmental mechanisms can be studied comparatively. But this current emphasis is largely a matter of practical considerations and not necessarily one of theoretical significance. Behavioral characters can as easily be considered as examples of phenotypic novelties. Within an Evo-Devo framework, focusing on behavioral novelties implies no change in theoretical perspective, as behaviors obviously also have genetic and developmental foundations. Rather, by emphasizing the analysis of behavior, we turn our focus toward the evolutionary dynamics of the emergence, subsequent stabilization, and further refinement of novelties (Müller and Newman 2003, 2005; West-Eberhard 2003; Müller 2007).

In the context of evolutionary models, behavior is the "integrator" of form and function; any new variant, whether it is a generic morphological or physiological variation or a novelty, has to be used by organisms in such a way that it contributes to fitness (e.g., Page and Amdam 2007). Because behavior tends to have higher degrees of plasticity, it can actually drive the subsequent acquisition and stabilization of morphological variation. For example, division of labor can first emerge as a behavior under certain circumstances—based, of course, on underlying physiological mechanisms— before the developmental system of these species "catches up" and stabilizes different castes and their specific morphological and behavioral repertoire (see West-Eberhard 2003, for many discussions on this subject). In short, in those systems in which it can be studied, behavior is indeed not only an integrator of form and function, but also the lens through which development and evolution can be integrated.

Recently researchers have discussed several potential mechanisms for the emergence of evolutionary novelties and conceptual frameworks for their analysis (Gottlieb 1992; Wagner 2001a; Carroll et al. Carroll, Grenier, and Weather bee 2005; Hall and Olson 2003; West-Eberhard 2003; Müller and Newman 2003, 2005). Some of these proposals also addressed behavioral novelties and the role of behavior in driving evolutionary innovation (especially Gottlieb 1992; West-Eberhard 2003). In the context of these discussions, it is clear that evolutionary novelties in most cases "do not fall from heaven," but emerge within the context of existing variation—genetic and, more importantly, developmental—of organisms, a phenomenon often captured by the notion of plasticity.

In her magnum opus, *Developmental Plasticity and Evolution,* West-Eberhard (2003) argued for the important role of developmental plasticity not only for understanding evolutionary novelties, but also as part of a more inclusive conception of evolutionary phenomena. Similarly, Müller and Newman (2005) suggested that several distinct phases can be distinguished in the process of phenotypic evolution, which, in turn, should be captured by different concepts. They argued that the term *evolutionary novelty* should designate the outcome of phenotypic evolution, for instance a new morphological or behavioral phenotype, whereas *innovation* should refer to those underlying processes that generate novel phenotypes within populations. In this view, novelty is primarily a comparative concept that allows us to distinguish differences among populations, species, or higher systematic groups, whereas innovation refers to the developmental and evolutionary processes that bring about evolutionary novelties. Furthermore, Müller and Newman proposed *origination* as a separate element in the process of phenotypic evolution. Origination captures a fundamental stage in phenotypic evolution and development that provides the basis for subsequent modification and further refinement of phenotypes, and that is determined more by physical-chemical and architectural principles than the specific features of the genetic architecture of development.

In our discussion, social insect examples provide additional support for these dynamic multistep models of evolutionary novelties. We do not want to enter semantic debates about proper designation of terms in this chapter; however, we want to propose that evolutionary novelties are indeed the product of more complex dynamic interactions, and should be distinguished from adaptations in a narrow sense, although they can be highly adaptive features. Our examples suggest that morphological and behavioral novelties first emerge as a consequence of developmental and epigenetic mechanisms interacting with specific environmental conditions through the behavior of individuals and colonies—a phenomenon best described in a general sense as plasticity—and that they are subsequently selected and canalized, which requires genetic variation for these traits within populations. This process of generating evolutionary novelties (innovation *sensu* Müller and Newman 2005) is frequently driven by behavioral responses that are predicated on developmental plasticity, which, in turn, is enabled by some fundamental elements of the developmental architecture, such as the reproductive ground plan. In that sense, the origination of evolutionary novelties is a system-level phenomenon; that is, novelties are the result of structural and regulatory changes of the whole genetic, developmental, epigenetic, social, and environmental system. Thus, it makes sense to distinguish them from other forms of variation that do not depend on such systemic responses and are therefore more localized. In this context it is interesting to note that modularity, the architectural principle that enables such quasi-independent forms of variation, can itself be seen as a fundamental novelty of metazoan evolution. The following are examples and possible research programs using ants as a model system for Evo-Devo.

Wing Polymorphism (Morphological
Novelty—Individual Level)

Ants (Formicidae) evolved from within the aculeate wasps. Most females in the aculaeate wasps are normally winged, though in multiple groups, females became wingless (e.g., Mutillidae). By comparison, ants are unique because both winged females (queens) and wingless females (usually workers) coexist, and whether an egg or larvae develops into a wingless worker or a winged queen is determined, in most cases, by the environment, usually by the amount and quality of food a larvae is fed. During the evolution of the Formicidae, some taxa, like the army ant genera *Dorylus* or *Eciton,* have abandoned winged females completely; in other ant genera we have species with winged or wingless queens; and in some species winged and wingless or short-winged queens coexist (for an overview see Table 27.2 in Heinze and Keller 2000).

The best example of a detailed study of the developmental genetics underlying a novel phenotype in social insects is the study by Abouheif and Wray (2002) on the genetic regulation of wing loss in worker ants.

Abouheif and Wray (see also Bowsher, Wray and, Abouheif 2007) used expression studies of genes known to be conserved in the wing development/patterning of insects to show that while these genes are normally expressed in winged queens, some are suppressed during the development of workers. While this was perhaps not surprising, they also found that different ant subfamilies, such as Formicinae, Mymecinae, and Dolichoderinae, vary in the genes that suppress wing development in workers, although wingless workers probably evolved once early in the evolution of ants. Abouheif and Wray hypothesized that "the simultaneous evolutionary lability and conservation of the network underlying wing development in ants may have played an important role in the morphological diversification of this group and may be a general feature of polyphenic development and evolution in plants and animals."

The North American harvester ant genus *Pogonomyrmex* is a good model system for further studies because it encompasses the complete range of winged (alate; e.g., *P. barbatus*), short-winged (brachypterous; *P. huachucanus*), or wingless queens (intermorphs; *P. imberbiculus*). In one species, *P. pima*, wingless and winged queens co-occur in the same population without any measurable genetic isolation (Figure 27.2 in Henze and Keller 2007; Johnson et al. 2007; Holbrock et al. 2007). Whether wingless or winged queens prevail in any given population of *P. pima* seems to be dependent on the environment, with harsher environments probably favoring wingless queens (Robert A. Johnson, pers. comm.). This example demonstrates how social insects would enable us to also connect and incorporate ecology to Evo-Devo (i.e., Eco-Evo-Devo). Termites evolved the same system of winged sexuals and wingless workers convergently; however, they are hemimetabolous insects and the workers are juveniles (nymphs) that are normally wingless. Hence, having wingless workers in termites would not be considered a novelty, as it is in ants. Nevertheless, most termite species do have replacement reproductives that take over the reproduction after the original queen or king has died, and will never develop wings (i.e., they become mature without developing wings, which could be considered a novelty that evolved through facultative neoteny). If a termite species no longer develops winged sexuals, wingless reproductives will become a fixed derived trait (a novelty).

Worker Polymorphism—Subcastes (Phenotypic Plasticity—Individual and Colony Level)

Conspicuously, the two most specious ant genera, *Pheidole* and *Campono-tus,* are characterized by polymorphic workers known as subcastes. Sub-castes in these two genera evolved convergently because the two genera are not closely related and their last common ancestor most likely did not have subcastes. Subcastes in ants refer to worker groups within a colony that differ in size and/or shape and fall, for example, into distinct clusters along a log-log plot for different morphological measurements, like head size versus body size (for a detailed review on ant subcastes, see Höll-dobler and Wilson 1990). The developmental basis of this intracolonial worker polymorphism is allometric growth during larval development and is mostly triggered by differences in larval nutrition.

The development of subcastes at least in some species is regulated by differences in hormone concentrations at specific developmental decision points (Wheeler 1991). However, other environmental factors such as temperature have been shown to influence caste determination, and it is likely that many more unknown factors contribute to caste determination. Additionally, it has been demonstrated that feedback regulations exist be-tween the extant worker force and larvae to ensure a specific ratio between subcastes in a given colony (Wheeler and Nijhout 1984), or to homeostati-cally restore a colony's characteristic subcaste ratio after distortions (John-ston and Wilson 1985). Yang, Martin, and Nijhout (2003) showed that the caste ratio of the ant *Pheidole morrisi* varies consistently between three different populations and, to add evolution and ecology to the mix, Passera et al. (1996) showed that ant colonies challenged with conspecific colonies produce more majors/soldiers—supposedly as a reaction to the potential threat of a conspecific colony.

Therefore, within just two ant genera that belong to different ant sub-families (Myrmicinae and Formicinae), we have the opportunity to com-pare species with a strong, weak, or even absent worker polymorphism or to try to understand how and why different populations of the same species show a different ratio between subcastes. Within the ant genus *Pogonomyrmex,* worker polymorphism evolved independently once in North America *(P. badius)* and once in South America *(P. coerctatus).* Thus, two interesting Evo-Devo questions arise: Do both species use the same developmental mechanisms? Did the same selective factor lead to the evolution of worker polymorphism in these two *Pogonomyrmex* species?

Queen Number (Social Phenotype—Colony Level)

Queen number per colony has major evolutionary impacts (Hölldobler and Wilson 1990), and the transition from monogyny (single queen) to polygyny (multiple queens) or the reversal is considered similarly significant as the evolution of sociality itself. There are two crucial points during the development of an ant colony when the number of reproductively active queens is decided. The first decision point involves the number of queens which found a colony, and the second is whether mated and fully reproductive queens are adopted to queenright colonies.

The ancestral status of queen numbers in ants is unknown, but during the evolution of different ant lineages queen number per colony went both ways and, in some lineages, queen number per colony is a very plastic trait. Queen number in ants can vary between zero and thousands in some unicolonial or highly polydomous species. Zero stands for species that have abandoned a queen phenotype completely or temporally, with workers taking over the reproduction (e.g., *Dinoponorea quadriceps*, *Pristomyrmex pungens* or *Harpegnathos saltator*; Hölldobler and Wilson 1990). We know that the queen phenotype was lost in these lineages because closely related species still have morphologically differentiated queens, or queens that are produced occasionally in a small percentage of colonies. Similarly, for many ant genera we can reconstruct the ancestral status of queen number. Queen number can change over the time of an individual colony life cycle. Colonies can begin haplometrotic (i.e., one queen founds a colony independently) or pleometrotic (i.e., multiple queens found a colony together). Pleometrotic foundress associations can either stay together, leading to primary polygynous colonies, or queens can kill each other until only one remains and the colony becomes monogynous. The fighting in these pleometrotic colonies usually starts after the first workers emerge; sometimes the workers eliminate the extra queens, sometimes only the queens fight and workers do not interfere. Multiple queen societies can also develop through adoption of queens into single queen colonies or through colony fission—the split of a mature colony into two or more colonies (for an overview and classification on the developmental pathways to queen numbers, see Hölldobler and Wilson 1977).

Can a change in queen number, which is a colony-level trait, promote speciation? In ants we know of many examples where sister taxa differ in queen number, for instance, *Formica rufa* and *Formica polyctena* (Gyllenstrand, Seppa, and Pamilo; Heinze and Foitzik, this volume 2004). This is circumstantial evidence that queen number can lead to speciation, but this correlation does not prove causality. One example, where we might actually see the evolution of a new species, is *P. californicus*. In *P. californicus*, Johnson (2004) described a polygynous population in this otherwise strictly monogynous species. This population does not seem to be genetically isolated but may be on its way to becoming a new species in the future. A study of the mechanism underlying the difference in this colony-level trait would allow us to understand how development changes a phenotype, and an evolutionary study on the fitness consequences of this shift could help to elucidate the evolutionary dimension. Laboratory studies on the proximate mechanism responsible for the switch have shown that it is essentially a change in queen colony-founding behavior. Queens of the polygynous population switched from haplometrosis to pleometrosis. The proximate mechanism for this switch is that founding queens from the polygynous population of *P. californicus* tolerate other queens, whereas queens from the monogynous population fight and kill each other when forced to found a colony together. The pleometrotic foundress association might enjoy a higher fitness in the polygynous population, because a higher colony density makes haplometrotic founding impossible. Or, since foundress queens in *P. californicus* need to forage, the pleometrotic association suffer less from parasitoids or brood raiding by other colony-founding queens than haplometrotic colonies.

Sociogenesis, Sociotomy, and Division of Labor

Wilson (1985) compared morphogenesis and sociogenesis, and defined *sociogenesis* as the "process at the level of the colony . . . by which individuals undergo changes in caste, behavior, and physical location incident to colonial development." Wilson treated the highly eusocial ant species, like leaf-cutter or army ants, as superorganisms to make his comparison between sociogenesis and ontogeny clear. He also outlined the fundamental questions of Evo-Devo concerning two important colony-level traits in social insects: What are the mechanisms that result in division of labor and the production of subcastes during colony development (Devo) and why are certain systems of sociogenesis more successful than others (Evo)? Some progress has been made in answering the first question. West-Eberhard (1987) developed a mechanistic framework for the evolution of division of labor in wasps with her ovarian groundplan hypothesis. This was further expanded and tested for the honey bee, and subsequently called the reproductive groundplan (Amdam et al. 2004, 2006 Page, Amdam, and Linksayer, this volume). Having a cohesive model about the developmental mechanisms leading to division of labor will allow us to do comparative studies between different evolutionary lineages and test whether the reproductive/ovarian groundplan hypothesis is generally applicable or a mechanism restricted to honey bees or wasps.

Another useful tool for understanding the regulatory mechanisms in the development of division of labor or sociogenesis is sociotomy. *Sociotomy* is defined as an experimental change in the task allocation pattern of a colony by removing age- or task-classes (e.g., Lenoir 1979; Tripet and Nonacs 2004). This is somewhat analog to a tissue transfer during embryogenesis and can yield very similar results concerning the regulatory mechanisms of colony-level traits during sociogenesis. Detailed studies in honey bees using expression microarrays yielded many significant differences between same-age foragers and nurses, but at the same time many genes changed age dependence in workers independent from task switching (Whitfield et al. 2006).

Our examples support the idea that behavior plays an important role in the emergence of evolutionary novelties, especially in those groups, such as the social insects, that have a sophisticated behavioral repertoire. The genotype-phenotype map is mediated by several contextual factors, resulting in various degrees of plasticity. Behavior, and especially social behavior, is one way that organisms can selectively constrain underlying developmental plasticity. We have seen that several developmental switches resulting in dramatic phenotypic changes, such as caste differentiation, are triggered by available food during critical periods of development. In the context of a colony, feeding is controlled by the behavior of the workers, which in turn restricts the expressed phenotypic plasticity of the larvae, contributing to the stabilization of a novel colony-level phenotype, such as the distribution of individual castes. Social insects are, therefore, an ideal system to further refine causal models of the origin of evolutionary novelties.

Conclusion

As we have seen, adopting social insects as new model systems allows us to tackle several central Evo-Devo questions and to close the "inference gap" (Wagner 2001b) in causal explanations of evolutionary novelties. While most of the current evidence in Evo-Devo focuses on the role of genetic and regulatory differences and similarities in explanations of large-scale phenotypic variation (often at the level of families and higher systematic groups), only a small number of studies on a limited number of species actually address developmental evolution within and between populations (e.g., Brakefield et al. 1996; Emlen et al. 2005; Nijhout 2003; Wheeler 1991; Wheeler and Nijhout 1984). And none of these cover all levels, from micro- to macro-evolutionary scales, within one closely related group of species.

The advantage of social insects, and ants in particular, is that they allow us to address questions of phenotypic evolution at many different scales and for a variety of different phenotypes, from morphological to behavioral characteristics. With social insects as a model system, we can study the emergence of evolutionary novelties as a continuous process—from epigenetically, developmentally, or environmentally induced intra-species variation (such as the queen-worker polymorphism), to increasing degrees of canalization and genetic assimilation of this variation between species and higher taxa. Another advantage of social insects (especially ants) is that we know quite a lot about the ecological causes and consequences of specific phenotypic characters, as well as the environmental, physiological, genetic, and epigenetic effects determining phenotypic variation and plasticity (Wheeler and Nijhout 1984; Hölldobler and Wilson 1990; Wheeler 1991; Toth and Robinson 2007; Page and Amdam 2007).

Comparative studies, like the ones of Abouheif and Wray (2002), ranging from within species, genera to between subfamily differences in phenotypes, will enable us to identify the relative importance of these different causes at each of these scales, thus allowing us to study phenotypic evolution and the emergence of evolutionary novelties as a continuous process, rather than just comparing the endpoints of different evolutionary lineages (the typological approach characteristic of many current studies in Evo-Devo). The emerging picture of phenotypic evolution is one in which initial phenotypic plasticity (often expressed by environmental and epigenetic factors, but ultimately based on genetic variation) is increasingly canalized by the developmental system and finally assimilated by underlying genetic variation (e.g., Abouheif and Wray 2002; Page and Amdam 2007).

The fact that social insect Evo-Devo incorporates different scales also helps us to bridge the gap between the macro-evolutionary orientation of much of current Evo-Devo theory (based on developmental genetics) and the micro-evolutionary focus of evolutionary biology. For example, we can study the developmental, physiological, and behavioral characteristics for the emergence of division of labor within groups of individuals, as well as the ecological and evolutionary consequences and mechanisms of this trait. The latter touches on arguments about inclusive fitness, kin selection, and group selection and brings one of the theoretically most advanced areas of evolutionary theory within the framework of Evo-Devo.

Even though the importance of life histories have been recognized within the Evo-Devo literature (e.g., West-Eberhard 2003), so far most attention has been placed on the earlier stages of embryonic development, which generally do not display the same rich behavioral characteristics as do later stages. In social insects, however, behavior often serves as the integrator of proximate and ultimate perspectives, thus providing us with a causal explanation that spans developmental and evolutionary time scales (Page and Amdam 2007). An emphasis on social insects as a model system for Evo-Devo brings with it a broadening of the range of phenotypes analyzed within this framework. And, as behavior crucially depends on environmental factors (as triggers and other context dependent causes), social insect Evo-Devo is always also Eco-Evo-Devo (Collins et al. 2007).

An Evo-Devo perspective on social insects affects the way we conceptualize the evolution and development of these species. The question of whether a colony should be seen as a superorganism, or at the very least an individual, needs to be addressed (see Hamilton, Smith and Haber, this volume). As we have seen, extending a developmental perspective to colony-level properties and studying the evolutionary effects of colony-level life histories is a major component of social insect Evo-Devo. It allows us to analyze the potential conflicts between different levels of selection, as well as the mechanistic causes for phenotypic differentiation among castes and between colonies and species. Depending on the explanatory context—evolutionary or developmental—we have two different types of cohesion relations. On the one hand, a colony acts as a unit of selection, which might explain the evolution of colony-level traits, especially in highly advanced eusocial ant species where workers have lost all of their reproductive potential. On the other hand, a colony shows developmental cohesion throughout its life history, from its founding to growth and reproduction. Together these two colony-level processes help to explain the major evolutionary transition of eusociality (Hölldobler and Wilson 1990; Maynard-Smith and Szathmary 1997).

Literature Cited

Abouheif, E., and G. A. Wray. 2002. "Evolution of the gene network underlying wing polyphenism in ants." *Science* 297: 249–252.

Amdam, G. V., A. Csondes, M. K. Fondrk, and R. E. Page. 2006. "Complex social behavior derived from maternal reproductive traits." *Nature* 439: 76–78.

Amdam, G. V., K. Norberg, M. K. Fondrk, and R. E. Page. 2004. "Reproductive ground plan may mediate colony-level selection effects on individual foraging behavior in honey bees." *Proceedings of the National Academy of Sciences USA* 101: 11350–11355.

Boveri, T. 1906. *Die Organismen als historische Wesen.* Würzburg: Königliche Universitätsdruckerei von H. Stürz.

Bowsher, J. H., G. A. Wray, and E. Abouheif. 2007. "Growth and patterning are evolutionarily dissociated in the vestigial wing discs of workers of the red imported fire ant, *Solenopsis invicta.*" *Journal of Experimental Zoology Part B Molecular and Developmental Evolution* 308: 769–776.

Brady, S. G., T. R. Schultz, B. L. Fisher, and P.S. Ward. 2006. "Evaluating alternative hypotheses for the early evolution and diversification of ants." *Proceedings of the National Academy of Sciences USA* 103: 18172–18177.

Brakefield, P. M., J. Gates, D. Keys, F. Kesbeke, P. J. Wijngaarden, A. Monteiro, V. French, and S. B. Carroll. 1996. "Development, plasticity and evolution of butterfly eyespot patterns." *Nature* 384: 236–242.

Cameron, S. A., H. M. Hines, and P. H. Williams. 2007. "A comprehensive phylogeny of the bumble bees *(Bombus).*" *Biological Journal of the Linnean Society* 91: 161–188.

Carroll, S. B., J. K. Grenier, and S. D. Weatherbee. 2005. *From DNA to diversity: Molecular genetics and the evolution of animal design.* Malden: Blackwell Publishing.

Collins, J. P., S. Gilbert, M. D. Laubichler, and G. B. Müller. 2007. "Modeling in EvoDevo: how to integrate development, evolution, and ecology." In *Modeling biology structures, behavior, and evolution,* M. D. Laubichler and G. B. Müller, eds. Cambridge, MA. MIT Press.

Danforth, B. N., L. Conway, and S. Ji. 2003. "Phylogeny of eusocial *Lasioglossum* reveals multiple losses of eusociality within a primitively eusocial clade of bees (Hymenoptera: Halictidae)." *Systems Biology* 52: 23–36.

Davidson, E. H. 2006. *The regulatory genome. Gene regularity networks in development and evolution.* San Diego: Academic Press/Elsevier.

Davidson, E. H., and D. H. Erwin. 2006. "Gene regulatory networks and the evolution of animal body plans." *Science* 311: 796–800.

Emlen, D. J., J. Hunt, and L. W. Simmons. 2005. "Evolution of sexual dimorphism and male dimorphism in the expression of beetle horns: Phylogenetic evidence for modularity, evolutionary lability, and constraint." *The American Naturalist* 166: S42–S68.

Fewell, J. H., and R. E. Page. 1999. "The emergence of division of labour in forced associations of normally solitary ant queens." *Evolutionary Ecology Research* 1: 537–548.

Gehring, W. J. 1998. *Master control genes in development and evolution: The homeobox story.* New Haven: Yale University Press.

Gottlieb, G. 1992. *Individual development and evolution: The genesis of novel behavior.* Oxford: Oxford University Press.

Gyllenstrand, N., P. Seppä, and P. Pamilo. 2004. "Genetic differentiation in sympatric wood ants, *Formica rufa* and *F. polyctena.*" *Insectes Sociaux* 51: 139–145.

Hall, B. K., and W. M. Olson. 2003. *Keywords and concepts in evolutionary developmental biology.* Cambridge: Harvard University Press.

Heinze, J., and L. Keller. 2000. "Alternative reproductive strategies: A queen perspective in ants." *Trends in Ecology and Evolution* 15: 508–512.

Hinchliffe, J. R. 2002. "Developmental basis of limb evolution." *International Journal of Developmental Biology* 46: 835–845.

Hines, H. M., J. H. Hunt, T. K. O'Connor, J. J. Gillespie, and S. A. Cameron. 2007. "Multigene phylogeny reveals eusociality evolved twice in vespid wasps." *Proceedings of the National Academy of Sciences USA* 104: 3295–3299.

Holbrook, C. T., R. A. Johnson, C. Strehl, and J. Gadau. 2007. "Single mating in the seed-harvester ant *Pogonomyrmex (Ephebomyrmex) pima:* Implications for the evolution of polyandry." *Behavioral Ecology and Sociobiology* 62: 229–236.

Hölldobler, B., and E. O. Wilson. 1977. "The number of queens: An important trait in ant evolution." *Naturwissenschaften* 64: 8–15.

———. 1990. *The ants.* Cambridge: Harvard University Press.

Johnson, R. A. 2004. "Colony founding by pleometrosis in the semi-claustral seed-harvester ant *Pogonomyrmex californicus* (Hymenoptera: Formicidae)." *Animal Behaviour* 68: 1189–1200.

Johnson, R. A., C. T. Holbrook, C. Strehl, and J. Gadau. 2007. "Population and colony structure and morphometrics in the queen dimorphic harvester ant, *Pogonomyrmex pima.*" *Insectes Sociaux* 54: 77–86.

Johnston, A. B., and E. O. Wilson. 1985. "Correlates of variation in the minor major caste ratio of the ant, *Pheidole dentate* (Hymenoptera, Formicidae)." *Annals of the Entomological Society of America* 78: 8–11.

Lenoir, A. 1979. "Feeding behaviour in young societies of the ant *Tapinoma erraticum* L.: Trophallaxis and polyethism." *Insectes Sociaux* 26: 19–37.

Maynard-Smith, J., and E. Szathmary. 1997. *The major transitions in evolution.* Oxford: Oxford University Press.

Moreau C. S., D. B. Charles, R. Vila, S. B. Archibald, and N. E. Pierce. 2006. "Phylogeny of the ants: Diversification in the age of angiosperms." *Science* 312: 101–104.

Müller, G. B. 2007. "Evo-Devo: Extending the evolutionary synthesis." *Nature Reviews Genetics* 8: 943–949.

Müller, G., and S. Newman. 2003. *Origination of organismal form: Beyond the gene in developmental and evolutionary biology.* Cambridge: MIT Press.

———. 2005. "The innovation triad: An evo-devo agenda." *Journal of Experimental Zoology* 304: 487–503.

Müller, G., and G. P. Wagner. 1991. "Novelty in evolution: Restructuring the concept." *Annual Review of Ecology and Systematics Biology* 22: 229–256.

Nijhout, H. F. 2003. "Polymorphic mimicry in *Papilio dardanus:* Mosaic dominance, big effects, and origins." *Evolution & Development* 5: 579–582.

Page, R. E. Jr., and G. V. Amdam. 2007. "The making of a social insect: Developmental architectures of social design." *Bioessays* 29: 334–343.

Passera, L., E. Roncin, B. Kaufmann, and L. Keller. 1996. "Increased soldier production in ant colonies exposed to intra specific competition." *Nature* 397: 630–631.

Toth, A. L., and G. E. Robinson. 2007. "Evo-Devo and the evolution of social behavior." *Trends in Genetics* 23: 334–341.

Tripet, F., and P. Nonacs. 2004. "Foraging for work or age-based polyethism? The role of age and previous experience on task choice in ants." *Ethology* 110: 863–877.

Wagner, G. P. 2001a. *The character concept in development and evolution.* San Diego: Academic Press.

———. 2001b. "What is the promise of developmental evolution? Part II: A causal explanation of evolutionary innovations may be impossible." *Journal of Experimental Zoology* 291: 305–309.

West-Eberhard, M. J. 1987. "Flexible strategy and social evolution." In Y. Itô, J. L. Brown, and J. Kikkawa, eds., *Animal societies: Theories and fact,* 35–51. Toyko: Japan Scientific Society Press.

———. 2003. *Developmental plasticity and evolution.* New York: Oxford University Press.

Wheeler, D. E. 1991. "Developmental basis of worker caste polymorphism in ants." *American Naturalist* 138: 1218–1238.

Wheeler, D. E., and H. F. Nijhout. 1984. "Soldier determination in ants: New role for juvenile hormone." *Science* 213: 361–363.

Whitfield, C. W., Y. Ben-Shahar, C. Brillet, I. Leoncinii, D. Crauser, Y. LeConte, S. Rodriguez-Zas, and G. E. Robinson. 2006. "Genomic dissection of behavioral maturation in the honey bee." *Proceedings of the National Academy of Sciences USA* 103: 16068–16075.

Wilkins, A. S. 2002. *The evolution of developmental pathways.* Sunderland: Sinauer Associates.

Wilson, E. O. 1985. "The sociogenesis of insect colonies." *Science* 228: 1489–1495.

Yang, A. S., C. H. Martin, and H. F. Nijhout. 2003. "Geographic variation of caste structure among ant populations." *Current Biology* 14: 514–519.

Acknowledgments

This volume was made possible only through the efforts and support of numerous people. We particularly want to thank Rob Page for his crucial role in initiating and sustaining the project. The volume's starting point was a series of workshops on social insects as models for social complexity arranged by Rob and sponsored by the Santa Fe Institute. However, the final author list extends far beyond the original working group, and the volume has gained significant substance as a result of their intellectual contributions. The chapter contributors make up the core of our team effort on this volume, both in providing chapters with the latest conceptual approaches to key areas in insect sociobiology and in intensively reviewing each other's work. Peggy Coulombe spent many valuable hours poring over the numerous chapters, helping to significantly improve formatting and flow. We additionally thank Ann Downer-Hazell, Vanessa Rossi, and the rest of the folks at Harvard University Press for working so flexibly with us. We both offer sincere thanks to our families, especially Mechtild Gadau and Jon Harrison, for their support and patience. Finally, the themes in this book reflect the far-reaching and profound impact that Bert Hölldobler has had on this field and on our own research. We thank him for the many lively discussions and thoughtful insights he has brought to this project.

Index